SOURCES OF POWER

SOURCES OF POWER

How Energy Forges Human History

VOLUME 2
THE OIL AGE AND BEYOND

Manfred Weissenbacher

Praeger Perspectives

PRAEGER
An Imprint of ABC-CLIO, LLC

A B C ≋ C L I O

Santa Barbara, California • Denver, Colorado • Oxford, England

Copyright 2009 by Manfred Weissenbacher

All rights reserved. No part of this publication may be reproduced, stored in a retrieval system, or transmitted, in any form or by any means, electronic, mechanical, photocopying, recording, or otherwise, except for the inclusion of brief quotations in a review, without prior permission in writing from the publisher.

Library of Congress Cataloging-in-Publication Data

Weissenbacher, Manfred.
  Sources of power : how energy forges human history / Manfred Weissenbacher.
    p. cm.
  Contents: v. 1. Before oil : the ages of foraging, agriculture, and coal—v. 2. The oil age and beyond.
  Includes bibliographical references and index.
  ISBN 978-0-313-35626-1 (set)—ISBN 978-0-313-35627-8 (set : ebook)—ISBN 978-0-313-35628-5 (v. 1 : hardcopy : alk. paper)—ISBN 978-0-313-35629-2 (v. 1 : ebook)—ISBN 978-0-313-35630-8 (v. 2 : hardcopy : alk. paper)—ISBN 978-0-313-35631-5 (v. 2 : ebook)
  1. Agriculture and energy—History. 2. Coal—Social aspects.
  3. Petroleum—Social aspects. 4. Civilization. I. Title.
    S494.5.E5W33 2009
    333.79—dc22   2009023895

13 12 11 10 09   1 2 3 4 5

This book is also available on the World Wide Web as an eBook.
Visit www.abc-clio.com for details.

ABC-CLIO, LLC
130 Cremona Drive, P.O. Box 1911
Santa Barbara, California 93116-1911

This book is printed on acid-free paper ∞

Manufactured in the United States of America

To Ingunn and Frank

Please visit the author's Web site at
www.manfredweissenbacher.com

Contents

Volume 1. Before Oil:
The Ages of Foraging, Agriculture, and Coal

Introduction   *xi*

Part I. Foraging Age

Bibliography to Part I   *11*

Part II. Agricultural Age

CHAPTER 1	What Is Agriculture?	17
CHAPTER 2	How Did Agricultural Technology Emerge?	23
CHAPTER 3	The Spread of Agriculture	37
CHAPTER 4	Fruit, Measles, and Backache	47
CHAPTER 5	Transformed Society	53
CHAPTER 6	Muscle Power: The Mammalian Machine	57
CHAPTER 7	Wind and WaterPower	77

CHAPTER 8	Biomass: Energy for Lighting, Heating, and Metallurgy	83
CHAPTER 9	Weapons Technology: Energy Used to Kill and Destroy	87
CHAPTER 10	The Rise and Fall of Grain-Fueled Empires	97
CHAPTER 11	The Early Wheat Empires	105
CHAPTER 12	The Rise of Europe	119
CHAPTER 13	The Opening of the World	127
CHAPTER 14	The Super-Agricultural Era	147

Bibliography to Part II — *173*

PART III. COAL AGE

CHAPTER 15	What Is Coal?	187
CHAPTER 16	The Emergence of Coal Technology	191
CHAPTER 17	How Coal Technology Spread	235
CHAPTER 18	Transitional Downturn: Pollution, Poverty, and Hard Work	257
CHAPTER 19	Longer Lives, Better Education, More Rights	261
CHAPTER 20	Knowledge and Technology	267
CHAPTER 21	Deadlier Weapons	277
CHAPTER 22	Improved Agriculture	283
CHAPTER 23	Increased Mobility and Trade: Global Economic Integration	289
CHAPTER 24	The Coal-Powered Empires	301

Bibliography to Part III — *347*

VOLUME 2. THE OIL AGE AND BEYOND

Introduction   *xi*

PART IV. OIL AGE

CHAPTER 25	What Is Crude Oil?	359
CHAPTER 26	The Emergence of Oil Technology	361
CHAPTER 27	How Oil Technology Spread	407
CHAPTER 28	Life in the Oil Age	413
CHAPTER 29	Technology and Knowledge	423
CHAPTER 30	Weapons	435
CHAPTER 31	Agriculture in the Oil Age	451
CHAPTER 32	Diverse Energy Mix	465
CHAPTER 33	Economic Expansion	477
CHAPTER 34	Population Development	491
CHAPTER 35	The Oil-Powered Empires	497

Bibliography to Part IV   *655*

PART V. BEYOND THE OIL AGE

CHAPTER 36	The Problem of Climate Change	677
CHAPTER 37	Transport Fuel for the Future	741
CHAPTER 38	Electricity for the Future	761
CHAPTER 39.	Extended Oil Age?—Energy Mix until 2030	773
CHAPTER 40	Who Got The Power?	779

CHAPTER 41	Nuclear Age or Second Coal Age?	797
CHAPTER 42	Energy for All	805

Bibliography to Part V   *819*

Epilogue   *837*

Index   *875*

Introduction

We have all asked these questions. Why are some of the world's regions so rich and others so poor? Why are European languages spoken dominantly on four of the world's six populated continents? Why is Africa's largest lake named for a British queen? Why were the Russians first to put a man into space? And how come the United States emerged as the world's sole remaining superpower at the end of the 20th century? However, answers that explain the big picture and the overall course of world history are rare. They are often designed to flatter the culture, religion, or intellect of those who are better off. Or they focus on environmental factors such as pronounced seasons, and on resistance to disease. And many historians reject ideas of orderly patterns in world history altogether: they consider the course of history chaotic, or at least emphasize the unpredictability of human behavior.

There is, however, a wide consensus on one issue: most people would agree that the emergence of agriculture has been the most critical milestone in human history. Agriculture triggered the rise of complex civilizations, as farmers were able to produce surplus food that sustained not just themselves and their families, but many other members of society who in turn specialized in newly emerging occupations to become full-time potters, weavers, miners, soldiers, priests, bankers, and so on. As it has long been known that agriculture emerged much earlier in Eurasia than in Africa and the Americas, it should thus not surprise us that Eurasians eventually wiped out or enslaved native Africans and Americans, and not vice versa. Eurasians simply capitalized on their agricultural head-start, which provided them with higher population densities, more advanced technology, and better weapons.

This leaves us with the question why agriculture actually *did* emerge earlier one place than another. Charles Darwin, among many others, pointed at climatic conditions to provide an answer. In *The Descent of Man* (1871), he explained that agricultural production is necessary to provide a basis for civilization, and speculated about the role of a temperate climate, though he concluded that "the problem, however, of the first advance of savages towards civilisation is at present much too difficult to be solved."[1] A more differentiated theory about how climatic conditions somewhat indirectly facilitated the emergence of agriculture has been popularized by Jared Diamond's 1997 book *Guns, Germs, and Steel*.[2] Diamond argues that the world's asymmetric endowment with species that lend themselves well to domestication is behind the delays in the onset of agriculture on different continents, and thus the course of world history. Research of the past decades has indeed shown that some regions are very rich, and others very poor, in terms of natural occurrence of species that agricultural people tend to eat. Most important were large-seeded grass species such as wheat, rice, and maize, which still now account for the bulk of the nutrition sustaining the global human population. As many as 32 of the world's 56 species of large-seeded grasses are native to the Mediterranean. The situation is similarly lopsided with respect to the world's large terrestrial mammal species. (These were not all that important in terms of food supply, but became indispensable for draft and portage, and decisive in warfare.) Worldwide, only 14 such mammal species proved suitable for domestication, with 13 of them descending from wild species found exclusively in Eurasia and North Africa.

The main problem associated with this theory is that it does not explain modern history. True, native Africans, Americans, and Australians may have been disadvantaged by the delayed onset of their farming, but there was not much time difference in the emergence of agriculture in western Eurasia compared to eastern Eurasia. Why isn't Chinese the main language spoken in the Americas today? Why did India become a colony governed by Europeans? Why did Britain rise to superpowerdom? And why did British global power decline at the expense of the United States of America and the Soviet Union? If we attempt to understand history beyond the 16th century, we certainly have to find answers that go beyond agriculture and its effects on societal development.

My personal quest for a general theory of world history began with a set of questions concerning the 20th century. Most importantly, I was curious how it was possible for the Soviet Union, a nation still languishing in technological backwaters in the early 20th century, to emerge as a superpower that controlled all of Eastern Europe and challenged the United States internationally. After all, the Soviet Union enjoyed little internal cohesion and practiced an economic system that did not allocate resources very well.

My approach to this question may have been quite conservative at first. My education was straightforwardly humanistic, and I studied Latin for six years, translating such classics as Caesar's *De Bello Gallico* (on the Roman conquest of what is now France) and Ovid's *Metamorphoses*. History had been presented to me as a long list of empires and battles, and each country in the world seemed to cherish the supposed brilliancy of its own respective generals and statesmen.

However, as I moved on to study first chemistry and then economics, my views and methods changed. After developing energy storage systems during my PhD years, I eventually took a position at the Stanford Research Institute, where one of my early assignments concerned the advantages and disadvantages of central governments to control energy prices. Around this time I began formulating a theory of world history that would apply to all epochs. Most notably, my consternation regarding Soviet superpowerdom disappeared like the Cheshire cat as soon as I saw the statistical figures of international oil production. Which two nations pumped the most oil by far cumulatively during the 20th century? The Soviet Union and the United States of America, the two principal superpowers of the 20th century. Coincidence? Hardly. In the century before, when coal was the most critical fuel, there was just one single country mining far more of it cumulatively than any other nation: It was Britain, the sole superpower of the 19th century and creator of the largest empire ever to exist on the face of the Earth. These observations in themselves prompted me to postulate that command of energy has probably always been the most important factor in the fate of societies and nations. What else could have been? After all, energy is formally defined as the ability or capacity to do work. But before I could claim this hypothesis to be universally correct, I had to check whether it applied to preindustrial times as well.

Let's go to the very beginning. Whenever I ask people what they think was the first major energy source commanded by humans, they tend to say: fire! This is a tempting choice indeed, but it is actually wrong. The first kind of energy that humans had access to, as trivial as it may sound at first, was the food that they found in nature. Accordingly, the human body itself was the first prime mover that people had at their disposal. And what a prime mover it was! Strong, endurable, and incredibly flexible! Useful to build shelters, to produce tools, and to find nutritional energy to fuel itself. And yet the notion of the human body being a prime mover was pushed to a whole new level as soon as slave-based agrarian societies emerged.

We have already heard that the emergence of agriculture is considered the most important milestone in human history. But why did agriculture make such a big difference and trigger the rise of complex civilizations? Because it was a source of energy! Think about it for a moment. What differentiates plants from animals is that plants can directly utilize sunlight to fuel their life. Animals (including humans), on the other hand, depend on the chemical energy that plants store in their tissue during growth. As humans happen to be unable to digest most of the cellulose-based plant material that tends to fill natural landscapes, original foraging societies had quite limited amounts of nutritional energy at their disposal. They had to collect edible fruits, nuts, seeds, and tubers from wide landscapes that were dominated by biomass people could not eat. Farmers, in contrast, removed natural vegetation with all its leaves, branches, bushes, and trunks, and directed sunlight towards the growth of edible plant material. Most importantly, they cultivated thin-stemmed grass species that store a lot of chemical energy in large digestible seeds.

By this stage, we have already identified the three principal fuels that have critically influenced the fate of human societies up until this day: grain, coal, and oil. Accordingly, we can divide human history into four main Energy Eras:

> Foraging Age
> Agricultural Age
> Coal Age
> Oil Age

Each of these eras is defined by its principal pair of fuel and prime mover. The work done by grain-fueled muscles dominated the Agricultural Age. Human workers (often slaves) were the principal engines of this era, and animal power was utilized where suitable beasts were available and fields (or natural landscapes) provided enough nutritional energy to feed them. In first approximation, the pairs of the succeeding eras were coal and the steam engine, followed by oil and the internal combustion engine. Coal-fired steam engines were initially used to pump water from mines, but soon powered all sorts of factory equipment. Once steam engines were fitted to boats and railroad cars, they also revolutionized human mobility and the transport of goods. However, oil-fueled internal combustion engines had an even bigger impact, as they powered motor vehicles and airplanes.

As societies took the three consecutive *energetic steps*—towards grain, coal, and oil—the amount of energy controlled by humans increased radically. Pick up a bag of rice and it will tell you that 100 grams of pealed rice contain about 350 kcal (kilocalorie) worth of energy. A man of average size needs to eat some 2,400 kcal worth of food energy a day to be able to function and to do work. In comparison, the amount of technical (inanimate) energy consumed daily per person living on the planet averaged around 2,100 kcal in 1860 and 18,000 kcal in 1950. By the end of the 20th century the figure was 41,000 kcal. In terms of food energy, 41,000 kcal per day would be sufficient to nourish 17 working men. Hence the average world citizen by now has a whole horde of personal slaves at her or his disposal in terms of technical energy.

But in reality there is no such thing as the average world citizen. At the beginning of the 21st century Americans consumed some 220,000 kcal, Latin Americans 35,000 kcal, and South Asians 11,000 kcal per day and person in technical energy. It is such differences in the command and consumption of energy that point at a fundamental understanding of world history: Those who commanded more energy have always been better off. They used energy in productive ways to increase their wealth, and they used energy in destructive ways to defeat their neighbors. Throughout history those who commanded more energy have killed, expelled, enslaved, or, at the minimum, politically controlled those who commanded less energy.

Although a one-sentence summary of this book would read "Command of energy is the most important determinant in world history," I am not going to claim that energy has fully determined every aspect of the fate of societies or nations. Rather, it laid out the broad patterns: energy resources set the outer limits as to what could be done within or by a society. Within their energetic boundaries people were free to make choices. There was thus space for all the curiosity, greed, love, hate, jealousy, folie de grandeur, and all the other human elements that may have influenced world history at one point or another. Energy realities can explain why the Aztecs did not conquer Spain, and why Germany lost World War II, but it would have been impossible to predict that Hitler would provoke a large-scale war (despite Germany's obvious lack of oil resources), and that the Spanish rather than the Portuguese would be first to sail to America. By the time the Soviet Union collapsed, it was the world's largest oil producer: Something had obviously gone wrong despite the availability of vast energy resources. (But the Siberian oil deposits are still there, and we can only guess what role the Russian Federation is going to play in the 21st century.)

Much energy technology served both peaceful and belligerent purposes. Horses, trains, ships, trucks, and airplanes—by increasing human mobility—improved productivity and trade as well as military operations. However, the principal imperative of warfare was the ability to kill from a distance: It increased from javelin to bow-and-arrow, to firearms, to bombs dropped from airplanes, and finally to intercontinental ballistic missiles. Societies seem to have invaded other societies throughout history simply because they had the means to do so. But the disposition or aptness to act aggressively was generally enhanced when the balance between command of destructive versus productive energy was out of proportion within a society. The mounted peoples of central Asia harassed settled Eurasia for centuries. They commanded lots of destructive energy as they had plenty of horses and superior compound bows, but their productive energy was minimal. Hence they kept being attracted by the riches of settled agrarian societies and attacked them regularly up until the time sedentary civilizations developed firearms.

An imbalance between productive and destructive energy command may also arise when destructive energy resources prevail while productive energy resources diminish. For various reasons some agricultural systems sustained empires literally for millennia, while other farming schemes collapsed after just a few centuries. (And what is happening in America right now? The United States rose to superpowerdom based on its enormous domestic oil resources, which are now rapidly declining.) A more obvious path towards such energetic imbalance is the spread of arms technology. The recent proliferation of weapons of mass destruction to relatively poor countries created an immense gap between destructive and productive energy commanded in these societies. Generally, the existence of weapons of such enormous destructive energy has decreased predictability and widened the bandwidth of possible futures: with enough nuclear warheads around to blow every person off the face of the Earth, who is to say what the future of humankind is going to be?

This set of books is divided into two volumes and five parts. The first four parts demonstrate how people proceeded towards more energy command; how their life changed in the process; and how political history unfolded as societies took the three energetic steps in different regions at different times. The fifth part is a look into the future.

Part I, titled "Foraging Age," very briefly presents what was by far the longest era in human history. Globally, it lasted from the time various protohumans roamed the Earth until about 10,000 years ago, when agriculture first emerged. This was the period when people were still spreading to all the continents; when there were hardly any differences between and within human societies; and when no artificial borders existed. The Foraging Age represents something of an energy baseline for human societies, an era in which people were sustained by natural energy flows. Gathering-hunting, the original way of life, provided no consistent oversupply in food energy, and the principal energy innovation of this era, the command of open fire, had somewhat limited applications. Mobile lifestyle and nutritional limitations kept population growth in check, and humans lived a lot like less intelligent animals do. Thus our distant ancestors seemed to represent a fairly inconspicuous species: it would have been hard to guess that these creatures would eventually reshape landscapes, overfish the oceans, and fly to the moon.

Part II, titled "Agricultural Age," covers the long time period from 10,000 years ago until the end of the 18th century C.E.,[3] when coal technology emerged. The start of this era marks an extraordinary turning point in human evolutionary development: Humans stopped adapting to natural environments (as animals normally do), and instead began to consciously shape the environment to their own needs on an ever vaster scale. This lacks comparison with any other species‡ and marks the point when humans truly began conquering the world. They became the dominant species on Earth and henceforth adapted to the artificial environment they had themselves created, rather than to nature as it once was.

The Agricultural Age proceeded in three periods. The first period concerned developments within continents only. Agriculture emerged independently in western Eurasia, eastern Eurasia, Mesoamerica, South America, and sub-Saharan Africa. People used quite different sets of domesticated species to capture sun energy in these regions, but the consequences were the same everywhere: as more nutritional energy was available, population growth accelerated, more human prime movers were around, and a lot more work could be done. Professional specialization promoted technological progress, and soon farming societies produced materials (most importantly hard metal) that do not occur in nature and were thus inaccessible to gatherer-hunters. Settled lifestyle and professional specialization triggered the emergence of social classes, while centralized governance and organized religions arose to manage agrarian communities.

Soon after the emergence of agriculture we see the first *energy campaigns*. Accelerated population growth prompted farming societies to expand further and further into gatherer-hunter regions in search for more nutritional energy and arable land. These expansions were bad news for indigenous foragers, as farmers commanded a lot more energy (more food and more people) and had better weapons. In this first wave of energy invasions agrarians committed something of a global genocide of gatherer-hunters, which irreversibly changed the genetic make-up of the human population. The only option for foragers to escape the fate of being killed, enslaved, or expelled was to rapidly adopt agriculture, or else to retreat to marginal lands.

Some of the early agrarian expansions are surprisingly well documented. At first it was rather loose groups of farmers migrating to new areas, but when empires emerged the campaigns into gatherer-hunter land became more organized. Empires kept expanding until they bordered seashores, areas unsuitable to agriculture, or other empires. (Sometimes they disintegrated into smaller entities, but the total area under agrarian administration then tended to remain the same.) And for a quite simple reason territorial expansion remained the name of the game for the entire Agricultural Age: The energy commanded by an empire was directly related to the field area under its authority. What is more, campaigns yielded slaves who served as additional prime movers. (The world's first empires appropriately emerged around the original centers of agriculture. Our whirlwind tour through political history will therefore start in Mesopotamia. From there we will quickly proceed to Egypt, Greece, Rome, and to the more northern regions of Europe.)

The Agricultural Age's second period was initiated by Europeans crossing the world's oceans and opening the world. In this period differences between continents became relevant. This era saw the near or full extinction of peoples native to the Americas and the beginnings of large-scale enslavements of West Africans. Wherever Europeans encountered people with inferior weapons and no resistance to Eurasian disease they slaughtered, infected, and enslaved them much the same way as was previously done by technologically advanced people on an intra-continental scale. And with natives disappearing from overseas fertile land, Europeans proliferated into many the world's temperate regions by introduction of Fertile Crescent domesticates. (The Fertile Crescent is the original center of agriculture in western Eurasia.) Hence, this was really a classic agrarian expansion, just on a global scale.

Compared to the Americas, the situation was entirely different in South and East Asia. In India and China European explorers encountered civilizations that were richer and more advanced than their own. Europeans nevertheless had two main advantages in terms of energy command, but these were limited to the open seas and did not enable them to actually conquer or occupy Asia's densely populated areas. First, Europeans had superior sailing vessels, which allowed them to harness wind energy for mobility as nobody else could. Second, they had the world's best naval cannons, thus commanding vast amounts of destructive energy. This combination enabled Europeans to become pirates of the Indian ocean, to terrorize Asian coastlines, and to control much of the intra-Asian maritime trade.

The third and last period of the Agricultural Age was what I call the "Super-Agricultural Era." It was characterized by the effects of New World domesticates spreading to the Old World and vice versa. As maize, potato, and sweet potato was grown in Eurasia and Africa, and wheat in the Americas, the world's arable land was better matched with domesticates suited for various local climates. (This was especially relevant as the global climate was changing around that time.) Hence, the worldwide flow of agricultural energy soared to unprecedented levels, and all the effects associated with the initial emergence of agriculture were accelerated. From around 1650 C.E. the global human population grew about ten times faster than before, and by 1750 C.E. some 750 million people populated the planet (compared to 300 million in the year 1000 C.E., and 500 million in 1650 C.E.). As more people will work, experiment, and think more, this was a period of fast technological progress. Europe experienced a scientific revolution that involved agricultural reforms and eventually climaxed into an industrial revolution in which waterpower drove complex machinery, transforming first the textile and then other industries.

Part III, titled "Coal Age," presents the time period that began in the late 18th century and lasted until the early 20th century. In this period global energy consumption skyrocketed as the development of steam engines allowed the chemical energy contained in coal to be translated into mechanical work. The ascent of coal began with its adoption as a fuel by certain industries in timber-lacking England. Most importantly coal was soon mined in large quantities to fuel the expansion of the iron industry, which had earlier exclusively relied on charcoal (that is, a biofuel). The coal-fueled steam engine was an entirely new type of prime mover, based on the expansion and condensation of water vapor. It remained crude and inefficient for a long time, but eventually matured to propel anything from production machinery to railroad trains and steam ships. An even more powerful prime mover, the coal-fired steam turbine, further increased the speed of steam ships and allowed for coal energy to be efficiently turned into electricity.

These developments redefined the energy mix that influenced political history. World energy (and thus political power) was no longer dominated by agricultural energy. The previous motivation for empires to expand diminished as coal deposits delivered energy that was unrelated to land area exposed to sunlight. Once nations had secured access to coal, the aim was to control large international (colonial) markets. These served as outlets for products delivered by efficient coal-fueled domestic industries. Britain (where coal technology first emerged) and other western European countries (where coal technology was rapidly adopted) competed fiercely in the international arena. And with nearly no delay the United States of America, while still expanding towards the Pacific coast, joined the contest.

Coal energy and the availability of unprecedented quantities of iron and steel revolutionized warfare and military power during the Coal Age. Coal-fired steamers were capable of penetrating rivers, the hitherto secure inland lifelines of agricultural societies. The Western powers could thus overcome the main shortcoming of their wind-powered oceangoing vessels, and finally made a break into Asia's interiors. India was made a colony, and China was firmly controlled. Africa was partitioned between all western European Coal Powers, and Britain snatched remote Australia and New Zealand. The United States began to engage in coal-fueled imperialism in Latin America as well as the North Pacific, where Hawaii was absorbed as a state and the Philippines acquired as a colony. What is more, U.S. cannon steamers forced Japan out of seclusion: Looking for fuel, they threatened Japanese coastal cities, demanding the construction of coaling stations and access to Japanese markets.

On land, coal-fired railroad locomotives redefined transport (and therefore military strategy). Most people had traditionally been living along shorelines and rivers. These regions were now being connected to the continents' interiors by railroad tracks. Much of the radical territorial expansion of the United States and Russia during the Coal Age, for instance, would have been impossible without steam trains. Meanwhile short-distance travel and agriculture was still dominated by animate power. The Coal Age saw a resurgence of slavery (predominantly in the U.S. South), and the world's horse population soared to unprecedented levels. (The cities were full of horse-drawn carts, and Californian harvester-thresher combines were being pulled by teams of up to 40 horses.)

As we would expect, the enhanced energy flows in Coal Age societies accelerated the effects observed following the emergence of agriculture. More work was taken over by machines, more people were released towards new tasks, and the rate at which knowledge accumulated gained more speed. But the shift towards coal-energy involved great hardships: Many of the traditional jobs disappeared, and scores of laborers and their families were pushed into urban poverty. In the longer run, though, the Coal Age saw major improvements for workers, a prolonged life expectancy for all, and an increase in the global rate of population growth. Chemicals extracted from coal tar were used to make both the first mass-produced pharmaceuticals and a new class of very powerful explosives (nitroglycerin, dynamite, TNT). The latter helped to increase the range of firearms, which were developed into more precisely engineered and deadly weapons (repeating rifles, machine guns).

Part IV, titled "Oil Age," covers the time period that began around World War I and continues until the present. Crude oil prospecting and refining techniques were developed from the mid-19th century to serve the lighting market. Only kerosene was sold as lamp fuel, while the other crude oil components (including gasoline, diesel, and asphalt) were initially discarded. Eventually gasoline and diesel were used to fuel internal combustion engines, whose pistons are moved directly by the expanding gases generated by the combustion of a fuel inside a cylinder. (In contrast, steam engines burn coal externally and direct water vapor into the cylinder.) These new lightweight engines revolutionized mobility as they allowed for the construction of off-rail motor vehicles and airplanes.

Global energy flows in the Oil Age reached levels ten times those of the Coal Age. Hence the world experienced an unparalleled push towards further technological progress and increase in wealth. Electrical appliances took over many household tasks; automobiles provided for unmatched individual mobility; and large passenger jets eventually opened extreme long-distance mobility for the masses. Urbanization proceeded even further, while oil-fueled tractors and harvesting machines increased agricultural output in combination with artificial fertilizers and pesticides. Pharmaceuticals improved the life-expectancy of Oil Age people, and the global population soared from two billion in 1930 to six billion in 1999.

Oil technology emerged mainly in western-central Europe and soon dominated military strategy in its manifestation as tanks and bomber planes. But western Europe lacked oil resources. European nations therefore used their powerful international position (read: their Coal Age might) to initiate policies whose consequences are still now influencing world politics. Most notably Britain, the principal superpower of the Coal Age, had no oil at all, none domestically (the North Sea reservoirs were discovered only after World War II), and curiously almost none in its huge global colonial empire. (The oil resources of Burma, then part of British India, were the one exception.) Nevertheless, Britain in 1912 decided to switch its navy from British coal to foreign oil. This risky step, taken to increase the speed of the British battle fleet, involved dependence on fuel shipped to Europe half way around the globe from Burma and Persia (Iran). World War I was fought right at the interface of the Coal Age and the Oil Age. Primitive airplanes and tanks entered the scene, but the conflict was generally dominated by coal energy. Germany, rich in coal but entirely without oil, eventually capitulated without losing a single decisive battle.

World War II, on the other hand, was a full-fledged Oil Age war. Germany lost it after failing to reach the oil fields of Azerbaijan, and Japan entered it for the sake of gaining control over the (then Dutch-owned) oil provinces of Indonesia. In the end, World War II was won through American oil. However, right after Germany's defeat the Russians developed the rich oil fields of the Volga and Ural basins. Consequently the world's two principal oil producers, the United States and the Soviet Union, emerged as superpowers from the global conflict and divided the world into two opposing zones of influence. Outside the United States and Russia the richest oil deposits were discovered in the Middle East (southwestern Asia). Western companies developed the oil fields of this region, but the area turned out difficult to control politically. The United States soon showed a special interest in the Middle East, as domestic oil discoveries wound down: American oil production started its long, slow decline in 1970.

Protected by large oceans to its east and west, the United States during World War II had taken over the safe-island position that Britain had enjoyed for centuries until it came into the reach of German zeppelins, airplanes, and missiles. However, the United States and the Soviet Union both perfected unmanned German missiles into, on the one hand, rockets that carried satellites and people into space, and on the other hand, intercontinental ballistic missiles that could reach every corner of every continent on the planet. These missiles soon carried nuclear warheads to deliver destructive energy of a magnitude that evades imagination.

Part V, titled "Beyond the Oil Age," takes a look into the future. When global oil production is starting to decline, what alternatives do we have in terms of liquid fuels or other sources of energy for transportation? And how should we produce electricity for those large parts of the world that are still waiting for their energy liberation: the adoption of a more energy-rich lifestyle with all the consequent benefits as experienced in the industrialized world? As a review of the most recent population projections shows, this is not going to be easy, especially because the issue is complicated by the perceived or real threat of global climate change. Futurists tend to say,

"The best way to foresee the future, is to create it." But in terms of climate change, this is a bit tricky. The climate record recovered from nature clearly indicates that we must expect to soon enter a new Ice Age, not a very attractive thought. On the other hand, fossil fuel burning has elevated atmospheric concentrations of carbon dioxide, a greenhouse gas that keeps the planet's surface warm, to unprecedented levels. If these amounts of carbon dioxide are not absorbed by the system (additional dissolution in the oceans; or increased plant growth as indicated by recent global greening[5]), or countered by means of technology (releasing dust in the atmosphere to cool it, adding iron to the oceans to promote algae growth, etc.), we need to expect a trend of global warming that, in extent and speed, outstretches the warming periods observed in recent history. If we look exclusively at the Coal Age and Oil Age, then global temperatures reached a record peak in 1998, while regional temperatures did not. (NASA recently had to revise its temperature record for the United States. The warmest year since the late Coal Age was not 1998, as previously claimed, but rather 1934, the year the infamous Dust Bowl devastated the Midwest. Five of the 10 warmest years were between 1920 and 1939, only one of the 10 was in the 21st century.) On global average, 1998 remained the warmest year in the period 1998–2008, with temperatures now being at about the same level as they were during the Middle Ages, when vineyards flourished in England, and Viking farmers settled in Greenland. But it remains unclear what is going to happen with the Earth's climate in the future, and just about everyone would agree that it is a bad idea for humankind to keep conducting an uncontrolled experiment on its own habitat. Unfortunately, the computer models currently employed remain weak in terms of predicting what the effects of global warming might be on a regional or local scale. This question is not to be taken lightly, as the consequences of global climate change have to be weighed against the benefits that the unhampered use of inexpensive fossil energy would provide for the world, and especially for the developing world. The only carbon-neutral energy alternatives that are currently competitive on a large enough scale are nuclear energy and waterpower. However, much of the world's waterpower potential has already been tapped, and nuclear (fission) energy remains controversial due to questions of security and final waste storage. The other nuclear energy technology, nuclear fusion, holds the potential of leading the world into a great new Energy Era, but currently remains illusive. Billions of dollars are being pumped into the development of a nuclear fusion reactor, but severe material limitations have yet to be overcome. Thus, the world seems to be heading into a Nuclear Age based on established nuclear fusion technology, or into a Second Coal Age, in which much of the energy needs is met by cheap, abundant, and fairly widely distributed coal, supplemented by various forms of renewable energy. Technology to capture carbon dioxide directly at the coal-fired plant already exists, but currently costs about a third of the energy gained. Similarly, coal liquefaction technology is in place (as it had been developed by Nazi Germany during World War II, and in South Africa during the international anti-apartheid embargo), but it is not yet economical. Thus, oil will in the near future remain the critical liquid fuel for transportation, and in terms of economic and military power. But how will this translate into global political developments?

We are now in the period that began with the collapse of the Soviet Union, but is dominated by the fact that the sole remaining superpower, the United States of America, is running out of oil. The traditional U.S. approach to this problem was to strive for control of the oil-rich Middle East, but the disastrous developments during the Iraq occupation for a moment seemed to change the attitude. At least in part under the disguise of a sudden interest in climate protection, an issue long marginalized by America to the despair of the rest of the world, the U.S. government seemed to aim for a new kind of energy self-sufficiency in terms of transportation fuel, notably by promoting the production of bioethanol. However, this was set in perspective when the United States announced in July 2007 that arms sales and military aid to U.S. allies Saudi Arabia, Israel, and Egypt were stepped up to arm the Middle East with more sophisticated weapons systems than ever before. It is questionable whether this will help the United States cling to global power. There are currently strong indications that relative or absolute U.S. decline will be accompanied by the rise of such populous Asian countries as China and India. But these countries do not have oil either. Energy-rich Russia will likely be able to improve its global strategic position, and has already reversed the trend of liberalizing and privatizing its oil and gas sectors. The Middle East will be in possession of increasingly larger shares of the remaining oil, and the whole picture has to be viewed in terms of the continuing global proliferation of weapons of mass destruction. To get a clear view of what is most likely to happen in the near future, we need to review the lessons learned from the previous Energy Eras and their transitions, and apply them to the current situation. And we need to assess how strong the energy theory of human history is in comparison to other theories explaining world history.

NOTES

1. Charles Darwin, *The Descent of Man, and Selection in Relation to Sex* (New York: Barnes & Noble Publishing, 2004), 113.

2. Jared Diamond, *Guns, Germs, and Steel: The Fates of Human Societies* (New York: W.W. Norton & Company, 1997).

3. C.E., or Common Era, replaces the religiously and culturally (more) biased term A.D. Accordingly, the now outdated term B.C. has been replaced by B.C.E., Before the Common Era. However, the new system indicates the same points in time, or time periods, as the old system according to the Gregorian calendar.

4. There are a few exceptions at a lower level. Semi-aquatic beavers, for instance, quite dramatically shape their environment through the construction of dams.

5. United Nations Environment Programme, *Global Environment Outlook, GEO Year Book 2003, International Environmental Agenda*, Box 4: Greening of the biosphere (Nairobi: UNEP, 2003), http://new.unep.org/geo/yearbook/yb2003/box7a.htm.

BIBLIOGRAPHY

Darwin, Charles. *The Descent of Man, and Selection in Relation to Sex.* New York: Barnes & Noble Publishing, 2004. (Original edition, London: John Murray, 1871, available online at Project Gutenberg, http://www.gutenberg.org/etext/2300).

Diamond, Jared. *Guns, Germs, and Steel: The Fates of Human Societies.* New York: W.W. Norton & Company, 1997.

United Nations Environment Programme. *Global Environment Outlook, GEO Year Book 2003, International Environmental Agenda*, Box 4: Greening of the biosphere. Nairobi: UNEP, 2003. http://new.unep.org/geo/yearbook/yb2003/box7a.htm.

PART IV

Oil Age

Given that it triggered the outbreak of a conflict that killed and wounded tens of millions of people, we should call this event nothing but sad. But the assassination of Franz Ferdinand, Archduke of Austria, was actually a somewhat bizarre affair. The designated successor of aging Austrian emperor Franz Joseph was cruising through the Bosnian capital of Sarajevo in an open Gräf & Stift motor vehicle that was part of a six-car convoy. Franz Ferdinand's automobile passed slowly through a cheering crowd, when a Serbian terrorist threw a bomb into the car. However, the Archduke, in a lucky reflex, lifted his arm and deflected the bomb towards the outside of the vehicle. Unfortunately it exploded right when the following car drove over it. The bomb injured two of the passengers and half a dozen spectators. Franz Ferdinand had the convoy stopped to take care of the victims, and it was decided that one of his companions had to be delivered to a nearby hospital. Meanwhile the bombthrower attempted to commit suicide. He swallowed potassium cyanide (but immediately had to vomit), jumped into the river (which, however, was not very deep at the site), and was nearly lynched by the crowd (but saved by his arrest). The Archduke allegedly commented, "This was some nutcase. We continue with our program." Thus the convoy proceeded towards the city hall as planned. From there, the party was supposed to move on to the next scheduled event, though along a route that was entirely different than the one previously publicized. However, Franz Ferdinand insisted on checking up on the hospitalized man before the next event, and on the way to the hospital his chauffeur took a wrong turn, right onto the original route. The driver immediately realized his mistake and backed off, but he stalled the engine.

And right then it happened. One of the 10 terrorists involved in the plot had bought himself a sandwich at the very corner the car came to a standstill. He was frustrated that the assassination attempt had failed, and had given up, but when he saw the Archduke's car, he instantly pulled out his Browning handgun and shot Franz Ferdinand and his wife Sophie. Both died in the car. The assassin, a Bosnian Serb, was arrested and sentenced to 20 years in prison. He escaped the death sentence as he was a few days short of his 20th birthday, but died four years later in his cell of tuberculosis. (All three principal assassins had been chosen by the head of the terrorist group, who was the chief of the Intelligence Department in the Serbian Army, because they were suffering from tuberculosis. They knew they did not have long to live and were ordered to commit suicide after killing the Archduke.)[1]

The fact that Franz Ferdinand was killed in a motor vehicle rather than a horse-drawn carriage suggests that the Oil Age had already begun when World War I broke out in 1914. What is more, some relatively reliable gasoline-powered airplanes had by this stage taken the skies, and the British had decided two years earlier, in 1912, to convert their navy from coal to oil. Nevertheless the world had not yet truly proceeded from the Coal Age. Coal still defined productivity (through stationary steam engines), mobility on

The Franz Ferdinand Vehicle   This is the 32-horse power gasoline-fueled motor vehicle in which Archduke Franz Ferdinand of Austria was assassinated on June 28, 1914. The event triggered the outbreak of World War I, a conflict waged at the interface of Coal Age and Oil Age. (Based on photograph "Gräf & Stift Double Phaeton" by Kadin2048 (Wikimedia Commons), licensed under Creative Commons Attribution 2.5 (http://creativecommons.org/licenses/by/2.5/), edited. The vehicle is permanently on exhibit at the Museum of Military History [Heeresgeschichtliches Museum] in Vienna, Austria.)

land (through steam trains), and mobility on the waters (through coal-fired steamers: not a single ship of the new class of oil-fueled British battle steamers had been completed by the start of the war.) Automobile technology was also still immature. (When U.S. President John F. Kennedy was assassinated half a century later, he was chauffeured in a car 10 times as powerful as the Archduke's.) Neither tanks nor capable oil-fueled tractors had yet emerged. Agricultural productivity, like individual land transport, was therefore still dependent on animal power, and the horse population was still swelling (to reach a maximum around 1920 in the United States, for instance[2]).

Perhaps during World War I, but definitely from the 1920s, the world fully stepped into the Oil Age. Coal had accounted for 93 percent of the world's commercial energy supply in 1900 as well as in 1910, but had passed its peak by 1920 (88 percent), and kept on sinking (to just under 80 percent in 1930) to continue its relative decline in favor of oil.[3] However, it was actually not before 1965 that oil passed coal in terms of contribution to the world's total energy mix. Both oil and coal contributed about 35 percent that year, thereafter coal stabilized a bit under 30 percent, and oil at about 40 percent. A share of 40 percent of all energy, more than any other energy source, may sound like a lot, but it actually understates the importance of oil in the Oil Age. Accounting for about 90 percent of the energy used for transportation, oil has become the lifeblood of industrial, urbanized societies, which depend on swift movement of raw materials, manufactured goods, food, and people. Oil fuels cars, trucks and airplanes, and powers tractors and other machinery for agricultural production. In the absence of other inexpensive liquid fuels crude oil has thus become indispensable for economic growth and military power. For a while oil was also used to produce a lot of electricity, but more importantly it turned out a critical raw material that was processed into anything from plastics (a whole new class of materials), asphalt, lubricants, paints, and pesticides, to medicines and thousands of other products.

The Oil Age's unprecedented energy command dwarfed even that of the Coal Age. The amount of technical energy consumed per person living on the planet averaged around 2,100 kcal per day in 1860, 18,000 kcal per day in 1950, and 41,000 kcal per day in 2000. Expectedly, commanding such huge amounts of energy dramatically accelerated the effects known from the previous Energy Eras. As industrial and agricultural mechanization continued, ever more people were freed from physical work, knowledge accumulated at extreme rates, and professional specialization advanced to new levels. The continuing electrification and the spread of household appliances helped to liberate women, who were in turn added to the pool of production, service, and, emerging later in the Oil Age, knowledge workers. Various technologies, including those in the areas of transport (cars, airplanes), weapons (nuclear bombs, unmanned intercontinental missiles), and communications (satellites, mobile phones, Internet), emerged with lightning speed compared to previous standards. Electricity-powered computers accelerated technological and economic development even more, as they helped people to store, process, and transmit information. Advances in agricultural sciences made it possible to feed a swelling human crowd, and advances in medical sciences allowed people to live longer than ever. The global human population more than doubled from less than a billion in 1800 to two billion in 1930, but in turn tripled within just 70 years to reach over six billion in 2000. Life expectancies reached some 80 years in industrialized regions towards the end of the 20th century, over twice as long as just a century earlier.

To be sure, the Oil Age had a somewhat difficult start. World War I undoubtedly promoted the development of oil technology, but shortly after the conflict, when countries were still recovering, the world slid into the global economic recession of the 1930s, which was immediately followed by World War II. Hence, the first half of the 20th century was mainly concerned with applying oil energy for military purposes. In the productive rather than destructive sense, the Oil Age fully took off from the 1950s. Many of the technologies developed under the competitive pressure imposed by military conflict now began benefiting non-belligerent life, but unfortunately it cannot be said that the second half of the 20th century was a peaceful time. The United States and the Soviet Union, the world's two principal oil producers and newly emerged superpowers, engaged in the Cold War and the Space Race, conflicts that yielded weapons of mass destruction that could be delivered to every corner of the world at the push of a button. Thus life in the Oil Age in a way became more unsafe, even though people enjoyed more energy, advanced technologies, and better health.

At the beginning of the 21st century it was unclear how long the Oil Age was going to last. It slowly became obvious, though, that this Energy Age might be fundamentally different from the previous ones. Never before had it been experienced that an Energy Era's defining fuel would show signs of exhaustion before a new principal energy source had emerged. All forecasts firmly put the peak of global oil production into the 21st century, some in the first and some in the second half.

NOTES

1. Friedrich Würthle, *Dokumente zum Sarajevoprozeß. Ein Quellenbericht*, Mitteilungen des Österreichischen Staatsarchivs—Ergänzungsband 9 (Horn: Verlag Berger, 1978), http://www.austria.gv.at/site/5212/default.aspx#a4; Michael Duffy, ed., "Who's Who: Gavrilo Princip," First World War.com, http://www.firstworldwar.com/bio/princip.htm. The Gräf & Stift motor vehicle in which Franz Ferdinand, Archduke of Austria, was assassinated is exhibited at the Vienna "Heeresgeschichtliches Museum" (Museum of Military History), www.hgm.or.at.

2. At the peak the U.S. horse and mule population counted about 26 million before going into steep decline. "Energy in the United States: 1635–2000," Energy Information Administration, http://www.eia.doe.gov/emeu/aer/eh/total.html.

3. United Nations Organization, "World Energy Requirements in 1975 and 2000," *Proceedings to the International Conference on the Peaceful Uses of Atomic Energy*, Geneva, 1955, Volume 1 (New York: UNO, 1956), quoted in Vaclav Smil, *Transforming the Twentieth Century: Technical Innovations and Their Consequences* (New York: Oxford University Press, 2006), 35. The figure for 1900 (93 percent) is mentioned in Stephen Hughes, "The International Collieries Study," ICOMOS (International Council on Monuments and Sites) and TICCIH (The International Committee for the Conservation of the Industrial Heritage), International Council on Monuments and Sites, Paris, France, 6, http://www.international.icomos.org/centre_documentation/collieries.pdf. In the United States, coal accounted for about 75 percent of total energy use at the end of World War I, and then began to decline. "Energy in the United States: 1635–2000," Energy Information Administration, http://www.eia.doe.gov/emeu/aer/eh/total.html.

CHAPTER 25

WHAT IS CRUDE OIL?

Crude oil, or petroleum, is a thick, greenish-brown, flammable liquid found underground in permeable rocks. It was generated in processes similar to those that yielded coal, but derives from marine organisms (chiefly algae, plus plant and animal plankton) that lived some 300 to 400 million years ago in the world's oceans. The leftovers of these organisms did not just decay, but sank to the ocean floor, where they were buried under gradually more sediment. This organic matter was thus cut off from oxygen supplies and decomposed with the help of bacteria under specific temperature and pressure conditions to yield energy-rich organic compounds, some gaseous, some liquid. Due to the mechanisms of tectonic movement and rock formation these liquids and gases may well have ended up at extreme depths, but owing to their low density, they began to move upward until being trapped, and accumulating, under formations of impermeable rock. To be sure, the formation of oil and natural gas, like that of coal, is an ongoing process. Large amounts of oil are probably being generated in the Black Sea right now, but the process will be completed only in about 100 million years.

Oil is typically found at a depth of between 7,000 feet (2,100 meters) and 15,000 feet (4,600 meters), but in some regions it made it much closer to the surface: there are even sites where oil has seeped into surface sediments. Oil always occurs in association with natural gas, but beyond 15,000 feet underground the temperature and pressure is so high that only gas and no oil can exist. Natural gas chiefly consists of methane (CH_4), plus some other relatively short hydrocarbons, including ethane (C_2H_6 or $H_3C\text{-}CH_3$), propane (C_3H_8 or $H_3C\text{-}CH_2\text{-}CH_3$), and butane ($C_4H_{10}$ or $H_3C\text{-}CH_2\text{-}CH_2\text{-}CH_3$). Crude oil itself is a liquid mix of hydrocarbons whose chain-lengths vary from just a few carbon atoms (such as dissolved propane, 3C) to over a hundred carbon atoms (asphalt). In fact, crude oil contains some 500 different hydrocarbons. Some are open carbon chains, others are cyclic.

It is relatively easy to separate the different crude oil components from one another. Short carbon chains are lighter and evaporate at low temperatures, while long carbon chains are heavier and evaporate at higher temperatures. Hence the crude oil mix may be gradually heated towards higher temperatures, with the gases that evaporate at different temperatures being captured and separately reliquified (condensed) by cooling. This process is called fractional distillation. (People have used the process of distillation since ancient times to separate alcohol from water to prepare intoxicating beverages.)

According to the lengths of their carbon chains, the obtained liquids have different properties and applications. Gasoline, the fuel used to power most present-day cars, contains mainly hydrocarbon chains ranging from 6 to 12 carbon atoms. (Octane, C_8H_{18}, is one of them.) Kerosene components consist of carbon chains of 10 to 16 carbon atoms. Kerosene was formerly used as lamp oil and is now best known as the fuel that keeps jet planes in the air. Diesel, also known as diesel oil, is used for diesel engines that power trains, boats, trucks, and busses. Diesel contains molecules of 15 to 18 carbon atoms, which is similar to light (domestic) heating oils. Viscous, heavy heating oils consist of somewhat longer carbon chains.

Other crude oil fractions are used for non-fuel applications. On the shorter end, benzene (5 or 6C) is used as a solvent to prepare dyes and paints, for instance. Long molecules of at least 22 carbon atoms constitute paraffin and other waxes. But there is more. Plastics, synthetic fibers, pharmaceuticals, detergents, fertilizers, insecticides, lubricating oils, explosives, and many other chemicals and materials that people are now taking for granted are being produced from compounds contained in the crude oil mix. Nevertheless the most important application of crude oil is as fuel. Like coal, oil owes its energy content to solar energy captured by plant growth hundreds of millions of years ago. But crude oil contains nearly twice as much energy per weight when compared to some common bituminous coals. And unlike coal (and even coke), the fuels distilled from crude oil burn without leaving ashes. What is more, a liquid fuel may be conveniently pumped from one place to another. On the downside, crude oil deposits are more rare than coal deposits, and a lot more difficult to find.

CHAPTER 26

THE EMERGENCE OF OIL TECHNOLOGY

Oil technology emerged towards the end of the 19th and the beginning of the 20th century in areas of central-western Europe and the United States. In these regions of mature coal technology the shortcomings of solid fuels became apparent, and the lack of appropriate liquid fuels obvious. The first crude oil component to be marketed was kerosene, which competed with whale oil on the lighting market. The tarrier crude oil compounds were at first discarded, but soon served the booming market for machine lubricants. Crude oil fuels were then used on ships to fire steam engines (and turbines), but most importantly they facilitated the development of an entirely new type of prime mover that would soon power cars, trucks, tanks, and airplanes.

ANCIENT USES OF OIL

Crude oil has been known to people since ancient times. It seeped out of underground reservoirs at different (global) locations, including the oil-rich areas adjacent to the Fertile Crescent. As early as the 4th millennium B.C.E. the nonvolatile, sticky components of crude oil have been used in Sumer (present-day Iraq) as mortar between bricks. Asphalt was also used as waterproof coating for boat hulls, baths, and pottery by some of the early civilizations of western Asia and the Mediterranean. Moreover, crude oil compounds were applied as axle grease for vehicles, for embalming Egyptian mummies, and for illumination purposes. When Alexander the Great entered Ecbatana (in present-day Iran), crude oil was spread onto the streets and ignited at night for lighting, and the thermae of Constantinople were known to use petroleum for heating purposes. (Petroleum, the other word for crude oil, literally means rock oil, derived from the Latin words petra, rock, and oleum, oil.)

The Chinese may have used oil lamps from the 4th millennium B.C.E. and later applied it in cooking stoves. The Chinese even drilled for underground oil as early as the third century B.C.E. They used metal drills to reach the oil and then pushed bamboo tubes into the holes to collect the oil at the surface. Interestingly, the Chinese also utilized natural gas at this very early time. The gas was delivered through bamboo pipelines to evaporate brines (for salt production) in enormous cast-iron pans in the landlocked Sichuan province. (Some natural gas was also used for lighting and cooking.) Chinese drilling technology was developed for the purpose of pumping brine from wells drilled into salt beds. The principal drilling technique, invented during the Han dynasty, was percussion drilling: two to six men jumped on a lever at rhythmic intervals to raise a heavy iron bit attached to long bamboo cables from a bamboo derrick.[4] The deepest recorded boreholes reached only 10 meters during the Han dynasty, but 150 meters by the 10th century. (The Xinhai well of 1835 was one kilometer deep.)[5] However, the use of oil and natural gas in ancient and medieval China was a rather isolated practice. Oil technology, as it paved the way into the Oil Age, emerged in the Western world.

OIL FOR LIGHTING

The modern oil industry had its starting point in the market for lighting agents. At the beginning of the Coal Age people had no sufficient means to illuminate their houses and workplaces. Light was provided by torches and candles (made from tallow), as well as lamps that burned oils rendered from animal fat. Coal gas (town gas), obtained during coke production, was used for lighting in industrial areas and in major cities from the early 19th century. Finally in the 1890s the first efficient electricity plants were installed to provide urban centers with electric illumination.

In the countryside, however, lighting remained a challenge. Whale oil, eventually used by 10s of millions of rural households around the world, emerged as an acceptable lamp fuel, because it burned with less odor and smoke than other fuels. Most of it came from the nose of the sperm whale, the animal named erroneously by early whalers thinking the semiliquid, waxy, milky-white substance found in the enormous, distinctively-shaped head of this whale species was indeed sperm. Known as spermaceti, this product soon became most popular as lamp fuel and for candle production, but other whale species were hunted as well to serve the fast growing markets for lamp fuel and machine (train) lubricant. The U.S. whaling fleet alone expanded from 392 ships in 1833 to 735 ships in 1846. Ten years later U.S. annual production of spermaceti had reached some five million gallons, and train oil production some eight million gallons. American whalers then accounted for 80 percent of the world's whaling fleet and killed over 10,000 whales per year. Some of the scarcer whale species were hunted to the brink of extinction, while the sperm whale population, which is distributed widely from Arctic waters to the equator, coped relatively well with the situation. Nevertheless whale oil remained expensive. What is more, whale oil had a poor shelf life and often spoiled before reaching its point of use on the countryside.[6]

When kerosene emerged as an alternative to whale oil, the idea of using crude oil for lighting was by no means new. However, unprocessed crude oil, as it had historically been used, develops a lot of smoke when burned. It is therefore impractical for indoor application, and it leaves a sticky residue that would rapidly render lamps unusable. The innovation about kerosene was that it was distilled from crude oil. Kerosene itself is a colorless, thin, flammable liquid that comprises between 10 and 25 percent of the total crude oil mix. The heavier parts of crude oil, such as bitumen and asphalt, were initially discarded, and the more volatile compounds, including gasoline, were initially burned off. Eventually various oil components were used to produce lubricants and paints, but all oil prospecting and processing was at this stage driven by the market for lamp kerosene.

Galicia

Commercial distillation of crude oil for the purpose of producing a suitable lamp oil was started in the Austrian monarchy, in Galicia. This region, which is now divided between Poland and the Ukraine, had become Austrian crownland in 1772 during the first partition of Poland. It counted only 2.8 million inhabitants in 1786, but (despite a wave of emigration from the 1880s due to famines) as many as 7.3 million in 1900. Roughly half of Galicia's population consisted of Poles, the other half mainly Ruthenians. Galicia included the northern slopes of the Carpathians, a mountain range that stretched through other parts of the Austrian Empire all the way south into Romania. The Carpathians had long been known for their oil seeps. In fact, Carpathian oil had been hand dipped in the 16th century to be burned in street lamps in the Polish town of Krosno.[7]

Between 1810 and 1817 Joseph Hecker and Johann Mitis were distilling Carpathian oil for lamp fuel production. Following a successful demonstration of an oil lighting scheme in the city of Prague in 1816, the Town Council ordered a large quantity of lamp oil from them. However, the business of Hecker and Mitis soon ran into problems and bankruptcy, mainly because the infrastructure for transporting their product from Galicia to the central regions of the Austrian Empire was inadequate in the days before the railroad. Hence, another four decades went by before Galicia truly became

Galician Oil Boom Galicia, in the Austrian monarchy, experienced an oil boom that started in 1854. Here, the world's first oil refinery went into operation in 1858, and soon the north Carpathian landscape was battered with derricks. Derricks were used for percussion drilling, also known as cable tool drilling, which involves the repeated dropping of a heavy metal bit onto the ground to break through rock. The picture shows a lithography of 1881. By this stage steam engines were used to lift the heavy drill bit.

somewhat of a birthplace of the modern oil era. In 1853 Ignacy Lukasiewicz, a Polish pharmacist living in Lemberg (now Lvov, in the Ukraine), together with Johann Zeh, traveled to Vienna to register the oil distillation process they had invented. Half a year earlier Lukasiewicz had been called to a local hospital to provide light from one of his kerosene lamps for an emergency surgery. Lukasiewicz's lamp gave a bright light without developing smoke, and the hospital administration was so impressed that it ordered several of these lamps and 500 kg (1,100 lbs) of kerosene. From 1855 many other hospitals in the Austrian Empire followed suit, and in 1858 Vienna's North Train Station installed kerosene lamps as well.

Lukasiewicz initially produced kerosene from seep oil obtained from hand-dug pits, but in 1854 he established what was referred to as an oil mine together with two business partners. This mine pumped crude oil from hand-drilled wells that were 30 to 50 meters deep. Well depth soon reached 150 meters, and the enterprise began attracting competitors. A true oil rush set in. Hundreds of prospectors moved into Galicia, battering the north Carpathian landscape with derricks. The world's first oil refinery went into operation in Galicia in 1858, and by 1873 about 900 firms employed some 12,000 workers in the Galician oil industry. (In later years mechanization and consolidation reduced the number of oil workers. Nevertheless there were still 285 companies with some 3,700 oil employees in 1890.)

The boom spread to the southern side of the Carpathians as well. The first Romanian oil wells were drilled at Bend, northeast of Bucharest, in 1857, and kerosene was immediately used to illuminate the streets of Bucharest. Subsequently the city of Ploiesti emerged as Romania's principal oil center. (The oil fields around this city were eventually going to fuel the war machinery of Nazi Germany.)

Pennsylvania

Meanwhile in North America, Canadian geologist Abraham Gesner, who had studied medicine in London, obtained kerosene by distilling cannel coal, a bituminous coal that contains much volatile matter. After moving to New York, Gesner in 1854 received three U.S. patents on a process for the production of lamp kerosene by distillation of "petroleum, maltha, or soft mineral pitch, asphaltum, or bitumen." (Four years earlier, in 1850, James Young had patented a process of distilling illuminating oils from petroleum springs and coal in Britain.) John Austin, a New York merchant, had come across an inexpensive type of oil lamp while traveling in Austria, and sold kerosene lamps from 1854, the same year Gesner received his patents. Three years later Michael Dietz, a producer of whale oil lamps, patented a clean-burning lamp for kerosene. And as demand for kerosene was increasing rapidly, over 50 companies were manufacturing kerosene from coal in the United States by 1859. A single large plant in Pennsylvania produced as much as six thousand gallons a day.[8]

North America was now ready to develop a crude oil (drilling and refining) industry. The continent could be reasonably expected to be quite rich in petroleum. Oil bubbled up along many streams in New York, Pennsylvania, and West Virginia, and even the earliest European explorers had observed seep oil in different locations. English navigator Walter Raleigh mentioned the Trinidad pitch-lake as early as 1595, and Russian traveler Peter Kalm included the oil springs of Pennsylvania in a map published with his work on America in 1748.[9] Northwestern Pennsylvania's Oil Creek runs over a very shallow oil deposit. Its banks are dotted by hundreds of artificial pits, 4 to 6 feet deep, 20 to 35 feet across, some of them shored up with timber. Whenever rain fills these pits, oil that seeps into the pits rises to the top where it can easily be harvested by skimming it from the water's surface. The construction wood of one of these pits has been dated to the early 1400s, suggesting that natives used petroleum in pre-Columbian times. Letters written by missionaries in the 18th century mention that Seneca Indians used oil as a salve, mosquito repellent, purge, and tonic. These natives traded Seneca Oil to European immigrants, who eventually began skimming oil from the surface of Oil Creek with wooden paddles or blankets. They used it to soothe aching joints and itching skin, and some burned crude oil in their lamps despite the black smoke and strong odor. Others used flour-thickened oil to grease wagon wheels and sawmill machinery.

However, petroleum was also an unwelcome by-product of salt wells drilled in Ohio, New York, West Virginia, Kentucky, and Pennsylvania. In these regions oil was contaminating several of the brine wells that supplied a booming salt industry in the early Coal Age. Samuel M. Kier bottled crude oil from his father's Pennsylvania salt wells, selling it as medicine around 1845, and soon thereafter distilled it into lamp fuel. Carbon oil, as he called it, was so successful that he built a refinery with a five-gallon still. By 1858 large quantities of carbon oil were sold in New York City for lamps, while crude oil (from northwestern Pennsylvania) became the chosen lubricant for America's booming textile industry.

When demand for crude oil pushed the price from 75 cents to two dollars a gallon, the stage was set to begin the deliberate search for underground oil deposits. In 1853 chemists at Dartmouth College evaluated a petroleum sample from an oil spring near Titusville in the Oil Creek valley. Their positive assessment caught the attention of George H. Bissell, a New York lawyer and Dartmouth graduate, who purchased the land where the oil sample had come from for $5,000 in late 1854. However, even though an oil boom was in full swing across the Atlantic in Austrian Galicia, Bissell had problems finding investors during the following years. He thus commissioned a report by Professor Benjamin Silliman, Jr. of Yale College to show the economic potential of petroleum, which finally enabled Bissell to persuade a group of New Haven speculators to finance oil prospecting efforts on his land. Edwin L. Drake, a 39-year-old retired train conductor who went by the nickname "Colonel," was hired in 1858 to organize the search. Drake then employed William Smith, an experienced salt driller, as technical supervisor. On August 27, 1859, one year after North America's first oil well had been drilled in Ontario, Canada, Drake's team struck oil at a depth of 69 feet (21 meters) below surface. This find at Titusville, Pennsylvania, launched the modern U.S. petroleum industry. The team had in fact discovered the world's first large oil field, and many more wells were immediately drilled along the creek. During the following four decades, until the East Texas oil boom of 1901, Pennsylvania accounted for half of the entire global crude oil production. Whale oil almost immediately disappeared, to be replaced by cheaper kerosene that did not spoil. (Nevertheless the whaling industry commenced again at the turn of the 20th century, when a new chemical procedure, hydrogenation, made it possible to process whale oil into soap and margarine.)

Launch of the U.S. Petroleum Industry   Retired train conductor "Colonel" Edwin L. Drake and his crew struck oil at Titusville, Pennsylvania, in August 1859. The picture, dating to 1861, shows Drake, in top hat, talking to his friend Peter Wilson, a Titusville pharmacist, in front of his oil well. This was the second derrick and engine house—the first one burned down in 1859. (Library of Congress image LC-USZ62-87910, edited.)

Azerbaijan

The Drake discovery touched off an international search for oil. There were several promising regions, but one area stood out, because it had been known for its oil seeps since antiquity: the Caucasus region of modern Azerbaijan, at the shores of the Caspian Sea. When Marco Polo in 1264 visited the city of Baku, the region's capital, he witnessed the mining of seep oil and observed "a fountain from which oil springs in great abundance." He also saw spectacular mud volcanoes and a flaming hillside, the "Eternal Fires of the Apsheron Peninsula," where burning condensate and natural gas that seeped through fractured shales had long been worshipped. Azerbaijan was part of the Ottoman Empire from 1516, part of Persia from 1618, and was annexed by Russia in 1806. During the following century Azerbaijan emerged as the world's largest oil producer.[10]

Azerbaijan's Absheron peninsula, northeast of Baku, counted about 120 hand-dug oil wells by 1821. Russian engineer F.N. Semyenov drilled the first modern oil well on the Aspheron Peninsula in 1848, but the Russian government retained a drilling monopoly, and nothing much happened until Drake's 1859 discovery revived interest in the area. In 1860 famous German chemist Justus von Liebig sent his scholar Eichler to Baku to establish refining technologies. Baku's first refinery was opened in 1863, but the industry did not grow much until the Baku fields were opened to foreign investment in 1872. A year after, some 80 kerosene refineries filled Baku and the surrounding area with smoke. Robert Nobel, brother of dynamite-famous Alfred, was one of the first foreigners to purchase oil properties. He had actually been sent to Azerbaijan to secure lumber supplies (which the Nobels' Russian munitions factory needed for rifle butts), but a few years later the Nobels owned the region's (and Russia's) largest oil company. In 1875 the Nobel brothers brought in several American drillers to introduce steam-assisted deep well drilling technology, and other concessionaires rapidly followed suit. Thus the hand-digging of wells became obsolete within a few years.

Transportation of oil from the wells to the refineries remained a challenge. Crude oil was filled in barrels at the well, which were loaded either on camels or carts. As carts often got stuck in the oily, sandy ground, the newly founded Nobel Brothers Association constructed oil pipelines to connect its oil fields to the distillation factories in Baku. The first of these pipelines was completed in 1878. It was 12 kilometers (7.5 miles) long and had a pipe diameter of 7.5 centimeters (3 inches), with steam engines forcing the oil through the pipe.

At the port of Baku, kerosene-filled barrels were loaded onto ships that steamed up the Caspian Sea to reach the mouth of the Volga, and further up the river into the heart of Russia. Initially, these steamboats were powered by costly Welsh coal, but in the 1860s two engineers (Shpavkosky from Russia and Aydon from Britain) developed an apparatus that foreshadowed the future use of crude oil as a fuel for prime movers rather than lamps. Their technology utilized a solid by-product obtained at the oil refineries. This refuse was pulverized and sprayed into a furnace to power steam engines. Mr. Kamensky, government engineer in charge of the port of Baku, used this fuel (which became known as astatki or mazoot) in 1869 to power a steamboat, and the Nobels soon adopted it for their fleet of steamers. (Also in 1869, the world's first oil tanker was launched in the United States to serve European markets.) In 1877 Ludwig Nobel ordered (in Sweden) a purpose-made steel-hulled tanker to improve the loading and unloading of kerosene. It was relatively small, but later versions could carry as much as 750 tons of liquid. The Nobels soon enjoyed a virtual monopoly on export of Baku kerosene into Russia, but water transport into central Russia was principally challenging because the Russian rivers and canals froze over during the winter, immobilizing the Nobel tankers during the five darkest months of the year.

Rothschilds Chasing Nobels

It was also difficult to transport kerosene from the landlocked Caspian Sea to international markets. This was especially problematic for the Nobels' competitors who did not have access to the Russian market. In the early 1880s, when Baku had over 100 refineries, they called for the construction of a Transcaucasian railway that would run from the Caspian Sea to the Black Sea, connecting Baku via Tiflis to the Armenian port of Batum (present-day Georgia). From there, kerosene would either be shipped to the major Black Sea port of Odessa, which was connected to the western European railway network, or straight into the Mediterranean, if the Ottomans would allow it to pass through the Bosporus. (From the Mediterranean kerosene would in turn be shipped into the Atlantic or through the Suez Canal to Asian markets.) The project became reality when the Baku kerosene producers mortgaged their properties to the Rothschilds to finance the railway. (The Rothschilds were a banking dynasty of German Jewish origin that had established operations across Europe. This particular deal was with the Parisian Rothschild branch.) Part of the line, between Tiflis and the Black Sea town of Poti, already existed, but the segment between Baku and Tiflis as well as Poti and Batum had to be constructed from scratch. When the railway opened at full length in 1883, it immediately pushed American kerosene off international markets. Some 100 tons of American kerosene were exported in 1883, two years later it was 30 tons. Another effect of the railway was that the Rothschilds became major players in the oil business. They received exclusive rights to their clients' oil as part of the financing deal and in 1884 established the Caspian and Black Sea Petroleum Company that rivaled the Nobel operations in size and controlled 42 percent of the export of Baku oil.

Due to Azerbaijani production the Russian Empire firmly replaced the United States as the largest oil producer in the world towards the end of the 19th century. In 1900 oil production in Azerbaijan amounted to 84 million barrels a year, compared to U.S. production of 62 million barrels per year. (1 ton of crude equals 7.33 barrels, based on worldwide average density. 1 barrel equals 159 liters.) Azerbaijan now had nearly 2,000 operational wells and more than 200 refineries, producing nearly 60 percent of the worldwide kerosene exports.

Downturn

Then things changed. First of all, production in the United States soared to new record levels with the East Texas oil boom of 1901 and the initiation of oil production at California's Kern River oil field, north of Los Angeles. (By 1903 the Kern River field alone produced 17 million barrels per year,

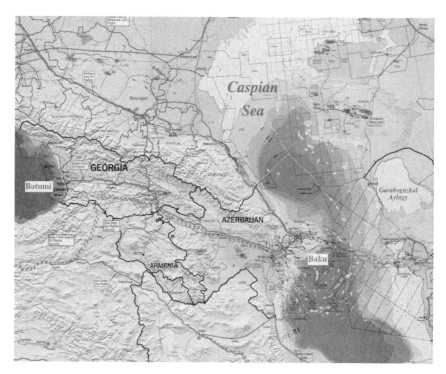

Azerbaijan was the world's top oil producing region in 1900. To transport Azerbaijani oil to international markets, the Transcaucasian Railway was constructed from the Caspian Sea to the Black Sea. Later on, operating from 1906, the world's longest pipeline ran parallel to the railway tracks at a length of 833 km from Baku to the Armenian port of Batumi (present-day Georgia). (The map has been created from a 2001 map: "Oil and Gas Infrastructure in the Caspian Sea Region." Compiled by the U.S. Central Intelligence Agency. Courtesy of the University of Texas Libraries, The University of Texas at Austin.)

and California was America's top oil-producing state.) But Azerbaijan also faced serious domestic problems. Baku's first major strike of 1901–02 was partly organized by young Georgian Dugashvili, the later Soviet dictator Stalin. Baku's oil production peaked at around 103 million barrels a year in 1904, when the disastrous Russo-Japanese War began. The economic crisis during this conflict further increased the tensions between two social groups working in the Azerbaijani oil industry: on the one hand, Muslim Azeris and Persians, and on the other hand, their Armenian employers, who were a Christian minority in the region and served as middlemen for Western factory owners. When strikes and riots erupted all over Russia in 1905, some 2,000 people were killed in Baku, oil fields were set on fire, and over half the oil installations were destroyed.

Nevertheless, Azerbaijan emerged from the tumults as the world's second largest oil producer (after the United States), well ahead of Austrian Galicia, where production peaked in 1909 at a distant 15 million barrels per year (or 4 percent of global production). Ludwig Nobel had proposed to build a pipeline parallel to the Transcaucasian railway, but the government had initially delayed the construction as it feared losing vital rail revenues. Nevertheless the pipeline was eventually constructed. Operating from 1906 at a length of 833 kilometers (518 miles), the Baku-Batum pipeline was the longest in the world. It employed 16 pumping stations; its pipe diameter was 20 cm (8 in); and its carrying capacity was 900 thousand tons per year.

Dutch East Indies (Indonesia)

Still in the Kerosene Era, oil was also found in the Dutch East Indies (present-day Indonesia). Active development began in Sumatra, Java, and Borneo in 1883, 1886 and 1896, respectively. A group of Dutch businessmen founded the Royal Dutch Company, which built its first refinery in Sumatra in 1892 and established its own tanker fleet four years later. In 1907 the Royal Dutch Company merged with Marcus Samuel's London-based Shell Company. Shell was a trading firm that Samuel had inherited from his father. (Among the top-selling imports were sea shells imported from the orient and used to adorn the decorative boxes and bowls then popular in Victorian parlors. Hence the name of the company.) Shell's first foray into the oil industry came in 1878, when Samuel began handling consignments of cased kerosene. In 1890, seven years after the Transcaucasian railway had opened, Samuel visited Batum and Baku together with a Rothschild representative. Thereafter he signed a contract according to which Shell would buy Baku kerosene from the Rothschild company to ship it (on coal-fired steamers) from the Black Sea through the Mediterranean and the Suez canal to the Far East. Samuel commissioned the first oil tanker in 1892, and many more were going to follow. They shipped kerosene to such East Asian markets as Singapore and Bangkok, and by 1913 over half of Baku's oil products were delivered to the Far East.

When Samuel expanded his oil-related business to the Dutch East Indies, he began to compete with the shipping operations of Royal Dutch. However, the two companies soon realized that their principal business activities, one drilling and refining oil, the other transporting oil products, complemented one another. They therefore decided to merge into Royal Dutch/Shell, which soon expanded radically, moving into Romania in 1906, Venezuela in 1910 (though oil was first found in 1914), and Egypt in 1911. In 1912 Royal Dutch/Shell began to operate in the United States. It also acquired the Russian Masut Company, which owned 11 percent of Baku's fields, and in 1913 Royal Dutch/Shell moved into Trinidad as well as Mexico, where oil had been found in 1901.

British Burma

The Burmah Oil Company was also founded in the days when oil prospecting was driven by the market for lamp kerosene. The firm was established to develop the Burmese oil fields as soon as Britain conquered Upper Burma in 1886. (Burma became part of British India. Now called Myanmar, the modern country of Burma borders India, Bangladesh, China, Thailand, and Laos.) Given the enormous size of the Coal Age British Empire, it is actually surprising that Burma was the first and only region under British rule where substantial amounts of oil were found in this period. Burmah Oil Company was created by Lord Strathcona, a Scottish millionaire who had financed the Canadian Pacific Railway. The firm with its head office in Glasgow expanded its operations greatly after 1890, but the kerosene era was coming to a close: At the beginning of the 20th century gas and electric lighting became prevalent even in the more rural areas. But the oil market was nevertheless going to expand: it was now dominated by fuels that powered prime movers to turn the chemical energy of oil into mechanical work.

OIL FOR SHIP PROPULSION

The steam engine, the Coal Age's principal prime mover, was very flexible in terms of fuel. As the task was simply to heat up water externally to generate steam, steam engines could be fired with coal, wood or just about any other combustible material. Oil was a viable option, too. Oil burners fired steam engines in Californian oil fields from 1885, for instance, and little later oil was used to fuel steam locomotives. As refining technology progressed, it became clear that oil might do the job even better than coal, simply because oil-derived fuels contained more energy per weight, burned without leaving ashes, and produced less smoke during combustion. Moreover, liquids can be conveniently stored in tanks and be delivered to the burning chamber through pipes. No human labor (in form of shovelers) was necessary for this operation, and the rather complex apparatus capable of moving solids became obsolete as well.

All these advantages were immediately obvious for steam-propelled warships. Refined oil products with twice the energy density of coal would instantly double the operating range of cannon steamers. Reduced smoke generation would decrease the chances of ships to be sighted from the distance by enemies. Flexibly located oil tanks allowed for more efficient ship designs, and pipes for refueling would eliminate the need for stokers and hence reduce manning. (Moving coal from shore to ship, and aboard ship, was dirty and hard work that often exhausted the ship's crew, robbing it of its brief period of rest during wartime.)

Spraying oil onto coal to improve combustion efficiency was an early step towards oil-powered ships. It was routine by the early 1900s, and soon the world's navies began to experiment with outright oil burners. The Italian navy had most of its torpedo boats equipped with oil burners by 1900, and Russia operated a warship that was solely oil-powered nearly as early. In the United States the use of oil as a standalone fuel was officially recommended in 1904, and the first oil-burning American destroyer, the USS *Paulding*, was commissioned in 1910. Germany was rich in coal, but had practically no oil, neither at home nor in any colony. Nevertheless the Germans built battle ships that could burn oil in addition to coal. The first series of German oil-fired battleships, the *Kaiser* class, was designed between 1907 and 1909. The S.M.S. *Kaiser* was completed in August 1912, featuring 16 coal-fired boilers with supplemental oil-firing. These boilers fed steam to three turbines that drove three propellers. The ship could hold fuel reserves of 984 tons of coal and 197 tons of oil under normal conditions, or a maximum fuel load of 2,952 tons of coal.

In Britain the development of oil-fired ships was promoted by Admiral John Fisher, First Sea Lord from 1904 to 1910.[11] He regularly received (incorrect) reports that the Germans were developing all-oil battleships and felt it was necessary for Britain to do the same in order to match the speed and power of the German fleet. Fisher had previously been instrumental in the development of the *Dreadnought* class. Released in December 1906, the heavily-armed *Dreadnought* was the first major warship powered solely by (coal-fired) steam turbines rather than steam engines. But Fisher's attempts to change the Royal Navy from coal to oil met fierce resistance. Britain had plenty of the world's best (Welsh) coal, but no domestic oil supplies. Britain had also built up an extensive global network of coaling stations. Moreover, coal supplemented armor by reducing damage from shells exploding in coal storage bins (unlike oil tanks that would explode when being hit).

Although Fisher did not succeed with his oil plans during his tenure as First Sea Lord, he managed to convince Winston Churchill, First Lord of the Admiralty from 1911, to switch the navy from British coal to foreign oil. (Churchill was later sacked for his role in the disastrous Battle of Gallipoli, but eventually became Prime Minister during World War II.) The decision was made in 1912, and Britain immediately began to construct an (all) oil-fired, steam-turbined class of capital battleship. The first of these ships, the *Queen Elizabeth*, was completed only after the start of World War I, in January 1915. Smaller ships were converted to oil as well. In essence, the British navy gave up fuel security to match the performance of German ships. Obviously this was a great risk. Britain's naval power now relied on fuel imported from colonial Burma (British India) or from other countries. Oil sourced from the other side of the Eurasian landmass meant that it had to be shipped either through the Suez canal and the Mediterranean, a highly risky undertaking during wartime, or around Africa, which was a very long stretch.

Persian Oil (Iran)

Transportation issues aside, Britain was at least hedging its bets when it came to securing access to oil. Still under Fisher, the British navy in 1904 signed a supply contract with Burmah Oil Company, which in turn grew into a sizable firm. However, the oil resources of Burma were actually quite limited, and Burmah Oil began to finance oil prospecting efforts in Persia (present-day Iran), where it had long been suspected that substantial oil reserves were to be found at the southern Caspian Sea, right next to the Russian Empire's oil-rich Baku region of Azerbaijan.

In fact, there had been a number of attempts to find oil in Persia before the turn of the century. Most notably German Baron Paul Julius de Reuter tried his luck, but failed twice, mainly due to political difficulties. (To be sure, Reuter did well in his other business. The founder of the famous news agency started a continental pigeon post in 1849, and in 1858 persuaded the press to use his news telegrams, which became an international service.) Reuter's oil prospecting activities caught the attention of the British ambassador in Persia, who in turn approached English-Australian lawyer William Knox D'Arcy, asking him if he would be willing to finance a search for oil in Persia. Knox D'Arcy, who had made a fortune in the Australian Mount Morgan gold field, was immediately interested, but it took substantial political maneuvering by the British before the shah of Persia would sign a concession agreement with Knox D'Arcy in 1901.[12] Russia protested the deal, but a relatively small payment to the Persian rulers made sure the British party received a 60-year concession to explore, produce, process, transport, and sell crude oil throughout Persia, except the five northern provinces that bordered Russia. This was acceptable, as it was expected that oil was also to be found in the more southern parts of Persia. However, years of unsuccessful prospecting went by, and Knox D'Arcy ran out of money. In 1905 he was forced to merge his oil operations into a partnership with Burmah Oil, but another three years went by, and the drilling efforts were just about to be abandoned. Then, in early 1908, the prospecting team discovered a substantial oil field a short 200 km from the Persian Gulf coast. This was of tremendous historic significance: The Persian Gulf region was going to turn out the world's most oil-rich area, which put it in constant focus of global politics for the entire Oil Age.

In April 1909 the Anglo-Persian Oil Company (APOC) was incorporated and took over ownership of the concession. (APOC was going to be known as BP, British Petroleum, from 1954.) APOC was established as a subsidiary of Burmah Oil, which held 97 percent of the shares, with Knox D'Arcy a director in the new company. By 1911 the company had constructed a pipeline to, and a refinery at, the (Persian Gulf) coast to serve international markets. In 1914, following the (1912) decision to switch the navy from coal to oil, the British government made a major investment in Anglo-Persian Oil to become the majority shareholder. The British government placed two directors on the company's board and signed a secret contract according to which APOC would provide Britain with oil under attractive terms for 25 years. Persian oil then fueled British ships during World War I, and the government's shareholdings were increased from 51 to 66 percent. (Burmah Oil held 23 percent of the company until the mid-1970s, and the British government decided to sell most of its shares in 1987.)

Iraqi Oil

Also in 1914, APOC became a 50 percent shareholder in the Turkish Petroleum Company, which had secured drilling rights in the Ottoman region of present-day Iraq. It was particularly difficult for Britain to get access to this potential oil area, because the Ottoman Empire had relatively close ties to Britain's rival Germany. In 1889 Deutsche Bank attained a contract to finance the construction of a railroad in Anatolia, and 10 years later this major German bank signed a deal for a complete railway from Berlin via Constantinople (modern Istanbul) to Baghdad (the capital of present-day Iraq), and further on to Basra at the Persian Gulf. To be sure, the railway from Turkey to Germany did not materialize as planned, but the Germans did indeed built a single-track railway through Iraq before World War I. Attached to this deal was a concession to mineral and oil rights over the land on either side of the Baghdad railway line by a convention dating to 1903.

Despite the German contract, numerous international groups kept on making bids for oil concessions in this area between 1900 and 1914. In 1912 the Turkish Petroleum Company was founded and was granted (or at least promised), under somewhat obscure circumstances, a concession to prospect for oil in the Baghdad and Mosum wilayets. Deutsche Bank owned 25 percent of Turkish Petroleum Company, Royal Dutch/Shell another 25 percent, and Turkish National Bank the remaining 50 percent. To be sure, Turkish National Bank was not the central bank of Turkey, but a private British-controlled bank set up in Turkey to promote British economic and political interests. Armenian millionaire Calouste Gulbenkian, a local of the Ottoman Empire, owned 30 percent of Turkish National Bank and organized the creation of the new Turkish Petroleum Company. (Hence, Gulbenkian came to own 15 percent of Turkish Petroleum.)

The British immediately pushed for a restructuring of Turkish Petroleum Company. In 1914, two years after the firm had been founded, Turkish National Bank's 50 percent share in Turkish Petroleum was transferred to the

Anglo-Persian Oil Company, of which the British government had just become the majority shareholder. Deutsche Bank and Royal Dutch/Shell both kept one quarter of the company. (More precisely, Anglo-Persian Oil and Royal Dutch/Shell each transferred 2.5 percent to Gulbenkian, who now owned 5 percent of Turkish Petroleum.) Iraq did indeed prove to be very rich in oil, but the Iraqi oil fields were not discovered before the 1920s.[13]

A Strategic Mistake?

All the political maneuvering, and the substantial investments, were part of Britain's attempt to maintain its role as the leading global naval power in a changing world that moved from the Coal Age into the Oil Age. The British navy did indeed gain a speed advantage, particularly since Germany did not develop oil-only steamers at this stage, but this improvement did not appear to be a decisive factor in World War I. During the 1916 Battle of Jutland, the only full-scale sea battle of World War I (and the last one in history without air support), the British lost more ships than the Germans despite the much larger size of the British fleet. (In terms of tonnage sunk as well as casualties British losses were twice those of Germany.) Superior German gunnery and armor was apparently more important than increased British speed. What is more, the British navy suffered from oil shortages, particularly in 1917. British ships were forced to stay in harbor for some time, and destroyers were restricted to move at relatively low speeds.

Hence, it may be argued that the British choice of switching the navy from coal to oil at this stage was a mistake. On the other hand, this decision instigated the policies that assured British access to international oil. This was going to be crucially important to fuel the tanks, submarines, and airplanes that represented the military might of the future. Besides, oil-firing technology developed for the world's navies rapidly spread to the commercial sector. Coal-fueled vessels soon became a rarity, as cargo vessels and passenger ships shifted to oil as well.

INTERNAL COMBUSTION ENGINES AND MOTOR VEHICLES

Steam engines and steam turbines are considered external combustion engines, because the fuel is burned outside the engine, in a separate unit, to create steam that acts upon a piston or fan blade on the inside. However, as soon as refined oil fuels that burned without leaving ashes became available, they opened the possibility to construct internal combustion engines. In such a setup, the fuel would be burned inside the engine, with explosively expanding combustion products acting directly upon a piston. (The combustion products are mainly carbon dioxide and water vapor. Air is a critical reactant, as burning or combusting implies a reaction of fuel with the oxygen contained in the air.) Internal combustion (IC) engines were developed from the 1830s. Practically all available hydrocarbons were tested as fuel, from the gases of very short carbon-chain length (methane, ethane) to the different crude oil fractions up to a chain length of 18 carbon atoms (C18). Gasoline, with a chain length of C6 to C12, emerged as the fuel that had just about the ideal properties for the most common type of internal combustion engine. Gasoline is slightly more volatile than kerosene (C10 to C16), but still liquid under normal ambient temperatures. Hence, gasoline can be conveniently stored in tanks, and has a much higher energy density per volume when compared to gases. Compact gasoline-fueled internal combustion engines were going to power passenger cars, trucks, and tractors. Thus, they allowed for oil energy to revolutionize human mobility, transportation infrastructure, and agricultural work, while horses (and oxen) retired from their millennia-long role in human societies.

Stationary IC Engines

The earliest internal combustion engines were developed for stationary applications. They were fueled by coal gas, which had become widely available by the mid-1830s. Austrian Christian Reithmann (from 1852) and Belgian Jean Joseph Etienne Lenoir are both credited with independently developing IC engines with electric ignition of the fuel. (These were two-stroke engines fueled with coal gas.) The stationary engine patented by Lenoir in Paris in 1860 was a commercial success: Some 400 units were sold within five years. Nikolaus Otto and Eugen Langen founded an engine factory in Germany to produce a similar stationary IC engine (also running on coal gas). Together with their licensees they built as many as 5,000 engines between 1867 and 1876. Such engines were purchased by various firms (which preferred them over coal-fueled steam engines) to power small items of machinery such as lathes or water pumps. (Electric motors did not exist by this stage.)

The construction of the world's first gasoline-fueled engine is usually accredited to Julius Hock of Vienna, Austria. His engine of 1870 was otherwise similar to that of Lenoir, who had presented an IC engine powered by liquid fuel as well. Neither of the two compressed the fuel, which was critically important, according to a patent by Frenchman Alphonse Beau de Rochas. Christian Reithmann apparently constructed a four-stroke engine in 1873, but it was Nikolaus Otto who, in 1876, built the first practical four-stroke piston engine powered by gasoline. His engine became known as Otto Cycle engine and served as the proto-type for all later gasoline engines. The four strokes of Otto engines are intake (of a mixture of air and vaporized fuel); compression (to about one-tenth of the original gas volume, which increases the temperature); power (ignition of the fuel-air-mix by an electric spark, which results in an explosion that creates high pressure and moves the piston); and exhaust (to force burned gases out of the cylinder). Otto's further development work yielded the invention of a magneto ignition system for low voltage ignition in 1884. The Otto engine model of 1885 was sold in large numbers as a stationary industrial prime mover, but more importantly Otto engines were soon going to power vehicles.

Gasoline Engines for Vehicle Propulsion

Much of the earliest development of internal combustion engines was actually directly intended for vehicle propulsion. Lenoir fitted an engine to a three-wheeled wagon that completed a trip of 18 km in 1863. The first four-wheeled gasoline-powered vehicle was built by Siegfried Marcus in Austria. Marcus invented a carburetor and an electro-magnetic ignition system in the 1860s and built his first motor vehicle in 1870.[14] Like the engine of Julius Hock, the engine of this vehicle worked without gasoline compression. Marcus then improved his engine, and his stationary models were produced by the machine works Schulz and Goebel in 1883, by Ganz & Co. in 1884, and by Heilmann & Ducomun of Mulhouse in 1885. In 1888–89 Marcus introduced his masterpiece vehicle, the Strassenwagen. It featured a true four-cycle gasoline engine, a throttle-control steering wheel, a clutch, a differential, a carburetor, a magneto ignition, and four wheels. This Marcus vehicle is still preserved at the Vienna Technical Museum, and it is still operable.

The breakthrough for motor vehicles came with the work of Germans Gottlieb Daimler and Karl Benz. Daimler had been working with Nikolaus Otto at the Otto-Langen engine factory for 10 years before developing an advanced Otto engine in 1885 together with his design partner Wilhelm Maybach, who invented the float-feed carburetor. Daimler fitted their small, lightweight engine to a bicycle to pioneer the motorcycle about the same time as Nikolaus Otto did. In 1886 Daimler created a four-wheeled motor vehicle by fitting his engine to a stagecoach. He then invented a V-slanted two cylinder and a four-stroke engine with mushroom-shaped valves. In 1889 Daimler presented an automobile with a four-speed transmission that maintained speeds of 10 mph, and the year after he founded Daimler Motoren-Gesellschaft to manufacture his designs. In later years the Daimler company delivered the world's first truck (1896) and the first car to be adopted as a motorized taxi (Victoria, 1897). The Mercedes name was first used for a Daimler car in 1901.

Karl Benz lived and worked quite close to Gottlieb Daimler, but the two actually never met. Benz constructed a lightweight, water-cooled, two-stroke gasoline engine and fitted it to a three-wheeled carriage in 1885. The following year Benz patented a four-stroke engine, and in 1891 he founded Benz & Cie to produce automobiles. In 1894 this firm built the world's first

The Strassenwagen of Siegfried Marcus (1888–89) This original Marcus vehicle is exhibited at the Vienna Technical Museum. It is the oldest preserved original motor vehicle in the world.

motorized omnibus, which seated eight passengers, and by 1900 Benz & Cie had emerged as the world's largest manufacturer of automobiles. In 1926 Benz merged with Daimler, but by this stage Otto's engine designs and car production technology had long spread from Germany to other Western countries and Japan. There were about 300 car makers in the United States in 1895, and following a wave of consolidation 109 were left by 1900, when France had more than 600 car manufacturers. On the eve of World War I Britain counted some 132,000 registered passenger cars, and the United States some 1.7 million.[15] (In addition the United States then had about 100,000 trucks.) Car manufacturing emerged as the leading industry in terms of product value in Western countries. Many other industries, including steel, rubber, glass, plastics, and oil refining, all became dependent on the motor vehicles industry.

Automobiles with four-stroke IC Otto engines truly revolutionized human transport. In terms of top speed, they allowed people to move faster than 130 mph by 1910, faster than 150 mph by 1919, faster than 200 mph by 1927, faster than 300 mph by 1935, and faster than 400 mph by 1947. Today's conventional Otto engines are still based on the original engine designs but have been radically improved in terms of efficiency. Towards the end of the 20th century, small passenger cars seating four people approached a fuel consumption of just three liters of fuel per 100 kilometers—about 78 miles per gallon.

Nevertheless, the average car engine still burns four times as much fuel as it would if it were as efficient as the laws of thermodynamics permit. The main losses of energy are due to incomplete fuel combustion and friction.

The Wankel Engine

A rotary variant of the four-stroke Otto engine was patented by German Felix Wankel in 1936. The idea was to increase engine efficiency by directly producing rotary power rather than translating the reciprocating (up-and-down) motion of pistons in conventional Otto engines via a crankshaft. (Rotary power is necessary to turn the vehicle's wheels.) In a Wankel engine a triangular rotor compresses the fuel towards the wall of an oval chamber and sweeps it along through all stages. The fuel is compressed, ignited, and exhausted, but all these stages happen simultaneously in different sections of just one single chamber that is separated into these sections by the particular shape of the rotor.

Rotary engines saw some commercial success when Wankel began collaborating with NSU, the German company that in 1955 was the world's largest motorcycle producer. (NSU started off in 1873 as a manufacturer of knitting machines, which is reflected in its name Neckarsulm Strickmaschinen Union.) In the 1960s NSU produced Wankel-engined motorcycles and automobiles in Germany, and licensed its technology to car manufacturers in other countries. Notably Mazda produced a large number of Wankel engine cars in Japan.

The Wankel engine is lightweight and compact, and simpler in construction than the four-stroke piston gasoline engine. Hence its main advantage is the high power to weight ratio. On the flip side, Wankel engines always tended to have problems with rotor seals, and the high surface to volume ratio in the Wankel combustion chamber is less thermodynamically efficient than conventional designs. Perhaps most significantly, Wankel engines met fierce resistance by many engine producers in light of decades of developing work invested in conventional Otto engines.

Diesel Engines

A quite different type of internal combustion engine has been named for German engineer Rudolf Diesel, who patented it in 1892. In this piston engine design, atmospheric air is compressed inside a cylinder, and fuel is thereafter injected to ignite spontaneously (without the use of a spark plug) due to the high temperature of the compressed air. In Britain Herbert Akroyd Stuart and Charles Richard Binney received a patent on a similar compression-ignition engine in 1890 and sold exclusive manufacturing rights to Hornsby, a producer of steam engines. Hornsby's kerosene-fueled low-compression IC engines in turn became a tremendous success for stationary applications despite the fact that they required the aid of a blow-lamp (to heat up the combustion chamber) to be started. The first two Hornsby-Akroyd oil engines were installed at the Great Brickhill Waterworks at Fenny Stratford in 1892. (They delivered 7 hp each and worked regularly until 1923.) Hornsby-Akroyd then sold over 32,000 engines. One turned an electricity generator to illuminate the Statue of Liberty in 1895, another powered Marconi's transmitter that sent the first wireless message across the Atlantic Ocean in 1901. Hornsby also diversified into nonstationary applications, releasing in 1896 both the world's first locomotive powered by an IC engine (10 hp) and a 20 hp tractor.

Diesel engines employed higher compression ratios than Hornsby's and were therefore more efficient. (Diesel compression ratios were between 14 and 24, compared to 7 to 10 for Otto engines.) Rudolf Diesel developed this engine type in 1893 at the Maschinenfabrik Augsburg (the later MAN) with financial contributions from the Friedrich Krupp company. In 1897 he demonstrated a 25-horsepower, four-stroke, single vertical cylinder compression engine. As in Otto engines, the four strokes were intake, power, compression, and exhaust, but Diesel engines are inherently more fuel efficient. (The best Diesel engines can surpass efficiency ratings of 40 percent, compared to 25 percent for the best Otto engines.) They are also more reliable, last longer, and are cheaper to operate than Otto engines. On the flipside, Diesel engines are more expensive and heavier than Otto engines: They deliver less power per weight unit.

Ships, Trains, and Trucks

Diesel made a true fortune by selling production licenses globally. Due to its higher weight, the Diesel engine competed more directly against the steam engine than against the Otto engine. The first applications were stationary, to power factory equipment and electricity generators. (Diesel engines are still now very important for electricity production, especially in remote areas and for back-up.) But before long, Diesel engines were also used for marine applications, where their weight did not matter much. They were installed in river-going boats from 1903, ocean-going ships from 1912, and entirely conquered the water-transportation sector from the 1920s. Diesel engines were much smaller than marine steam engines of comparable power and were first adopted by small to medium ships. Larger ships initially preferred steam turbines, but later in the 20th century even the largest oil tankers and container ships used Diesel engines. (The Diesel engines that drive modern supertankers deliver up to 100,000 hp.)

Another kind of marine application was for submarines. These vessels, capable of operating under water, have few practical applications outside the military. In the later Coal Age advanced steam-powered as well as battery-powered underwater boats (with electric motors) were constructed in various countries. In 1900 the U.S. Navy purchased a submarine launched by Irish immigrant John P. Holland. This vessel had a gasoline engine for surface locomotion and a battery-powered electric motor for submerged cruising. Holland's design became the principal prototype for the submarines built during the following years. From 1905 Germany began to develop a submarine with real fighting properties and produced the first diesel-powered Unterseeboot (U-boat). By the eve of World War I all major navies had diesel-electric submarines, and in World War II the Germans introduced such innovations as the snorkel, which supplied fresh air to the diesel engine without having to surface the boat. (Hence, the submarine could remain under the surface while the Diesel engine, running an electric generator, recharged the storage-battery, whose limited capacity restrained the submerged operating range of submarines.)

Diesel engines made a difference for land transport as well. Diesel still witnessed successful test runs of a Diesel train in 1913, but he drowned later that year. (Diesel had earlier collapsed under the burden of being responsible for 30 companies, but it was rumored that he committed suicide, because the overworked man had at this stage lost nearly all his fortune.) Diesel engines then almost entirely replaced steam engines on rails, at least as far as nonelectrified lines were concerned. (By 1960 steam trains had disappeared almost everywhere around the world.) Off rails, Diesel engines turned out the right choice for heavy vehicles such as trucks, buses, tractors, and earth-moving machinery: Virtually every heavy commercial vehicle in the world is now equipped with a Diesel engine. From the 1950s lighter Diesel engines were also used for passenger cars. Partly for tax reasons, these have become tremendously popular in western Europe. (More than one in three of new automobiles registered in western Europe in 2000 had a Diesel engine.)

Various Fuels

The crude oil fraction that became the preferred and principal fuel for Diesel engines also came to carry Rudolf's name: diesel, or diesel oil, contains molecules of ca. 15 to 18 carbon atoms. Diesel is generally simpler to refine than gasoline and hence tends to cost less, even though it contains almost a fifth more energy per unit of volume than gasoline. (This adds to the already greater efficiency of diesel versus gasoline engines.) But Diesel engines are actually quite flexible when it comes to their fuel. The original engine demonstrated by Rudolf Diesel in 1897 was running on unrefined crude oil, and in 1911 Diesel delivered a speech, saying, "The Diesel engine can be fed with vegetable oils and would help considerably in the development of agriculture of the countries which use it." He further predicted that "the use of vegetable oils for engine fuels may seem insignificant today, but such oils may become in the course of time as important as petroleum and the coal tar products of the present." Diesel demonstrated that his engine could run on peanut oil, for instance, though it was soon optimized for the C15 to C18 fraction of crude oil. (Nevertheless even conventional Diesel engines can run on vegetable oil derived from canola, soy, sunflower, hemp, coconut, etc., which can be easily transferred into biodiesel in a process called transesterification.)

Tractors and Tanks

It was quite obvious that motor vehicles that could operate off rails would make a crucial difference in warfare and hence translate into military and political power. The development of automatic weapons towards the end of the Coal Age also made clear that vehicles used in warfare would have to carry heavy armor. Thus tractors used for earth-moving and field work evolved into a new type of combat vehicle. Tractors were initially powered by steam engines, but soon employed internal combustion engines. The construction of the first gasoline-powered tractor is accredited to John Froehlich, who built his vehicle in Iowa in 1892, and about a decade later the tractor started to revolutionize farming, gradually eliminating work horses from temperate zone agriculture.

A key element of emerging tractor technology were endless, or continuous, metal belts looped over the wheels. Vehicles traveling on such tracks distribute their weight more evenly over the surface than ordinarily wheeled vehicles do, which allows them to move over terrain where other wheeled vehicles of the same weight would get stuck. One of the early pioneers of crawler tractor development was Thomas S. Minnis of Pennsylvania, who constructed an eight-ton vehicle with fixed endless chain tracks and in 1867 was granted a U.S. patent for "Improvement in Locomotives for Ploughing." The Hornsby company built the world's first fully-tracked oil-fueled tractor in Britain in 1905. The vehicle featured a 20 hp kerosene-fueled Akroyd engine and tracks that had been patented by David Roberts, Hornsby's chief engineer. Hornsby also fitted tracks to a 40 hp Rocket-Schneider motor car and eventually achieved tractor speeds of 25 mph (40 km/h). All these vehicles could easily pass through most difficult terrain and were demonstrated to the military, but the British army did not catch on, and Hornsby sold the patent for the tracks to the Holt company of Stockton, California, right before World War I.

Benjamin Holt had produced his first horse-drawn combine, which required 20 horses, in 1886. In 1904 the Holt company demonstrated a gasoline-fueled track-laying tractor, which however was not fully-tracked. (It had a tiller wheel in the front.) Holt's most serious competitor in California was the

Plowing with Gasoline-fueled Track-laying Tractor (1912)   This picture was taken in Oxnard, California, in 1912, but similar Holt tractors had been around since 1904. Vehicles of this kind evolved into both compact agricultural tractors that replaced horses in fieldwork during the Oil Age, and fully tracked and armored tanks that came to dominate warfare on the ground. They were typically fitted with Diesel rather than gasoline engines. (Library of Congress image LC-DIG-npcc-30912, edited.)

company of Daniel Best, who had built a gasoline-fueled tractor in 1896. The Best company sued Holt for patent infringements around 1908, the year Daniel Best died, but in turn the two companies decided to cooperate and in 1928 fully merged under the name of Caterpillar Tractor Company

Tank Development during World War I

World War I saw the adoption of tractor technology for warfare. Shortly after the start of the conflict the western front turned static. The French and British on the one side, and the Germans on the other side, defended their trenches with heavy machine guns, and neither party was able to advance through the heavy gunfire. Hence it was suggested to employ an armored track-laying vehicle that would be capable of moving across trenches. The British ordered a Holt tractor in February 1915, but the War Office remained unimpressed. In turn William Foster and Co., an English manufacturer of agricultural equipment, was contracted to develop two vehicles that were later dubbed 'Little Willie' and 'Big Willie.' One hundred Big Willie-type vehicles were ordered in February 1916, and 50 were delivered to France in August that year. As the story goes, early secret shipments of these vehicles were labeled in Russian script, and it was claimed that they were mobile water tanks produced for delivery to Russia. The name stuck and vehicles of these kind have been called tanks ever since.

Big Willie tanks had the enormous weight of 28 tons and employed 105 hp of engine power to reach a speed of 4 mph. They typically seated a crew of eight; some were equipped with 57 mm guns, others with smaller machine guns. Renamed Mark I, the tank made its debut at the First Battle of the Somme on September 15, 1916. Seventeen of 49 of these tanks broke down before they reached the starting line, and only nine made it across the no man's land to the German lines. The Mark I was underpowered and lacked climbing skill. It was unable to cross the larger trenches, but it nevertheless demonstrated the potential of such tracked, armored vehicles. Thus the British built some 2,000 tanks during World War I with gradually improved designs.

Meanwhile the French pioneered light tanks. The Renault FT-17, of which over 3,000 were built from 1917 to 1918, weighed under seven tons and was equipped with a 35 hp engine. One man was handling the single heavy machine gun on the rotating turret, the other was driving. This vehicle was adopted by all allied nations. The U.S. M1917 and the Italian Fiat 3000 were both based on the Renault FT-17 design. The Germans responded by developing their own tank, the truly massive A7V, which weighed 32 tons and featured two 100 hp machines. It was manned by a crew of 18. Short in resources towards the end of the war, the Germans managed to build only 20 of them.

Post–World War I Development

Tank development continued after the war. One notable innovation was J. Walter Christie's suspension chassis. Christie built a series of fast tanks in America, but the U.S. army turned him down, and he ended up selling his prototypes to the Soviet Union, where they were developed into the T-34 that gave the Soviets a military advantage in World War II. The Germans were prohibited from producing tanks under the peace treaty that ended World War I, but eventually resumed development work under the cover of agricultural tractor systems. The power-to-weight ratio kept on increasing, from 3.3 hp/ton for the early British models, to 12 hp/ton in the 1930s in Germany, to above 20 hp/ton at the end of the 20th century. The Western armies generally contracted automobile manufacturers for tank development. The U.S. army worked with Chrysler, for instance, and the German army with Daimler-Benz and MAN. The largest tank ever built was the Maus, designed in 1942 by Ferdinand Porsche under direct order from Adolf Hitler. Weighing 188 tons, the Maus was covered with up to 240 mm (9 inch) of armor and armed with a 128 mm (5 inch) cannon and a coaxial 75 mm (3 inch) gun. Only two prototypes were ever built. Modern battle tanks normally weigh over 50 tons and achieve road speeds of about 40 mph (70 km/h). Their armor-piercing projectiles are typically fired from a 120 mm (4.7 inch) gun. Generally tanks have been equipped according to advances in weapons technology. Some are now armed with Anti-Tank Guided Missiles, for instance. However, the military importance of tanks was somewhat kept in check due to the development of airplanes.

INTO THE AIR: FROM ZEPPELINS TO JET PLANES

Light but powerful internal combustion engines also opened the skies for human mobility: the energy of oil finally helped people to realize the ancient dream of controlled flight. There were two principal pathways. One was the development of lighter-than-air rigid airships, the other the construction of heavier-than-air airplanes.

The lighter-than-air strategy has its roots in the balloon experiments of the early 18th century. The French Montgolfier brothers discovered that heated air collected in a lightweight bag would cause the bag to rise, demonstrating a balloon that reached 3,000 ft (1,000 m) in a 10 minute flight in 1783. Later that year they sent a sheep, a duck, and a rooster up as passengers; thereafter they successfully completed the first untethered manned balloon flight. This worked because warm air is lighter (has a lower density) than cold air of the same volume, which makes a hot-air balloon rise when it is surrounded by colder, denser air. Later on it was discovered that gases such as helium and hydrogen, which are lighter than air even when they are cold, can be used for ballooning as well.

Balloons reached a certain military importance during the Coal Age as they served as aerial observation sites. Frenchman Henri Giffard in 1852 constructed and flew the first engine-powered, propeller-driven, steerable airship that already had the cigar-shape that was to become typical for later airships. Giffard's airship was filled with hydrogen and equipped with a 3hp steam engine that drove a single three-bladed propeller. (The idea of a propeller was similar to that of the ship screw.) The rudder, located at the rear, was like a sail that could be pulled into either direction. On its first flight Giffard's ship traveled 27 km (16.78 miles) from Paris to Trappes at a speed of 9 km/h (5.6 mph). However, this vehicle could only be steered in calm weather and the high weight of the steam engine prompted Giffard to add a second balloon to support the engine. On the next flight gas escaped and the balloon became deformed. The nose tilted up, the balloon escaped from the net and burst, and Giffard and his passenger were slightly injured. Plans to continue the experiments with an even larger balloon never materialized for cost reasons.

A steam-engined airship that required supplies of coal and water had obvious limitations, but the idea of the steerable, propelled airship was picked up again in Germany as soon as small powerful gasoline engines became available. Ferdinand von Zeppelin's company constructed the world's first and most successful untethered rigid airships.[16] Zeppelin had been working in the United States from 1863, serving as a military observer for the Union army during the Civil War. Later on he explored the headwaters of the Mississippi River, making his first of several balloon ascensions at St. Paul, Minnesota. The first of Zeppelin's own massive airships took off the ground in Germany with five passengers in July 1900. It had a cloth-covered aluminum structure that was 420 feet (128 meters) long and 38 feet (12 meters) in diameter. It was powered by two 15hp Daimler gasoline engines, each turning two propellers. A generator provided electricity for 17 hydrogen cells that used electric current to split the molecules of liquid water (H_2O) into oxygen (O_2) and the hydrogen gas (H_2) that lifted the airship. Zeppelin's airship designs were a big success. The *Deutschland* launched in June 1910, became the world's first commercial airship, and the *Sachsen* followed in 1913. By 1914 Zeppelins had flown 107,208 miles (172,535 km) and had carried 34,028 passengers.

During World War I the Zeppelins were used to drop bombs over France and England. However, the German airships had to fly at altitudes above the range of British and French fighter planes, which made it difficult to hit

A Zeppelin Airship: The *Deutschland* of 1910 An early manifestation of oil energy fueling mobility in the air. Zeppelin airships took the skies from 1900, using gasoline engines to turn propellers. (Library of Congress image LC-DIG-ggbain-09494, edited.)

targets accurately. Soon fighter planes were improved to reach higher and intercept the huge Zeppelins. Airships therefore had limited strategic importance. Germany had 10 Zeppelins in 1914, constructed another 57 during the war, and had 16 left by 1918. Most had either been lost in bad weather or shot down. After the war Germany was not allowed to keep the remaining airships or to build new ones, but the Zeppelin company stayed in business, as the U.S. military commissioned it to build a huge airship. This Zeppelin, named *Los Angeles*, was delivered in 1924. It could accommodate 30 passengers and had comfortable sleeping facilities. The *Los Angeles* made some 250 flights, including trips to Puerto Rico and Panama. Americans attempted to construct their own airships as well. The *Akron* and *Macon*, modeled on the Zeppelins, were built at enormous cost for the U.S. Navy, but both these helium-filled rigid airships crashed with less than two years of service, one in 1933 the other in 1935. (The combined death toll was 154.)

With the restrictions lifted, airship construction was reassumed in Germany as well. The *Graf Zeppelin* was launched in 1928 and the year after flew around the world within just 12 days. This airship provided a transatlantic flight service before any airplane did, and from 1932 flew regularly to South America. (Within 10 years the *Graf Zeppelin* completed 590 flights including 144 ocean crossings.) In 1936 the Zeppelin company released the *Hindenburg*, the largest rigid airship ever constructed. It was 804 feet (245 meters) long, had a maximum diameter of 135 feet (41 meters), and traveled at a top speed of 82 miles per hour (132 kilometers per hour), due to the power delivered by four 1,050-hp Diesel engines delivered by Daimler-Benz. The *Hindenburg* provided the first scheduled air service across the North Atlantic, between Frankfurt am Main, Germany, and Lakehurst, New Jersey. One trip took between 50 and 60 hours, and accommodations included a lounge with a grand piano, a dining room, and a library. The *Hindenburg* shipped over 1,300 passengers plus thousands of pounds of mail and cargo per year, but disaster struck in May 1937, when the airship went up in flames during the landing operation at Lakehurst on its 11th round-trip between Germany and the US. A third of the 97 people on board lost their life as the *Hindenburg*'s hydrogen ignited. Most likely the *Hindenburg* caught fire from a static discharge spark that initially ignited the highly flammable impregnant of the fabric covering. This episode pretty much ended the era of commercial airships, at least for the remainder of the 20th century. Instead, winged oil-fueled airplanes gained increasingly more importance.

Airplanes (with Piston Engines)

Heavier-than-air flight has its roots in the development of unpowered gliders. One of the numerous early pioneers was Briton George Cayley, who researched aerodynamics and briefly flew with a full-size glider in 1853. However, the most prominent and influential figure in glider development was German Otto Lilienthal, who completed over 2,000 flights before dying in a gliding accident in 1896.[17] Lilienthal received a degree in mechanical engineering from Berlin University in 1870 and opened his own workshop to manufacture boilers and steam engines. He had been interested in flight from early age and now devoted himself to the study of aerodynamics. In 1889 he published *Der Vogelflug als Grundlage der Fliegekunst* (Bird Flight as the Basis of Aviation), which was translated in many languages and read by virtually all later flight pioneers. The book examined the details of bird wings and applied the study of wing area and lift to the problem of human flight. From 1891 Lilienthal presented 18 gliders, 15 monoplanes, and 3 biplanes, which carried him up 1,150 feet (350 meters) through the air after take-off from a hillside. Photographs visually documenting his flights were distributed internationally, together with Lilienthal's writings, inspiring many to participate in the development of aircraft. In principle winged aircrafts can fly because the angle and shape of the wings provide an upward lift as soon as the aircraft moves forward. And as soon as Otto-type internal combustion engines had become small and powerful enough, they were mounted onto gliders. As in airships, the chemical energy of the oil-fuel was translated into forward movement through propellers.

First Motorized Flights

The world's first successful motor-powered airplane flight was once firmly ascribed to the Wright brothers, but is now associated with German Gustav Albin Weisskopf.[18] Weisskopf was born in 1874 and received training as a mechanic before being shanghaied (forcefully put aboard a ship) in Hamburg in 1888. Weisskopf spent several years at sea and in 1897, the year after Lilienthal's deadly accident, settled in the United States, where he anglicized his name to Whitehead. Apparently he immediately began building and flying gliders for the Boston Aeronautical Society, and reportedly ended a short motorized flight in 1899 by crashing into a wall. In 1900 he moved to Bridgeport, Connecticut, where he built motors and aircrafts. Weisskopf supposedly completed the world's first stable engine flight on 14 August 1901, traveling in his model Number 21 over 800 meters. (The Bridgeport *Herald*, the *New York Herald*, and the *Boston Transcript* published Weisskopf's own report about his flight experiments that day.) In January 1902, Weisskopf reported that he had flown 10 km (7 miles) over the Long Island strait in his improved Number 22. Both Number 21 and 22 were monoplanes, the former powered by a 20hp engine, the latter by a 40hp engine. During the starting phase the engine powered the front wheels, like an automobile, and after take-off the engine power was switched to drive a propeller. Weisskopf's monoplanes, with wheels, an enclosed fuselage and two propellers up front came very close to modern designs. They featured (lightweight) aluminum in engines and propellers, ground-adjustable propeller pitch, and individual control of propellers. However, Weisskopf's planes have only been photographed while on the ground, and the lack of hard evidence made it easy for those interested in dismissing Weisskopf's claims once his engine-building business had gone bankrupt.

There were quite a few early pioneers of motorized flight in other Western countries as well. Frenchman Clément Ader managed to hop 200 meters with a steam-powered flying machine in 1891. Richard Pearse reportedly flew a motor plane in early 1902 in New Zealand, and German Karl Jatho completed air jumps in mid-1903. In the United States Wilbur and Orville Wright were inspired by Otto Lilienthal's writings and tutored by French-American flight pioneer Octave Chanute. On December 17, 1903 the

Weisskopf Airplane (ca. 1901) Weisskopf's monoplanes with wheels, an enclosed fuselage, and two propellers up front, looked a lot more modern than those of the Wright brothers, which they predated. However, like several other flight pioneers, Weisskopf did not enjoy a full technological and commercial breakthrough, and his firm disappeared. (Courtesy of Fritz Majer & Sohn KG, Flughistorische Forschungsgemeinschaft Gustav Weisskopf, http://weisskopf.de/.)

Wright brothers completed four successful liftoffs in a motorized aircraft in the dunes near Kitty Hawk, North Carolina. Their biplane aircraft depended on strong (at least 20 mph) headwinds to overcome the ground drag and managed to stay in the air for less than a minute, which arguably made it a substantial hop, rather than a real flight. As their wheel-less craft with the propellers behind the wings got damaged that day, further trials had to be postponed until the following year. But the Wright brothers did not give up. They constructed a light four-cylinder internal combustion engine with an aluminum body and a steel crankshaft. Starting from a rail, Wilbur Wright in September 1904 completed the first circular flight, and Orville flew for 33 minutes and 17 seconds in October 1904.

Other flight pioneers were busy improving their motor planes as well. Brazilian Santos-Dumont, who was best known for constructing airships, completed a public 200 meter flight in a motorized airplane in Paris in 1906. (At this time the Wright brothers had still kept their flights secret.) Frenchman Henry Farman achieved the world's first city-to-city flight, from Bouy to Reims (27 km [17 miles] in 20 minutes) in October 1908, and L. Blériot crossed the English Channel from Calais to Dover with his front-propelled monoplane *Blériot XI* in July 1909.[19]

In 1910, two years after Orville Wright had completed a one-hour flight, Igo Etrich in Austria presented his front-propelled, distinctly shaped monoplane *Taube* (Dove). This aircraft reached special status, as 54 manufacturers under license agreements produced over 500 of it between 1910 and 1914 in 137 different configurations.[20] Shortly after Otto Lilienthal's deadly 1896 accident Etrich's father had purchased a Lilienthal glider, and Igo constructed tail-less wing-only gliders from 1903, when he was 24 years old. Etrich based the shape of his extremely stable gliders on the zanonia tree seed pod, which drifts through the air over great distances. (He found out about the qualities of this seed through publications by German naturalist Frederick Ahlborn.) In 1906, one of his zanonia flying wings took Etrich 900 meters through the air, but soon it was time to add a gasoline engine and a dove-like tail to the zanonia-shaped wings. The Etrich Taube of 1910 was gasoline-powered by a 105 hp engine designed by Ferdinand Porsche, who in 1906 had succeeded Gottlieb Daimler's son Paul as the technical director at the Austro-Daimler works in Wiener Neustadt. The Taube soon became known for its stability and safety, and was produced in large numbers in Austria and Germany. In 1911 the Etrich Taube apparently became the first airplane used as a bomber. During the Italian invasion of Libya, which was then part of the Ottoman Empire, an Italian pilot purportedly dropped four 4.5 lb (2 kg) grenades on a Turkish camp at Ain Zara from an altitude of 600 ft (185 m). The same year explosives were reportedly thrown off a Taube in the Balkans, and in 1914, right after the outbreak of World War I, three-kg-bomblets as well as propaganda leaflets were dropped from a Taube over Paris.

Etrich *Taube* in Flight The Etrich *Taube* (dove) of 1910 featured a 105 horsepower gasoline engine designed by Ferdinand Porsche. It was widely considered the best airplane at the start of World War I, but was soon outperformed due to the rapid airplane evolution during the war.

Airplane Development in World War I

No war party had lots of aircraft at the beginning of World War I. Most were still biplanes, the maximum range, load, and flight altitude was limited, and no real bombers had been developed. The principal function of airplanes in World War I was artillery observation and reconnaissance, though newly developed light machine guns were mounted on planes from 1915. These became especially viable once the Germans invented interrupter equipment that enabled the pilot to fire the gun through the aircraft's propeller blades. The Taube, the principal airplane on the German-Austrian side at the start of the war, was removed from the front lines after just six months, as it turned out too slow and became an easy target for rapidly improving enemy planes. By the end of the war, competition in the air had led to the construction of fighter airplanes that were equipped with 200 horsepower engines and reached speeds of 200 km/hr (124 mph).

What is more, all war parties constructed special bomber airplanes. The first genuine bomber to be used in combat was the French *Voisin*, a biplane that bombed the Zeppelin hangers at Metz-Frascaty in August 1914. However, a German plane of 1916 emerged as the most infamous bomber of World War I. The *Gotha G.V.* had two Daimler engines and a wingspan of over 77 feet (23 meters). It was strong enough to carry more than 1,000 pounds (454 kilograms) of bombs. On 23 May 1917 a fleet of 21 Gothas bombed the

English coastal town of Folkestone, killing 95 people. Another 26 Gotha raids were flown, and by the end of the war the English reported 835 killed and 1,990 wounded in aerial attacks.[21] This is a minimal number when compared to the losses in ground warfare and the number of people that were going to lose their life in the bombings of World War II, but World War I was critical for aircraft development. France had less than 140 aircraft in 1914, but produced as many as 68,000 during the war. (A staggering 52,000 of them were lost.) Similarly, Britain had no more than 154 airplanes in August 1914, but turned out planes at a rate of 30,000 units per year in mid-1918.

Post-War Development

After the war the initiation of airmail service was a major stimulus for the aviation industry. In Germany, where the peace settlements prohibited military flights, the availability of scores of trained pilots also promoted passenger transport. Passengers were initially sitting in the open, covered in blankets, but the Junkers company soon offered transport in full metal airplanes instead of the usual canvas-covered wood constructions. Hugo Junkers had founded the Junkers Motor Works in Magdeburg in 1913, producing large diesel engines for ship propulsion. Convinced that aircraft of the future would be monoplanes built entirely of metal, he crafted wings of corrugated sheet iron in 1911, and presented an all-metal monoplane J1 in 1915. Lightweight aluminum was in short supply in Germany during the war, but the government brought him into a partnership with Anthony Fokker, a Dutch designer who was building warplanes for Germany. Once Junker had access to aluminum, he promptly presented the *J3*.

After the war Junkers focused on commercial aviation, designing the famous Junkers F 13, which seated six passengers and pinpointed the start of airplane passenger transportation. To stimulate sales Junkers promoted the formation of airlines in Germany and other countries. His largest airline, Junkers Luftverkehr, merged with a competitor in 1926 to form Lufthansa, Germany's principal national carrier. In 1928 Lufthansa introduced flight attendants on board the Junkers G31 and became the world's first airline to serve food and drinks to its passengers. The Junkers G38 had a dining room and seated passengers inside its thick wings, offering dramatic views. However, it was the three-engined Junkers JU-52 of 1932 that became Lufthansa's principal aircraft. Nearly 5,000 of them were built and some continued their service for years after World War II. (During the war the JU-52 served as troop carrier, cargo transporter, bomber, flying ambulance, and tug that pulled troop-carrying gliders.)

Stable all-metal airplanes eventually allowed for long-distance flights. The immediate challenge after World War I was to cross the Atlantic. The first nonstop transatlantic flight, from Newfoundland to Ireland, was completed

Junkers JU-52 of 1932 The aviation industry was taking shape during the 1920s and 1930s. German airplane producer Junkers pioneered full metal airplanes, and German airline Lufthansa the serving of food and drinks to passengers. The Junkers JU-52 became the principal aircraft of Lufthansa. (Reproduced from H. J. Cooper, O. G. Thetford, and D. A. Russell, *Aircraft of the Fighting Powers*, Volume I, Harborough Publishing Co, Leicester, England, 1940.)

by British aviators Alcock and Brown in a Vickers Vimy bomber in 1919. Eight years and 13 transatlantic flights later, Charles Lindbergh became famous for crossing the Atlantic non-stop from New York to Paris by himself. His *Spirit of St. Louis*, a custom-made monoplane, was powered by a 220-hp 9-cylinder Wright engine. (The flight took 33.5 hours.) An Atlantic crossing from east to west, against the eastward headwinds, was finally achieved by a Junkers airplane in 1928. Thereafter long-distance flights became part of commercial passenger transportation. Lufthansa's 1935 service from Berlin to Bangkok took five days and required four stopovers. In the United States, Pan American Airlines asked Boeing to construct a long-range, four-engine flying boat for its transoceanic routes. Boeing was making aircraft for the military and had developed America's first true heavy bomber, the XB-15, in 1934. Drawing upon the wing design of this bomber, but adding new powerful 1,500 hp engines, Boeing delivered to Pan American Airlines the Model 314, nicknamed the Clipper. With a 149-foot (45-meter) wingspan it was larger than any winged aircraft ever built, and started to make routine flights across the Pacific Ocean in 1939.

At this stage the internal combustion engines in use had reached such high standards that there was little space left for improvements that could have further advanced airplane performance. But right then, in 1939, German engineer Ernst Heinkel, who had once been hired by Igo Etrich, sensationally presented the Heinkel He 178, the world's first airplane that was powered by jet engines rather than propellers.

Jet Planes (Gas Turbines)

Jet engines were introduced as an entirely new prime mover. In principle they work similarly to Otto piston engines, but they operate as continuous internal combustion engines (rather than intermittent combustion engines). In jet engines both air and fuel flow steadily into the engine, where a stable flame is maintained for continuous combustion. (Otto engines, in contrast, burn discrete quantities of fuel and air that are periodically ignited.) All of the jet engine's parts are situated on a centrally rotating spindle, pointing towards the front of the airplane. A continuous stream of air enters into the rotating centrifugal compressor that forces the air backwards, compressing it more and more until it reaches a high (typically circa 30 times the initial) pressure and a correspondingly high temperature. (The compressor simply consists of a series of blades mounted on rotating disks. A fan in front of the compressor sucks in the air.) At the end of the compressor the air enters the combustion chamber where fuel is injected. (When the engine is started, the air/fuel mix is ignited by igniter plugs, after which the combustion is continuous.) The high pressure generated by the combustion of the fuel-air mix is acting upon a turbine that is located right behind the combustion chamber. In principle, the technology of steam turbines could be applied at this point. (It is irrelevant if the turbine blades are driven by high-pressure steam or the gases released by the fuel explosion.)

In true jet engines (turbojet engines) the turbine only takes minimal pressure out of the exhaust gas stream, just enough to drive the compressor as well as other accessories that sit on the same central axis. Most of the pressure generated by the exhaust gases is directly expelled out of the engine towards the back of the aircraft, which pushes the craft forward. (The discharge of a jet of exhaust gases propels an aircraft according to Newton's third law of motion: action and reaction are equal and opposite.) Most commonly the hot exhaust gas that leaves the turbine is directed through a nozzle, which is a tapering casing that accelerates the jet stream and increases the thrust.

An alternative to direct jet propulsion are the more recent turboprop engines. In this design the turbine takes as much pressure as possible out of the exhaust stream in order to drive a propeller that sits on the same axis (shaft) in front of the jet engine. This setup was introduced because propeller propulsion is more efficient than jet propulsion at lower speeds. There is also a mix of the two designs. In turbofan engines the size of the first-stage compressor is increased to the point where it acts as a ducted propeller (or fan) that blows only part of the captured air into the central core that hosts the combustion chamber and turbine, while the remainder of the captured air is bypassed around the core and meets the hot exhaust gas at the end. Turbofan engines run best at speeds between 400 km/h to 1,000 km/h (250 mph to 650 mph) and have therefore become by far the most widely used type of engine in aviation.

Jet engines revolutionized flying because they are simpler in their construction, directly deliver rotary power, and provide more power per weight. Even though the mass-to-power ratio of internal combustion engines had declined by two orders of magnitude in less than 50 years, jet engines eventually carried these improvements by almost another two orders of magnitude. Jet engines are also a lot more fuel efficient because they can suck in vast amounts of air, which is especially relevant in the thin air of high altitudes. Kerosene, the crude oil fraction that due to its application as lamp oil had been in high demand even before the Oil Age had truly started, turned out the preferred fuel for jet engine propulsion.

World War II Jet Planes

The jet engine was independently invented by Pabst von Ohain in Germany and Frank Whittle in England. As soon as Ernst Heinkel heard that Ohain had received a patent on the jet engine, he hired him to develop it into production status in a private venture. The first aircraft to fly powered by an Ohain engine took off near Rostock in August 1939, while the first British jet-powered aircraft was in the air in 1941. However, the only jet plane to be used in combat during World War II was the German Messerschmitt ME 262 A2, developed by the Junkers works to be mass-produced from mid-1942. (Wilhelm Messerschmitt had earlier designed the Messerschmitt 109, which set a speed record of 775 km/hr (481 mph) and was produced at a count of 35,000, which makes it the most-produced fighter in history.) At a maximum speed of 870 km/h (540 mph), compared to about 650 km/h (404 mph) achieved by piston-engined propeller planes of the day, the jet-engined ME 262 was bound to make a difference. However, Germany had problems securing

Heinkel He 178 of 1939: The World's First Jet Plane Jet propulsion pushed aviation towards new limits. German engineer Ernst Heinkel had once worked for Igo Etrich (of Etrich *Taube* fame) and now presented the world's first airplane without propellers.

supplies in cobalt, nickel, and chromium during the war, and the Junkers engineers thus had to redesign their jet engine in order to avoid these metals. This resulted in serious production delays. When some 5,000 jet engines were finally delivered, it was too late for the ME 262 to affect the outcome of the war.

Post-War Development

After World War II Britain and the United States both scraped their own jet engine designs and copied the German ones instead. What is more, Pabst von Ohain was brought to the United States by Operation Paperclip in 1947 and was employed by the U.S. Air Force at Wright-Patterson Air Force Base. In 1956 he became the Director of the Air Force Aeronautical Research Laboratory. (Paperclip was the codename for the operation under which the United States imported and nationalized 700 members of the Nazi scientific community, including specialists in airplanes, rocketry, missiles, chemical weapons, and medicine.) By the 1970s such strike fighters as the F/A-18C Hornet had been developed. Powered by two turbofan engines, this aircraft reached top speeds of over 1,800 km/h (1,100 mph) at high altitudes.

Britain introduced the first jet engine passenger plane, the De Havilland Comet, in 1952, but it was grounded in 1954 after three of them had crashed. Hence the American Boeing 707 of 1958, with 200 seats, emerged as the world's first successful passenger jet. (920 of this four-engine aircraft were sold by 1977.) In 1960 several different commercial jet aircraft were in service and the era of mass air travel had definitely begun. The wide-bodied Boeing 747 made its first commercial flight in 1969. This humped aircraft, with the characteristic two-deck configuration (the small upper deck usually being reserved for business-class passengers), seats about 380 passengers in a typical mixed-class layout. The modern 747–400 flies at a speed of just over 900 km/h (560 mph) and features an operating range of about 13,600 km (8,400 miles), sufficient to fly non-stop a third around the globe. (A Boeing 747 burns about 5 tons of fuel to take off and over 100 tons to fly across the Atlantic.) At the beginning of the 21st century the Boeing 747 remained the largest aircraft in commercial service, but Airbus developed the even larger A380, which made its maiden flight on 27 April 2005. At a length of 73 meters, and a wingspan of 80 meters, this aircraft may seat up to 656 passengers on two full-length decks. Equipped with four kerosene-fueled jet engines, its maximum take-off weight is 560 tons.

To be sure, jet engines have also been adopted for stationary applications. Such gas turbines are used in steel mills, for instance, but more generally for centrifugal compressors and electricity generators. However, gas turbines have not completely replaced steam turbines, which are the better choice for (coal-fueled) electricity production and have ranked as the most powerful prime mover ever since 1910. What made the gas turbine indispensable in aviation is its unrivaled power per weight ratio. To a small extent gas turbines have also been used in automobiles, an application that may gain importance later in the 21st century.

PROSPECTING, TRANSPORT, AND REFINING TECHNOLOGY

One integral part of the emergence of oil technology was continual progress in oil prospecting and refining techniques. The first operators of Galicia were drilling with simple spring poles: a springy wooden pole was stuck in the ground at an angle and supported by a fulcrum, which was usually a forked heavy limb fixed in the ground not too far from the pole's base. At the head of the pole, which rested right above the bore hole, a cable with a heavy metal drill bit was attached. Also attached at the head of the pole were lines with stirrups. During operation the spring pole would lift the drill slightly above the ground, and some three operators would bounce up and down on the stirrups, causing the bit to literally chop a hole into the ground.[22] (The same drilling method was used in North America.) Before long, steam engines were used to lift the heavy drill bit on a lever (a walking beam) and let it fall on the ground over and over again. Galician operators applied steam power from the late 1860s and soon reached 200 meters beneath the surface. In the 1870s much drilling technology was developed in Canada, around the southwestern Ontarian city of Petrolia.[23] This place is sometimes credited with being the starting point of the world oil industry, because its technicians traveled to many places around the globe, spreading the so-called Canadian rig, which suspended the drilling tools from a series of linked hardwood rods. Among these technicians was William Henry MacGarvey, who came to Galicia in 1883 and a year later bored holes to a depth of 1,000 meters. Rotary drilling, the technique that remained the industry standard until this day, was for the first time successfully used by Anthony F. Lucas at the Spindletop gusher in Beaumont, Texas, in 1901. In 1909 Howard Hughes introduced the rolling cutter rock bit, the device that assured the continuing success of rotary drilling. (Nowadays these drills are powered by Diesel engines.)[24]

As oil usually does not exist below 15,000 feet (4,600 meters), oil drillers had a natural and limited goal to their efforts.[25] Drilling rigs capable of penetrating to 15,000 feet became available as early as the late 1930s, and in 1946 the world's deepest oil well (in Louisiana) bottomed at 13,800 feet. However, subsurface drilling technology kept advancing, not least because natural gas is found in greater depth. By the 1980s some gas wells were drilled that reached 9,000 meters (30,000 feet) below the surface. The world's record gas exploration bore holes of Krivoy Rog (1985) reached to 12,260 meters. (For comparison, the center of the Earth is located more than 6,000,000 meters beneath the surface.) A more recent development is horizontal drilling and multibranch wells to reduce the number of wells by extending the range of a single vertical well.

Offshore Drilling

As two thirds of the planet's surface are covered by water, offshore drilling technology was bound to emerge as well. This development actually started very soon, because some of the earliest oil fields were discovered close to coastlines. In Summerland, California, south of Santa Barbara, the first offshore wells were drilled from piers around 1896, after it had been realized that wells at the beach produced better than those further inland. The longest wharfs stretched over 1,200 feet (400 meters) into the Pacific ocean. In the Caspian Sea, offshore drilling near Baku started around the same time. The first true offshore drilling platform, (almost) out-of-sight of land, was built in 1947 in shallow waters 19.3 km (12 miles) off the coast of Louisiana in the Gulf of Mexico. Such (submerged) drilling rigs rested permanently on the ocean ground, but in the 1950s jack-up rigs were introduced to operate in water depths up to about 100 meters. These drilling platforms are barges (boats) from which legs are lowered to the ocean bottom to raise the platform above the water at the drill site. To drill in deeper waters, semisubmersible rigs were developed in the 1960s. These floating rigs owe their stability to large columns with water tanks that extend far below the ocean surface and submerge the rig deep into the water. They are first towed to the drill site, then the water tanks are flooded, and thereafter the rig is anchored. Semisubmersible rigs could potentially drill in water depths of 150 meters (500 ft) in the early 1960s, 610 meters (2,000 ft) in the early 1970s, and 1,500 meters (5,000 ft) in the 1980s. Drilling offshore is, however, expensive, and the actual offshore production records lagged behind: 300 meters in 1978, just over 1,000 meters in 1994, and 1,800 meters in Brazil in 1998. (Drilling at such ocean depths costs about seven times as much as producing oil in the Saudi desert.) For ultra-deep water depths between 2,000 meters and 3,000 meters drillships are currently used. These have to carry enormous loads including ultra-deepwater risers, anchoring systems, and the attendant mud column. (Mud refers to a mixture of water, clay, and chemicals used in drilling for lubrication and cooling of the bit, and for flushing of rock particles to the surface.)

Prospecting Methods

Oil prospectors also developed methods to identify likely locations of oil fields without even having to drill a test hole. They soon figured out which geologic elements are necessary for oil or gas fields to occur, recognizing the structural traps under which fluids and gas accumulate: under nonporous arches, along faults and folds (created by the movement of the Earth's crust),

Drilling Offshore: Semisubmersible Rig Being Transported Semisubmersible rigs could potentially drill in water depths of 150 meters (500 ft) in the early 1960s, 610 meters (2,000 ft) in the early 1970s, and 1,500 meters (5,000 ft) in the 1980s. The picture shows the *MV Blue Marlin* carrying BP's 59,500-ton Thunder Horse, the largest semi-submersible rig in the world, from Okpo, South Korea, to Corpus Christi, Texas, on September 23, 2004, for completion. The rig is designed to process 250,000 barrels of oil per day and 200 million cubic feet per day of natural gas in challenging deep water environments of the Gulf of Mexico. It is now located 150 miles southeast of New Orleans. (Courtesy of Dockwise Shipping B.V.)

or near subsurface salt domes. With the advent of airplanes (and later satellites), geologists started to examine land surfaces from the air, especially when difficult terrain was concerned. The German-born Schlumberger brothers introduced a method to chart the Earth's subsurface by measuring the electrical resistivity of rock formations. They were able to reliably locate subsurface deposits of iron ore, copper, and oil reservoirs, and could even map the shape of oil reservoirs, the quantity of oil they held, and the optimum location from which to drill. Their company signed its first major contract to search for oil in 1929 with the Soviet Union, was hired by Shell in 1932 to look for oil in California and on the Texan coast, and grew into the world's leading oil field services provider. Gravity meters and magnetometers (on board airplanes) were increasingly used to measure subtle changes within the Earth's magnetic and gravitational fields that might signal the presence of oil traps. On the surface, sound waves are sent underground to measure the time it takes for subsurface rocks to reflect them back up. Gas detectors can find traces of gaseous hydrocarbons escaping from subsurface accumulations, and data from all such measurements are fed into the world's largest and fastest computers to produce (since the 1970s) accurate three-dimensional underground images that identify the best location and trajectory for drilling wells. Nevertheless, most exploratory wells are actually dry. The average U.S. wildcat well (that is, an exploratory well drilled a mile or more from existing production) currently has a 1 in 10 chance of striking hydrocarbons, and a real wildcat well, drilled in an unproven frontier area, stands a 1 in 40 chance.[26]

Transport Systems

To get the oil from the drill site to refineries, and on to final markets, required building up a whole oil infrastructure. The best choice to transport large quantities of liquids or gases over long distances is through pipelines. In the United States the construction of an extensive pipeline system began as early as the 1870s, and a century later, during the 1970s, the world's longest oil pipeline was built from western Siberia to central Europe. For transport on the oceans specialized tankships were designed. The Nobel oil company's first tanker was steaming across the Caspian Sea from 1878, and the first modern ocean-going tanker, the British-made *Glückauf*, was launched in 1886. (This steamer had a deadweight of 2,740 tons.) The year after, 17 steam-powered tankships operated between Europe and North America. Tanker size gradually increased to reach around 16,000 tons of deadweight at the time of World War II, during which the United States built numerous T2 standard vessels to deliver fuel to Europe.

Up until World War II, it was primarily processed oil products that were shipped in tankers; thereafter refineries were built in the consuming industrialized areas, and thus chiefly crude oil was transported over long distances. Oil tankers kept on growing, and transport costs per unit of volume were tumbling. Record deadweight tonnage increased to 100,000 in 1959, 200,000 in 1966, and 300,000 in 1968. By 1971 over 200 tankers with over 200,000 tons deadweight were in service. The *Batillus* of 1976 was the first tanker to exceed half-a-million tons in deadweight, but only six of those were built until the end of the 20th century, because their enormous size touched upon the limits of practicality and safety. The 1980 *Seawise Giant*, the largest tanker of all, is 460 meters (1,504 foot) long and by many standards the largest human-made structure ever created. But even tankers with only 200,000 tons deadweight are too large to enter many of the world's major ports. They are loaded and discharged far out at sea, at pipeline terminals at piers or loading buoys. (Alternatively they can be discharged into smaller shuttle tankers.) Several severe oil tanker accidents, including the one of the *Atlantic Empress* in the Caribbean in 1979, and the one of the *Exxon Valdez* off the coast of Alaska in 1989, resulted in disastrous environmental pollution. From 1990, newly-built large oil tankers were required to be fully double-hulled.

Refining Technology

From its humble beginnings, when crude oil served only the lamp kerosene market, it had to be processed in order to separate the different components of the crude oil mix. Practically all of these components gained some importance (as petroleum products are processed into anything from detergents to plastics, synthetic rubber, and lubricants), but from the 1920s demand for crude was chiefly driven by demand for gasoline (and diesel) for automobiles (and trucks). The focus was therefore on the lighter fractions of crude oil, with shorter carbon chains. So-called cracking processes were developed to break up the long-chained constituents of crude oil into shorter molecules by applying heat, pressure, and certain catalysts. This technology is closely associated with the work of German chemist Fritz Haber. Thermal cracking (at high temperatures) was introduced around 1913, while less expensive catalytic cracking (at lower temperatures) was introduced in 1935 (to yield high-octane gasoline). Nowadays a barrel of crude oil, which is 42 U.S. gallons (or 159 liters), yields some 19 gallons of gasoline, 9 gallons of fuel oil, 4 gallons of jet fuel and 10 gallons of other products, including lubricants and asphalt.

Increase in Oil Production

The amount of crude oil produced, transported, and refined has increased sharply during the Oil Age. Annual world crude oil production rose from 21 million tons in 1900, to 210 million tons in 1929, to 765 million tons in 1955, to 2,340 million tons in 1970, to 3,500 million tons in 2000.[27] Much of the increased demand derived from the huge number of registered motor vehicles, which in the United States alone rose from 9 million in 1920 to 221.3 million in 2000. (Of these, 133.6 million were cars, the remainder were buses and trucks.) Worldwide some 10,000 cars, trucks and buses existed in 1900, compared to some 70 million motor vehicles on the roads in 1950, and 750 million in 2000. All these vehicles depended on an efficient oil prospecting, extracting, transporting, and refining industry: Towards the end of the 20th century, only some 0.005 percent worth of the energy contained in Middle Eastern oil was invested in finding and producing it. Thus, the overall amounts of energy invested were often negligible in comparison to the energy gained, which allowed the Oil Age to become such a high energy era, and to achieve such fast technological advances.

NOTES

4. Vaclav Smil, *Essays in World History: Energy in World History* (Boulder, CO: Westview Press, 1994), 167. *Wellspring Africa* offers an account of how the percussion drill works. Percussion drilling gets its name from the action of its drill which rises and

falls to beat upon the earth and chop up the soil and rock. It is more popularly known today as Cable Tool Drilling since the more modern drilling rigs use steel cable. It has many variations, like the Chinese springboard, the American springpole, and the walking beam, but all employ the same basic means of cutting the earth and clearing out the hole.

The drill involves a heavy steel bit attached to a rope which is lifted, either by hand or by machine, and then dropped to cut the earth. As the bit chops the earth, water is added to the well hole so that the bit makes mud out of the earth it has cut. After the hole is filled with several feet of mud, the heavy bit is withdrawn and a tool called a bailer is attached to the rope and lowered into the hole. The bailer is a hollow tube with a door at the bottom. The door, called a flap valve, opens when it hits the mud to allow the mud to fill the bailer, and then closes to trap the mud inside the tube so that the mud can be lifted to the surface. The tube is emptied at the surface and the procedure is repeated until the hole is clear. The bit is then reattached to the rope and the above process begins again. Wellspring Africa, "Hand Powered Percussion Drill," http://www.wellspringafrica.org/drildesc.htm.

5. Hans Ulrich Vogel, "The Great Well of China," *Scientific American* 268 (1993): 116–21, http://www.sciamdigital.com/index.cfm?fa=Products.ViewIssuePreview&ARTICLEID_CHAR=169EA6F2–2CA5–4C50–8EBE-114F3AF2991.

6. James L. Coleman, "The American Whale Oil Industry: A Look Back to the Future of the American Petroleum Industry?" *Natural Resources Research* Vol. 4, 3 (1995), http://www.springerlink.com/content/7178248713276401/fulltext.pdf.

7. Alison Fleig Frank, *Oil Empire: Visions of Prosperity in Austrian Galicia* (Cambridge: Harvard University Press, 2005). Alison Fleig Frank, "Galician California, Galician Hell: The Peril and Promise of Oil Production in Austria-Hungary," *Bridges* 10 (2006), http://www.ostina.org/content/view/1172/506/; Valerie Schatzker, "Petroleum in Galicia," Drohobycz Administrative District, http://www.shtetlinks.jewishgen.org/drohobycz/history/petroleum.asp; "The History of the Oil Industry," San Joaquin Geological Society, http://www.sjgs.com/history.html.

8. This section is largely based on "Edwin L. Drake and the Birth of the Modern Petroleum Industry," Pennsylvania Historical and Museum Commission, http://www.phmc.state.pa.us/ppet/edwin/page1.asp?secid=31.

9. "Petroleum," *Encyclopaedia Britannica*, 11th Edition (1911), http://www.1911encyclopedia.org/Petroleum.

10. This section is mostly based on: Sarah Searight, "Region of Eternal Fire—Petroleum Industry in Caspian Sea Region," *History Today* Vol.50, Issue 8 (2000): 45–51, http://www.historytoday.com; Find more on this topic in Daniel Yergin, *The Prize: The Epic Quest for Oil, Money & Power* (New York: Free Press, 1991). Wolfgang Sartor, "International and Multinational and multilateral Companies in the Russian Empire before 1914: The Integration of Russia into the World Economy," XIII Economic History Congress, Buenos Aires, July 2002, The International Economic History Association, http://eh.net/XIIICongress/cd/papers/43Sartor202.pdf.

11. Erik J. Dahl, "Naval Innovation: from Coal to Oil," *Joint Force Quarterly* 27 (Winter 2000–01): 50–56, http://www.dtic.mil/doctrine/jel/jfq_pubs/1327.pdf. Alternative site: http://www.findarticles.com/p/articles/mi_m0KNN/is_2000_Winter/ai_80305799.

12. J. H. Bamberg, *The History of the British Petroleum Company: Volume 2: The Anglo-Iranian Years, 1928–1954* (Cambridge: Cambridge University Press, 1994).
13. Daniel Yergin, *The Prize*; J.H. Bamberg, *The History of the British Petroleum Company*.
14. Austrian Press and Information Service (Washington, DC), "Siegfried Marcus-Mechanic and Inventor (1831–1898)," *Austrian Information*, Vol. 51, No. 8, (1998), http://www.austria.org/oldsite/aug98/marcus.html; Samuel Kurinsky, "Siegfried Marcus, An Uncredited Inventive Genius," *Fact Paper* 32-I, Hebrew History Federation, http://www.hebrewhistory.info/factpapers/fp032–1_marcus.htm.

Marcus was actually better known for many of his other various inventions. He invented the telegraph relay and a portable field telegraph, for instance, and patented an apparatus for automatic picture-making in 1876 and an electric lamp in 1877, two years before Edison and Swan. He also improved gas and petroleum lamps (which was related to the invention of the carburetor), and developed famous detonators (related to the invention of the magneto ignition) that could activate explosions from the distance and were adopted for sea mines. He also constructed a hand-held firearm capable of 30 shots per minute and, for the Austrian Arctic Expedition of 1872–4, a special whaling knife or spear. He also installed an electric bell system for Empress Elisabeth in the Hofburg and tutored the Austrian royal family in science. In short, he was largely preoccupied with other projects and left the further development of the motor vehicle to others.

15. J. Bradford DeLong, "Slouching Towards Utopia?: The Economic History of the Twentieth Century, VII. The Pre-World War I Economy," University of California at Berkeley and NBER (National Bureau of Economic Research), January 1997, http://econ161.berkeley.edu/TCEH/Slouch_PreWWI7.html/
16. "The Zeppelin," US Centennial of Flight Commission, http://www.centennialofflight.gov/essay/Lighter_than_air/zeppelin/LTA8.htm.
17. "Lilienthal—The 'Flying Man'," US Centennial of Flight Commission, http://www.centennialofflight.gov/essay/Prehistory/lilienthal/PH6.htm; Otto Lilienthal Museum, http://www.lilienthal-museum.de; Otto Lilienthal, "Practical Experiments for the Development Of Human Flight," *The Aeronautical Annual* (Editor James Means), 7–20, 1896, http://invention.psychology.msstate.edu/i/Lilienthal/library/Lilienthal_Practical_Exp.html.
18. "Gustav A. Weißkopf," Flughistorische Forschungsgemeinschaft Gustav Weißkopf (FFGW), http://www.weisskopf.de/; "Gustave Whitehead's Flying Machines," http://gustavewhitehead.org/; William J. O'Dwyer, "The 'Who Flew First' Debate," *Flight Journal Magazine*, October, 1998, http://www.flightjournal.com, http://findarticles.com/p/articles/mi_qa3897/is_199810/ai_n8815811.
19. Find information on various flight pioneers in "Born of Dreams—Inspired by Freedom, Centennial of Flight Day: December 17, 2003," US Centennial of Flight Commission, http://www.centennialofflight.gov/.
20. E. T. Wooldridge, "Early Flying Wings (1870—1920)," http://www.century-of-flight.net/Aviation%20history/flying%20wings/Early%20Flying%20Wings.htm.
21. "Bombing During World War I," US Centennial of Flight Commission, http://www.centennialofflight.gov/essay/Air_Power/WWI_Bombing/AP3.htm.
22. "Petroleum in Galicia," Drohobycz Administrative District, http://www.shtetlinks.jewishgen.org/drohobycz/history/petroleum.asp

23. "The Town of Petrolia Online," http://town.petrolia.on.ca/; "Canadian Petroleum Pioneers," The Petroleum History Society, 2001, http://www.petroleumhistory.ca/archivesnews/2001/01feb.html.

24. Most oil wells in the United States are drilled by the rotary method that was first described in a British patent in 1844 assigned to R. Beart. Todd M. Doscher, "Petroleum," Microsoft Encarta Online Encyclopedia, http://encarta.msn.com/encyclopedia_761576221_2____6/Petroleum.html#s6.

25. K. N. Deffeyes, *Hubbert's Peak: The Impending World Oil Shortage* (Princeton: Princeton University Press, 2001).

26. "A Petroleum Prospecting Primer," Chevron Learning Center, http://www.chevron.com/products/learning_center/primer.

27. Find annual oil production figures since 1960 by nation as well as global, in: "World Crude Oil Production, 1960–2005 (Million Barrels per Day)," Energy Information Administration, http://www.eia.doe.gov/aer/txt/ptb1105.html.

CHAPTER 27

HOW OIL TECHNOLOGY SPREAD

Oil technology emerged in Western countries where coal technology had matured. Again, the technological advantage enjoyed by people of European descent may ultimately be traced to the population density and agricultural head start of (western) Eurasia. As the world was thoroughly globalized towards the end of the Coal Age, linked by steam ships, railroads, internationally active companies, and empires that stretched beyond continents, oil technology spread as soon as it emerged.

WORLDWIDE OIL PROSPECTING

The worldwide search and discovery of oil suggests a rapid global diffusion of drilling technology. In the second half of the 19th century large-scale oil prospecting was started in Galicia, the United States, the Caspian region, and the Dutch East Indies (Indonesia). These efforts continued in Latin America in the early 20th century, while Middle Eastern oil production began in the 1930s. In central Russia, large oil fields were developed towards the end of World War II, and on the African continent substantial oil deposits were discovered in 1956 in Algeria and Nigeria, followed by Libya in 1959. Alaskan oil was first found in 1957, and in 1968 the supergiant Prudhoe Bay oil field was discovered on Alaska's arctic North Slope. Oil prospecting in Western Europe finally paid off when the oil reserves in the North Sea offshore Scotland were found in 1969. North Sea oil production began in 1975, and both Britain and Norway turned out rich in oil. The only new major oil province discovered after 1980 was the one in the northeastern Caspian Sea. Production of oil in the ultradeep waters off the coast of Brazil and Angola began towards the end of the 20th century.

However, for a long time it was almost exclusively European and U.S. firms that stood for international prospecting and drilling. Eventually some of the oil-rich but non-industrialized countries decided to nationalize the foreign oil infrastructure on their territory, but those nations typically saw their oil technology fall behind, unless they kept cooperating with Western oil corporations.

SPREAD OF AUTOMOBILE TECHNOLOGY

The spread of IC engine and automobile technology from its German source to the rest of the world is quite well documented. In principle it was relatively straightforward to fit Otto-type engines to experimental vehicles: all kinds of wheeled vehicles were already common, gasoline and kerosene were available, and steam car engineers had solved many of the transmission problems. On the other hand, the early spread of practical gasoline engines was somewhat stalled by the patents that Nikolaus Otto had obtained in 1876–77 in different countries. Otto put a high price on royalties, and patent protection expired only in 1894, three years after Otto's death. Soon thereafter several car manufacturers emerged in Germany next to Daimler (Mercedes) and Benz. Opel, for instance, produced bicycles from 1885 and cars from 1898. NSU produced bicycles from 1886, motorbikes from 1901 and cars from 1906, but eventually (1969) merged with Audi, founded by August Horch, a former Benz employee who constructed cars from 1899.

In Austria, brothers Franz, Heinrich and Carl Gräf founded a workshop in Vienna that in 1898 constructed the world's first gasoline automobile with front-wheel-drive (patented 1900).[28] Ferdinand Porsche built automobiles for Lohner in Vienna from 1898, and in 1906 was assigned technical director at the Austro-Daimler works, where he designed aircraft engines. In 1934 Porsche's own company began developing the Volkswagen (VW) Beetle in Germany, while the first Porsche sports car was manufactured in Austria in 1948. Meanwhile Puch produced engines from 1901 and cars from 1904. Merged into Steyr-Daimler-Puch in 1934, this Austrian manufacturer is now best known for its military vehicles and the Puch G 4-wheel drive, which is also sold under the Mercedes-Benz (G) name.

In Britain, Daimler set up a subsidiary to assemble cars with German-made engines in Coventry from 1896.[29] The "English Daimlers," as they were known, became truly British cars when this subsidiary was restructured in 1904, the same year car-maker Rolls Royce was founded, and Rover started to produce automobiles. Morris was founded in 1913, Austin Martin in 1914, and Bentley in 1919. (Rolls Royce absorbed Bentley in 1931.) In France, the Peugeot brothers initially used Daimler engines to produce cars from 1890.

Renault employed about 5,000 workers to make 4,200 vehicles per year in 1913, but it was André-Gustave Citroën's company that became Europe's first true mass-producer, making low-priced automobiles from 1919. Meanwhile Italian Fiat, founded in Turin in 1899, set up a production facility near Vienna in 1907 (Austro-Fiat), and started to build Europe's largest contemporary automobile factory in Italy in 1916.

In the United States, the owner of the Steinway & Sons piano factory, William Steinway, bought Daimler manufacturing rights in 1888 and set up the American Daimler Motor Company (in Hartford, Connecticut), which from 1891 produced gasoline engines for tramway cars, carriages, fire engines, and boats. However, the first American automobile was built by the Duryea brothers in Springfield, Massachusetts, in 1892. Henry Ford constructed his first car in 1896 and offered his first motor vehicle for sale in 1903. Ransom Eli Olds produced cars from 1899. He established the first mass production of gasoline powered automobiles in the United States to supply his hugely successful Curved Dash Oldsmobile, of which he sold 600 in 1901, and 2,500, 4,000 and 5,000 in the following years. Ford sold a grand total of 15 million of his Model T, which was manufactured solely by assembly-line methods from 1908. That same year General Motors (GM) was founded in Flint, Michigan. Some 200 garage-sized companies were bundled into GM, which acquired Oldsmobile, Pontiac, Cadillac, Chevrolet, and the German Opel (1929). GM actually grew into the world's largest corporation. Chrysler Corporation was founded in 1925 by Walter Percy Chrysler, self-made son of a German immigrant, and a major railroad industry figure. Chrysler eventually included such brands as Plymouth, Dodge, and (from 1987) Jeep, and arose as one of the Big Three U.S. car manufacturers, next to Ford. (In 1998 Chrysler merged with Daimler-Benz into German-based DaimlerChrysler AG, but in 2007 Daimler AG sold Chrysler to an American equity firm.)

In Japan, automobile pioneers started to make cars in 1902, initially with imported engines. Japan was by this stage fully integrated into the global Coal Age economy, and the first all-Japanese motor vehicle was produced in 1907. That same year the military imported trucks from Germany and France, and in turn oversaw the production of the first domestic military truck at the Osaka Artillery Factory in 1911. Mitsubishi Zosen manufactured 22 Mitsubishi Model A cars in 1914, while the Kaishinsha Motor Works manufactured their Dattogo during the following years. The first Mazda was made in 1931, and true mass production of Japanese automobiles begun in 1935 with a Datsun passenger car, produced by the Jidosha Seizo company, a predecessor of Nissan. (The Datsun pioneers had actually begun building cars as early as 1911.) Toyota started to mass-produce soon thereafter, in 1937, and following gradual growth had emerged as the world's largest automaker in terms of revenue, production, and profit by the early 21st century. (By the 1980s the cars produced annually in Japan had outnumbered those produced in the United States.) In South Korea, the first gasoline engine was constructed in 1973 by KIA, the firm that introduced its first passenger car the year after. Meanwhile Hyundai manufactured a Ford model from 1968, and a Mitsubishi model from 1975. In Russia, the Soviet government decided in 1966 to mass-produce a passenger car. Technical know-how was to be acquired through cooperation with a Western producer, and Fiat was chosen due to the success of the new middle-class Fiat 124. Based on this Fiat model, the first 200,000 Lada vehicles were manufactured in 1971, with production rising to 600,000 units per year by 1973.

Towards the end of the 20th century, India and China emerged as serious car manufacturers as well. Tata Motors, India's largest car producer, manufactured light-duty commercial vehicles from 1954, and heavy-duty commercial vehicles from 1983, through a collaboration with Daimler-Benz. The first fully Indian-made passenger car was produced in 1998. (In 2005 the company sold some 245,000 commercial vehicles and 209,000 passenger cars.) In China nearly all the large automobile manufacturers set up production joint-ventures, but by 2005 there were also over 100 home-grown car carmakers, copying Western designs closely. Three different vehicles made in China by local firms for export to Europe were displayed at the Frankfurt Motor Show in 2005.[30] Such Chinese cars were set to eventually conquer market share in the United States as well, but the Chinese and Indian car markets actually held more potential: In fact, it might be argued that India and China had not fully entered the Oil Age by the turn of the century. In 2000 the United States had 790 motor vehicles per 1,000 citizens, Japan 570, Germany 580, the UK 530, and Brazil 90, while there were no more than eight motor vehicles per 1,000 people in China, and seven per 1000 people in India. However, in 2004 car sales in China exceeded five million, thus overtaking the German market, and was set to replace Japan as the second-largest national market for cars after America. Notably, the motorization of populous India and China may cause the number of motor vehicles produced in the first two decades of the 21st century to surpass the number of motor vehicles produced in the entire 20th century.

SPREAD OF AVIATION TECHNOLOGY

Much of modern aviation technology has its origin in the German-speaking countries as well. Otto engine technology was critical in the early years, but other German engineering know-how, including all-metal monoplane construction, also spread quickly to various countries. Germany was not allowed to build aircraft engines after World War I, but once again held the technological lead by the end of World War II. German jet engine technology and German technicians were in turn taken to Britain, the United States, and the Soviet Union to initiate further development. However, various other governments also managed to lure some of the famed German aviation engineers of World War II into their countries. Kurt Tank, for instance, first moved to Argentina to design the Pulqui II jet, and afterwards to India to develop the subcontinent's first jet aircraft, the Hindustan Aeronautics' HF-24 Marut.

American, British, French, and Soviet producers spread jet engine and fighter technology by selling their products or production licenses worldwide. Japan introduced its first locally developed fighter jet, the Mitsubishi F-1, in 1978. (It resembled the Anglo-French SEPECAT Jaguar of 1973.) China's fighter aircrafts Shenyang J-8 and Chengdu J-7 entered full-scale production and service in the 1980s. These models were based on the Russian (Mikoyan-Gurevich) MiG-21, introduced in the Soviet Union in 1959. (The MiG-21 itself used jet engines that were reverse-engineered from Rolls Royce engines, which were based on German technology from the 1940s).

In terms of passenger transport, much growth in the aviation (as the automobile) industry shifted from Western countries to Asia towards the end of the 20th century. The number of passengers carried by Chinese airlines doubled to 138 million between 1999 and 2004, and was expected to double again during the five consecutive years. (Chinese airborne freight volume grew by as much as 20 percent in 2004 alone.) China thus emerged as a major customer for airplanes, buying 219 units from Airbus (one-fifth of Airbus' global orders) in 2005, while Boeing estimated that China will need another 2,600 new planes through 2025.[31] To be less dependent on imports, various Chinese aviation companies formed a (government-controlled) consortium in 2002 to develop a Chinese passenger jet that would seat about 80 people. Its body showed a strong resemblance with the DC-9 series of aircraft (as China has produced the McDonnell Douglas MD-90 under a license agreement), and nearly 20 Western companies contributed to the project. General Electric delivered the jet's engines, for instance.

AN ONGOING PROCESS

Compared to the global spread of agriculture, which took several millennia, the worldwide diffusion of oil technology was lightning quick. Nevertheless, many of the world's regions had not fully entered the Oil Age by the beginning of the 21st century: Numerous countries in Africa and Asia, including India and China, remained overwhelmingly rural, while urbanization in true Oil Age countries plateaued above 70 percent, a level of approached by Latin America towards the end of the 20th century. To be sure, various mechanisms of technology spread have long been well established, and all the world's regions adopted some aspects of oil technology (by importing trucks and military jets, for instance). Many skipped the Coal Age to become rapidly dependent on imported (or less often domestic) crude oil. But they did not yet become full-fledged Oil Age societies. Both global industrialization and the Oil Age are continuing, and large parts of the world population have yet to experience the full-blown consequences of oil energy command.

NOTES

28. "Gräf & Stift," AEIOU, the Austrian cultural information system of the Federal Ministry for Education, Science and Culture (BMBWK), http://aeiou.iicm.tugraz.at/aeiou.encyclop.g/g638817.htm. The site quotes M. Minnich, "Der Fall Gräf & Stift" (diploma thesis, Vienna, 1992).
29. Jonathan Glancey, "Jaguar Halts Production of Venerable Daimler," *The Guardian*, March 2004, http://arts.guardian.co.uk/critic/feature/0,1169,1177372,00.html.
30. "The Global Car Industry: Extinction of the Predator," *The Economist*, September 8, 2005, http://www.economist.com/business/displaystory.cfm?story_id=E1_QPGJSGN.
31. "Chinese Aviation: On a Wing and a Prayer," *The Economist*, February 23, 2006, http://www.economist.com/business/displaystory.cfm?story_id=E1_VVVSQRT.

CHAPTER 28

LIFE IN THE OIL AGE

Oil Age people enjoy a way of life that is unbelievably energy-rich in comparison to all previous standards. During the Agricultural Age the amount of work a well-fed man could do in a day was a good measure to describe productive energy. As an average sized man needs to eat about 2,400 kcal a day, a slave owner able to provide food worth that energy had a man work for him full-time. (One kilogram of rice contains some 3,700 kcal, for instance.) During the Coal and Oil Age fossil energy began to replace increasingly more human muscle work. (One kilogram of crude oil contains about 10,000 kcal of energy.) Soon a lot more work was done by machines than all humans living on Earth could possibly do. Hence the amount of technical energy consumed on average per person living on the planet (despite the enormous global population growth and despite substantial regional disparities) climbed to some 2,100 kcal per day per person on average in 1860, and thereafter to 18,000 kcal in 1950, and to some 41,000 kcal in 2001. This is equivalent to every person on the planet having about 17 energetic personal slaves at her or his disposal all day long.[32] Life thus became much easier for Oil Age people in terms of physical work. Motorized cranes, trucks, elevators, conveyor belts, and forklifts began to move loads; earth moving machinery eased construction work and mining; stationary equipment crushed, hammered, rolled, and stretched materials; and electric motors in hand-held tools (such as drills) even eased minor tasks for mobile workers. Accordingly, human muscle power by the end of the 20th century contributed less than 0.2 percent of the total work done in the United States, for instance.

People also started to walk less. Affordable family cars (starting with such automobiles as Ford's Model T, the VW Bug, Citroen 2CV and Fiat Topolino) redefined mobility during leisure time, but even more important, people began driving cars to work and no longer had to live close to factories or other sites of employment. This greatly increased the quality of life for families, and induced urban sprawl. Urbanization itself was further promoted by the spread of oil-powered tractors and other mechanized agricultural machinery that eliminated rural jobs. In the United States, rural labor fell from 15 percent in 1950 to 2 percent in 1975. Similarly, over 80 percent of Britain's and Germany's populations is now living in cities. The number of people concentrated in very large cities increased worldwide as well. There were only 13 urban areas with more than 1 million inhabitants in 1900, compared to well over 300 by the year 2000.

In urbanized living, electrification made all the difference in terms of household work. General Electric offered irons, vacuums, immersion water heaters, and cookers by 1900, but electric household appliances were first sold in large numbers in Europe and North America from the 1930s, when the electrical grid had grown, and electrical wiring of households had become standard.[33] Electrolux of Sweden introduced the first compact built-in electric refrigerator in 1930, and refrigerator ownership rose sharply in America during 1940s, and in Europe during the 1960s. Electric (and gas) ovens were followed by microwave ovens (invented in 1947), and eventually washing machines, dryers, and dish washers became standard household equipment. (By the mid-1970s, about 60 percent of French households had washing machines, for instance.) Thus, the energy-rich lifestyle began to save people a lot of time. The washing of clothes turned from a day-filling task into a background operation, and the new heating and cooking systems needed a lot less attention than previous (ash-producing) wood- and coal-fired ovens. Refrigerators for food storage eliminated the daily trip to the market, and cooking activities were partly outsourced as eating out became more common, and ready-made frozen (or otherwise preserved) meals entered the diet of Oil Age people. In turn, much of the saved time was spent with such electricity-powered items as radios, TV sets, and eventually home computers.

As household work was traditionally the work of women, the extra time won through the increasing electrification helped to bring about a societal revolution that led women towards equal standards in terms of political rights and education. In Britain, for instance, fully equal women's suffrage was won in 1928, and later on in the Oil Age many Western countries saw a larger number of women than men graduate from universities. Higher education became generally more important, as knowledge creation and professional specialization were accelerated in comparison to the previous Energy Eras, and traditional production jobs were increasingly eliminated. Literacy quickly approached 100 percent in fully industrialized Oil Age countries, and eventually office jobs outnumbered all other jobs. In the United States, the number of so-called white collar jobs exceeded the number of blue collar jobs for the first time in 1956, and Oil Age societies increasingly turned into Knowledge Societies.[34] (Only 16 percent of all American jobs are now in industry.) But just as coal and oil had earlier eliminated jobs in production, inanimate energy in the form of electricity soon began to take over much work in the administrative, service, and other white collar sectors. Computers became faster and smarter, data networks provided a new communications infrastructure, and voice recognition systems began to ease communications between humans and machines. Hence computers increasingly replaced people in various office jobs, eliminated much supervising personnel in factories, turned pilots into bystanders even during the starting and landing maneuvers, and interpreted X-ray pictures for doctors, for instance. In short, electricity, mainly a manifestation of fossil fuels, replaced brain power, just as coal and oil had earlier replaced muscle power.

The shift towards higher energy consumption benefited all members of Oil Age societies. The wealthy urban middle class was growing, physically straining work was on the decline, and people generally worked less than ever before. In fact, the substantial amounts of leisure time enjoyed in energy-rich Oil Age societies soon approached those of the Foraging Age. The Fair Labor Standards Act of 1938 mandated worker protections, including a maximum 8-hour workday and 40-hour workweek in the United States, for instance, and in many western European countries people began to enjoy workweeks below 40 hours plus five or six weeks of vacation per year from the first time they were employed. Add up to 15 paid holidays per calendar year (in Austria, for instance) to work-free Saturdays and Sundays (and in some countries Friday afternoons), and it becomes clear that people in many western European countries now work little over two days per day off.

LONGER, HEALTHIER LIVES

Oil Age people also began to live a lot longer than people in the previous Energy Eras could have possibly imagined. In fact, life expectancies in some industrialized countries nearly doubled between 1900 and 2000. In Britain, average life expectancy was 47 years in 1901, but 77 years in 2000. In 1840 the world's longest average life expectancy, which was then 45 years, was enjoyed by Swedish women. In 2002 the world's longest average life expectancy, 85 years, was enjoyed by Japanese women. (Japanese men, for comparison, could expect to live until 78. Women generally tend to live longer than men.) The rise in average life expectancy at birth during the first half of the 20th century was mainly due to improved child survival rates, while the rise in the consecutive decades reflected the increased chances of survival for people older than 65 years. The trends that caused better chances of survival had already begun in the Coal Age, but were now enhanced and accelerated. One trend was increased personal hygiene. Less than 60 percent of all French households had running water in 1954, only one-quarter had an indoor toilet, and only one in 10 had a bathroom. Two decades later 75 percent of French households had toilets and 70 percent had bathrooms. Water taps, especially for warm water, were generally a novelty of the Oil Age. (Coal Age people applied muscle power to receive cold water from hand pumps.) Another trend was better nutrition for all, achieved through energy-intensive agricultural output gains. But most important were medical advances that helped to control what had traditionally been the worst killer: infectious diseases.

In the beginning of the Oil Age it did not look as if this was possible. The global influenza epidemic of 1918, right at the end of World War I, killed perhaps 50 million people, easily more than died in combat in World War I (nine million) and World War II (15 million) combined.[35] (Apparently this pandemic started with a fairly mild strain in early March 1918 at a military camp in the United States, but American troops then carried the virus to western Europe and returned half a year later from the front with a deadly influenza strain that was responsible for half of the 60,000 World War I casualties of American soldiers, and an estimated 675,000 deaths in the United States. The disease spread to nearly all populated areas of the world. In Germany, over 400,000 civilians died in the second half of 1918, and in India up to 16 million people may have died between June 1918 and July 1919.)

By the 1930s viral diseases such as influenza were fully understood, but their treatment remained difficult. The situation looked a lot better in terms of bacterial diseases. They were easier to identify, and real treatments emerged in form of antibacterial drugs from 1909 (with Paul Ehrlich's Salvarsan). German Gerhard Domagk discovered the antibacterial action of the sulphonamides in 1932, and German firm IG Farben marketed the first general-purpose bacteria-killing drug, Prontosil. German Ernst Chain and Australian Howard Florey isolated (and produced in quantity) penicillin from Penicillium mold in 1939, over 10 years after the substance's potency had been coincidentally discovered by Scotsman Alexander Fleming in 1928. Penicillin was produced in large quantities from the early 1940s and proved vital for the treatment of World War II casualties. Penicillin turned out to affect the bacteria that cause strep throat, spinal meningitis, gas gangrene, and syphilis, and soon more antibiotics were identified to render previously life-threatening infections harmless. (Selman Waksman, a Ukrainian biochemist working in the United States in 1944, isolated streptomycin, the first specific agent effective in the treatment of tuberculosis.) Vaccination programs proved effective to prevent both bacterial infections (typhoid, whooping cough, tetanus, diphtheria, tuberculosis) and viral disease[36]. The smallpox virus, for instance, had killed an estimated 500 million people during the 19th century, but was globally eradicated by the mid-1970s following a vaccination campaign promoted by the World Health Organization (WHO). Similarly, the incidence of poliomyelitis (caused by the poliovirus) dropped in the United States from 38,000 per year in the 1950s to less than 10 per year by the early 21st century. A preventive vaccine against influenza was developed after World War II as well.

While Oil Age people forgot the constant fear of epidemics that had been part of human life in the previous Energy Eras, the primary causes of death shifted to non-communicable diseases, with cardiovascular diseases being the number one killer: Worldwide cardiovascular diseases now account for about one third of all deaths, and in mature Oil Age countries for a substantially larger share. Coronary heart disease (or ischemic heart disease) is the most frequent cause of death globally, followed by cerebrovascular disease (stroke).[37] Both are associated with risk factors that reflect the energy-rich lifestyle of the Oil Age, including lack of physical activity and extensive meat (as well as egg and dairy product) consumption. Other risk factors include smoking, high blood pressure, diabetes mellitus, and obesity, which are in themselves serious health problems. Lung cancer, associated with tobacco smoking, was responsible for 2.1 percent of all deaths globally in 2002. (All cancers combined accounted for 12.6 percent.) Worldwide over one billion adults (over one sixth of the total population) are overweight, and at least 300 million are clinically obese. Obesity reflects an imbalance between ingested food energy (which is available in excess quantities in Oil Age societies) and physical activity (as the human body hardly serves as prime mover any longer). Obesity now kills some 300,000 people per year in each North America and western Europe. In America nearly two thirds of the adult population is overweight, and almost a third is obese. Another cause of death typical for the Oil Age are road traffic accidents: these are globally the leading cause of death for men aged 15 to 44 years.[38]

SMALLER FAMILIES

The observed improvement of child survival, combined with prolonged education of women, contributed to a further decline in family size. In many Oil Age countries families decided to have fewer children, and fertility rates, at least temporarily, fell below replacement level. (Less than two surviving children on average per couple, in order to replace both parents, will cause a population to shrink, unless immigration is allowed to offset the effect.) Children were no longer viewed as an additional pair of hands to do work (as was the case still in the Coal Age in the countryside), and wealthy urban Oil Age parents generally do not regard their children as some sort of insurance or pension provision for the last part of their life. The availability of birth-control pills, in addition to other contraceptives, supported the trend from the 1960s by providing parents with a practical means to restrict family size.

In turn families became smaller, and women gave birth later and to fewer children than ever before.[39] In Britain, for instance, the average age of first-time mothers had risen above 30 years by the beginning of the 21st century.

TRANSITIONAL PROBLEMS

The Oil Age generally spared people negative impacts of the magnitude experienced in the transitional periods between the previous Energy Eras. The emergence and spread of agriculture led to the enslavement of people, to hard work, and to epidemics thriving on higher population densities; coal culture is associated with pollution and the emergence of a poor, urban working class. Oil culture, however, was deeply embedded in coal culture: the transition was relatively smooth and did not create large groups of winners and losers in society at a fast pace. That is not to say that there weren't certain professional groups that were negatively affected. Horse breeders and farmers found themselves diminished in numbers, and railroad workers and coal miners saw the importance of their profession dwindling. But since societies had matured, these groups were often supported by central governments to ease the transition. Coal mining remained heavily subsidized in western Europe, for instance. A quite radical and late transition occurred in Britain in the mid-1980s, when most coal was used for electricity production and iron smelting. The coal miners had reached an agreement with the government in 1974, right after the first oil price shock, but in 1975 British North Sea oil production began. In 1984 the conservative government under Margaret Thatcher attempted to close 20 of Britain's 170 collieries, which translated into the loss of 20,000 jobs. In response, up to 165,000 British coal miners went on strike during 1984–85 in what was Oil Age Britain's most bitter industrial dispute. In 1985 alone, 25 pits were closed, and by 2005 only 8 active British collieries employing 4,000 miners were left. (British electricity production was shifted to nuclear energy.)

NEGATIVE IMPACTS

Environmental pollution due to fossil fuel burning remained a problem as well. As late as December 1952, sulfur dioxide and particulates during the Great Smog caused 4,000 excess deaths within five days in London, where foul air now kills about 1,000 people a year, for the most part by exacerbating lung diseases.[40] On a less regional scale, it also became clear towards the end of the 20th century that the accumulation of carbon dioxide gas released by fossil fuel burning might have a serious negative impact on the world's climate. (Besides, cars and their infrastructure also take up a lot of space. By the mid-1970s two thirds of Los Angeles's downtown area was covered by roads and parking lots.[41])

On the other hand, air quality greatly improved during the second half of the 20th century in the West's major cities. Oil fuels burn cleaner than coal, and air pollution caused by gasoline- and diesel-fueled motor vehicles was countered by various legislative measures that introduced mandatory exhaust gas filters and catalytic converters, for instance. Cities were increasingly supplied with electricity and warm water from power plants located quite far away from population centers. The electricity and gas infrastructure directly helped to reduce the incidence of respiratory diseases by eliminating traditional open-fire stoves fueled by solids. In the less electrified parts of the world respiratory diseases now rank as the second most common cause of death (after coronary heart disease). Indoor air pollution, including contributions from cigarette smoke, is estimated to cause globally 36 percent of all lower respiratory infections, 28 percent of chronic obstructive pulmonary disease, 22 percent of tuberculosis, 11 percent of asthma, and about 3 percent of lung cancers.[42]

Lifestyle changes caused by oil (or generally large) energy command negatively impacted people's health as well, as many became obese due to the oversupply in food and reduced physical activity. Even the new global mobility and infrastructure had negative aspects to it, as diseases could now potentially spread very fast around the globe. The influenza epidemic of 1918 spread quickly to Africa, Brazil and the South Pacific, but today such diffusion would be even faster due to frequent intercontinental passenger flights. Extensive international shipping in many places has introduced aggressive foreign species that threaten indigenous ecosystems or human health. Tires transported in 1985 from Asia to Texas on a containership carried larvae of the Asian tiger mosquito, which transmits dengue fever and other tropical diseases. Within five years, Asian tiger mosquitoes were living in 17 states.[43] Generally, people had to learn not to abuse the vast amounts of energy they commanded: the application of oil energy led to large-scale deforestation and overfishing, for instance, and new mining machinery allowed humankind to become a major geological agent. Similarly, people had to learn not to use the enormous amounts of destructive energy that they now commanded (in the form of weapons of mass destruction) against one another.

SOCIAL DISPARITIES

Increased amounts of energy commanded in the Oil Age also meant increased wealth, and increased wealth may easily lead to major disparities between regions or social groups. As Oil Age nations became richer, the distance to less industrialized regions increased. The poorer regions became poorer in relative terms, as they stayed at the level of previous Energy Eras, but sometimes even in absolute terms, when population growth outpaced economic growth. Such regions, even though they were now independent (rather than colonial holdings), received direct support by rich Oil Age nations, which was an unprecedented development. What is more, many Oil Age nations had internally matured enough in terms of their social organization to counter directly exaggerated asymmetric wealth accumulation in society. Well-functioning democracies, with one vote per adult, can easily opt to redistribute wealth, but redistribution measures also have their risks. Some western European countries introduced high income taxes (at least on income above average) and high estate taxes (death duties), which works to slow extreme accumulation of wealth over generations but may also be counterproductive in terms of economic growth and entrepreneurial spirit. Free market economies are expected to create wealth for societies as a whole, but the system needs to provide incentives for individuals to work hard and take risks in exchange for the prospect of acquiring personal wealth. Individuals also demand to be allowed to share their wealth with their family and children.

Oil Age societies generally saw their wealthy middle class grow, but the United States experienced the emergence of a two-tier labor market, in which those at the bottom lack the education and the professional/technical skills of those at the top. Since 1975, practically all the gains in U.S. household income have gone to the top 20 percent of households.[44] Generally the richest 20 percent of the population in wealthy Oil Age countries tend to earn 40 to 50 percent of the total income, while the other 80 percent of the population are sharing the other half of the total income. The disparity looks even more extreme when the very wealthiest people are compared to the poorest people. When Microsoft chairman Bill Gates' net wealth was $40 billion, this was more than the combined net wealth of the poorest 110 million U.S. Americans, or about 40 percent of the American population.[45] (By 1998, Gates was worth $59 billion; a year later, he was worth $85 billion.) While such figures definitely call for more wealth redistribution in America, the problem might also be overstated: There is a limit to what one person can consume, be it in terms of champagne, toilet paper, or miles flown or driven. In western Europe, the distance between the richest and the middle class is much smaller than in the United States, but the distance between the middle class and the poor is actually quite similar. Western Europe has higher standards in terms of general health care, vacation time, and paid holidays when compared to the United States, but it seems to have problems removing some of the societal structures that have been inherited from previous Energy Eras. The House of Lords still existed in Britain in the early 21st century, though both Houses of Parliament in 1999 finally voted to strip most hereditary peers and peeresses of their right to a seat in the House of Lords. Similarly, many European nations are at this stage still granting their monarchs a birthright to represent their countries. (And many parents are still sympathetic with the system, installing a positive attitude towards monarchy into their children.)

Nevertheless, Oil Age societies saw their increasingly educated people make predominantly rational decisions to improve social organization.

NOTES

32. The commercial energy consumption of 40,740 kcal per person on global average is the 2001 amount based on oil, coal, natural gas, hydropower, and nuclear energy calculated with data from the BP Statistical Review of World Energy, June 2002. BP's current Statistical Review of World Energy can be downloaded at www.bp.com/statisticalreview. If we take into account that perhaps a third of the 2,400 kcal worth of food intake is used by the working man to keep his own metabolism going and to cover some non-work activities, he can deliver perhaps 1,600 kcal worth of work. However, engines and machines are not fully efficient either. The consumption of 41,000 kcal per day per person in technical energy would thus not provide the whole amount as useful output either.

33. Vaclav Smil, *Essays in World History: Energy in World History* (Boulder, CO: Westview Press, 1994), 212.

34. Generally, a knowledge society is considered a society in which knowledge is a primary resource of production, more important than capital and labor. The terms knowledge workers, knowledge economy, and knowledge society were coined by Peter Ferdinand Drucker in the 1950s. Drucker published over 30 books, including: *The Landmarks of Tomorrow* (New York: Harper, 1959), *The Age of Discontinuity* (London: Heinemann, 1969), *Management: Tasks, Responsibilities, Practices* (New York: Harper & Row, 1973), *The Post-Capitalist Society* (New York: HarperBusiness, 1993).

35. The death toll of the 1918 influenza epidemic has been repeatedly revised upward. The following paper suggests that it was of the order of 50 million, though it may have been 100 million. The higher figure would be even well above the total death toll of both World Wars, which was perhaps 75 million. (This figure is uncertain as well.) N.P. Johnson and J. Mueller, "Updating the Accounts: Global Mortality of the 1918–1920 "Spanish" Influenza Pandemic," *Bull Hist Med.* 76(1) (2002):105–15, http://www.ncbi.nlm.nih.gov/pubmed/11875246?dopt=Abstract.

36. While diseases caused by bacteria may be treated with antibiotics, vaccination programs are usually the only effective tool to counter viral infections by providing resistance to them. The treatment of viral infections is difficult, as it will typically also damage the host cell in which the virus reproduces. To be sure, a lot of research has gone into the development of antiviral drugs, but with limited success. A typical strategy is to trick the virus into incorporating nucleoside analogues, that is, fake non-functioning DNA building blocks, into its genome during replication. See, for instance: H. Kapeller, C. Marschner, M. Weissenbacher, H. Griengl, "Synthesis of Cyclopentenyl Carbanucleosides via Palladium(0) Catalysed Reactions," *Tetrahedron* 54(8) (1998): 1439–56, http://www.sciencedirect.com/science?_ob=ArticleURL&_udi=B6THR-3SH5D6B-1C&_user=10&_rdoc=1&_fmt=&_orig=search&_sort=d&view=c&_acct=C000050221&_version=1&_urlVersion=0&_userid=10&md5=a1e2f00d273c69e99ede1a89bbaaab92#bibl1.

37. World Health Organization, *Cardiovascular Diseases* (Geneva: WHO, 2007), http://www.who.int/mediacentre/factsheets/fs317/en/index.html.

38. "Global Burden of Disease and Injury Series: Executive Summary," Burden of Disease Unit, Center for Population and Development Studies at the Harvard School of Public Health, http://www.hsph.harvard.edu/organizations/bdu/summary.html, http://www.hsph.harvard.edu/organizations/bdu/gbdsum/gbdsum2.pdf; World Health Organization, "The World Health Report 2002: Reducing Risks, Promoting Healthy Life" (Geneva: WHO, 2002), http://www.who.int/whr/en/; World Health Organization, "WHO Statistics, Mortality Database," "Table 1: Numbers of registered deaths, United States of America—1999" (Geneva: WHO, 2000), http://www3.who.int/whosis/mort/table1.cfm?path=mort,mort_table1&language=english; World Health Organization, "The World Health Report 2002: Reducing Risks, Promoting Healthy Life," "Annex Table 2 Deaths by Cause, Sex and Mortality Stratum in WHO Regions, Estimates for 2001" (Geneva: WHO, 2002), http://www.who.int/whr/2002/whr2002_annex2.pdf; Center for Disease Control and Prevention, National Center for Health Statistics, http://www.cdc.gov/nchs; World Health Organization, "The Top 10 Causes of Death" (Geneva: WHO, 2008), http://www.who.int/mediacentre/factsheets/fs310/en/; "We're Living Longer, Getting Fatter," *CBSNEWS.com*, September 12, 2002, http://www.cbsnews.com/stories/2003/03/10/health/main543315.shtml.

39. The trend of increasingly later births may have contributed to the increased longevity of people even before the Oil Age. Research showed that later births are associated with changes in telomeres (that is, the structures that cap the end of chromosomes) in a way that delays aging at a basic molecular level. "Ageing: Parents, Telomeres and Lifespan," *The Economist*, July 12, 2007.

40. Peter Brimblecombe, *The Big Smoke: A History of Air Pollution in London since Medieval Times* (London: Methuen, 1987). Peter Brimblecombe and László Makra, "Selections from the History of Environmental Pollution, with Special Attention to Air Pollution. Part 2: From Medieval Times to the 19th Century," *Int. J. Environment and Pollution* 23 (2005): 354, http://www.sci.u-szeged.hu/eghajlattan/makracikk/Brimblecombe%20Makra%20IJEP.pdf; "Air Pollution: Driving Polluters from London," *The Economist*, February 1, 2007, http://www.economist.com/world/britain/displaystory.cfm?story_id=E1_RGQGQSV.

41. Richard White, *It's Your Misfortune and None of My Own: A History of the American West* (Norman: University of Oklahoma Press, 1991); Martin V. Melosi, "Automobile in American Life and Society: The Automobile Shapes The City," http://www.autolife.umd.umich.edu/Environment/E_Casestudy/E_casestudy.htm.

42. Lower respiratory infections kill more people in developing countries than HIV/AIDS, for instance. World Health Organization, "The Top 10 Causes."

43. Chris Bright, *Life Out of Bounds: Bioinvasion in a Borderless World*, (New York: W.W. Norton, 1998).

44. "United States," CIA—The World Factbook https://www.cia.gov/library/publications/the-world-factbook/geos/us.html.

45. Ralph Nader, "Open letter to Bill Gates," July 27, 1998, http://www.nader.org/index.php?/archives/1893-An-Open-Letter-to-Bill-Gates.html. Nader quotes a study by Edward Wolff of New York University, whose calculations included home equity, pensions, and mutual funds, but excluded personal cars.

CHAPTER 29

TECHNOLOGY AND KNOWLEDGE

Knowledge accumulated at unprecedented rates during the Oil Age, and technological advance was accordingly fast. More people than ever before were freed from productive tasks (that were now handled by machines), and many spent a substantial part of their life studying and researching. Besides, there were a lot more people around. Two billion people (in 1930) will potentially produce twice as many ideas as one billion (in 1810), and six billion people (in 2000) may potentially produce three times more knowledge than two billion people (in 1930).

Many of the Oil Age's most critical technological advances were in the areas of productivity and mobility. Even small current earth-moving machines are more powerful than thousands of humans (or hundreds of horses) combined. One billion tons of coal were mined worldwide in 1910, but 5 billion in 1990 (even though coal was no longer the dominant fuel). People traveled at maximum speeds of about 20 kilometers per hour (12 mph) in horse-drawn coaches, while trains and cars soon reached top speeds of well over 100 km per hour (60 mph), and today's modern passenger jet planes fly at speeds of 800 to 1,000 km per hour (500 to 600 mph). U.S. railroads carried passengers some 35 billion miles per year in 1913, while U.S. airlines carried passengers some 400 billion miles a year in 1987. Trucks began to have an impact in America from the 1920s, carrying farm products cheaper and faster to the urban centers. Trucks also became indispensable for the transport of raw materials and manufactured goods, but bulk transport remained cheapest on the water. Some of today's supertankers carry 500,000 tons of oil, while the largest cargo planes carry around 100 tons of goods.[46] However, high energy inputs and the transition to the knowledge society also delivered major advances in material and natural science, as well as communications and information technology.

MATERIALS: PLASTICS, NEW METALS, AND CERAMICS

The material world that surrounded people in their everyday life radically changed during the Oil Age. First of all, the trend towards more use of metal continued. During the Coal Age people got acquainted with large metal structures, an effect of coal-fired mass production of iron. In the Oil Age metals remained the principal material for the construction of buildings and bridges, and for engines and tools. Iron continued to be the most important metal due to its availability, material characteristics, and low production cost. Annual steel production increased from some 28 million tons in 1900, to 70 million tons in 1913, to 170 million tons in 1950, to 875 million tons in 2000. The Linz-Donawitz process, developed in Austria in 1948, now accounts for nearly 70 percent of all steel produced worldwide. (It is also known as the basic-oxygen process, as it blows a jet of high pressure pure oxygen through a pipe or lance into the molten metal.) Electric arc furnaces, which maintain precise temperatures, became dominant in the making of high-quality steel towards the end of the 20th century. Japan had emerged as the most innovative iron maker by the 1960s. In 1973 a Japanese blast furnace reached the unprecedented size of 32 meters in height and 16 meters in width.[47] Modern continuous blast furnaces produce 10,000 tons of metal a day, using 8 million tons of raw materials and nearly 2.5 million tons of coal per year. As of 2003, China produced 220 million tons of the world's total 965 million tons of steel per year, with Japan being a distant second at 110 million tons. (U.S. production was 92 million tons.)

Global annual cast iron production is about 10 times larger than combined production of 5 other leading metals: aluminum, copper, zinc, lead, and tin. Aluminum was first isolated in 1824, but an economic process for large-scale production was first devised in 1886, when C.M. Hall in the United States and P.L.T. Heroult in France independently electrolyzed aluminum oxide.[48] Aluminum production consumes huge amounts of energy, and aluminum smelting grew only slowly even after large electricity plants had been introduced. The aviation industry used aluminum as a replacement for wood and cloth during the 1920s, and after World War II aluminum became a substitute for steel wherever designs required both lightness and strength (including parts for automobiles).

Plastics

The Oil Age also saw a whole new class of materials emerge and surround people during every minute of their life. Clothing, shoes, mattress fillings, blankets, carpets, furniture, cooking utensils, telephones, TV sets, computers, paints, plumbing fixtures, pipes, medical instruments, dental fillings, optical (contact) lenses, packaging materials, building walls, boats, cars, and airplanes all consist fully or to a large extent of plastics, which are generally lightweight, waterproof, and chemical resistant materials that can be formed in virtually any shape and be produced with properties that range from hard as stone, strong as steel, transparent as glass, light as wood, and elastic as rubber. (Thus, the volume of plastics produced in the United States passed the volume of domestically produced steel in 1979.)

A few plastics had already been made from natural materials in small quantities in the late Coal Age. Shellac, a resin secreted by a tiny scale insect (Laccifer lacca) and harvested from trees in southern Asia, was used to make phonograph records, buttons, and knobs. Celluloid, a highly flammable tough plastic, was initially developed by American John W. Hyatt as a substitute for ivory billiard balls. (Celluloid is made from nitrocellulose and camphor, a waxy solid obtained from the wood of the camphor laurel.) The first fully synthetic polymer, now known as phenolic resin, was produced by Belgian chemist Leo Baekeland in the United States in 1909 (from formaldehyde and phenol). This artificial resin was used to make such products as telephones and pot handles.[49]

The foundation for the great expansion of the plastics industry in the Oil Age was laid by German chemist Hermann Staudinger, who received a Nobel Prize for proposing (in 1922) and demonstrating the nature of plastics as long-chain molecules. The quality to produce such chains is quite unique to the carbon atom. (The only other atom with a somewhat comparable quality is silicon.) Long carbon chains are observed in nature and are critical for the living world. (Proteins and wood are built up by long molecular chains, for instance.) Oil Age chemists took various hydrocarbons out of the crude oil mix and rearranged them into similarly long carbon chains. Usually this is done by producing short units (monomers) first, which are in turn reacted into polymers. Polyethylene, for instance, is produced by reacting ethylene gas (CH_2CH_2) in presence of a catalyst into extremely long molecules that each contain over 200,000 carbon atoms.

German chemical companies extended their (Coal Age) lead from dyes and pharmaceuticals to the polymer sector. Among the plastics I.G. Farben synthesized between 1920 and 1932 were polystyrene (either a foam used to make such items as insulating coffee cups or protective packaging for shipping fragile items, or a solid used to make throw-away cutlery, yogurt containers, compact disc casings, etc.), polyvinyl chloride (PVC— either rigid, for water pipes, plumbing fittings, etc., or flexible, for toys, hoses, shower curtains, imitation leather). Bayer in 1937 introduced the highly versatile polymer polyurethane, used either as a flexible foam (mattresses, packaging), a rigid foam (insulation in refrigerators and homes), elastomer (shoe soles, medical equipment), fiber (Spandex fibers used for textiles), surface coating (sealants for machine parts, linings for tanks and pipes), or adhesive (general-purpose waterproof glue). I.G. Farben in 1939 introduced polyepoxide (epoxy), a resin that hardens when a catalyzing agent is added to it. Epoxy is used to make adhesives, surface coatings and composite laminates. Composites are plastics that are reinforced by fibers. Strong and affordable glass fibers, commercially available since 1936, are the most common reinforcing material. (They are produced by extruding molten glass through extremely small nozzles.) Many aircraft parts, including wings, are made from epoxy composites. Other applications for reinforced plastics include car body panels, boat hulls, windmill rotor blades, skis, tennis rackets, and golf clubs. The development of synthetic rubber was of special economic and strategic importance. By the end of the Coal Age natural rubber had become indispensable for insulating underwater telegraph cables (gutta-percha), and for tires and washers. German chemists developed synthetic rubbers (Buna) in the early 1930s, and several countries established major synthetic-rubber plants during World War II, when natural rubber shipments from the plantations of Southeast Asia were disrupted.

Oil Age people also began to dress in oil-derived materials. Rayon (produced in the United States from 1910) and acetate (produced in the United States from 1924) foreshadowed this development, but both these fibers were derived from plant cellulose. Nylon, on the other hand, was a fully oil-derived fiber. Different types of nylon, which is similar to silk, were developed in the United States (DuPont, 1935) and Germany (I.G. Farben, 1938). Acrylic, a wool-like fiber, was marketed in the United States from the 1950s. Polyester fibers, made from PET (polyethylene terephthalate) were increasingly used for clothing in the 1960s and 1970s. Synthetic fibers eroded the dominant position that cotton had held in the textile industry in the Coal Age. Nearly 70 percent of the materials processed in U.S. textile mills are now artificial fibers. (Yet the textile industry only accounts for something in the range of one percent of crude oil consumption.)[50]

Advanced Ceramics

Another type of Oil Age material that was unknown in previous Energy Eras are advanced ceramics. People are less familiar with these than with plastics, as they are hardly ever seen in consumer products. However, industrial ceramics are a lot lighter and yet stronger and harder than metals, and thus have countless applications. Some ceramics are resistant to abrasion and just about every chemical, others are stable at extremely high temperatures, and others again have a variety of useful electronic properties. Like traditional pottery ceramics, the advanced types are prepared from pliable, earthy materials. These are dug from the ground, dried, crushed, and ground into fine powders. (Some are dissolved and precipitated from solutions to obtain a highly pure product.) The powder is then pressed into the desired shape, often with the addition of a wax or plastic as a binding agent that allows the mix to be molded or extruded. The resulting object is in turn sintered at a high temperature, which causes the powder particles to coalesce, providing for the internal structure that results in high density and great strength. Zirconium dioxide (ZrO_2), for instance, is used as a refractory material (to make linings for furnaces, etc.). Silicon ceramics (silicon carbides, silicon nitrides) are used in such in high-temperature applications as gas turbines, jet engines, and diesel engines. Silicon nitrides, like alumina (aluminum oxide, Al_2O_3), are used to cut iron and other metals. The metal surface of refrigerators, stoves, and washing machines is often coated with ceramic enamel to make it harder and more corrosion resistant. Inert ceramics are also used inside the human body, to replace bone, dental caps, and hip joints. In electronics, chromium dioxide works as conductor, alumina as isolator, and silicon carbide as semiconductor. Barium and strontium titanate serve as miniature capacitors in computers and televisions. German physicist Johannes Georg Bednorz and his Swiss colleague Karl Alex Müller in 1987 received the Nobel prize for discovering that certain (copper-oxide-based) ceramics become superconductive at temperatures much higher than previously thought possible. Superconductivity, the ability to conduct electricity with no resistance, had formerly been observed only in metals cooled to ultra-low temperatures that are difficult to maintain. Certain ceramics, in comparison, act as high-temperature superconductors at just minus 148°C (-234°F), a temperature easily sustained by inexpensive liquid nitrogen. This opened a host of (now economically feasible) applications based on superconductor-made electromagnets generating strong magnetic fields with no energy loss. These include Magnetic Resonance Imaging (MRI), a medical diagnostic method providing spatial images without any surgical intervention on the patient; high resolution Nuclear Magnetic Resonance spectroscopy to determine detailed molecular structures of complex molecules such as DNA (this facilitates the development of new drugs.); and magnetic levitation trains that levitate above the track, which eliminates friction and allows for high speeds. Ultimately superconducting ceramics may allow for the development of computers of unprecedented speed and, above all, the more efficient generation and transmission of electricity.[51]

NATURAL SCIENCE

Advances in material science were largely based on a better understanding of nature and matter in general. Electron microscopes, initially invented in Germany in the 1930s, eventually made atoms visible, confirming that matter is indeed constituted of tiny particles. (But as it turned out, atoms are not the indivisible particles they had been proposed to be. Even the protons and neutrons that constitute the core of atoms are made up by yet smaller particles known as quarks.) German physicist Albert Einstein showed in 1905 in his theory of relativity that matter is principally equal to energy ($E=mc^2$). In biology, the rediscovery of the work of Austrian Coal Age scientist Gregor Johann Mendel made a major difference. His 1866 paper "Experiments in Plant Hybridization" had been sent to more than 120 scientific organizations and universities in Europe and America, but Mendel's laws of inheritance became famous only after 1900. These laws furnished Oil Age plant and animal breeders with a scientific foundation, and completed Darwin's theory of evolution by explaining the mechanisms that provided for the variation among offspring upon which the environment's selective forces could act.

One of Mendel's main conclusions was that heredity is transmitted by a large number of independently inheritable units (which are now called genes), but relatively little was known about the chemistry behind inheritance until after World War II. German physicist Max Delbrück (whose mother happened to be a granddaughter of famous Coal Age chemist Justus von Liebig) theorized that the genetic information might be stored in the form of a code, using the Morse code as an example of such code. Austrian chemist Erwin Schrödinger, who is best known for founding quantum mechanics through the fundamental Schrödinger equation of 1925, popularized Delbrück's theory in a 1944 book called *What Is Life?*, in which he speculated that the code of inheritance was to be found in a huge molecule.[52] This widely-read publication inspired the work of British physicist Francis Crick, who together with American biologist Jim Watson, in 1953 proposed the double-helix structure of DNA after studying the X-ray pictures taken by British crystallographer Rosalind Franklin. (High energetic X-rays had been discovered in 1895 by German physicist Wilhelm Roentgen, who produced an image of the bones of his own hand and immediately realized the importance of his discovery for medical diagnosis.) DNA (deoxyribose nucleic acid) had been isolated (from pus and many other types of cells) as early as 1869 by Swiss chemist Friedrich Miescher, who named the material nuclein, as he found it exclusively in the nucleus of cells. Miescher correctly proposed that this was the principal substance of inheritance. As it turned out, DNA is an extremely long molecule with an equally long sequence made up by just four chemical "symbols" attached to its backbone. These four "symbols," the four bases adenine (A), thymine (T), guanine (G), and cytosine (C), are arranged into groups of three (that is, triplets), which code for 20 amino acids. (The first triplet was decoded by German biochemist Johann Heinrich Matthaei and his American colleague Marshall W. Nirenberg in 1961. GGC codes for an amino acid by the name of glycine, for instance, while AAG stands for lysine.) According to the sequence of these triplets on the DNA backbone, cells link (20 different) amino acids into proteins consisting of chains of several hundred amino acids. (This is like threading beads of twenty different colors on a long string. It allows for countless different combinations.) The various species found in nature are characterized by their genetic material and thus the specific proteins found inside them: almost every organic molecule in a living creature is either itself a protein or the result of a protein's activity. The DNA coding system, indicating which triplets stand for which amino acid, turned out universal: it is the same in the smallest single-celled bacteria as it is in plants, animals and humans. This allowed scientists to conclude from DNA samples how species are related and how higher species evolved over time from primitive single-celled species. The *unity of life* also led to the emergence of the science of genetic engineering. A single gene, that is, a DNA section that encodes a single protein, can be transferred from one species to another and will be expressed. An Arctic fish's gene encoding an anti-freeze protein has been introduced into crops to make them frost-resistant. Two genes from a flower (wild daffodil) have been introduced in rice to make its seeds rich in provitamin A, whose absence in the diet of rice-based societies causes a marked incidence of blindness and premature death in children. Human genes were inserted into bacteria (and later plants and animals) from 1977. This served the production of pharmaceuticals. It had been known from 1922 that insulin, a vital protein expressed in insufficient quantities in diabetes patients, could be used to treat such patients. The human gene coding for insulin was therefore engineered into a strain of the bacterium Escherichia coli, which can be grown (multiplied) in a fermentor, and insulin isolated from the batch. Since its approval for the U.S. market in 1982, such human insulin has been available in pure form in large quantities to diabetes patients. Other human proteins have been expressed in various plants, cow milk, and chicken eggs, for instance. However, direct gene therapy, by inserting normal or genetically altered genes into a patient's cells to replace nonfunctional or missing genes, had not been substantiated by the beginning of the 21st century.

COMMUNICATIONS

Advances in communications technology were promoted by increased Oil Age (long-distance) mobility. Transatlantic telegraph cables from the Coal Age maintained their importance well into the Oil Age, but (expensive) radio-based telephone service was available from 1927, and in 1956 the first transatlantic telephone cable was finally laid. The eighth transatlantic phone cable, laid in 1988, carried 40,000 simultaneous channels, compared to 36 channels of the 1956 cable, and 4,000 channels of the 1978 cable. Accordingly, the cost of a three minute phone call between London and New York decreased from $245 in 1930 to $53 in 1950, to $32 in 1970, and to $3 in 1990 (expressed in constant US$). The 1988 cable was the first transatlantic cable made of optical fibers. Thin transparent fibers of glass (or plastic) transmit digitized light pulses through their length by internal reflections with extremely low loss (which translates into high transmission rates) and do not interfere with one another (in the way electricity cables made of copper wire do). This allows for a thousand fibers or more to be strung into a single optical cable. Already in 1992 a new transatlantic telephone cable doubled the capacity of the 1988 cable, and soon data transfer between America and Europe was practically unlimited. Meanwhile communication via satellites matured as well. The first truly successful active communication satellite was the American TELSTAR I of 1962, and the next year NASA launched SYNCOM II, the first successful geostationary communication satellite. In 1969 a series of satellites was placed over the Atlantic, Pacific, and Indian oceans to provide for worldwide commercial satellite communications. The Intelsat 600 series, launched in 1989, reached a capability of 80,000 simultaneous voice channels. In the 1970s, telephone networks also connected fax (facsimile) machines, which scan a piece of paper and convert its printed text or graphics into a digital code that is transmitted and reproduced (printed) by the receiving fax machine. Such machines soon became indispensable in commercial activities, as they communicated contracts, signatures, pictures, and so on at the speed of a phone call.

From the 1980s, mobile phones (cellular telephones) became available. These transmit data via radio waves to stationary antennas, which direct them into landlines (or to other mobile phones). Each stationary antenna covers an area, a cell, around it, and many such overlapping cells provide for continuous mobile coverage. Cellular phones emerged out of the mobile rig technology developed for use in vehicles (such as police cars and taxis), but by the 1990s they had shrunk to pocket-size. At the beginning of the 21st century, market penetration was practically complete in Europe and the wealthier parts of Asia, while the United States lagged behind (66 percent of the population in 2003).[53] Mobile phones soon featured phone-to-phone texting capabilities (SMS, Short Message Service), which became tremendously popular as an inexpensive, fast, mobile, non-intrusive, silent form of communication that remains private even in public places. (In 2000, about 17 billion SMS messages were sent, in 2001 it was 250 billion, and in 2004 over 500 billion worldwide. In some Asian countries the average user in 2003 sent over 2,300 SMS messages, in Britain and Germany some 450, and in the United States 160.) The high penetration of mobile phones in Asia is in part explained by the relative low cost of building up a wireless infrastructure in areas that had not previously been penetrated by a landline network. What is more, mobile phones have been equipped with built-in digital cameras that can transmit live pictures through the telephone network.

Radio and Television

In terms of one-way communication, the Oil Age saw the rise of radio and television broadcasting. Regular radio broadcasts started in 1920, and in

1936 the BBC (British Broadcasting Corporation) offered the first scheduled television service. The basis for television cameras and TV sets had been laid by Karl Ferdinand Braun's 1897 invention of the cathode-ray (vacuum) tube, and color TVs were introduced in the 1950s, followed by cable TV in the 1960s, and (extensive) satellite broadcasting in 1970s. People used radio and television for entertainment and information, and governments used them from the start to expose people to propaganda.

Dutch electronics producer Philips in the 1960s pioneered cassette tapes to record radio programs, and in the early 1980s introduced compact discs (CDs) to replace the long-playing records that had been on the market since 1948. Sony of Japan in 1969 introduced VCRs (videocassette recorders) to record broadcast TV programs for later viewing and to play commercially recorded cassettes. In the 1980s, when VCRs started to be popular, watching television had become the third most time-consuming commitment (after work and sleep) for many Oil Age people. In the 1990s, the development of large flat-panel displays (liquid crystal display and plasma screens) to replace Braun (picture) tubes allowed for television sets to become a lot less bulky. However, the initial downsizing of radios and TVs, which made them accessible worldwide, was owing to the development of transistors to replace vacuum tubes as the amplifier elements in the 1950s. (Transistors practically do the same thing as vacuum triodes (that is, they amplify, control, and generate electrical signals), but they are small, have low power requirements, and generate little heat. Transistors consist of layers of different semiconductors, which are made from a base material (usually silicon) with addition of impurities (such as arsenic or boron).)

INFORMATION TECHNOLOGY (COMPUTING)

The substitution of transistors for vacuum tubes was also behind the downsizing of previously room-filling computers in the 1950s. Computers have their origins in the mechanical calculators of Frenchman Blaise Pascal (1642) and German Gottfried Wilhelm Leibniz (1673), as well as the data storage systems of the early mechanized weaving industry in France. Joseph Marie Jacquard improved on earlier similar systems in 1804 when he developed his loom that applied interchangeable punched metal cards (connected in an endless belt) that controlled the weaving of a cloth so that any desired pattern could be automatically obtained. The Jacquard loom spread worldwide, and from 1884 American Herman Hollerith used Jacquard-type punched cards as an input medium for his census machine. Punched cards were in turn used to feed data into electromechanical, and later into electronic computers. (In electromechanical computers an electric motor drives a mechanical calculator. Such computers required just seconds to complete multi-digit multiplications or divisions, for instance.)

Computers principally communicate and calculate through a binary two-letter-language, with the two letters being denoted as 1 and 0. Words in languages (codes) with few symbols are necessarily longer than those in languages with more symbols. An English sentence written in the Morse alphabet would be quite long, for instance, as there are only four symbols: short and long signals and breaks. (The letter C is encoded as "-. -.," for instance, the cipher 9 by "- - - -.") The same English sentence written in the Latin alphabet (26 letters, ten ciphers) will be much shorter. Two-symbol binary computer language is thus necessarily very long, but data machines were able to process (and store) large amounts of data, with 1 and 0 being represented by "hole" and "no hole," or by "current flowing" and "no current flowing" (or by different magnetic polarities on a magnetic disk).

Portable electro-mechanical rotor cipher machines (known as Enigma machines) were developed in the 1920s and 1930s in Germany, where Konrad Zuse in turn built the world's first real and functional program-controlled computer, the Z3 of 1941. This third of Zuse's home-made computers was destroyed in a bomb attack in 1945, but he founded a company the year after to build the Z4, the world's first commercial computer, leased in 1950 to ETH Zürich (Swiss Federal Institute of Technology Zürich). The venture capital to build this computer was raised through ETH Zürich and a 1946 IBM option on Zuse's patents. (IBM, International Business Machines, was a U.S. firm that had introduced an electromechanical machine in 1943.) By 1967 the Zuse company had built 251 computers, and was in turn sold to Siemens AG.

In the United States, the army financed the development of a general-purpose computer during World War II, but the ENIAC (Electronic Numerical Integrator And Computer), a room-filling structure of 19,000 vacuum tubes, was completed only in 1946. After the transistor was invented in 1947, and the first commercially available computer to use transistors in its circuitry delivered to the United States Air Force in 1956, computers became smaller and more wide-spread. Texas Instruments had earlier introduced the first commercially produced silicon transistor, and in 1958 demonstrated that it was possible to integrate various components such as transistors, capacitors, and the necessary wiring onto a single piece of silicon: Such computer chips became known as integrated circuits (ICs). The number of components placed on single chips grew exponentially, and in 1974 the U.S. Intel Corporation presented the first microprocessor: an IC chip featuring a central processing unit (CPU). Developed for use in a calculator, it incorporated 4,800 transistors, but the number increased to 3.2 million transistors on Intel's Pentium chip of 1993. A microprocessor is essentially a computer on a chip, and has thus found various applications in consumer electronics, telecommunications, and manufacturing. (Microprocessors are found in anything from watches, pocket calculators, telephones, and refrigerators, to cars and airplanes.)

The decrease in microprocessor prices promoted the spread of generic computers that helped to store and communicate the information and knowledge that accumulated at unprecedented rates during the Oil Age. By the end of the 20th century most households in the Western World had at least one personal computer, and these were many times as powerful as the huge government computers of the 1950s. Like businesses before, people started to use their computers for a variety of everyday tasks, including the writing and printing of letters, the tracking of budgets, and so on. And with the advent of the Internet, people also began using their personal computers to send and receive e-mail, to access their bank accounts, to read newspapers, to share information, to shop, to download computer programs, music, and movies, to place phone or video calls, and more.

The Internet

The Internet started out as small in-house networks that connected the computers in larger companies, universities, and other research organizations, mainly to share information between individual computers or to have access to the most powerful in-house computer from different terminals. By the late 1960s time-shared computing was a well-established technology, with terminals either being directly wired to a shared (out-of-house) computer, or connected via a telephone line. IBM was at this stage firmly committed to the mainframe rather than the personal computer, and in 1968 offered time-shared computing using a ten character per second Teletype terminal at about $40 per hour. However, the definite starting point of the Internet is given as the 1969 ARPANET (Advanced Research Projects Agency Network), which connected the computers of the Stanford Research Institute, the University of California at Santa Barbara, the University of California at Los Angeles, and the University of Utah. It was financed by the Department of Defense (DOD) with the goal of developing technology for a robust computer network to interconnect the main computers at the Pentagon and other locations. ARPANET quickly expanded in the United States as academics and researchers in different fields began to use it to exchange and access information, and in 1973 international connections to ARPANET were made from Europe. At this stage e-mail, developed in 1971, represented most of the traffic on ARPANET.

In 1986 the National Science Foundation established the NSFNET, a distributed network of networks capable of handling far greater traffic, and in 1990 ARPANET ceased to exist. One ongoing problem was to find a common language for all the different computers and networks that became part of what was growing into the Internet. In 1991, the European Organization for Nuclear Research (CERN) released the World Wide Web based on the communications protocol called HyperText Transfer Protocol (HTTP) developed at CERN two years earlier to standardize communication between computer servers and clients. When Mosaic, a browser with a graphical interface, was released in 1993, the World Wide Web began to exceed the popularity of earlier information-exchange services of the Internet. The first commercial dial-up access to the Internet was offered in 1990, and in 1995 the NSFNET reverted to the role of a research network, leaving Internet traffic to be routed through commercial network providers rather than NSF supercomputers. The Internet can now be accessed via mobile phones, wireless modems, cable-television lines, satellites, fiber-optic connections, and the public telephone network.[54]

NOTES

46. Vaclav Smil, *Essays in World History: Energy in World History* (Boulder, CO: Westview Press, 1994), 238.
47. Ibid., 179.
48. Ibid., 181.
49. "New Materials: Plastics," *Making the Modern World*, The Science Museum, http://www.makingthemodernworld.org.uk/stories/the_second_industrial_revolution/05.ST.01/?scene=6&tv=true.
50. This section is based on Terry L. Richardson, "Plastics," Microsoft Encarta Online Encyclopedia, http://encarta.msn.com/encyclopedia_761553604/Plastics.html.
51. This section is based on "Ceramics," Microsoft Encarta Online Encyclopedia, http://encarta.msn.com/encyclopedia_761565098/Ceramics.html.
52. Erwin Schrödinger, *What Is Life?* (Cambridge: Cambridge University Press, 1944)
53. "Je ne texte rien," *The Economist*, July 8, 2004, http://www.economist.com/displaystory.cfm?story_id=2908047.
54. "Internet," Encyclopædia Britannica, http://www.britannica.com/EBchecked/topic/291494/Internet.

CHAPTER 30

WEAPONS

The increased command of energy in the Oil Age also had horrifying consequences. Never before had people used such vast amounts of destructive energy, never before had combat casualties been so high, and never before had so many civilians been killed during wartime. The World War I Battle of the Somme, which lasted for four and a half months, cost over one million lives in 1916. Casualties at the World War II Battle of Stalingrad, which lasted about as long in 1942–43, surpassed 2.1 million. During World War II some 40 million civilian deaths accounted for more than 70 percent of the 55 million total. In Japan about 100,000 people died in less than two weeks in March 1945 during nighttime raids by American B-29 bombers, which leveled about 83 sq km (32 sq mi) of four principal Japanese cities,[55] and in Germany nearly 600,000 people, including 70,000 children, died, and almost 900,000 were wounded, when Allied airplanes dropped 80 million incendiary bombs over urban areas.

Incendiary bombs, or firebombs, contain chemicals such as white phosphorus to incinerate wide target areas. Napalm bombs, first used by the United States in 1944, actually contained gasoline as the active ingredient, which was mixed with a solid thickener to yield a gel that would tenaciously stick to surfaces (including the human body) and continue to burn. (It was during the Vietnam War that the American use of napalm was first considered controversial.) However, the most relevant advance in bomb technology in the early Oil Age was the Haber-Bosch process. Strangely, this German technology to fix atmospheric nitrogen served humanity tremendously well and tremendously badly. On the one hand, it eliminated the shortage of

Haber Bosch Plant Without this German technology perhaps half the people now populating the planet would not be able to survive due to food shortage. However, the fixation of atmospheric nitrogen not only served the purpose of fertilizer production; it also facilitated the mass production of explosives. Large scale ammonia and fertilizer (ammonium sulfate) production started in 1913 at BASF's new Oppau plant near Ludwigshafen. In 1921 a fertilizer storage facility exploded at the plant, killing over 500 people—Germany's largest industrial accident to date.

nitrogen fertilizer and allowed the human population to grow to current levels. On the other hand, it also allowed for the production of large amounts of those nitrogen-rich high explosives that had been developed in the late Coal Age. The Haber-Bosch process itself yields ammonia gas (NH_3) by reacting hydrogen gas directly with nitrogen gas. Hydrogen gas, H_2, was obtained from coal gas or natural gas, and nitrogen gas, N_2, could be obtained from the atmosphere, thanks to German refrigeration pioneer Carl von Linde, who in 1895 developed a technique to liquefy large quantities of air from which nitrogen could be separated. The Haber-Bosch chemical reaction itself is quite obvious ($3H_2 + N_2 \rightarrow 2NH_3$), but chemists were unable to carry it out economically until Fritz Haber conceived his method in 1909, and Germany built the world's first ammonia plant in 1913. (Nitrogen gas and hydrogen gas need to be mixed and subjected to extreme pressures and moderately high temperatures in the presence of a catalyst, iron oxide, which was identified by Carl Bosch.) Germany rapidly built several more ammonia synthesis plants and was therefore independent of Chile saltpeter during World War I. Britain and the United States tried to build Haber-Bosch process plants during the war as well, but they failed, because they lacked the skills to build high-pressure plants (and because the German patents omitted many vital technical details). However, when the war was over Haber-Bosch technology spread, and from the 1920s the United States and several European countries built plants that supplied nitrogen for nearly all the explosives used in World War II.

NUCLEAR BOMBS

World War II also saw the development of a new type of explosive that dwarfed all the world's nitrogen-rich high-explosives. These were nuclear or atomic bombs, which essentially release the kind of energy that powers the sun (and all other stars). Nuclear energy is the energy that holds together the particles (protons and neutrons) that are found in the nucleus (core) of an atom. The energy between these particles varies depending on the size of an atom's nucleus and the ratio between protons and neutrons.

Most atoms are very stable, but some occur in nature in unstable, radioactive versions that release nuclear energy in the form of radiation while they are falling apart. The reason for their instability is usually that the ratio between their protons and neutrons in the core is not correctly balanced. A carbon atom, for instance, is defined by having six protons in its core, but in addition to the common C-12 atoms (with six protons and six neutrons in the core), there are also some C-14 atoms (with six protons and eight neutrons in the core). C-14 is unstable and decays at a slow rate. (Every 5,730 years half of the original amount of C-14 atoms has fallen apart.) Accordingly, very little energy is released by this decay. This implies that if radioactive elements radiating off useful amounts of energy were to be found in nature, they would not be there very long, as they would decay at a high rate. Scientists therefore searched for atoms that could be made unstable (by investing a bit of energy) and would in turn fall apart and release a lot of nuclear energy. The target of choice were large, heavy atoms, and the strategy was to bombard them with neutrons to ruin the stable ratio between protons and neutrons. Uranium, the heaviest of all atoms found in nature, contains 92 protons. It is rare and occurs almost exclusively as U-238. However, a bit of uranium (0.7 percent) occurs in nature as U-235, and it is this version that turned out to be feasible for use in atomic bombs (and nuclear power plants): The decay of U-235 releases vast amounts of radiation energy, but it also releases neutrons, which in turn cause a chain reaction by penetrating and destabilizing further U-235 atoms.

Nuclear fission of uranium by bombardment with neutrons was achieved in early 1939 by Germans Otto Hahn and Fritz Strassmann. Austrian Lise Meitner, professor of physics at the University of Berlin from 1926 and long-term co-worker of Hahn, had relocated to Sweden shortly before the experiment was carried out to escape restrictions imposed by Nazi Germany against researchers of Jewish background. However, Meitner received the test results from Hahn, and it was she, together with her nephew Otto Frisch, who interpreted the nuclear reaction as such and coined the term nuclear fission. Meitner and Frisch quickly traveled to Copenhagen to inform Nils Bohr, who was just about to leave for a scientific conference in the United States, where he spread the news to an international audience. Most researchers in turn concluded that a uranium-based nuclear bomb would be impractical, simply because it would have to weigh over 50 tons to contain enough of the actual reactant, U-235. Otto Frisch, however, together with German Rudolf Peierls, calculated the critical mass (minimal amount) of uranium 235 needed for an atomic bomb to be as small as one kilogram. (This was later revised to 10 kg or 22 pounds.) Hence, they concluded that it was possible to build such bomb by separating U-235 from U-238, and they submitted this information to the British government.[56]

Not knowing that the Germans had shelved the idea of producing a nuclear bomb (because it would have diverted too many resources from other war efforts), Britain then pushed for the development of an atomic bomb, which was achieved in the United States at enormous expense. In total the effort involved 175,000 workers, and the plant at the Oak Ridge National Laboratory in Tennessee alone, where U-235 was separated from U-238 on the basis of the slight weight difference between these two isotopes, consumed more electricity than all of New York City. (The plant was actually the largest building that had ever been built.) At another major plant, at Hanford, Washington, U-238 was bombarded with neutrons, as it was found that uranium would then turn into plutonium (Pu-239), which was shown to be a fissile material suitable for a bomb as well. U-235 and Pu-239 were then shipped to Los Alamos, New Mexico, where the bombs were put together. Austrian Otto Frisch was the head of the Critical Assembly Group, and the project was joined by many scientists who were either German native speakers or had studied or worked at German institutions. Among the latter was Enrico Fermi from (Germany's ally) Italy, who had relocated to the United States in 1938, and produced the first controlled nuclear chain reaction in 1942. Constructing the U-235 bomb was relatively straight-forward, but the plutonium bomb was complex and had to be tested. The world's first nuclear explosion, the Trinity test, took place in the desert of New Mexico on July 16, 1945. To be sure, this was a couple of months after Germany had capitulated on May 7, 1945. However, there was still time to demonstrate the new weapon in combat: a month after the Trinity test, the United States dropped two nuclear bombs on Japan in a three-day interval, one on the city of Hiroshima, with about 340,000 inhabitants, the other on the city of Nagasaki, with about 240,000 inhabitants. The Hiroshima bomb was uranium (U-235) bomb and released energy equivalent to the explosion of about 12,000 tons of TNT (trinitrotoluene); the Nagasaki bomb was a plutonium (Pu-239) bomb that was equivalent to about 20,000 tons of TNT (trinitrotoluene). The blast, heat rays, collapsing houses, fires, and radiation poisoning killed about 90,000 to 140,000 people in Hiroshima, and 60,000 to 80,000 people in Nagasaki, immediately or within four months of the bombing. (Thousands more died of cancer due to the longer term effects of radiation. Excess deaths due to leukemia were soon noticed, while an increase in deaths caused by other types of cancer was noticed from about 10 years after the bombing and continues until the present.)

The next step was to develop a hydrogen bomb. Such a thermonuclear bomb is based on nuclear fusion rather than fission, capable of releasing energy equivalent to the explosion of millions of tons of TNT. Hydrogen bombs are made of deuterium or tritium, which are the heavy (neutron-containing) versions of hydrogen. When these atoms merge (fuse) to yield helium, enormous amounts of energy are released. (This is truly the process that occurs in the sun.) The effect is great heat, enormous shock waves, high winds, and deadly radiation in the form of gamma rays and neutrons that destroy living matter and contaminate soil and water. Several of the nuclear scientists working in the United States were deterred by the effects of the original nuclear bombs, but others chose to build a hydrogen bomb that was tested in a remote region of the Pacific Ocean, on Enewetak Atoll, on November 1,

Unprecedented Destructive Energy The depicted Hiroshima explosion killed between 90,000 and 140,000 people, and the Nagasaki bomb between 60,000 and 80,000 people immediately or within four months. This added to the enormous civilian casualties produced by the other U.S. raids on Japan, particularly the firebombing of Tokyo and other major cities in March 1945. (Library of Congress image LC-USZ62-39852.)

1952. (Its explosive force was about 500 times greater than the Hiroshima or Nagasaki bombs.) Just a little later, in August 1953, the Soviet Union detonated a hydrogen bomb as well. This bomb is considered the work of Andrey Dmitrievich Sakharov, while the Soviet fusion bomb, demonstrated in 1949, may have been built with information received directly from Los Alamos. (German physicist Klaus Fuchs, who worked on the American fission and fusion bombs, admitted in 1950 to have passed information on to the Soviets.) On October 30, 1961 the Soviet Union tested the most powerful bomb ever detonated. This hydrogen bomb was equivalent to 58 million tons of TNT. Less than 15 months later the Soviets revealed that they had built a hydrogen bomb equivalent to 100 million tons of TNT.

ROCKET-PROPELLED AIRPLANES

The means to deliver such bombs became more sophisticated as well. The development of bomber airplanes was clearly a continuation of the historic trend to kill people from increasingly longer distances, and by the end of World War II, bomber ranges had surpassed 6000 km with up to 9 tons of bombs. However, even these ranges were soon exceeded by (unmanned) intercontinental ballistic missiles.

The whole development started in Germany, where engineers developed both rocket-propelled airplanes and unmanned rocket-propelled missiles. Rocket propulsion refers to forward propulsion through combustion of a fuel with backward discharge of a rapidly expanding exhaust gas stream. The Chinese had used gunpowder-filled solid rockets in warfare by the 13th century, while modern German rocket engines were somewhat similar to conventional jet engines. The main difference was that ordinary jet engines take in and use air for the combustion of their fuel, while rocket engines carry both fuel and oxygen, or an oxygen-rich substance, with them. A rocket engine thus consists of nothing more than a combustion chamber and an exhaust nozzle: it is the only prime mover that can deliver more power per unit of weight than turbojet engines (gas turbines).

German aeronautical engineer Alexander Lippisch designed the *Ente* (Duck), which in 1928 became the first aircraft to fly under rocket power when two gunpowder rockets were attached to it. (This experiment was in part financed by German car manufacturer Opel, which during the economic depression of the 1920s undertook a variety of publicity stunts involving rocket-powered vehicles. One Opel was powered by 24 solid rockets to reach a speed of 230 km/h [143 mph] in 1928.) German engineer Hellmuth Walter argued that an engine powered by a fuel that is itself rich in oxygen would have obvious advantages for powering submarines, torpedoes, and airplanes. In 1925 he patented liquid hydrogen peroxide (H_2O_2)

as the fuel of choice, as it would decompose into water steam (H_2O) and oxygen gas (O_2) when it comes into contact with certain metal catalysts. This would create a rapid expansion of the fuel's volume, while another fuel could be injected into the hot mixture of gases to be combusted by the available oxygen to generate additional thrust. Based on Walter's system, the Heinkel He 176 in 1939 became the world's first aircraft to be propelled solely by a liquid-fueled rocket. However, the army was not interested in this small experimental plane, and the Heinkel company discontinued the project. So it happened that Germany presented the world's first operational rocket-propelled fighter, the Messerschmitt ME 163, only towards the end of World War II, in 1944. Designed by Alexander Lippisch, this delta-wing aircraft reached a top speed of 960 km/h (596 mph) and climbed to 40,000 ft (12,000 m) in just three minutes. No other fighter came close to this speed, but the ME 163 was not sufficiently developed. It had to start from a dolly, and after just eight minutes of powered flight glided back to the ground. Besides, it lacked a targeting system to approach much slower-flying bombers safely. Another rocket-propelled plane, the Messerschmitt ME 263 (also known as Junkers Ju 248, as Junkers did part of the developing work), overcame some of the ME 163's shortcomings. It featured a second cruising combustion chamber for longer range, as well as a conventional retractable undercarriage. Designed to achieve a top speed of 1,000km/h (620 mph), this aircraft never reached series production, because Soviet forces overran the Junkers factory where it was about to be manufactured. Factory materials and staff were taken to the Soviet Union, where an enlarged copy of the ME 263 appeared as the Mikoyan-Gurevich I-270 interceptor in 1947. Meanwhile Hellmuth Walter was taken to Britain together with his colleagues and all his research materials, and was forced to work for the British military. (With Walter's cooperation, one of the German submarines using his engine, the U-1407, was raised from where it had been scuttled and recommissioned as HMS *Meteorite*.) After three years in Britain Walter was allowed to return to Germany, and in 1960 relocated to the United States. Alexander Lippisch was taken to the United States immediately after the war under Project Paperclip, living and working there until his death in 1976.

A U.S.-built rocket-propelled experimental plane, the Bell X-1, was supposedly the first aircraft to break the sound barrier, reaching the speed of sound in 1947. (According to a disputed claim a German pilot achieved this in 1945 in a ME 262.) However, due to advances in nonrocket jet engine technology, and the difficulties in handling (or moderating) the high speeds of rocket planes, rocket-propulsion for airplanes was ultimately abandoned. Their initial function, as interceptor, was instead assumed by unmanned surface-to-air missiles.

MISSILES AND SPACE PROGRAMS

The intercontinental ballistic missiles and space rockets developed by the United States and the Soviet Union after World War II were based on German technology as well. A key figure in this context was Wernher von Braun, who shook the hand of Hitler as well as of five United States presidents (Truman, Eisenhower, Kennedy, Johnson, and Nixon). Von Braun developed missiles for the German military from 1932, and in 1936, at the age of just 24, was made technical director of a newly founded rocket development facility at Peenemünde, at the Baltic coast, west of the current German-Polish border. In 1938 von Braun became a party member of the NSDAP, 18 months later he received an officer's commission in the SS (Schutz-Staffel) elite corps, and in 1943 he was promoted to a higher rank within the SS. Von Braun met Hitler on several occasions, and he was photographed in his SS uniform during an official visit by SS chief Himmler. Nevertheless von Braun was later described as a career scientist rather than ardent Nazi.[57]

By mid-1943 von Braun's team had developed and tested rockets with a pre-set guidance system, capable of carrying one ton of explosives to a target up to 340 km away. These rockets were fueled by liquid hydrogen or ethanol. Hydrogen can be obtained from coal gas or natural gas, while ethanol (CH_3-CH_2OH), which is best known as the active ingredient in alcoholic beverages, is traditionally made by fermenting organic compounds such as grain or sugar cane. (Nowadays industrial ethanol is also produced from crude oil, after cracking, via ethylene.) The German military pushed for the use of ethanol, because Germany was notoriously short in fuel, and ethanol could be produced from potatoes, for instance.

Hitler immediately ordered mass production, which was moved to the Harz mountains when Peenemünde was bombed in August 1943. Two types of missiles were launched against Britain, France, and Belgium in 1944 and 1945. One was the smaller V-1, which was the world's first cruise missile. Equipped with a guidance system, it flew essentially as an unmanned airplane, powered by gasoline through a pulse-jet engine. Over 8,000 V-1s were launched against Britain, but it was possible to intercept them with airplanes. The other type was the larger V-2, which was the world's first ballistic missile.[58] V-2s were guided in the ascent of a high-arch trajectory to freely fall in the descent. They carried ethanol as fuel, and liquid oxygen to burn it. Hydrogen peroxide served as auxiliary fuel to generate steam according to the Walter-system. The steam drove a turbine that pumped ethanol and oxygen into the rocket's combustion chamber. Over 1,100 V-2 rockets were launched against Britain, and due to their speed there was no defense against them at the time, except for destroying the launch pads in Germany.

By the end of the war Americans, Britons, and Russians rushed to the German rocket production sites to capture parts, engineers and documents. The

R-7 Semyorka The missile and space programs of both the United States and the Soviet Union were based on German technology adopted after World War II. Never have people been able to kill from further away. The Soviet R-7 Semyorka, the world's first intercontinental ballistic missile, stood 34 meters high. It could deliver a warhead over a distance of more than 8,000 km, nearly a fifth of the equator. The Russians used this type of rocket to put Sputnik I, the world's first artificial satellite, into orbit in October 1957. (Photograph "Semyorka Rocket R7" by Sergei Arssenev (Wikimedia Commons), licensed under Creative Commons Attribution 2.5 (http://creativecommons.org/licenses/by/2.5/), edited.)

Operation Paperclip: German Technology Spreading to America At the end of World War II, the United States took a large number of Nazi Germany's scientists and military staff to America, where many of them were eventually naturalized. Perhaps the best-known example is the Wernher von Braun Rocket Team, posing here at the Redstone Arsenal in 1950. (According to NASA, the picture was, however, taken at Fort Bliss, Texas.) Their contract with the U.S. Army Corps was part of Operation Paperclip. The team developed guided missile designs based on the German V-2 rocket, and was eventually transferred to a newly established field center of the National Aeronautic and Space Administration (NASA). (NASA image NIX MSFC-8915531, and, in similar form, Library of Congress image HAER AL-129-A-34.).

Soviets snapped, plenty of German rocket workers and V-2s, but they actually also had some previous experience from a Soviet rocket program of the 1930s. By 1948 the Soviet Union was testing its R-1 missile, a knockoff of the V-2, and under chief designer Sergey Korolyov emerged as the global leader in missile and space technology. The world's first intercontinental ballistic missile, the R-7 Semyorka, was tested in 1957. Filled with kerosene and liquid oxygen, this missile stood 34 meters high and weighed 280 tons. It could deliver a warhead over a distance of more than 8,000 km. In October 1957 this type of rocket carried Sputnik I, the world's first artificial satellite, into orbit. (Sputnik I was an aluminum sphere of 84 kg and 58 cm in diameter. It was capable of sending radio signals and performing simple measurements, including the temperature.) Sputnik II, which carried a dog into space, was launched a month later. (After orbiting Earth, the small female terrier Laika died on November 4, 1957, some seven hours after launch, due to stress and overheating.) So-called Vostok rockets then carried the first man (1961) and the first woman (1963) into space. (These people were safely returned to Earth.) The Soviets also explored the Moon. In 1959 Luna 2 impacted on the Moon's surface, and between 1966 and 1970 the Soviet Union landed several (unmanned) space probes on the Moon that carried exploration robot vehicles, took photographs, drilled into the surface, and safely returned lunar soil samples to Russia. What is more, the Soviets pioneered the landing of a spacecraft on another planet's surface: Planetary probe Venera 1 was launched in 1961 and crash-landed on the surface of Venus in 1966. Then, in 1971, the USSR landed a space probe on planet Mars and established the first orbital space station.

The United States started somewhat more slowly into the Space Age, even though it captured more (and more crucial) German scientists, and V-2 materials. (In fact, the United States had already begun launching V-2s in America in 1945.) Most importantly, von Braun together with Walter Dornberger, who had directed the construction of the V-2 rocket, were taken to the United States with 126 of their staff members. (All received American passports.) The German scientists trained American personnel, and von Braun became chief of the U.S. Army's missile program. Under his leadership the Redstone, Jupiter-C, Juno, and Pershing missiles were developed. In 1958 von Braun and his group used a Jupiter-C missile to launch the first American satellite, the Explorer I, while an Atlas rocket, tested in 1959, was the first successful American intercontinental ballistic missile. In 1960 von Braun and his development team were transferred from the Army to NASA, the National Aeronautics and Space Administration, founded in 1958. Here, von Braun led the development of the giant Saturn rockets, which launched manned Apollo spacecraft into the Earth's orbit and to the Moon. (The scientific significance of the first successful manned lunar landing on 16 July 1969, as well as the other five lunar landings, seems rather doubtful. It may be viewed as a very expensive publicity stunt in context of the space race against the Soviet Union, which the United States appeared to lose. Unmanned space probes of the kind the USSR had landed on the Moon earlier on, might just as well have delivered all the desired information.) Saturn V technology was also used in 1973 to launch the Skylab space station, two years after the Russians had put a station into the Earth's orbit. Thereafter Saturn rockets were retired in favor of the Space Shuttle, a partially reusable, manned, rocket-launched vehicle developed by NASA to go into Earth orbit from the early 1980s. The Space Shuttle emerged from the Air Force/NASA project Dyna-Soar, in which Walter Dornberger participated. (Ultimately, the concept for the Dyna-Soar and Space Shuttle can be traced back to Austrian Eugen Sänger's Silverbird.) The best known launch vehicle of the European Space Agency, founded in 1974 for strictly civilian purposes, is the (expendable) Ariane rocket. First launched in 1979, this series turned out the most reliable of all commercial rocket systems.

As increasingly more satellites were being lifted into space, they became an integral part of everyday life in the Oil Age. Some are communication satellites (for voice, data, TV), others are crucial for weather forecasting and Earth resources management, and many are critically important for military reconnaissance. Equipped with powerful optical lenses, satellites can photograph every little detail on the Earth's surface (unless it is cloudy). The precise Global Positioning System (GPS), with a fleet of more than 24 communications satellites, was initially developed for the U.S. military as well. It is now available to the general public with receivers commercially available and used for ship (and later on automobile) navigation, for instance.

COUNTERING THE NUCLEAR THREAT

The existence of long-range missiles combined with nuclear warheads potentially made the world a much more dangerous place than it had ever been. Literally every corner of the world could be reached by missiles launched either from land-based silos or from submarines. Had the United States and Soviet Union engaged in a limited thermonuclear exchange in the late 1980s, and targeted not cities, but exclusively strategic facilities, tens of millions of people would have immediately lost their lives. (One major problem in nuclear warfare is the interception of missiles, because their explosion high up in the atmosphere may have serious global consequences through nuclear fallout.) Besides, developing and maintaining nuclear capacity came at an enormous economic, environmental, and social cost. The United States from 1940 spent nearly $5.5 trillion (in 1996 dollars) on its nuclear weapons and weapons-related programs. To test the effects of radiation both the United States and the Soviet Union have deliberately exposed their soldiers to radioactivity without their knowledge or consent. U.S. nuclear tests exposed people living downwind from the Nevada Test Site (NTS) to external radiation, while radioactive iodine (I-131) concentrated first in cow milk and then in human thyroid glands. (Such exposure caused some 120,000 extra cases of thyroid cancer.) Within "Project Sunshine," initiated at the University of Chicago, some 6,000 dead babies were snatched from hospitals in Australia, Britain, Canada, Hong Kong, the United States, and Latin American countries over a period of 15 years without parental consent to be shipped to the United States to test the impact of nuclear fallout on body parts. A lot worse, thousands of civilian Americans have been deliberately exposed to radiation in some 3,000 to 5,000 secret tests without their knowledge or consent between 1944 and 1974. At one hospital people stopping by for various reasons were injected with radioactive plutonium. At Vanderbilt University in Nashville, Tennessee, 829 pregnant women in the time right after World War II came to the clinic thinking that they were getting vitamins to drink, when in fact they were served radioactive iodine to study how fast it would pass into the placenta. Scientists from Harvard, MIT, and other universities conducted radiation experiments on children up until the 1960s. (In one such experiment 73 disabled children were spoon-fed oatmeal that had radioactive calcium mixed into it.) In Oregon a group of prisoners had their testicles irradiated in a project conducted for NASA to learn about the effects of space radiation on astronauts.[59]

Deterred by the vast destructive effects of nuclear explosions and radiation poisoning, many countries began in the early 1960s to negotiate limitations on the production and use of nuclear weapons and fissile materials. Over 180 nations eventually signed the Nuclear Nonproliferation Treaty of 1968, pledging not to acquire nuclear weapons or distribute nuclear weapons technology. At the beginning of the 21st century, the United States, Russia, Britain, France, China, India, and Pakistan were the only nations to officially possess nuclear weapons. Israel very likely has some, but does not admit (nor deny) it. South Africa previously constructed, but then voluntarily dismantled, six uranium bombs. North Korea and Iran have sought to develop nuclear weapons.

The United States and the Soviet Union somewhat reduced their nuclear arsenals through agreements negotiated in the 1980s. Following the disintegration of the Soviet Union in 1991, a supplementary agreement obligated the Ukraine, Belarus, and Kazakhstan to destroy their nuclear weapons or to hand them to Russia. However, subsequent U.S. efforts to develop an anti-missile defense system put strains on the arms control regime. Besides, in a more recent trend, military planners promoted the downscaling of nuclear weapons to make them more acceptable for use on the battlefield. The United States in 2003 began developing a new atomic weapon known as a mininuke. Designed to deliver a blast comparable to some 5,000 tons of TNT, this weapon would be used to penetrated underground bunkers protected by steel and concrete. The neutron bomb, developed in the US from the 1960s but dismantled since, somewhat fit this trend as well: its blast and heat are confined to a small radius (under a mile), while its massive generation of neutrons and gamma radiation are extremely hazardous to living tissues in a confined area and on a limited time-scale.

CHEMICAL WEAPONS

Another new arms threat of the Oil Age were chemical weapons, typically delivered by artillery shells bursting in midair, or directly released from airplanes. The Germans had a general lead in chemicals production and in World War I began using chlorine gas, tear gas (xylyl bromide), and the choking agent phosgene ($COCl_2$, carbonic dichloride). Mustard gas, causing skin burns and deep blisters, was rarely lethal, but was widely feared because the gas masks of the day offered little protection against it. Britain and France followed suit, and by the end of the war perhaps 100,000 had been killed by toxic gas, mostly by phosgene. This gave rise to the Geneva Protocol of 1925, which banned the use of chemical agents in war. (Most major powers supported it right away, though the United States refused to sign it until 1975.) Nevertheless, chemical weapons were used by the Spanish against Morocco (1923 to 1926), by the Italians against Ethiopia (1935 to 1936), by the Russians against China (1930), by the Japanese also against China (1938 to 1942). During World War II chemical weapons were fortunately hardly used at all, probably because of fear of retaliation and the risk of exposing one's own troops to them. However, chemical weapons killed over 50,000 people during the Iran-Iraq War (1980–1988), with both sides violating international law.

Research in turn continued, because the Geneva Protocol banned the use, but not the production and stock-piling of chemical weapons. Tabun, sarin and soman, which are organophosphorus nerve agents, were all developed (but not used) by Germany during World War II. The United States and the Soviet Union after World War II produced huge amounts of these, plus the newer and more powerful agent VX. (Even exposure to a minimal amount of such nerve agent will result in almost immediate death.) The dismantling of these chemical arsenals finally began under the terms of the 1993 Chemical Weapons Convention, which prohibits all development, production, acquisition, stock-piling, or transfer of such weapons. However, not all nations signed the convention, and chemical weapons production may be difficult to prevent. (In 1995 a religious sect used sarin gas in the Tokyo underground to kill 12 people and to injure thousands.)[60]

BIOLOGICAL WEAPONS

Biological weapons may be even more difficult to control. Biological agents include such toxins as botulinum (produced by bacteria) and ricin (found in castor beans), and a variety of bacteria (anthrax, brucellosis, typhus) and viruses (smallpox). Biological warfare has a long history. The English, for instance, handed blankets from smallpox patients to American natives sympathizing with France. Earlier on, in the 14th century, Eurasians threw plague-infected cadavers into each other's camps. The Geneva Protocol of 1925 banned the use of both biological and chemical weapons, but the Japanese between 1938 and 1942 dropped plague and other bacteria from airplanes over several Chinese towns. The U.S. Army has admitted to 339 open-air tests of biological weapons in the United States in the 1950s and 1960s (including the release of the whooping cough agent in Sebring and Palmetto, Florida), and terrorists have occasionally chosen to use biological agents.[61] Members of a religious group placed salmonella bacteria in the salad bars of several restaurants in Oregon in 1984, for instance, which caused 750 people to fall sick. In 2001 letters containing anthrax bacteria were sent to five media offices and two U.S. senators, which resulted in 24 infections and five deaths.[62] Various countries have attempted to develop anthrax as a weapon of biological warfare, because it is very lethal when inhaled, because it spreads easily through the air, and because it is an extremely hardy agent that withstands extreme heat (as developed in explosions). The 1972 Biological and Toxic Weapons Convention banned the development, production, stockpiling, and use of microbes or their poisonous products, but it allowed for "protective and peaceful research." Besides, biological weapons programs are easy to conceal. The Soviet Union, for instance, likely weaponized various pathogens up until the 1990s.

The advent of genetic engineering technology lifted the potency of biological weapons to a whole new level. The United States, Russia, Britain, and Germany have all genetically engineered biological weapons agents. Gene coding for toxins can easily be transferred to other organisms. By 1986 U.S. researchers had inserted the gene for the lethal factor of Bacillus anthracis, into Escherichia coli, a normally harmless gut bacteria that can be added to water supplies and all kinds of foods. Similarly, any gene could be engineered into the Rhinovirus that causes common cold.[63] (There are also commercial applications for such research. In 1991 the toxin gene from the North African scorpion was transferred into baculovirus, which was then sprayed onto plants and immediately paralyzed infected parasitic caterpillars.)[64] Another strategy is to modify established bioagents in a way that they become resistant to known treatments.[65] Antibiotic resistance genes were introduced into anthrax in the late 1980s at the University of Massachusetts, for instance. Typically, a lethal strain would be modified, and a vaccine would be developed to render one's own troops immune.

Ethnic weapons target genetic features linked to specific populations. The U.S. Army in 1951 at the Naval Supply Depot in Mechanicsville, PA, released a non-lethal mutant of the agent that causes Valley fever, known to affect African blacks more than other ethnic groups. (This was to test the threat of anyone using such strategy against African American soldiers, who represent much of U.S. troops.) As increasingly more ethnic genetic markers are being identified, genetic engineers can turn previously indiscriminate bioagents into weapons targeted towards specific enemies. Besides, better knowledge of the human genome will allow for the creation of pathogens that specifically target certain functions of the human body such as vision, memory, fertility, among others. Genetic engineers have also developed anti-material agents designed to destroy fuels, camouflage colors, asphalt, and so on.

HIGH- AND LOW-ENERGY DEADLINESS

Much of what the Oil Age witnessed in terms of weapons development was a continuation of the historical trends towards command of more destructive energy and belligerent use of technology used for mobility. This included more powerful conventional, as well as nuclear, bombs; newly developed bomber planes; helicopters (the first fully controllable one was developed in Germany in 1934, while extensive combat use started with the Korean War, 1950–53); intercontinental ballistic missiles; shorter range missiles (anti-aircraft, anti-ship, anti-tank); fast, heavily-armored tanks; armed submarines; purpose built aircraft carriers (from 1922); light assault rifles with accuracy ranges of 300 to 500 meters (1,000 to 1,600 feet); grenades with high-explosive charges that fragment the shell's iron casing; and booby traps and land mines. Biological (and to some extent chemical) weapons, on the other hand, are different. They are weapons of knowledge rather than weapons of energy and material. A single scientist may grow a tiny culture of harmful pathogens into a large amount of biological weapons agent using standard materials found in laboratories around the world. Even more worrying, it has been demonstrated that off-the-shelf chemicals can be used to make a whole virus culture from scratch.

NOTES

55. Vaclav Smil, *Essays in World History: Energy in World History* (Boulder, CO: Westview Press, 1994), 216.
56. Richard Rhodes, *The Making of the Atomic Bomb* (New York: Simon & Schuster, 1986).
57. The famous photo of von Braun in his SS uniform can be viewed at http://www.v2rocket.com/start/chapters/vb-009.jpg.
58. Find an Internet site dedicated to the V-2 at this address: http://www.v2rocket.com.

59. Eileen Welsome, *The Plutonium Files: America's Secret Medical Experiments in the Cold War* (New York: Dell Publishing, 2000); Amy Goodman, "Interview with Eileen Welsome," May 5, 2004, http://www.democracynow.org/article.pl?sid=04/05/05/1357230.

60. Leonard A. Cole, "Chemical and Biological Warfare," Microsoft Encarta Online Encyclopedia, http://encarta.msn.com/encyclopedia_761558349/Chemical_and_Biological_Warfare.html.

61. Find information on admitted biological weapons tests by the U.S. Army conducted in America in Conn Hallinan, "Of Mice & Men," *San Francisco Examiner*, May 11, 2001, http://homepages.ihug.co.nz/~dcandmkw/ge/micemen.htm#top.

62. Leonard A. Cole, "Chemical and Biological Warfare," Microsoft Encarta Online Encyclopedia, http://encarta.msn.com/encyclopedia_761558349/Chemical_and_Biological_Warfare.html.

63. Most of the examples of genetically engineered bioagents are taken from the Web site of "The Sunshine Project," an international non-governmental organization committed to inform about the dangers of new weapons deriving from advances in biotechnology. http://www.sunshine-project.org/. Find more on this topic in The Sunshine Project, "US Armed Forces Push for Offensive Biological Weapons Development," News Release, May 8, 2002, http://www.sunshine-project.org/.

64. Stephen Nottingham, *Eat Your Genes: How Genetically Modified Food Is Entering Our Diet* (London: Zed Books Ltd., 1998).

65. A.P. Pomerantsev et al., "Expression of Cereolysine Ab Genes in Bacillus Anthracis Vaccine Strain Ensures Protection against Experimental Hemolytic Anthrax Infection," *Vaccine* 15(17–18) (1997):1846–50.

CHAPTER 31

AGRICULTURE IN THE OIL AGE

Agriculture changed entirely during the Oil Age. The combination of the Haber-Bosch process providing for virtually unlimited amounts of nitrogen fertilizer, and Mendel science being applied in breeding programs, resulted in extra ordinary yield-gains. New chemicals protected crops against pests, vast irrigation schemes were installed, and oil-fueled machinery was employed. Thus, world cereal production rose from 650 million metric tons in 1950 to 1,193 million tons in 1970, to 2,109 million metric tons in 2001.[66] (The 2001 harvest included 615 million tons of maize, 598 million tons of rice [paddy], and 591 million tons of wheat: well over half of all food energy consumed by people worldwide comes from these three cereals: wheat, rice and maize.) Expansion of land under cultivation was a relatively minor factor in this increase. A little over one third of the world's arable land of nearly 1.5 billion hectare was first cultivated in the 20th century, but the area under cultivation has remained quite constant since World War II. What made the difference was output per hectare: 1.35 tons of cereals were harvested per hectare on global average in 1961, compared to 3.12 tons per hectare in 2001. In western Europe total cereal production between 1961 and 2001 increased from 93 to 205 million tons, in the United States from 164 to 325 million tons, in China from 110 to 398 million tons, in India from 87 to 243 million tons, and in Africa from 46 to 116 million tons. In developing countries the introduction of dwarf varieties of wheat and rice was critically important: These overcame the problem that heavily-fertilized crops in hot countries grew too tall and fell over. Nevertheless, there was still ample space for improvement at the beginning of the 21st century: Myanmar harvested

1.2 tons of wheat per hectare per year in 2001, compared to 3.6 tons wheat per hectare per year in Japan. Some 70 countries remained below the world average of 3.9 tons of rice per hectare, with Africa averaging just 2.2 tons of rice per hectare.

Much of the remaining relative inefficiency is due to inadequate agricultural management, but some regions lack water (or energy to pump it). In fact, much of the sensational yield gains of the Green Revolution was due to the installation of irrigation schemes on which fast-growing crop varieties depend. The area under irrigation increased from about 94 million hectare in 1950 to 272 million hectare in 2000, when irrigated land accounted for some 17 percent but yielded about 40 percent of the world's cereal output. About half of all irrigated land depends on pumped water, and about 70 percent of all irrigated land is located in Asia.

Mechanization on the fields varies regionally well. Oil energy made a true difference in agriculture, ending the era of the work horse. Oil-fueled tracklaying tractors emerged by 1904 and soon became widespread in America. The population of U.S. draft animals peaked shortly before 1920. By 1927 tractors provided half of the total draft power applied in agriculture, and by 1963 America's tractor power was nearly 12 times the record draft animal capacity of 1920. By the 1980s one worker with a powerful tractor pulling three plows, each with five moldboards, could plow 110 acres in a 10-hour day, accomplishing the work that once required 55 workers and 110 horses. Thus, the U.S. countryside was emptied of people and horses. Rural labor accounted for nearly 40 percent of the total in 1900, while it was 15 percent in 1950 and 2 percent in 1975.[67] The trend was similar in Europe, where tractors were widely adopted from 1950, while typically two-thirds of the population of Asian nations, including the three largest economies, China, India, and Indonesia, remained rural still towards the end of the 20th century. Nevertheless the number of tractors in use in non-industrialized Asia increased dramatically from 186 thousand in 1961 to 2.0 million in 1980 to 4.9 million in 2001. (In Asia's rice fields small two-wheeled tractors are common.) The increase in tractors in use in sub-Saharan Africa, where the population remained overwhelmingly rural, was comparatively meager: It rose from 52 thousand in 1961 to 161 thousand in 2001.

Fertilizer application reached the highest levels in western Europe in the 1980s (over 300 kg of nitrogen per hectare.) The use of nitrogen fertilizer remained generally low prior to World War II, as the ammonia gas (NH_3) obtained from rapidly spreading Haber-Bosch plants was diverted to ammunitions factories for war preparations. However, between 1950 and 2000 worldwide fertilizer application increased from 14 million tons to 137 million tons. With high fertilizer inputs China managed to pass the United States around 1980 in terms of wheat yield per hectare (though output remained twice as high in such western European countries as France and the Netherlands).[68]

Double cropping, made possible by the combination of fast-maturing crops with high fertilizer inputs, pushed the overall grain yield into new spheres: The North China Plain combined winter wheat (4 tons per hectare annually) and corn (5 tons per hectare annually) to emerge as the world's leading grain producer (9 tons of grain per hectare per year.) Similarly, winter wheat is combined with soybeans as a summer crop in the United States, while northern India added a rice harvest to the traditional wheat harvest on the same fields within a year. (To be sure, high fertilizer inputs may also have negative consequences. Any oversupply may be washed into rivers, lakes and oceans, where fertilizer promotes overgrowth of aquatic vegetation (algal bloom), which deprives water of oxygen and causes fish and other aquatic life to suffocate.)

Highly productive monocultures began attracting a lot of pests, but the problem was generally solved by stepping up the arsenal of agricultural chemicals: Herbicides against weeds, fungicides against molds, insecticides against insects. The German Bayer company marketed the first synthetic organic pesticide in 1892. It also proved to be the first selective herbicide, that is, it harmed the weed but not (so much) the crop. Tens of thousands of different pesticides have since been developed, and their use has grown exponentially since World War II to prevent crop losses. Unfortunately, these chemicals also involve problems. DDT $(C_1C_6H_4)_2CH(CCl_3)$, released in 1944 as the first large-scale insecticide, persisted in the environment and had toxic effects on various bird and fish species. What is more, many species of insects by the 1960s had become resistant to DDT. (Nevertheless, many countries kept on using DDT in its initial application to control insect-borne diseases: the chemical's effectiveness in malaria control seemed to outweigh the environmental problems.)

Vast amounts of chemicals are also used to process plants directly on the fields before they are harvested. Defoliants, for instance, strip the leaves of such plants as cotton, tomato and soybean. Cotton is the single most chemical-intensive crop, consuming over half of the world's chemical input in agriculture. (Its oil seeds aside, cotton is not grown for food consumption, which allows for higher acceptable levels of chemical residues on the plant. In the 2001/02 season world cotton production was 22 million tons, with China producing 5.32 million tons and the United States 4.42 million tons.)

GENETICALLY ENGINEERED CROPS

The advent of genetic engineering technology started a new era in agriculture. Traditional plant (and animal) breeders produced new varieties by cross-breeding existing varieties of the same or similar species. (They would often identify a naturally-occurring mutant variety with a beneficial trait, trying to introduce the trait into the otherwise already optimized crop.) Genetic engineering technology, on the other hand, allowed scientists to mix the genetic material of entirely different species, and to identify exactly which genes are responsible for which trait. This opened up opportunities to manipulate crops for anti-aging, herbicide resistance, increased sweetness, higher oil content, resistance against frost, flooding, or drought, and so on. The first transgene (a gene from another species) was successfully inserted into a plant (tobacco) in 1983, and by 1995 60 plant species had been manipulated through genetic engineering. The first genetically modified vegetable to reach the market (in 1994) were tomatoes. These did not include the trait of another species, but had one of their own genes silenced. (This can for instance be done by introducing a transgene that interferes with the expression of the target gene: these tomatoes were durable, as they did not express the chemicals that normally start the conversion of solid plant tissue into softer tissue.) Potatoes were engineered to increase their starch content by 20 percent using a gene from a starch-producing strain of the bacterium E. coli. (More starch translates into less water content, which speeds up the frying process and results in less fatty fries.) Potatoes have also been engineered to express thaumatin, a protein that occurs naturally in the West African katemfe plant and displays many times the sweetness of sucrose. (Similarly, high-sweetness strawberries and other fruits have been developed.) Oilseed rape and canola have been genetically altered to contain nutritionally healthier types of oil.[69]

The most common trait engineered into crops during the early period of genetic engineering was herbicide resistance. The idea was to make the chemicals applied onto fields less indiscriminate: they would now kill weeds without harming the crop. (Target genes are found in soil bacteria that break down the non-persisting agrochemicals now in use.) Soybeans engineered to withstand glyphosate, the world's biggest-selling herbicide, were introduced in the United States commercially in 1996. Herbicide resistant maize and cotton followed soon thereafter. By 1997 nearly a quarter of all U.S. cotton crop was grown from transgenic seed; a few years later it was over half. Similarly, transgenic maize was planted on a third of the U.S. maize fields by the year 2000. Cotton and maize were engineered for both herbicide resistance and insect resistance. The latter is achieved by insertion of a gene that expresses a protein that is toxic to insects if they should try to eat the plant. (Lots of target genes are found in various plant species that naturally have a chemical arsenal to protect themselves against insect attacks.) In result, pesticide application can be avoided or reduced. Transgenic tobacco resistant to the tobacco hornworm, tomatoes resistant to lepidopteran larvae, the Russet Burbank potatoes resistant to the Colorado potato beetle, and maize resistant to the European corn borer, were early commercial applications. Meanwhile such related species as squash (zucchini), pumpkin, cucumber, and watermelon have been engineered for virus resistance. About half of Hawaii's papaya production consisted of genetically engineered fruit soon after resistance to the Papaya Ring Spot Virus had been introduced in 1998.

In the longer run, it may be much more important for humankind to produce crop varieties that are tolerant to various climate extremes or poor soil conditions. A high-yield rice variety has been made tolerant to high salinity using a gene from a wild rice strain found in salty mangrove swamps in Bangladesh. Genes from a salt-tolerant yeast have been engineered into tomatoes, melons, and barley. Genetic engineers have identified which gene allowed an Indian rice variety to survive total immersion for several weeks. Drought-resistant plants have been engineered by insertion of a gene from baker's yeast. Frost resistant tomatoes have been developed by inserting a gene of the Arctic winter flounder. (An even more potent gene has been identified in the Canadian mealworm beetle, an insect that survives minus 30 degrees Celsius due to an antifreezing protein in its blood.)

Ultimately, genetic engineers would increase agricultural output by improving the photosynthetic efficiency of plants, which would make them utilize sun energy better, capture more carbon dioxide, and grow faster. However, this is quite complicated, as it involves several genes. Scientist have also attempted to turn normal C_3 rice into C_4 rice. (Most plants convert carbon dioxide into sugars that contain three carbon atoms, C_3. But in warmer climates certain plants have evolved the C_4 mechanism, which is more efficient above 25 C.)[70] Similarly, energy-expensive Haber-Bosch nitrogen production would become unnecessary if the capability of certain bacteria to capture nitrogen straight from the atmosphere could be transferred into the genome of crops. (However, nitrogen fixation involves at least 17 bacterial genes.)

In non-food applications crops can be engineered to yield pharmaceuticals. Human genes have already been inserted into bacteria to produce insulin, human growth hormone, and alpha interferon, for instance. Plants cannot be grown as fast as microorganisms, but they may deliver pharmaceuticals in a package that can be readily consumed. (Vaccines and therapeutic drugs have already been produced in bananas and cowpeas.) Transgenic plants may also deliver specialty fibers (plastics). But again, slow plant growth is a problem. Transgenic bacteria are currently preferred to produce such chemicals as adipic acid (a component of nylon).

To be sure, there are various problems associated with current genetic engineering technology and products. These include the areas of human health (allergies), ecological risks (such as gene escape and increased herbicide use), and ethical concerns (against animal or human genes in plants, or the consumption of transgenes in food stuffs in general).

ANIMAL HUSBANDRY

Animal husbandry underwent a transition that was similar to that of crop farming during the Oil Age. As world meat production increased from about

47 million tons in 1950 to 248 million tons in 2002, global average meat consumption per person increased from 17 kilograms a year to 40 kilograms a year despite rapid population growth. In nonindustrialized countries meat consumption rose fivefold, and milk consumption threefold, between 1971 and 1997 alone, while beef prices (in real terms) sank by two thirds.

Increase in meat production was directly related to increase in crop production. Surplus grain had historically been scarce, but since the early 1970s around 40 percent of the grain harvested on the world's fields has been fed to animals. (Due to the losses of nutritional energy involved in meat consumption, meat nevertheless provides no more than about 7 percent of the total nutritional energy eaten by people.) As in crop farming, application of the Mendel Laws to create fast-maturing varieties was critical. Such varieties were not fed with grain alone, but with all kinds of supplements that would promote meat growth, milk productivity, and egg-laying efficiency. Cows, which are naturally plant-eaters, were soon put on the same protein diet as pigs and chicken. All these farm animals were essentially turned into cannibals, as the meat industry began to recycle animal parts considered unfit for human consumption back into animal feed.[71] Blood meal, bone meal, tankage (dried animal residues), and similar products worked remarkably well to boost the growth of cows, pigs, and poultry. In the United States, but not in the European Union, animals are also treated with growth hormones, which are steroid-type chemicals that aid in feed efficiency and growth rates. (U.S. cows treated with Bovine Growth Hormone may produce 60,000 pounds of milk per year, which equals about 7,000 gallons or 26,000 liters.) Broiler chicken in 1925 were marketed when 15 weeks old at a weight of 2.8. In 1990 they were sold after 6.4 weeks at a weight of 4.5 pound. A U.S. hen laid about 70 eggs a year in 1933, 175 eggs in 1950, and 275 eggs a year in 2002. Similarly, an average U.S. cow produced some 4,200 pounds of milk per year in 1925, while it was 5,000 pounds in 1940, and about 18,000 pounds of milk in 1996. Meanwhile the feed-efficiency was improved to just over 7 kilograms of grain per 1-kilogram gain in live weight during cattle fattening, 4 kilograms of grain per kilogram of pork weight gain, and just over 2 kilograms of grain per kilogram of poultry weight gain. Animal farms turned into major factory operations. By 1990 over 90 percent of U.S. broiler production came from halls with more than 100,000 birds under one roof. Egg factories had about the same size, while pig farms reached 500,000 animals, cattle feedlots 100,000 head, sheep feedlots 50,000 head, and milking farms some 12,000 cows.[72]

The mass production of meat, milk, and eggs in the later Oil Age has undoubtedly overstretched compromises in animal welfare. The animal domesticates of the Fertile Crescent have been bred into creatures with metabolisms fully focused on meat growth. Many broiler chicken suffer from broken bones prior to their slaughter, for instance, and farmed turkey now has such extreme breast that it literally falls over to the front when attempting to walk. In many parts of the world nearly all chicken eggs come from birds that spent their entire life in small cages, while broilers are also kept under close confinement on a sawdust-covered concrete area in halls without windows. Pigs, too, are spending all of their five-month-short life indoors. (Pigs have been shown to be more intelligent than dogs, and may suffer more than other farm animals under confinement.) Diary cows are now carrying huge oversized udders and are prone to develop mastitis, a painful udder infection. Generally the poor health of farm animals, the high levels of medication to keep them alive under close confinement until slaughter, the crowded transport conditions (during which animals often nearly suffocate or die with thirst), and inadequate slaughter methods have outraged animals' rights activists. In addition, the high meat consumption comes at huge environmental costs. It takes 12,000 gallons of water to produce a pound of beef, compared to 100 gallons to produce a pound of wheat. It also takes about eight times as much (fossil) energy to produce animal protein than plant protein, in addition to overgrazed pasture or increased soil loss (as far as grain fodder is concerned.)[73]

GENETICALLY MODIFIED ANIMALS AND CLONING

Farm animals have also been modified by genetic engineers. This is achieved by injecting foreign DNA into fertilized egg cells. The success rate is low, but surviving founder animals can in turn be bred in traditional ways. First transgenic animals were created in the 1980s, and by the mid-1990s hundreds of thousands were born per year. Most were developed for biomedical research, including the patented transgenic mouse of 1988, which easily develops cancer to facilitate the screening for cancer cures. Humanized pigs have been engineered with key human genes, to turn them into organ donors whose organs will not be rejected by the human body. Transgenic cows, goats, and sheep express additional proteins (usually pharmaceutical drugs) in their milk, and transgenic hens express them in their eggs.[74]

This kind of research for medical purposes spilled over to agricultural production. Before long, farm animals were genetically engineered to grow faster, to deliver more milk, to contain less fat, and to show increased disease resistance. Inserted genes usually express growth hormones. It is also attempted to produce pigs and poultry that are more docile and therefore better suited to intensive rearing units. Featherless chicken and self-shedding sheep have been the target of research as well.

Since the initial production of transgenic animals is a hit-and-miss affair, founder animals are extremely expensive. However, the offspring of these animals, like all offspring, show variation and hence include high as well as low producers of the desired proteins. Thus, substantial efforts have been directed towards the cloning of founder animals to produce animals that are exact genetic copies. Cloning is quite an established technology as far as embryos are concerned. The eggs of especially milk-productive (or otherwise superior) cows are routinely mass-fertilized with semen from superior bulls, and the resulting embryos are either transferred into less superior cows for breeding, or frozen and sold. Such embryos may also be cloned by splitting them into two or more parts, all of which will develop into genetically identical offspring. However, in case of transgenic animals it is impossible to predict which ones will develop into healthy, productive founder animals before they have grown up. It is therefore necessary to clone them from cells taken from the adult animal. This was achieved in 1996 at the Scottish Roslin Institute with cell material extracted from the udder of a 6-year-old sheep. In 1998 Dolly bore her first lamb, proving that clones are able to produce healthy offspring. Adult goats, cows, and pigs have since been cloned with similar methods, and again the technology spilled over to agricultural production. Livestock breeders began to clone superior adult animals simply because they wanted (genetically) identical offspring rather than offspring with the usual variations. (In October 2000 a cloned calf was for the first time auctioned off in the United States. It was the genetically identical copy to a diary cow that produced 3,300 gallons [ca. 12,500 liters] of milk per year, about double the average. It was sold to a dairy before its birth for $82,000.)

FISHING AND FISH FARMING

The application of oil energy led to an enormous increase in fish consumption as well. Global fish production was below 20 million tons per year in 1950, but rose to 65 million tons in 1970, and to about 129 million tons of fish and seafood in 2001. Nearly a third of this tonnage is now provided by fish farms. Fishing itself is actually a relic from the Foraging Age: fish are the last wild animals hunted on a large scale. The actual global catch peaked in 1989 at 100 million metric tons per year, and is now close to its maximum sustainable yield, at roughly 90 million tons. About 85 percent of the global catch comes from the oceans, the other 15 percent from freshwater. In terms of numbers of fish, most fish caught in the oceans are actually being fed to cattle, pigs, poultry, and sheep as a protein supplement. In terms of weight, only about one third of the captured fish is processed into feed. (It is smaller fish that are processed into feed while more of the larger species are directly consumed by people.)[75]

The echo sounder and sonar made schools of fish easy to find even in deepest waters, and the advent of light plastic filaments allowed for the production of huge nets, up to 30 miles long by the 1970s. However, nets longer than 1.5 miles were eventually banned due to environmentalist pressure. Such driftnets are typically floating through the sea during the night (stretching from the surface to about 40 feet below), and are recovered with the help of radio beacons the next day. Groundfish, on the other hand, are caught by bottom trawlers, which drag a heavy metal bar with a net behind it across the ocean floor. These methods were so effective that they led to the collapse of whole fisheries. In addition, the problem of by-catch arose. All vessels, no matter how large, have limited storage capacity, which they attempt to fill with the commercially most valuable species. Hence other species (or small specimens of the target species) are sorted out and thrown back into the ocean, dead or dying as they had been caught by their gills in driftnets or damaged in other ways by the gear. Perhaps as much as 30 percent of the fish caught around the world are thrown back in the sea, but these fish are not accounted for by the annual production statistics. (Bottom trawling produces the most by-catch. Perhaps as little as 16 percent of the initial catch by shrimp fishers are shrimp and prawns. But in terms of absolute amounts, the larger fisheries such as herrings, sardines, and anchovies produce much more discarded by-catch.) Fishing boats have turned into outright factory ships that process target species on the open sea with filleting machines and store the product in immense on-board freezers. However, such large vessels, in combination with fishing quotas introduced to prevent the collapse of further fisheries, have sharply reduced the manpower and number of boats needed on the oceans. Governments have therefore heavily subsidized this ailing sector. The UN's Food and Agriculture Organization estimated in 1993 that subsidies were 80 percent larger than the worth of fish caught globally.

As the global catch approached its sustainable production limit, fish farming began to expand rapidly. Aquaculture accounted for a mere eight percent of the global fish production in the mid-1980s, but 29 percent (or 37 million tons) in 2001. (In terms of fish for human consumption the aquaculture share is actually much larger.) Roughly 60 percent of aquaculture production (about 22 million tons in 2001) comes from inland waters, roughly 40 percent (about 15 million tons in 2001) from marine waters. In the United States, catfish is the principal fish species farmed, with production in ponds tripling from 200 million pounds in live weight in 1985 to 600 million pounds in 2002. Norway still produces about half of the world's farmed salmon, while Chile emerged as the second largest producer. However, China alone produces nearly 70 percent of all globally farmed fish. Its output is quite evenly divided between freshwater and seawater farming, with carp dominating the inland production, and shellfish (mostly oysters, clams, and mussels) the coastal production. Carps eat just about everything, and their farming is considered an efficient way to produce protein. Quite the opposite is the case in regards to such marine species as shrimp and salmon, whose farming results in a net loss of fish protein. (Salmon is a carnivorous species, and shrimp are fed fish meal made from lower-value captured fish. Some 15 percent of all fishmeal goes into aquaculture feeds, the rest is used for terrestrial farm animals.) Coastal fish farming is also associated with the destruction of natural habitat (mangroves, for instance), water pollution, and oxygen depletion (algal blooms). Moreover, aquaculture exhibits similar problems in regards to diseases as terrestrial mass rearing does. Farmed fish is fed large amounts of antibiotics and other drugs, but viral and bacterial outbreaks have nevertheless wiped out the shrimp farm industries of entire countries.

Fish offer a number of advantages for genetic manipulation due to their high fecundity and external fertilization and development. Coho, one of the five Pacific salmon species, has been engineered to incorporate growth hormone genes from other salmonids. The transgenic coho grows into an enormous fish nearly 40 times as heavy as the original, but the over-production of growth hormone turned out to be detrimental to the health of Atlantic as well as Pacific salmon. Nevertheless, development efforts continue as these fish are of great commercial interest. Other species, including bass, trout, carp, catfish, and tilapia have also been engineered to incorporate additional growth hormone genes from other fish, insects, and humans.

ENERGY SINK

Overall, agriculture had a somewhat strange fate during the Oil Age. This great source of energy, which had been behind the rise of complex civilizations and the knowledge accumulation that led humanity into consequent Energy Eras, was turned into an energy sink. In traditional agriculture the (sun) energy captured in harvested produce was about ten to 30 times higher than the (nutritional) energy used to do the fieldwork and process food. However, in industrialized Oil Age countries far more (fossil) energy is used in agricultural production than is contained as nutritional energy in the diet of people. In the United States, for instance, the energy input is ten times as high as the energy harvest. Fertilizer production consumes the largest share of all energy spent in agriculture, accounting for about half of the global total. Production of field machinery and their fuel requirements plus spare parts account for about 40 percent, and pesticide synthesis and irrigation for about five percent. (In addition, a lot of energy is used to ship agricultural output to the urban centers, to cool storage facilities, and to transport food from supermarkets to homes.)

The reason why fertilizing accounts for the lion's share of agricultural energy inputs is the Haber-Bosch process. It consumes a lot of energy because air has to be liquefied (and slowly regasified) to separate nitrogen from the mix, and because hydrogen is obtained either from glowing coal (and steam, $3C + O_2 + H_2O \rightarrow H_2 + 3 CO$), or from natural gas at about 800°C ($CH_4 +$

$H_2O \rightarrow 3\ H_2 + CO$). The Haber-Bosch reaction itself ($3\ H_2 + N_2 \rightarrow 2\ NH_3$) also uses a lot of energy as the reactants have to be compressed and the reaction temperature is relatively high (450 to 600°C).

To be sure, the energy cost of different agricultural products varies widely. Staple crops such as wheat, rice and maize actually have a positive energy balance: they may contain about twice as much food energy as is invested in terms of technical energy to grow them. (This remains true even though the energy used to produce a ton of maize in the United States, for instance, has increased four-fold during the second half of the 20th century.) The two highest-yielding crops in terms of energy are maize in temperate regions and sugar cane in the tropics. However, the energy balance looks a lot worse as soon as pumped irrigation water is used. (Energy inputs are equivalent to less than 150 kg of oil per hectare for Canadian wheat, but 10 times that amount for heavily fertilized and irrigated Nebraska corn.[76]) Vegetables and fruits, due to their large fertilization and irrigation needs, may contain as little as 1/10th in food calories compared to the energy invested (if tomatoes are grown in heated greenhouses, for instance).[77]

The energy balance looks generally bad for meat and other animal products. Milk is the least energy-intensive animal food, ahead of eggs. (Yet chicken and cows eat about 4.5 calories of plant material per calorie contained in their eggs or milk.) In terms of meat, broilers are more efficient than pigs, while beef production is the most energy intensive. Over 30 times as much energy may be invested to produce beef than is contained in the final meat product. Deep ocean fishing is wasteful in terms of invested versus obtained energy as well. Generally it requires more than eight times as much (fossil) energy to produce animal protein than plant protein.

The global crop harvest increased more than sixfold during the 20th century, but energy inputs increased 80-fold. Half of all food calories eaten worldwide derive from Haber-Bosch fertilizer.[78] Hence half the world's population would not survive without high energy inputs. Humans are clearly dependent on technical energy for their survival. Yet there is little concern, because the percentage of energy spent in agriculture is small in comparison to energy spent in other sectors. In 1990 the global total of agricultural energy inputs was equivalent to about 300 million tons of oil, or less than 5 percent of the total worldwide primary energy consumption.[79] In case of a severe energy crises, there would thus still be enough energy to be used for food production. On the other hand, it would be impossible to go back to traditional agriculture based on animate energy inputs. If the United States were to match the existing power of its tractors in terms of horses, a draft stock of at least ten times its record numbers of the 1910s would be required. Some 300 million hectares, or twice the total area of U.S. arable land would be needed to feed these animals.

NOTES

66. Find agricultural output data in FAOSTAT (FAO Statistical Databases), *Agricultural Data*, Food and Agriculture Organization (FAO) of the United Nations, http://apps.fao.org/faostat/collections?version=ext&hasbulk=0&subset=agriculture. A wide variety of (historical) agricultural statistics (fertilizer consumption, pesticide use intensity, tractors in use, water use intensity, total cereal area harvested, cattle stocks etc.) can be obtained at the World Resources Institute's (http://www.wri.org/) Agriculture and Food searchable database at http://earthtrends.wri.org/searchable_db/index.php?theme=8.

67. Vaclav Smil, *Essays in World History: Energy in World History* (Boulder, CO: Westview Press, 1994), 189

68. Ibid., 191

69. Stephen Nottingham, *Eat Your Genes—How Genetically Modified Food Is Entering Our Diet* (London: Zed Books Ltd., 1998).

70. "Genetic Modification: Filling Tomorrow's Rice Bowl," *The Economist*, December 9, 2006.

71. The public was informed about cannibalistic feeding practices when BSE (Mad Cow Disease) turned into a health problem for humans. See, for instance: Manfred Weissenbacher, "BSE Requires Stricter Feed Ban," *Toronto Star*, August 7, 2003, http://www.healthcoalition.ca/tstar-bse-aug7.pdf.

72. Manfred Weissenbacher, *Rinderwahnsinn: Die Seuche Europas* (Vienna: Böhlau, 2001). Robert E. Taylor, *Scientific Farm Animal Production: An Introduction to Animal Science* (Upper Saddle River: Prentice Hall, 1995); Ray V. Herren, *The Science of Animal Agriculture* (Albany: Delmar Publishers Inc., 1994).

73. PETA (People for the Ethical Treatment of Animals), http://www.peta.org/; Juliet Gellatley, "Murder, She Wrote," Viva! USA, http://www.vivausa.org/activistresources/guides/murdershewrote1.htm#fowlplay; David Pimentel, Laura Westra and Reed F. Noss, eds., *Ecological Integrity: Integrating Environment, Conservation and Health* (Washington: Island Press, 2000).

74. Stephen Nottingham, *Eat Your Genes*.

75. Most information in this section is based on: Michael Berril, *The Plundered Seas, Can the World's Fish Be Saved?* (San Francisco: Sierra Club Books, 1997). Statistical data are available at the World Resources Institute's (http://www.wri.org/) searchable database "Coastal and Marine Ecosystems" at http://earthtrends.wri.org/searchable_db/index.php?theme=1. For instance: "Fishery Production Totals (Aquaculture and Capture): Total for All Species," http://earthtrends.wri.org/searchable_db/index.php?theme=1&variable_ID=834&action=select_countries. World Resources Institute, United Nations Environment Programme, United Nations Development Programme, World Bank, "Diminishing Returns: World Fisheries Under Pressure," in *World Resources: A Guide to the Global Environment-Environmental Change and Human Health* (New York/Oxford: Oxford University Press, 1998), 195.

76. Vaclav Smil, *Essays in World History*, 219.

77. Ernst Ulrich von Weizsäcker, Amory B. Lovins, and L. Hunter Lovins, *Faktor vier. Doppelter Wohlstand—halbierter Naturverbrauch* (München: Droemersche Verlagsanstalt Th. Knaur, 1995, 1996). English language edition: Ernst Ulrich von

Weizsäcker, Amory B. Lovins, and L. Hunter Lovins, *Factor Four. Doubling Wealth—Halving Resource Use* (London: Earthscan, 1997).

78. Synthetic nitrogen now supplies about half of the nutrient used annually by the world's crops. Because about three-quarters of all nitrogen in food proteins come from arable land, at least one-third of the protein in the current global food supply is derived from Haber-Bosch ammonia synthesis. Vaclav Smil, *Essays in World History*, 190.

79. Vaclav Smil, *Essays in World History*, 219.

CHAPTER 32

DIVERSE ENERGY MIX

Though agriculture turned from being humankind's principal energy source into an energy sink, the total energy mix generally became more diverse when people proceeded through the Energy Eras. New sources of power were added, while old ones remained, but not all regions adopted all new energy sources, or all aspects associated with them. Some societies essentially skipped one Energy Era, moving from the Agricultural Age straight into the Oil Age. Others decided to forego nuclear energy for electricity production due to security concerns.

The rise in energy consumption during the Oil Age reflects both enormous population growth and availability of more, and more diverse, energy sources. (The one main factor mitigating the trend towards more fuel consumption was efficiency gains.) Biomass fuel consumption increased globally from some 700 million tons in 1700 to over 1.8 billion tons towards the end of the 20th century. By 1950 commercial energy provided about five times as much useful energy as biomass, by 1990 it was 20 times as much. Total global energy consumption rose perhaps 70-fold between 1800 and 2000. Coal consumption kept expanding until the 1980s, when it began to decline slightly. Crude oil production (extraction) increased about 300-fold between the 1880s and the 1980s, and natural gas production 1000-fold. Less than 10 percent of all fossil fuels were converted to electricity in 1945, but the share increased to a quarter towards the end of the 20th century. Electricity is produced from coal, natural gas, nuclear energy, hydropower, and to a smaller extent from non-hydro renewable sources such as wind, solar, and geothermal energy. In total about one third of the world's commercial energy is currently used as electricity, which can be transported over long distances and is versatile to the extent that it can be converted either into light, heat, or mechanical energy. Electricity powers the compressors in Haber-Bosch ammonia plants, the refrigerators in supermarkets and households, the sewage pumps of municipalities, countless industrial electric motors, as well as personal computers and electric light bulbs. (About half of the world's electricity is used by industry, the other half by commercial activity and homes.) On the downside, the production and consumption of electricity involves energy losses, as energy conversions are never 100 percent efficient. Often about two thirds of the original (primary) energy is lost during conversion into electricity, with only one third of the remaining third being actually available after the conversion of electricity into mechanical work or heat. However, modern steam (turbo)generators turn over 40 percent of the energy contained in the fuel into electricity. The efficiency of steam turbines in electricity plants is especially important because they operate more than 5,000 hours a year. (The world's prime mover capacity in form of internal combustion engines (cars, trucks, tractors, combines, etc.) is much larger than that of steam turbines, but vehicles are often parked and operate perhaps 500 hours a year.)[80]

Fossil fuels currently provide for about 87 percent of the world's commercial energy, with oil contributing 38 percent, coal 25 percent, and natural gas 24 percent. The remaining 13 percent are equally split between hydropower and nuclear energy.[81] (The small contributions of commercial renewable energies other than hydro energy, which, combined, amount to about 2 percent of the total energy consumption, are not included in these figures.) If smaller commercial fuels (such as solar energy and wind power) as well as non-commercial, traditional fuels (such as fuel wood, animal and vegetal wastes) are taken into account, the contribution of fossil fuels to meet world energy needs amounts to about 77 percent. Oil accounts for 32.5 percent, coal for 26.5 percent, natural gas for 18 percent, nuclear energy for 5 percent, hydropower for 6 percent, biofuels (biomass, wood, charcoal) for 11.5 percent, and solar and wind energy combined for 0.5 percent.[82]

About 38 percent of the world's commercial energy is used for industrial processes; another 38 percent is consumed by the commercial and residential sector (for space heating, air conditioning, lighting, and running small appliances such as refrigerators, TV sets, etc.); and about 24 percent of the world's commercial energy is used for transportation.[83] Both the relatively small percentage of energy used in transportation, and the share of oil below 40 percent, are understating the importance of oil. Transport and mobility on land, on the waters, and in the air are crucially important for Oil Age societies, and practically the entire sector (save electric trains) depends on oil. Just to keep the world's fleet of motor vehicles running consumes about half the global oil production. In comparison the other main fuels, coal, natural gas, nuclear energy, and hydro, can all substitute for one another, as they are all mainly used to produce electricity. (Coal, natural gas, and nuclear energy through steam turbines, hydropower through water turbines.)

COAL

Coal has now two main applications. Electricity production accounts for about 60 percent of all coal consumption, while most of the remainder goes into steel manufacturing, where coal continues to be indispensable. (Earlier on in the Oil Age, coal was also very important for heating houses. In the United States coal heated 35 percent of all homes in 1950, but less than 1 percent in 2000.[84]) About 38 percent of the world's electricity was produced from coal at the beginning of the 21st century. From the 1920s electricity plants began injecting pulverized coal into the combustion chamber (rather than burning lump coal) to achieve higher efficiencies. In the 1950s utilities began to build power stations close to the mines rather than to consumers due to concerns about air pollution. (Sulfur dioxide [SO_2] and particulates generated by coal combustion killed some 4,000 people in five days during the Great Smog of London in December 1952.[85] The sulfur in coal derives from a type of amino acid present in living organisms. Once SO_2 gas is released into the atmosphere it interacts with moisture to form tiny droplets of sulfuric acid [H_2SO_4] that fall down as acid rain and damage the natural environment as well as buildings.) Rather expensive installments are necessary to clean the exhaust gases of coal-fired electricity plants, but despite these installments energy derived from coal is very cheap compared to oil and gas. (Varying with actual energy prices, producing a unit of energy from coal may cost about half as much as producing it from gas, and a third compared to oil. A ton of coal cost between $30 and $40 per ton in 2001 on different markets in the world.)

Coal consumption resurged after the first oil price shock of 1973. (In 1976 the Soviet Union was the largest producer in the world, extracting over 630 million tons annually, while China produced 415 million tons, Poland 185 million tons, and Czechoslovakia 113 million tons.) However, by the 1980s the growth in coal consumption was stagnating, though coal continued to be a main fuel in non-iron industries in such populous and coal-rich countries as China and India. In 2001 the United States accounted for 26 percent of world coal production and China for 24 percent. India was the third largest producer. Europe as a whole (excluding the former Soviet Union) accounted for 10 percent, but production had decreased by over a third during the 1990s (when oil was rather cheap). The areas of the former Soviet Union produced 9 percent of the world's coal in 2001, with production picking up again from the late 1990s.

In 2001 the world's proved commercial coal reserves totaled 984 billion tons, including anthracite and bituminous coal (combined 519 billion tons), and sub-bituminous coal and lignite (combined 465 billion tons). If production were to continue at the current level, these proved reserves, which are equivalent to approximately 3,560 billion barrels oil,[86] would last for 216 years. (This compares to 1,050 billion barrels of oil in proved 2001 reserves, which would last for 40 years at current production rates, and to natural gas reserves equivalent to 975 billion barrels of oil equivalent, which would last for 62 years at the current natural gas production rate.) The United States (25 percent), Russia (16 percent), China (12 percent), India (9 percent), Australia (8 percent), and South Africa have the largest share in the world's coal reserves. Despite the lower price of coal, its contribution to the world's commercial energy consumption has been projected to decrease from 25 percent in 2001 to 20 percent in 2020 at the expense of the more versatile fuels oil and natural gas (whose consumption growth is substantially larger than that of coal). Nevertheless, the importance of coal will undoubtedly resurge, simply because coal exists in much larger quantities, and is more equally distributed throughout the world than the other fossil fuels.

NATURAL GAS

Natural gas is usually found in conjunction with crude oil. During the early Oil Age the gas was usually flared off, and only in the United States were significant amounts produced and consumed prior to World War II. During this time natural gas was essentially a substitute for coal gas, and indeed natural gas is still now important for cooking and heating in homes and offices. The first gas-powered electricity turbine began operation in Switzerland in 1940, and eventually natural gas became very important for electricity production. Natural gas is a rather clean fuel that leaves no ashes and generates only water and carbon dioxide (CO_2) when combusted. What is more, natural gas burns very efficiently: the boilers of large power plant and household gas furnaces alike may be up to 95 percent efficient. In modern, so-called combined-cycle natural gas power plants, gas turbines are directly driven by burning natural gas, and exhaust heat is utilized to produce steam for (conventional) steam turbines. Such combined-cycle technology can increase the thermal efficiency from about 40 percent (as is achieved with modern steam turbines) to about 50 to 60 percent. These technological changes, combined with environmental and economic realities, have helped natural gas to become the fuel of choice for new power plants in many regions. In the United States, almost 95 percent of the new electric capacity installed in 2000 was natural-gas-fired (while large coal or nuclear power plants had been the choice of most electricity providers during the 1970s and 1980s.)

Some 41 percent of the natural gas consumed in the United States in 2003 went to residential and commercial establishments, 35 percent to industry, and 24 percent to electricity production.[87] In industry, natural gas serves as fuel and critical feedstock for the chemical industry. It is used in the production of pulp and paper, metals, (petro)chemicals, stone, clay and glass, plastic, and processed food, but it is also the chief source of hydrogen.[88] This works because methane, the chief constituent of natural gas, can quite easily be transferred into synthesis gas in a process known as steam reforming. ($CH_4 + H_2O \rightarrow CO + 3\ H_2$: syngas consists of hydrogen gas and carbon monoxide.) The obtained hydrogen is then used for such essential applications as the Haber-Bosch ammonia synthesis, critical for fertilizer production to feed the world. (The price of natural gas is thus quite critical for the cost of agricultural production. During the 2000 planting season, ammonia fertilizer cost around $100 per ton in the United States, while farmers faced ammonia prices of $350 or more per ton during the 2003 growing season, mainly due to the higher price of natural gas.) Syngas can also be reacted into liquid fuels, so-called Gas-to-Liquids fuel.

The rise of natural gas as fuel and feedstock depended on means to transport it. In various regions natural gas extraction only became feasible when turbine-driven compressors and large diameter pipelines entered the scene. Dense networks of gas pipelines were built in North America after 1945 and in Europe after 1960. The pipelines running from western Siberia to Europe, some as large as 2.4 meters in diameter, stretch over a distance of nearly 6,500 kilometers. For overseas deliveries liquefied natural gas (LNG) tankers have been developed from the 1960s. Japan initiated this trend and due to its island position and lack of fossil fuels emerged as the largest importer of LNG. Natural gas (methane) is liquefied by cooling it to minus 161 degrees Celsius (minus 259 degrees Fahrenheit), which consumes energy. At this temperature the gas turns into a clear, colorless, odorless, nontoxic and noncorrosive liquid that weighs less than half the weight of water and occupies only about 1/600 of its gaseous-state volume. Handling LNG is relatively hazardous, since it explosively increases its volume if it is exposed to ambient temperatures. LNG tankers transport their load in insulated double-hulled tanks. At the delivery terminals, LNG is usually pumped into tanks and thereafter carefully regasified (by warming) and fed into continental gas pipeline systems. In 1990 just 65 liquid natural gas tankers existed (compared to 2,600 oil tankers), while 136 LNG tankers were operational by 2003, and 224 LNG tankers were in service in March 2007 (and another 145 on order).[89] (Tankers that transport LPG, liquefied petroleum gas, are similar. LPG consists of liquefied butane and propane.)

World consumption of natural gas more than doubled between 1970 and 1990, and rose by another 20 percent between 1991 and 2001 (to ca. 2,400 billion cubic meters or ca. 15.1 billion barrels oil equivalent).[90] This makes natural gas the fastest growing fossil fuel in terms of consumption, which is expected to double again between 2001 and 2021. Worldwide natural gas production will peak perhaps around 2030, but this is almost certainly after oil has peaked: natural gas is therefore expected to play an increasingly critical role in the near future. An optimistic estimate for the world's ultimate natural gas resources (including production to date, known reserves, and reserves yet to be discovered) is 2,600 billion barrels of oil equivalent; however, significant portions of the world's natural gas resources lie in remote locations or are found in small accumulations.[91] Proven natural gas reserves rose from 82.44 trillion cubic meters in 1981 to 123.97 trillion cubic meters in 1991 to 155.08 trillion cubic meters in 2001. Generally, countries endowed with oil also tend to have natural gas, but natural gas reserves are even more regionally concentrated than oil reserves. The former Soviet Union and the Middle East each account for a bit over a third of the proven natural gas reserves, while just three countries, Russia, Qatar, and Iran, combined control almost 60 percent. Russia, with over a quarter of the total, is especially rich in natural gas, while Iran and Qatar have similar shares. The tiny Gulf state of Qatar (slightly smaller than Connecticut) is in possession of the world's largest single gas field and determined to become the top LNG exporter. In February 2005 Qatar signed deals worth approximately $20 billion with ExxonMobil and Royal Dutch/Shell for the production of liquefied natural gas (LNG) to be exported to Europe and North America. France's Total, not to be left out of the boom, immediately invested $3.5 billion in LNG production in Qatar as well. However, Russia remains the main supplier of natural gas to the European Union. In 2001, the EU imported 43 percent of its gas requirements from Russia, and this dependency is expected increase to some 75 percent by 2020. The United States is importing increasingly more natural gas from Canada, while Japan receives most of its LNG from Indonesia and the Middle East.

HYDROPOWER

Hydroelectric schemes have been around since the late Coal Age, and hydropower supplied about 17 percent of the world's electricity at the beginning of the 21st century, though the contribution varied widely by region.[92] Austria, with large alpine rivers of significant gradient, covers 78 percent of its electricity needs with hydropower, Norway even 99 percent. Similarly, large rivers provide for most of the electricity in some Latin American and African countries: around 96 percent in Brazil, and some 97 percent in the Democratic Republic of the Congo.

The initial investment in large hydroelectric schemes is substantial, but the life of hydro plants is long (about 100 years), and the operating costs are low. In the United States and in the Soviet Union massive dams with large generators were constructed during the 1930s with state support under the New Deal's Tennessee Valley Authority and as part of Stalin's industrialization program. During World War II demand for electricity soared in the

United States to supply aluminum plants, which delivered metal for airplane and bomb production. The Tennessee Valley Authority then engaged in an enormous hydropower construction program, employed 28,000 people, and had 12 hydroelectric projects under construction by 1942. Thus it emerged as the nation's largest electricity supplier from the war.

The largest hydroelectric scheme of the United States, and about the third largest in the world, is the Grand Coulee plant (6,500 megawatt capacity) in Washington State. It is by far exceeded in capacity by the Itaipu installation (12,600 megawatt capacity) at the Brazilian-Paraguaian border (officially dedicated in 1982). In 1993 China started to build the largest hydroelectric scheme in the world, the Three Gorges project (18,200 megawatt capacity). It involves harnessing the Yangtze River (Asia's longest river) through a dam that is 185 meters high and spans 2.3 km across. Worldwide there are some 40,000 large dams (more than four stories high), and hundreds of thousands small hydropower units, many of which generate the energy for just one farm or a small village.

NUCLEAR ENERGY

Nuclear energy, like hydropower, is almost exclusively used for electricity production, and both contributed about the same share (17 percent) to the world's electricity needs in the beginning of the 21st century.[93] But again there were substantial regional differences. France supplied 75 percent of its electricity needs with nuclear energy in 2002, South Korea 43 percent, Japan 28 percent, the United States about 20 percent, Russia 16 percent, and China 1.5 percent, while countries such as Australia, Austria, Indonesia, and Italy operated no nuclear electricity plants at all. As of 2005, 443 nuclear power plants were operating or operable in 31 countries worldwide. Of these 443, 104 were located in the United States, 59 in France, 56 in Japan, 31 in Russia, 23 in the UK, 20 in South Korea, 18 in Canada, 17 in Germany, 15 in India, 15 in the Ukraine, and 10 in Sweden.[94]

Nuclear power plants use the same raw materials as nuclear bombs, and the two technologies are closely related. During World War II Germany worked on a nuclear electricity plant, while an atomic bomb was built in the United States. However, the technology of the first operable nuclear electricity plants derives from a nuclear reactor developed by the U.S. military for submarine propulsion.[95] In nuclear-powered submarines a nuclear reactor generates heat, which is used to generate steam, which is used to turn both propulsion turbines (which drive the propeller) and electricity generators (which supply the ship with electricity). Unlike the diesel-electric submarines of World War II, nuclear-powered submarines do not rely on oxygen (air) for power generation, and hence can remain submerged for very extended periods of time. The first operable nuclear-powered submarine, the

USS *Nautilus*, was launched in 1954 and crossed the North pole beneath the Arctic ice cap in 1958, the year the Soviet Union put its first nuclear-powered submarine into service. Soon such submarines had the capability to launch intercontinental ballistic missile with nuclear warheads. The United States currently has over 100 submarines in operation, almost all of them nuclear-powered. However, submarine propulsion is the only significant application of nuclear power besides stationary electricity plants (and atomic bombs). The pressurized water reactor (PWR) developed by the U.S. military for submarine propulsion was directly adopted for large-scale stationary nuclear electricity generation (and weapons-grade plutonium production). The first such electricity plant began operation in 1956,[96] and though the PWR was not a superior design, its early adoption provided it with a market position that was quite easily defended against later nuclear reactor types. (By the end of the 20th century PWRs accounted for nearly 60 percent of all commercial nuclear stations.)

Inside a nuclear reactor uranium is bombarded with neutrons in order to start the nuclear reaction. Once it is started, the decaying nuclei release free neutrons, which, in a chain reaction, split further uranium atoms which release further neutrons and energy. This chain reaction is kept in check with control rods that absorb neutrons. (Cadmium and boron are strong neutron absorbers and are the most common materials used in control rods.) Another important component of the reactor is the moderator, a material that slows down (moderates) neutrons, which makes collisions with the fuel more likely. (Either graphite blocks or water work as moderator. If water is used, it functions as a coolant at the same time.) The thermal energy released by the nuclear reactions is used to heat up water to turn steam turbines that are connected to electricity generators.

Like large hydropower schemes, nuclear power plants are very expensive to build. Up to 75 percent of the lifetime costs (excluding decommissioning) of a nuclear plant are incurred up front (compared with ca. 25 percent for a gas-fired plant). National governments are typically involved in the financing of nuclear energy, and more than half of the subsidies (in real terms) ever lavished on energy by Western countries have gone to the nuclear industry. After the 1973 oil price shock there was a major increase in orders for nuclear reactors, but Pennsylvania's Three Mile Island (1979) and the Ukraine's Chernobyl (1986) accidents helped to reverse the trend. (At Chernobyl 135,000 people were permanently evacuated the day after the explosion at the nuclear plant from within an 18-mile radius of the accident. Huge amounts of radioactive material entered the atmosphere and were washed down by rain all over Europe, but mainly in the north. About 15,000 relief workers engaged at the site later died and another 50,000 were left handicapped.) But it was not alone these accidents that brought orders for new nuclear plants to an almost complete end in the Western world in the 1980s. High construction costs, certain technical weaknesses, and not least the unresolved problem of long-term disposal of radioactive waste contributed as well. After decades of running nuclear power plants, no country has figured out what to do with its radioactive waste. In the United States, the Department of Energy (DOE), under the guise of research, has spent $5 billion to built a full-scale system of tunnels into Yucca Mountain in Nevada (90 miles northwest of Las Vegas) for long-term storage of U.S. nuclear waste. However, at the beginning of the 21st century the legal battle between the people of Nevada and the federal government continued, and spent fuel rods were kept at the nuclear power plants.

And then there is the problem that the materials that occur in nuclear power plants can be used to build atomic bombs. The fuel cycle starts with uranium ore, which consists to 99.3 percent of U-238. From the mines the ore is typically shipped to processing plants, which concentrate it into a useful nuclear fuel that contains about three to five percent of U-235. (Most uranium concentrate is made by leaching the uranium from the ore with acids, but sometimes the concentrate is produced *in situ* underground, without removing the uranium ore.) At the end of the process uranium (in form of uranium oxide) is produced into rods, which are bundled together into fuel assemblies. Such fuel rods, which are low-enriched with uranium-235, can be concentrated into highly enriched uranium containing enough U-235 to construct a nuclear bomb. But even more problematic are heavy water reactors. These use natural uranium, because heavy water moderates neutrons enough to react with U-238 rather than U-235. Such reactors produce considerable amounts of bomb-usable Plutonium-239.

Despite these concerns nuclear energy production has been growing considerably even after the Chernobyl accident. In the decade between 1987 and 1997 this increase was led by Japan, where nuclear electricity output grew by 70 percent, France (50 percent increase), and the United States (40 percent increase). However, over half of the world's 34 nuclear power plants under construction at the beginning of the 21st century were located in Asia. Eight were under construction in India, three each in China and Japan, two each in Taiwan and South Korea, and one each in North Korea and Iran. Six new nuclear power plants are being installed in Russia. Meanwhile the German government in June 2000 concluded an agreement with German utilities for an early phase-out of their 19 operating nuclear power plants.

WIND ENERGY AND OTHER RENEWABLES

The only source of renewable energy (called so to differentiate it from energy that comes from limited underground fuel deposits) to compete economically with fossil fuels in the 20th century was hydropower. Of the other renewables, which include solar energy, biofuels, and geothermal energy, most government support went to wind energy. By the beginning of the 21st century, tall, slim windmills with large propeller-type rotors about 50 to 90 meters in diameter had become a quite common feature in some areas of North America and western Europe, and were on the rise in Asia. These mills directly drive a built-in generator that feeds electricity into the grid. Germany is not especially windy, but a legal environment was created that turned the country into the world's number one producer and consumer of wind power by the start of the 21st century, ahead of the United States and Spain.

NOTES

80. Vaclav Smil, *Essays in World History: Energy in World History* (Boulder: Westview Press, 1994), 229.

81. BP Statistical Review of World Energy June 2002. BP's current Statistical Review of World Energy can be downloaded at www.bp.com/statisticalreview.

82. Ernst Ulrich von Weizsäcker, Amory B. Lovins, and L. Hunter Lovins, *Faktor vier. Doppelter Wohlstand—halbierter Naturverbrauch* (München: Droemersche Verlagsanstalt Th. Knaur, 1995, 1996). English language edition: Ernst Ulrich von Weizsäcker, Amory B. Lovins, and L. Hunter Lovins, *Factor Four. Doubling Wealth—Halving Resource Use* (London: Earthscan, 1997). Find an overview of renewable and non-renewable energy sources in World Energy Council, "2007 Survey of Energy Resources (Executive Summary)," http://www.worldenergy.org/documents/ser2007_executive_summary.pdf.

83. Michael D. Morgan, Joseph M. Moran and James H. Wiersma, *Environmental Science: Managing Biological & Physical Resources* (Dubuque: Wm. C. Brown Publishers, 1993); N. Stern, *The Stern Review on the Economics of Climate Change* (Cambridge: Cambridge University Press, 2006), http://www.hm-treasury.gov.uk/stern_review_final_report.htm.

84. Michael D. Morgan, Joseph M. Moran and James H. Wiersma, *Environmental Science: Managing Biological & Physical Resources* (Dubuque: Wm. C. Brown Publishers, 1993).

85. Peter Brimblecombe, *The Big Smoke. A History of Air Pollution in London since Medieval Times* (London: Methuen, 1987). Peter Brimblecombe and László Makra, "Selections from the History of Environmental Pollution, with Special Attention to Air Pollution. Part 2: From Medieval Times to the 19th Century," *Int. J. Environment and Pollution* 23 (2005): 354, http://www.sci.u-szeged.hu/eghajlattan/makracikk/Brimblecombe%20Makra%20IJEP.pdf.

86. Different types of coal contain different amounts of energy. According to the BP Statistical Review of World Energy, June 2002, global coal reserves amounted to 984 billion tons. The report also states this amount in terms of million tons of oil equivalent, which was converted to billion barrels oil equivalent by using a factor 7.33, based on worldwide average gravity (metric tons of oil by 7.33 = barrels).

The BP-Report quotes The World Energy Council as the source of coal reserves data. Historical data can be obtained from World Resources Institute's "Energy and Resources" searchable database at http://earthtrends.wri.org/searchable_db/index.php?theme=6. The World Resources Institute quotes the BP reports (BP Statistical Review of World Energy, http://www.bp.com/statisticalreview) for coal reserves data.

87. "U.S. Natural Gas Consumption by End Use," Energy Information Administration, http://tonto.eia.doe.gov/dnav/ng/ng_cons_sum_dcu_nus_a.htm.
88. "Natural Gas: Uses In Industry," Natural Gas Supply Association, http://www.naturalgas.org/overview/uses_industry.asp.
89. "Introduction to LNG," Energy Economics Research at the Bureau of Economic Geology, University of Texas at Austin, http://www.beg.utexas.edu/energyecon/lng/documents/Cee_Introduction_To_Lng_Final.pdf.
90. BP Statistical Review of World Energy, June 2002.
91. In addition to that, there are also deposits of methane in the form of methane hydrates on the sea floor, where cold temperatures and high pressures cause the methane gas to be trapped with water in an icelike state. The global size of these deposits may be very large, perhaps in the range of 10,000 Gt of methane carbon, but it would be concentrated enough to be harvested only in few regions. Even there it is questionable, whether such undertaking could be viable. Alexei V. Milkov, "Global Estimates of Hydrate-bound Gas in Marine Sediments: How Much is Really Out There?," *Earth-Science Reviews* Vol. 66, Issues 3–4 (2004): 183–97, http://www.sciencedirect.com/science?_ob=ArticleURL&_udi=B6V62-4BMTJ4G-1&_user=10&_rdoc=1&_fmt=&_orig=search&_sort=d&view=c&_version=1&_urlVersion=0&_userid=10&md5=c4a913455b5e27ed6935405cc476fced.
92. According to estimates by the Energy Information Administration, hydroelectricity in 2005 accounted for 16.7 percent of world net electricity generation, while nuclear energy contributed 15.1 percent, and "Geothermal, Solar, Wind, and Wood and Waste" added 0.3 percent. "World Net Electricity Generation by type," Office of Energy Markets and End Use, International Energy Statistics Team, Energy Information Administration, Table posted on September 17, 2007, http://www.eia.doe.gov/emeu/international/RecentElectricityGenerationByType.xls.
93. Ibid; "International Energy," Annual Energy Review, Energy Information Administration, http://www.eia.doe.gov/aer/inter.html; "International Energy Outlook 2002, Highlights," Energy Information Administration, http://www.eia.doe.gov/oiaf/archive/ieo02/index.html.
94. International Energy Agency, *World Energy Outlook 2006*, 347, http://www.iea.org/textbase/nppdf/free/2006/weo2006.pdf.
95. Federation of American Scientists (FAS), "Nuclear Propulsion," http://www.fas.org/man/dod-101/sys/ship/eng/reactor.html.
96. The B Reactor at Hanford, Washington, completed in September 1944, was the world's first large-scale plutonium production reactor. The first nuclear reactor to provide electricity to a national grid opened in England on the Sellafield site in 1956: Calder Hall was primarily used to produce weapons-grade plutonium rather than electricity until 1964, when focus shifted towards electricity production. However, it was not until 1995 that the British government announced that all production of plutonium for weapons purposes had ceased at this site. Calder Hall's four cooling towers were demolished in 2007 as part of the decommissioning of the plant. U.S. Department of Energy, "B Reactor-Hanford, Washington," http://www.energy.gov/about/breactor.htm; "Sellafield Towers are Demolished," *BBC News*, September 29, 2007, http://news.bbc.co.uk/2/hi/uk_news/england/cumbria/7019414.stm.

CHAPTER 33

ECONOMIC EXPANSION

With unprecedented amounts of energy available to humans, the Oil Age generated extraordinary wealth. When societies entered the Oil Age there seems to have been no real transitional downturn of the kind experienced at the beginning of the two previous Energy Eras. Perhaps the transition into the Oil Age was smooth because oil technology was rooted deeply in the Coal Age and emerged quite slowly. However, the Oil Age did in fact have a somewhat difficult start that may in part reflect transitional problems between the two Energy Eras. During the 1920s the world was still recovering from the effects of World War I, and in the 1930s it slid into a great economic recession. This recession helped Adolf Hitler to come to power in Germany, which ultimately led to the outbreak of World War II (1939–1945). Thereafter, between 1950 and 1973, the world experienced an unprecedented period of rapid and widespread economic growth. This was a time of steadily declining crude oil prices that ended in late 1973 with a sudden quadrupling of the oil price. The sum of all countries' GDPs, the gross world product (GWP), increased from about $1 trillion in 1900 to nearly $4 trillion in 1950 to $14 trillion in 1973 (expressed in constant 1990 U.S. dollars). Per capita GDP, which is a country's GDP divided by the number of its inhabitants, is an approximate measure of a country's overall wealth. The U.S. per capita GDP was the highest in the world in 1950, but it was another 60 percent higher by 1973. Meanwhile West Germany's per capita GDP more than tripled, and Japan's more than sextupled. Despite experiencing rapid population growth, China and Mexico managed to double their per capita GDPs, and Brazil did even better.[97] After the global economic slowdown of the early 1980s, which was accompanied by record inflation and high unemployment, the world economy picked up again and rapidly grew during the 1990s, a decade of low oil prices. In 1992 the GWP was about $23 trillion, but by 2003 it had reached $51.5 trillion (in current terms, adjusted for purchasing power parities).

Even though a country's economic output will generally be higher the more energy it consumes, maturing industrialized economies tend to consume diminishing amounts of fossil fuel per unit of GDP. This has two reasons. First, maturing Oil Age societies improve their industrial efficiency and manage to produce more per unit of energy invested. Second, maturing Oil Age societies tend to move from production to service to knowledge economies, which are increasingly less energy-intensive. (Much of the heavy industries has been moved from Western to developing countries.) Britain reached its peak in energy input per unit of GDP as early as around 1850, the US around 1920, Japan in 1970, and China in the late 1970s. Total energy intensity of the world economy as a whole has been declining since the 1920s, while the consumption of electricity per unit of GDP was still growing in the 1990s.[98] (However, in some industrialized countries even the latter measure started to decline slightly for the first time during the 1980s.) The oil-intensity of industrialized countries decreased radically after the oil price surge of 1973–74, as these nations began diversifying their energy mix (largely by shifting electricity production from oil to coal and nuclear). In 2003 Britain used 3.8 barrels of oil per $10,000 of GDP (at purchasing power parity), for instance, which was a third less than 20 years earlier. In comparison, the United States in 2003 used 6.8 barrels of oil per $10,000 of GDP, Japan 5.7 and Germany 4.2.[99]

INTERNATIONAL TRADE

International trade increased even more in the Oil Age than economic output. The gross world product (GWP) increased nearly six-fold between 1950 and 2000, while world merchandise trade rose by a factor of about 17. Foreign trade was worth less than 5 percent of GWP in 1900, but about 15 percent towards the end of the 20th century. Between 1950 and 2000 alone the ratio of world exports to GWP more than doubled, and by 2002 total exports were worth $6.4 trillion.

The enormous rise in trade volume was facilitated by improved oil-powered transport systems in form of trucks, ships, and airplanes. Transport on the water is still the cheapest option for large loads, and from the 1950s specialized ships for such widely traded commodities as ores, grain, lumber, and chemicals were introduced. (Transport costs for dry bulk cargoes such as iron ore, coal, phosphates, bauxite, and grain on carriers of more than 100,000 tons deadweight are often less than one U.S. cent per ton per mile.)

Meanwhile manufactured goods are typically transported in standard (ISO 20 ft and 40 ft) containers, with literally thousands of them fitting on today's enormous container ships.

In the Oil Age the trend continued that wealthy industrialized nations were mainly each other's trading partners and customers. (The largest trading partner of the United States, for instance, has traditionally been Canada, while well over half the exports from France, Germany and Italy go to other EU countries.) These nations often sold very similar things to one another: Cars made in France were exported to Germany while German cars were shipped to France. What is more, Western Europe, for the first time since the 1860s, stopped being a cereal importer in 1983 and in turn started to produce major surpluses. Industrialized countries also reduced their import dependencies in rubber and fertilizer due to the development of synthetic products in the Oil Age. In the 1930s industrialized nations were still generally 97 percent self sufficient in terms of raw materials, but the picture changed significantly from the 1950s, most notably because much of the industrialized world started to rely on oil imports. Owing to the wide availability of coal and America's incredible oil riches, the West had a surplus in commercial energy up until the 1940s, and a deficit of only four percent by 1950. However, in the mid-1950s crude oil became cheaper than coal for the first time, which had the effect that imported oil was substituted for coal in electricity production (in addition to assuming its critical function for transportation and mobility). In western Europe, where domestic oil production was entirely absent, coal accounted for 100 percent of the energy used for thermal electricity production in the 1950s, but the share sank to 48 percent in 1973, when oil accounted for 36 percent of thermal electricity production. (The remainder was natural gas.) At its peak in 1973, western Europe's foreign energy dependency amounted to 58 percent of consumption. In the United States oil consumption exceeded domestic production from 1948, despite America's huge domestic oil supplies. However, U.S. coal was inexpensive to extract, and the share of oil in American thermal electricity production reached no more than 20 percent in 1974 (compared to coal, 59 percent). Nevertheless, the United States emerged as the largest oil importer in the world. In terms of metals the situation was somewhat similar. The industrialized world developed a six percent deficit in iron ore trade only in 1960, but the deficit grew to 33 percent of the iron ore consumption by 1970. (Japan in particular emerged as major iron ore importer.) In terms of bauxite (aluminum ore) the industrialized world had a deficit starting in the 1970s.[100]

For developing countries trading patterns with wealthier industrialized nations changed significantly during the Oil Age. In terms of agricultural goods, exports slowed due to increasing domestic demand following rapid population growth, plus slowing demand for food from the Western world. The picture of the developing world being an exporter of raw materials to the industrialized world did not last very long either: the share of manufactured goods in export from developing countries increased from 20 percent in 1970 to 60 percent in 1990. As the globalized infrastructure allowed for an international division of labor that favored some production to be moved to less-developed regions of low labor costs, developing countries from the 1980s on increasingly exported labor-intensive low-tech products (such as textiles) and intermediate goods (such as iron ingots), and by the 1990s they also exported many goods of the kind that were previously only made in the industrialized world. (Mexico exported refrigerators, for instance, Thailand computer peripherals, and China CD players.) As the communications infrastructure (phone lines, satellites, Internet) also allowed for the quick and inexpensive transport of data, Western corporations had by the end of the 1990s even outsourced parts of the service sector to developing countries. India was ideally positioned in this respect, as it offered a rather inexpensive but well-educated work force, and as it had inherited English, the preferred international language of the Oil Age, as its schooling language from its former mother country, England. First, Western companies moved parts of their software development (the writing of computer programs) to India; then entire customer-support call-centers and e-mail support; and soon Indian doctors started to analyze foreign X-rays sent over the Internet to provide remote diagnosis of patients in other countries. By 2002, financial services firms had moved many of their back-office jobs to India as well, and some 3.3 million white-collar American jobs (500,000 of them in information technology) were expected to be transferred offshore to countries such as India by 2015.

Trade Barriers

Trade and international integration tends to help less wealthy countries to enter the Oil Age and industrialize. China, India, and the four Tiger states (Hong Kong, Singapore, Taiwan, and South Korea) all have done much better economically ever since they decided to integrate into the global economy. Supposedly, export-oriented open economies with focused investments and a good education system would do best. Their industries would have to compete with international price and quality standards, and hence evolve rapidly.[101] (This is in sharp contrast with Coal Age industrialization, when Western countries closed their borders to have their industries mature before they would face foreign competition.) Classical trade theory (as well as practical evidence) suggests that trade increases the wealth (or maximum possible consumption) of both the richer and the poorer country involved in trade, but the absolute wealth gap between the two may increase in the case of an equal percentage gain by both trading partners.[102] That might create an even less equal world, but the benefit of lifting the poor country out of poverty undoubtedly weighs heavier. It is thus a problem that various rich, industrialized countries, though they often officially promote free trading schemes, put up trade barriers whenever it serves their own interest. Such countries typically protect strategic industries to maintain a certain degree of autarchy (The United States slapped safeguard tariffs of up to 30 percent on foreign steel in March 2002, for instance), or they use their industrial wealth to subsidize their agricultural sector, which artificially decreases prices in a sector where developing countries might otherwise be able to compete.[103] As of 2003, the United States paid some $3 billion a year to some 25,000 American cotton farmers, whose overproduction impoverished millions in West Africa who rely on cotton for their livelihood. (On the other hand, it may be argued that the developing world as a whole profits from lower cotton prices at the expense of U.S. tax payers: The developing world has been a net cotton importer since 1980, and exports textiles to the industrialized world. But this translates into shifting wealth away from the world's poorest nations, in Africa, to the less poor Asian textile producers.) Similarly, the European Union is the world's second largest sugar exporter, even though the cost of producing sugar from beet in the EU is six times higher than producing sugar from cane in Brazil. (Official EU subsidies to its sugar producers currently amount to about €1.3 billion ($1.5 billion), but there are hidden subsidies worth over €800 million in addition to that.)[104]

RICH AND POOR IN TERMS OF ENERGY AND MONEY

As true Oil Age societies proceeded towards fabulous wealth levels, while regions that kept relying on more traditional energy systems remained unchanged, the trend towards more inequality between different global areas (which had already been witnessed in the Coal Age) continued for most of the 20th century. At the beginning of the 21st century regional consumption of energy varied widely between areas of different economic output. In 2001 Americans consumed some 221,000 kcal of commercial energy per person per day on average, compared to EU citizens, 107,000 kcal per day, and Latin Americans, 35,000 kcal per day. Meanwhile people in less industrialized countries, such as India, consumed only some 11,000 kcal per day on average. In such areas noncommercial traditional fuels, including firewood, animal dung, and plant debris, still comprised a substantial percentage of the total energy consumed, especially at the residential level. In fact, over a quarter of the world's population relied wholly on traditional fuels and lacked access to commercial energy in any of its forms at the start of the 21st century. Without electricity or gasoline, scores of people had no means to refrigerate food or medicine, turn on an electric light, make a phone call, pump well water, or power a tractor. Four out of five people without electricity lived in

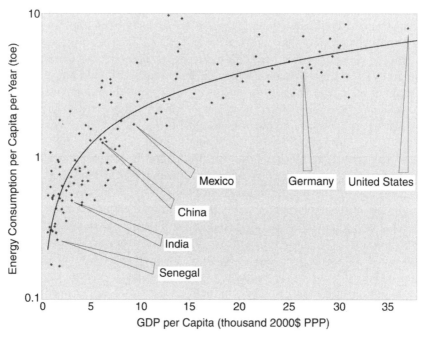

The Correlation between Energy Consumption and Wealth Creation The correlation between energy consumption and wealth creation is strong. Countries in which large parts of the population still live on the countryside tend to use less energy and create less wealth, while fully industrialized nations consume vast amounts of energy and are a lot richer. (Note the log-scale on the left axis, which improves the illustration of energy consumption differences in low-income countries, but makes the difference in energy consumption between poor and rich countries appear smaller than it is. Based on data in International Energy Agency, "Key World Energy Statistics 2007," http://www.iea.org/textbase/nppdf/free/2007/key_stats_2007.pdf.)

rural areas, mainly in South Asia and sub-Saharan Africa, where human and animal labor remained dominant. About 40 percent of the world population, or 2.4 billion people, depended on primitive biomass for cooking and heating. Many of them used much of their time collecting firewood, crop residues, or animal dung, while the richest 10 percent of the global population used over a third of the world's commercial energy. But though it remained substantial, the difference in energy consumption between the rich and the poor world actually became smaller during the second half of the 20th century.[105]

In terms of money, disparities in the wealth of people living in different regions is expressed by per capita GDPs. In 2003 per capita GDPs (at purchasing power parity, PPP) varied from $37,800 for the US to about $27,700 for Germany, France, and Britain, $5,000 for China, $2,900 for India, $1,200 for Mozambique, and $500 for Sierra Leone. Hence, the per capita GDP of

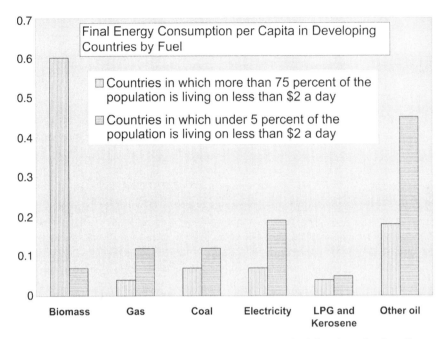

Not Fully in the Oil Age: Energy Mix in Poorer and Richer Developing Countries The energy mix in wealthier and less wealthy developing countries reflects the stage of their *energy development*. In less wealthy countries well over half of the total energy mix is based on biomass, while declining poverty is associated with increasing consumption of liquid, oil-based fuels. Just as it was in the very start of the Oil Age in Europe and North America, the first small step into the new Energy Era in current developing countries is often the use of kerosene. Back then it was used as lamp oil, while crudely refined kerosene now serves as inexpensive cooking fuel. (Based on a graph in International Energy Agency, *World Energy Outlook* 2004, www.iea.org/textbase/nppdf/free/2004/weo2004.pdf.)

the U.S. was about 76 times that of Sierra Leone, the poorest country in 2003. Only some 15 percent of the global population lived in high-income countries in 2000, generating 55 percent of the world gross product (at PPP). Some 11 percent lived in upper-middle income countries, some 34 percent in lower middle-income countries, while some 41 percent of the global population lived in low-income countries, generating merely 11 percent of the world's economic output. About 1.1 billion people were living on less than a dollar a day (at purchasing power parity). The largest share of these were living in south (432 million) and east (261 million) Asia, the second largest share in sub-Saharan Africa (323 million).[106]

Despite the Oil Age's fabulous wealth levels in many regions, scores of people do not even have access to adequate nutrition. At the beginning of the

21st century hunger and malnutrition were the number one risks to global health, killing more people than AIDS, malaria, and tuberculosis combined. Hunger often kills by increasing susceptibility to disease, and underweight contributes to about 60 percent of all child deaths in developing countries. At this stage about 840 million people (more than the combined populations of the United States, Europe, and Japan) do not have enough to eat, with hunger and malnutrition claiming some 25,000 lives every single day. About 70 percent of the hungry live in rural areas, where neither the step into the Coal Age nor the Oil Age has been fully taken. (In 2000 roughly 524 million of the hungry lived in south and east Asia, 185 million in sub-Saharan Africa, 34 million in Arab States, 53 million in Latin America, and 33 million in Central and Eastern Europe, including all of the former Soviet Union.) As with poverty, the situation in regards to hunger is worst in sub-Saharan Africa, where the largest percentage of the population lacks an adequate diet.[107] Meanwhile more than one billion adults worldwide are overweight, and at least 300 million clinically obese. Well over a third of all the grain grown on the world's fields is fed to livestock. The grain fed to livestock in the United States alone could fully nourish nearly 800 million people. Oil Age countries with about 20 percent of the world's population in 2002 produced 897 million tons (or 47 percent) of the 1.905 billion tons of cereals produced globally.

Trends

Much of the poor regions' inadequate situation in terms of both per capita GDP and food available per person is closely related to the extreme 20th century population growth that was concentrated in the developing world. The peak in global population growth in percentage terms was reached in the late 1960s (at 2.1 percent), while the peak in absolute numbers of people added per year was reached in the late 1980s (at ca. 87 million people per year.) But world cereal production grew even faster than the human population, doubling over the 30 year period from 1960 to 1990. Global cereal production per person thus increased until 1984; thereafter population growth slightly (and temporarily) outpaced cereal output growth. Roughly 1.6 billion people were added to the planet between 1980 and 2000, yet the number of chronically undernourished declined from about 920 million to 830 million. Unfortunately, this trend disguised an actual reversal during the second half of the 1990s: the number of hungry people increased by 18 million between 1995 and 2000.[108] (On the other hand, since population growth remains high, the proportion of undernourished people keeps declining, from 20 percent in 1991 to 17 percent in 2002. The United Nations in 2006 thus reported that progress continued towards the goal of halving the percentage of undernourished people between 1996 and 2015.[109])

In principle, it is not necessary for a country to be self-sufficient in terms of food supply, if only it is wealthy enough to buy grain on international markets. (Neither Japan nor the oil-exporting Persian Gulf countries grow enough food to sustain their populations.) In countries where many go hungry, population growth has typically outpaced economic growth, leaving the region impoverished. However, East Asia, followed by South Asia, began narrowing the wealth gap to Western industrialized countries from the early 1980s, and due to extreme economic growth in populous China and India, per capita GDP growth in the developing world as a whole outpaced that of the industrialized world for the first time during the 1990s. Per capita GDP in real terms more than doubled from $2,000 to $4,200 in the developing world between 1960 and 2000, while life expectancy increased from 46 to 63 years, and mortality rates for children under five years of age more than halved. Adult illiteracy nearly halved between 1975, when one of every two adults in the developing world could not read, and 2000. And even though the global population rose by 1.6 billion people between 1980 and 2000, the absolute number of people living on less than $1 per day declined by some 375 million people.

Nevertheless, the picture remained grim. The number of people living worldwide on less than $2 per day actually increased, from 2.45 billion in 1981 to 2.74 billion in 2001, and almost no positive trend was reported from sub-Saharan Africa, where population growth remained high, and economies stagnated. (Economic growth in sub-Saharan Africa was actually negative between 1981 and 2001[110], and the number of extremely poor Africans rose from 164 million [41.6 percent of the population] to 316 million [46.9 percent of the population].) Though still about two-thirds of the world's extreme poor lived in Asia at the beginning of the 21st century, poverty was more (and increasingly) concentrated in Africa. The distance between the national incomes of the world's poorest countries (now located in Africa) and the world's richest countries kept on widening as well. Per capita GDP of western European nations was less than 30 percent above that of India, China, or African countries at the start of the Coal Age, while the per capita GDP of the richest compared to the world's poorest nations was about 10:1 in 1900, more than 20:1 in 1950, and nearly 40:1 by 1973. At the end of the 20th century it was more than 60:1. (To be sure, the World Bank has also published studies that conclude that international income inequality has decreased continuously from the late 1960s to the late 1990s, even if incomes in the countries comprising the richest and poorest quintile of the world population are compared.[111])

Problems Faced by Poor Countries

Typically, countries of low per capita income have not yet fully entered the Oil Age, and hence have not experienced all the consequences of such energy transition. While rural labor in the United States had fallen to 2 percent by 1975, it still accounted for two-thirds of the labor force in Asia's three largest economies, China, India, and Indonesia, and typically above 20 percent in Latin American countries in 1990. People living in rural areas often lack access to commercial fuels, electricity, and education. Life expectancy may be under half that of the planet's wealthy regions. Lack of clean water is a major health issue, and drought is the main cause of food shortages in poor countries. (Irrigation would boost crop yields by up to 400 percent.)

Clearly, there has been a major change in the attitude of the planet's wealthy towards the planet's poor during the Oil Age. Powerful energy-rich societies generally no longer attempt to kill those who are worse off. (Industrialized Oil Age nations inflicted most deaths upon each other rather than on people living in poor countries of the developing world. Notable exceptions include Japanese atrocities in China and other parts of Asia, and those of Europeans suppressing insurrections in African colonies such as British Kenya.) What is more, the United Nation's General Assembly adopted the Declaration of Human Rights in 1948, containing a provision prohibiting slavery. (Slavery was officially abolished in Saudi Arabia in 1963 and in Mauritania in 1980.) Most significantly, the richer nations began making direct payments to poorer nations simply to support them. But even though this is a spectacular trend in comparison to the previous Energy Eras, aid spending is not large enough when viewed with mature Oil Age eyes. Just one week of subsidies given to farmers in the developed world would cover the annual cost of food aid. And the wealthy nations' military expenditures of some 900 billion dollars a year hardly look favorably when compared to their combined foreign aid spending of 50 billion dollars.

What is more, many developing countries are facing a serious debt problem. Some are owing amounts twice as high as their yearly national income, and in various countries this debt problem was not entirely home-made. It has its roots in the oil price surge of 1973, which provided record earnings for oil exporters who deposited their money in commercial banks in the industrialized world. As the high oil price slowed down economic growth in the industrialized world, banks turned to the developing world, where they peddled loans at bargain prices. These loans made the borrowers vulnerable, as they were due in relatively short term. And since the high oil price caused inflation, the Western governments increased interest rates. Developing countries therefore had to renew loans at high interest rates, while export earnings to repay debt failed to rise correspondingly, often because commodity prices decreased. (Between 1975 and 1992 the price of minerals and food commodities kept on falling in real terms; only the price of timber increased.) In turn, African countries, especially, had problems getting out of their debt trap. (Thus, the West decided at the beginning of the 21st century to write off billions of dollars owed by African countries.)

To be sure, not all problems faced by the world's poor nations are imposed on them from the outside. Many are simply the product of poor governance and false allocation of resources. Poor countries have purchased arms (from the West, the Soviet Union, and China, for instance) while their populations were starving. China is the largest receiver of foreign aid and yet spends huge amounts on developing arms systems. (China even put a man into space in 2003.) Corruption also contributes to the false allocation of resources. Leaders (and officials at all levels) in the developing world often escape the kind of scrutiny that governments in mature democracies are subjected to. Between 26 billion and 130 billion U.S. (2003) dollars in World Bank credits have been diverted or alienated in the second half of the twentieth century, according to different estimates. Such diversions cost economic growth and human lives, while various rulers of poor countries made it on the list of the world's most wealthy people.[112]

NOTES

97. Vaclav Smil, *Essays in World History: Energy in World History* (Boulder, CO: Westview Press, 1994), 205.

98. Ibid., 206.

99. Energy intensity is now falling by about 1.5 percent a year. The price of energy is a main factor: America's energy intensity was falling by 0.4 percent until the oil shock of 1973. It is now falling by 2% a year. "Energy Use: Less Intense," *The Economist*, May 8, 2008, http://www.economist.com/daily/chartgallery/displaystory.cfm?story_id=11332762; "Energy Efficiency: The Elusive Negawatt," May 8, 2008, *The Economist*, http://www.economist.com/displaystory.cfm?story_id=11326549.

100. Paul Bairoch, *Economics and World History. Myths and Paradoxes* (Chicago: University Of Chicago Press, 1993).

101. Daniel Cohen, *Fehldiagnose Globalisierung* (Frankfurt/New York: Campus Verlag, 1999).

102. The following paper, however, argues that global inequality (among citizens of the world) has declined, modestly, since the 1980s, reversing a 200-year-old trend toward higher inequality: David Dollar, "Globalization, Poverty, and Inequality since 1980," *World Bank Study* 3333 (2004), http://econ.worldbank.org/external/default/main?pagePK=64165259&piPK=64165421&menuPK=64166093&theSitePK=469372&entityID=000112742_20040928090739.

The following paper comes to a similar conclusion. It investigates the view that globalization creates winners and losers, and thus leads to greater inequality, as supported by some reports published by, among others, the UNDP (United Nations Development Programme). The study concludes that global inequality between countries has decreased during the period from the 1960s until 1998 in terms of income gaps and some indicators of living standards.

Arne Melchior, Kjetil Telle and Henrik Wiig, "Globalisation and Inequality: World Income Distribution and Living Standards, 1960–1998," Norwegian Institute of International Affairs (NUPI), Royal Norwegian Ministry of Foreign Affairs, Studies on Foreign Policy Issues, Report 6B: 2000, October 2000, http://www.nupi.no/pub likasjoner/boeker_rapporter/2004_2000/globalisation_and_inequality_world_in come_distribution_and_living_standards_1960_1998.

103. UNCTAD (United Nations Conference on Trade and Development), "The Least Developed Countries Report, 2004," http://www.unctad.org/Templates/WebFlyer.asp?intItemID=3074&lang=1.

104. "Oh, Sweet Reason: A Report Counts the Cost of Europe's Sugar Subsidies on Poor Countries," *The Economist*, April 15, 2004, http://www.economist.com/display-story.cfm?story_id=2599003.

105. Find more on the link between energy use and poverty in International Energy Agency, *World Energy Outlook 2002*, "Energy and Poverty" (2002), 365, http://www.iea.org/textbase/nppdf/free/2000/weo2002.pdf.

106. The World Bank in 2001 classified gross national incomes per capita as follows. Low-income: $745 or less, lower-middle-income: $746 to $2,975, upper-middle: $2,976 to $9,205, and high-income $9,206 and above per capita, http://www.worldbank.org. A better measure for wealth or poverty is the Human Development Index (HDI). It focuses on three measurable dimensions of human development: living a long and healthy life, being educated, and having a decent standard of living. Thus it combines measures of life expectancy, school enrolment, literacy, and income to allow a broader view of a country's development than does income alone. "Human Development Report 2004-Cultural Liberty in Today's Diverse World," United Nations Development Programme, 2004, http://hdr.undp.org/reports/global/2004/, http://hdr.undp.org/en/media/hdr04_complete.pdf.

107. Food and Agriculture Organization (FAO) of the United Nations, "Undernourishment Around the World: Depth of Hunger—How Hungry Are the Hungry?," The State of Food Insecurity in the World 2000, http://www.fao.org/docrep/x8200e/x8200e03.htm.

108. Food and Agriculture Organization (FAO) of the United Nations, "Undernourishment Around The World; Counting the Hungry: Latest Estimates," The State of Food Insecurity in the World 2003, FAO Corporate Document Repository, http://www.fao.org/docrep/006/j0083e/j0083e03.htm.

109. Food and Agriculture Organization (FAO) of the United Nations, "The State of Food Insecurity in the World 2006, Eradicating World Hunger—Taking Stock Ten Years After the World Food Summit," FAO Corporate Document Repository, http://www.fao.org/docrep/009/a0750e/a0750e00.htm.

110. To be sure, the All-of-Africa GDP from 2002 for several years grew at five percent per year, that is, faster than World GDP. Seven out of the world's 20 fastest growing economies were African in recent years, but much was due to rising commodity prices. Nigeria had enormous oil revenues, for instance. The Africa-7, South Africa, Egypt, Nigeria, Kenya, Botswana, Ghana, and Morocco, attracted substantial foreign investments, much of it from China, which is eager to secure access to minerals. However, financial services, telecommunications, and the production of consumer goods also contributed to the economic growth.

111. David Dollar, "Globalization, Poverty, and Inequality since 1980," *World Bank Study* 3333 (2004), http://econ.worldbank.org/external/default/main?pagePK=6 4165259&piPK=64165421&menuPK=64166093&theSitePK=469372&entityID=

000112742_20040928090739; The World Bank is a specialized agency within the United Nations system, http://www.worldbank.org/.

112. Corruption is a common problem in various developing countries, but becomes especially obvious in low-income countries that are rich in resources. Venezuela had incredible oil revenues. When the oil price rose during the 1970s, savings accounts were created, which were soon directed to private accounts by people close to the government. Moreover, ambitious industrial projects were initiated without paying attention to efficiency criteria. They were approved to maximize bribes and commissions. Unnecessary aluminum and steel factories were built with government subsidies. Per capita GDP in 1990 was lower than in 1970, and productivity lower than 20 years before. Similarly, in oil-rich Nigeria, oil money was used to build the new capital Abuja from scratch, while urgent problems on the countryside remained unsolved. The pattern is often the same. Nonsense projects are approved in order to channel public funds into private pockets.

However, sudden natural-resource booms may curiously hinder economic growth for reasons other than corruption. This is known as the Dutch disease, because it was first observed in the Netherlands, where the development of natural gas deposits in the 1960s caused the manufacturing industry to shrink. After the oil price surge of 1973, oil-rich Venezuela, Ecuador, and Nigeria became rich overnight, but their agriculture and manufacturing industry languished as oil revenues pushed up the real exchange rate, rendering most other exports uncompetitive. By the end of the 1970s, Nigeria's non-resource economy was 29 percent smaller than it would have been without the oil boom. Daniel Cohen, *Fehldiagnose Globalisierung* (Frankfurt/New York: Campus Verlag, 1999), 30–31; Jonas E. Okeagu, et al., "The Environmental and Social Impact of Petroleum and Natural Gas Exploitation in Nigeria, *Journal of Third World Studies* (Spring 2006), http://findarticles.com/p/articles/mi_qa3821/is_200604/ai_n17179247/pg_1?tag=artBody;col1.

CHAPTER 34

POPULATION DEVELOPMENT

Many of the problems faced during the Oil Age in less industrialized countries were rooted in, or aggravated by, rapid population growth. When the global human population reached one billion (1,000 million) in the early 19th century, this was considered an enormous number. But a bit over a century later, in 1930, another billion people had been added to the planet. And thereafter a true population explosion set in, as the global population grew to three billion by 1960, four billion by 1974, five billion by 1986, and six billion by 1999. With other words, the time span to add an additional billion people to the planet decreased from, say, 100,000 years to a century, to 30 years, to 14 years, and to 12 years (and then slightly increased to 13 years). At the beginning of the 21st century, the human population kept on growing, adding an astounding 76 million people to the planet per year.[113]

BETTER HEALTH, MORE FOOD

Most of the enormous population growth during the Oil Age occurred in developing countries, where some 80 percent of all people are now living. It was mainly the result of decreasing death rates with a delay of a corresponding decrease in birth rates. Death rates decreased because health-related knowledge and technology that accumulated in Coal Age and Oil Age societies was exported to the rest of the world. Most importantly vaccines eliminated the threat of such diseases as smallpox, measles, whooping cough, neonatal tetanus, polio, and diphtheria. In addition, drugs to cure various diseases including tuberculosis and malaria were introduced, antibiotics healed infections, and

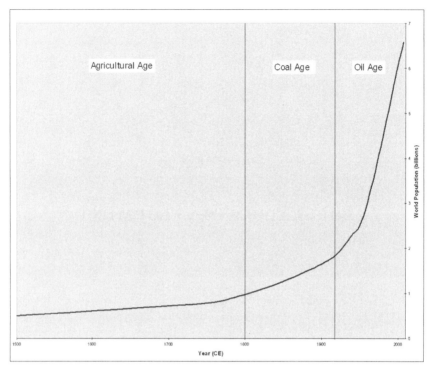

Global Population Development The graph shows the number of humans living on the planet at different times. The energy-rich Oil Age witnessed unprecedented population growth in both percentage and absolute terms. High energy inputs in agriculture sustained this population growth, which was concentrated in developing countries, while true Oil Age societies maintained the size of their populations. (Based on data from the Population Reference Bureau, "FAQ: World Population," http://www.prb.org/Journalists/FAQ/WorldPopulation.aspx.)

synthetic pesticides (such as DDT) were applied to control insect populations that spread infections such as malaria, yellow fever, typhus, and elephantiasis. DDT (dichlorodiphenyltrichloroethane), isolated in 1874 in Germany and recognized for its potency as nerve poison on insects by Swiss chemist Paul Müller in 1939, reduced malaria from 75 million cases to fewer than 5 million cases in India within a decade. (Though it was shown from the early 1960s that DDT caused severe environmental problems, several countries decided not to ban its use due to its major potency in disease control.) What is more, public health measures, including safer water supplies and ways of segregating and treating human wastes, were introduced in more and more countries to help death rates to decline in Asia, Africa, and Latin America.

The consequent population growth was supported by the achievements of the Green Revolution that kept agricultural output growth well ahead of population growth. The global population never grew faster than 2.1 percent per year, a peak that was reached between 1965 and 1970, while grain production grew at an annual rate of 2.9 percent between the mid-1960s and the mid-1980s. Hence, grain production per person was climbing from 251 kilograms in 1950 to 344 kilograms in 1984 despite the rapid population growth. This increased food supply contributed to people's better health as well.

While life expectancy at birth had been below 45 years almost everywhere in the world in 1900, the world average life expectancy at birth was some 65 years at the beginning of the 21st century. However, the global average disguises large disparities. In fact, life expectancy in the least industrialized countries was still below 50 years, while average men could expect to live about 76 years, and women 80 years, in the industrialized world. Sorted by region, life expectancy at birth in 2002 was about 76 years in North America, 73 years in Europe (as well as Australia and New Zealand), 69 years in Latin America and the Caribbean, 66 years in Asia, and 50 years in Africa. The European average includes eastern European countries where life expectancy is relatively low, at about 68 years. Similarly, life expectancy in East Asia, where it is about 70 years, is higher than the total Asian average. Meanwhile, life expectancy at birth was still below 45 years in several sub-Saharan African countries at the beginning of the 21st century.

Global Fertility Transition

Global population growth eventually slowed despite the continuing increase in life expectancy, because people decided to have fewer children. The world population grew at about 0.9 percent per year on average from 1900 to 1950, 2.0 percent on average per year from 1950 to 1975, and 1.6 percent on average per year between 1975 and 2000. Though the peak in percentage terms was reached in the years between 1965 and 1970 (at 2.1 percent per year), the absolute number of people added per calendar year peaked between 1985 and 1990 (at about 87 million people.) The most critical factor to slow the growth of a healthy population is whether women on average give birth to more, or less, than two children that survive to sexual reproduction age: two children are sufficient to replace both parents, less than two will cause population decline, more than two population growth.[114]

Though with a delay, the world's less industrialized regions witnessed a remarkable reduction of fertility levels, with average total fertility falling from 6 to 3 children per woman within just 50 years. Many countries seem to have gone the same path towards reduced fertility. Before the transition, both the long-term average birth and death rate were high, resulting in a population growth around zero. Next, death rates fell, while birth rates remained high or even increased: this resulted in higher population growth. Next, in a period referred to as fertility transition, birth rates fell as well, while death rates remained low, and population growth slowed. Thereafter, death and birth rates were both low and nearly equal. Growth rates were then very low or even negative, but the population size had become much larger than it was before the transition began.

As this sort of transition has been experienced in Western Oil Age societies, it was assumed that it is the result of an energy-rich lifestyle, with industrialization, economic growth, urbanization, increased access to education, and so forth. However, a fertility transition may also be caused by factors disconnected from energy consumption and economic development. Global population growth in percentage terms came off its peak rate in the early 1970s, as birth rates began to fall in such populous countries as China, India, Brazil, Egypt, Indonesia, South Korea, Mexico, and Thailand. (In addition, the baby boom, the high birth rates in prosperous Western countries after World War II, came to an end after the 1960s.) China's decreasing fertility rate contributed much to the slowdown of world population growth, as it dropped radically from an average of 6.5 children per woman in 1968 (total population: ca. 800 million), to about 2.2 children per woman in 1980 (total population: ca. 980 million). It then decreased further to 1.8 children per woman in 1999 (total population: ca. 1.24 billion). China first prohibited youthful marriage and then, in 1979, introduced a one-child policy (applying to ethnic Han Chinese living in urban areas) that would have been impossible to implement in any democracy. Fines, pressures to abort pregnancies, and even forced sterilization, accompanied second or subsequent pregnancies. In 1999 China finally decided to weaken its population control regulations, after the rule had reduced population growth in China by perhaps as much as 300 million people during just twenty years.

During the 1980s and 1990s, global population growth did not sink as much as it had during the 1970s. For one thing, India backed off from family-planning programs viewed as too vigorous. (India's population grew from 357 million in 1951 to one billion in 2000 and is set to pass that of China to become the largest in the world.) More generally, many countries had simply completed their fertility decline, and many of the numerous children born during the 1960s (when infant and childhood mortality rapidly fell in poor countries while rich countries experienced the baby boom) entered their childbearing years in the 1980s, when the average annual global population growth rate remained at 1.8 percent.

In Africa population growth and fertility remained high throughout the 1990s. No more than 220 million people lived in Africa in 1950, but it was 800 million in 2000. Average fertility was still 5.3 children per woman in the late 1990s, and the African population grew at 2.6 percent annually, that is, faster than the global population as a whole had ever grown. Fertility also remained high in the Middle East. In the 1990s, the average completed family size in Muslim countries was about six children. Arab fertility rates were twice the world's average, and infant death rates far higher than those in Western countries. Even in wealthy oil-exporter Saudi Arabia mortality rate among babies up to one year old was 8.3 percent, compared to ca. 0.9 percent in western Europe.

THE GLOBAL POPULATION AT THE BEGINNING OF THE 21ST CENTURY

Rapid Oil Age population growth in developing countries radically altered the make-up of the global human population. The share of Europeans decreased from 25 percent of the global population in 1900 to 12 percent in 1999, while Latin America doubled its share to 9 percent, and Africa increased its share from 8 percent to 13 percent. Asia slightly increased its share to 61 percent, while that of the United States/Canada remained stable at about 5 percent of the world population. In 2001, the global population was about 6.1 billion, growing at about 1.21 percent annually, adding about 75 million people to the planet per year. (A population growing at a constant 1.21 percent per year will double within 57 years.) This growth rate came about through an average birth rate of 2.09 percent (20.9 births per 1,000 people per year) and an average death rate of 0.88 percent (8.8 deaths per 1,000 people per year). The global average fertility was 2.82 children per woman. Less developed regions (that is all regions but Europe, North America, Australia, New Zealand, and Japan) accounted for 80 percent of the world's inhabitants and about 95 percent of the 2001 population growth. Just six countries accounted for half of the world's annual percentage growth: India 21 percent; China 12 percent; Pakistan 5 percent; Nigeria 4 percent; Bangladesh 4 percent, and Indonesia 3 percent.

Most rich, industrialized Oil Age countries, where less than 20 percent of the entire world population lived, experienced zero or slightly positive or negative population growth at the beginning of the 21st century, because survival rates were high while many parents decided to have less than two children. Meanwhile average fertility in the developing world as a whole was about three children per woman. However, there were still over 20 African countries that had total fertility levels of 6 children per woman. Birth rates in these countries remained high, as death rates remained high. Worldwide, more people are killed by lung cancer (together with trachea and bronchus cancer) than by malaria, but 90 percent of malaria cases occur in tropical Africa, and most victims are children. In addition, much of sub-Saharan Africa is suffering from the impact of the human immunodeficiency virus (HIV), the cause of acquired immunodeficiency syndrome (AIDS). AIDS killed 20 million people between 1981 and 2002, 70 percent of whom were Africans. The virus may have emerged, but initially remained contained, in rural communities in central Africa in the 1930s. In the early 1980s it reached epidemic proportions, with Uganda and Rwanda as the epicenter of occurrence, and then spread southward, with the highest rates of infection occurring in Botswana and South Africa, where HIV/AIDS has become the prime cause of death.[115]

NOTES

113. Joel E. Cohen, *How Many People Can the Earth Support?* (New York: W.W. Norton & Company, 1995). (This is a principal source used for historical population issues and figures.); "Population Growth," Population Reference Bureau (PRB), http://www.prb.org/Educators/TeachersGuPopulation Reference Bureau (PRB), ides/HumanPopulation/PopulationGrowth.aspx.

114. Joel E. Cohen, *How Many People Can the Earth Support?*; Paul R. Ehrlich, Anne H. Ehrlich, *The Population Explosion* (New York: Touchstone, 1991); "The World at Six Billion," United Nations, http://www.un.org/esa/population/publications/sixbillion/sixbilpart1.pdf.

115. "AIDS Epidemic Update," UNAIDS/WHO, Joint United Nations Programme on HIV/AIDS (UNAIDS), World Health Organization (Geneva: WHO, 2006), http://data.unaids.org/pub/EpiReport/2006/2006_EpiUpdate_en.pdf; Joint United Nations Programme on HIV/AIDS, http://www.unaids.org/; Sabin Russell, "Turning Point for HIV Infection Rate in Africa," *SF Chronicle*, November 21, 2006, http://sfgate.com/cgi-bin/article.cgi?f=/c/a/2006/11/21/BAGLTMHB0C12.DTL&type=health.

CHAPTER 35

THE OIL-POWERED EMPIRES

Oil Age societies commanded vast amounts of energy that could be used to control regions with less energy command. But fortunately, one major Coal Age trend by and large continued: those who commanded more energy no longer expanded by killing people outright in other regions (as agrarians did). Oil Age Powers rather dominated less technologically advanced regions by dictating international rules (of trade, for instance) and threatening trade embargos or military intervention. Even the right of national independence of tightly controlled overseas regions was eventually acknowledged, and the empires built up during the Coal Age were formally dissolved. Western Europe could accept this, as it was well on its way to achieve self-sufficiency in terms of food supply (and thus no longer depended on grain imports from colonial regions), and one of the Coal Age's main reasons for empire-building, access to markets, could now be achieved without formally ruling over foreign economies. Besides, the energy-rich countries no longer sold so much to less developed regions, but had become each other's main customers.

The one problem that remained, and led to plenty of wars, was access to oil, the Oil Age's principal fuel and key to wealth and military dominance. Oil is not nearly as widely distributed over the globe as coal is, and most notably, the western European Coal Age Powers had practically no domestic oil resources at all. Austria and Germany, where much oil technology emerged, had access to some Galician and Romanian oil in their near neighborhood, but France and Britain were far away from any known oil reserves. (North Sea oil was discovered only in 1969 and produced from 1975.) Curiously, the British could not even find oil anywhere in their vast global Coal Age empire with the exception of the Burmese part of British India. Nevertheless, the British decided shortly before World War I to convert their navy from coal to oil. This implied that Britain had to rely on oil from Persia, which was not a British colony, and for the remainder of the Oil Age Western international policy had much focus on securing access to foreign oil.

Conflicts generally became more international during the Oil Age, simply because mobility had increased and weapons systems had become more far-reaching. (Aircraft carriers and long-range bombers were eventually followed by nuclear-powered submarines and intercontinental missiles carrying nuclear warheads). The progression of World War II clearly reflected the amount of oil available to the various combatant nations. Germany lost the war after failing to reach the Azerbaijani oil fields by a hairbreadth. Japan entered the war to gain access to the (Dutch) oil fields of Indonesia. and America won World War II after pouring an astounding six billion barrels of oil into it. (For comparison, the United States produced 2.6 billion barrels per year in 2001.) This saved Britain and France, but all western European nations emerged seriously weakened from the war: they had now definitely lost the powerful international status they had carried over from the Coal Age. (Arguably, Britain ceased being a superpower on August 15, 1947, when 400 million people, the inhabitants of British India, were released from the British Empire.) The Soviet Union successfully defended Azerbaijan and developed its vast West Siberian oil fields at the end of World War II. Consequently, the world's principal oil producers, the Soviet Union and the United States of America, emerged as the world's two principal superpowers. (Up until 1995 the United States produced 171 billion barrels of oil cumulatively, and the (former) Soviet Union 125 billion. Saudi Arabia was a distant third at 74 billion barrels, mainly produced for export, and no other nation had even reached 50 billion barrels.) But even though they had been fighting on the same side during World War II, the Soviet Union and the United States opposed one another immediately after the war and created two power blocks that divided the world during the second half of the 20th century.

However, both these superpowers had problems. The Soviet Union showed little internal cohesion, as it was a non-democratic, multinational state practicing an inefficient economic system imposed by a totalitarian central government. Hence, the Soviet Union disintegrated in 1990 despite then being the world's largest oil producer. (It then reemerged as a major power in the form of the Russian Federation.) The United States, too, had to learn that even oil is not a guarantee to success, when it poured as much as five billion barrels of oil into the Vietnam War only to lose it. But more importantly, the United States became a net oil importer shortly after World War II (in 1948) and saw its domestic oil production peak in 1970. (That year confidence in the U.S. dollar deteriorated, and funds moved to the financial centers of Europe and Japan. Hence, the dollar-based Bretton Woods exchange rate system was abandoned the year after, in 1971.) Much of U.S. international policy and warfare after World War II was thus focused on securing access to international oil, and especially Middle Eastern (Persian Gulf) oil, as it is the world's most plentiful. This policy slowed the international spread of democracy, as it involved installing and maintaining repressive dictatorships, which worked well in some countries, but led to violent regime changes in others. New regimes then promoted anti-Americanism, which in Arab countries was also fueled by U.S. support for Israel. However, the United States was quite invincible, maintaining its position as the world's chief economic and military superpower, especially after the downfall of the Soviet Union. On the other hand, the United States imported over half of its oil by the early 21st century, and it became questionable if its military (and economic) might would be sufficient to keep controlling those foreign regions that had enough oil to export it. The situation was bound to get more difficult, as the world's remaining oil reserves were increasingly concentrated in the Middle East, and because competition for oil was getting tougher due to rapid industrial development in some of Asia's populous regions. (Notably, China became a net oil importer in 1993.)

WORLD WAR I (1914–1918)

World War I was fought at the interface between the Coal Age and the Oil Age. The outbreak of the war was triggered by the June 1914 assassination of Austrian Archduke Franz Ferdinand, who was shot in Bosnia-Herzegovina by a Serbian nationalist while being chauffeured in a gasoline-powered automobile. However, the armies of the day had very few motor vehicles and based nearly all their on-land transportation on steam locomotives and horses. By the beginning of the war tanks had not yet been invented, and motor-powered flight was barely a decade old. The small number of existing aircraft were mainly used for reconnaissance, and the fronts turned static in trenches, with all sides employing heavy machine guns that rendered the opponent's cavalry useless.

Nevertheless oil had already become a strategic resource at the beginning of the war, because most Western navies had experimented with oil-powered turbines for ship propulsion. Germany had built battleships that could flexibly run on either coal or oil, and Britain had decided in 1912 to base much of its navy entirely on oil. But the strategic importance of oil was going to reach entirely new levels as the war progressed. The number of conventional motor vehicles used by the armies grew exponentially, and from 1916 Britain, Germany, and France even introduced track-laying tanks. These had a relatively minor impact as they were not yet sufficiently developed, but their numbers by the end of the war were quite impressive: Britain had produced some 2,636 tanks, France some 3,870 (mainly light) tanks, and Germany, reluctantly, 20 tanks. The production of aircraft soared in a similar manner. France had less than 140 airplanes in 1914, but produced some 68,000 during the war, of which well over 50,000 were lost. (By the end of the war in 1918 France had 4,500 functioning aircraft, more than any other combatant nation.) Britain, too, initially had just 154 airplanes in all of its military forces, but cranked out aircraft at a rate equivalent to 30,000 units per year by mid-1918. From 1915 newly developed light machine guns were mounted onto planes, and France, Germany, Britain, Russia, and Italy all constructed special bomber airplanes in an attempt to target supply lines, factories, and cities. Yet, even after a total of 27 raids by the infamous German *Gotha G.V.* bombers the English reported just 835 killed, an extremely small number compared to the bomb victims of World War II. Neither tanks nor airplanes were truly decisive in World War I, which was still mostly coal-powered. Nevertheless, the relatively small amounts of oil energy applied through gasoline-fueled aircrafts and diesel-powered submarines had a disproportionate impact, especially towards the end of the war. France's war-time prime minister Georges Clemenceau is quoted as saying at the start of the war that if he wanted oil, he would find it at his grocer's, while he said in 1917 that every drop of oil is worth a drop of blood.

The Troubled Balkans

It took Austria some three weeks to figure out how to react to the murder of Franz Ferdinand, heir to the throne of the Austrian-Hungarian Monarchy, Europe's largest empire except for Russia. The Russians had expanded significantly into the Balkans at the expense of the Ottoman Empire during the Coal Age, and successfully fought the last in a series of Russo-Turkish Wars in 1875 following an anti-Ottoman uprising in Bosnia and Herzegovina. The Congress of Berlin (1878) recognized Serbia, Montenegro, and Romania as independent states, while Bosnia-Herzegovina was assigned to Austria for administrative and military occupation, though these two provinces remained officially associated with the Ottoman Empire. Thirty years later, in 1908, the Young Turks staged a revolution in the Ottoman capital of Constantinople, and Austria decided to annex Bosnia-Herzegovina formally to avoid the new Turkish regime's attempt to regain control over these provinces. Russia formally approved of this annexation (in return for the promise of Austrian support for the opening of the Dardanelles to Russian warships), but the Russians also promoted Pan-Slavism, a movement intended to elevate the political or cultural unity of all Slavs, including those living in the Balkan Peninsula and Austria-Hungary. Russia's (fellow Greek Orthodox) ally Serbia was outraged. The Serbs wanted to expel Austria-Hungary from the region to create an all-Slav state. In 1912–13 an alliance of Serbia, Bulgaria, Montenegro, and Greece, mediated by Russia, fought a short war against the Ottoman Empire, stripping it of all its European possessions except the area adjacent to the European side of Constantinople. During this war Serbia gained access to the Adriatic Sea, which both Austria and Italy would not accept, fearing that a Serbian port would ultimately become a Russian port. Hence, the Great Powers in 1913 created an independent Albania of fair size, cutting Serbia off from the sea. In turn the Serbian government stepped up its anti-Austrian propaganda in the region, and allegedly financed and equipped the terrorist organization behind the murder of Franz Ferdinand and his wife. In the aftermath of the assassination Austria thus decided to send a list of demands to Belgrade, asking the Serbian government to arrest the terrorists behind the plot, to prevent illicit trafficking in arms and explosives into Bosnia, and to officially condemn anti-Austrian propaganda and remove it from public education. Serbia made a conciliatory reply, but in its entirety this ultimatum was unacceptable for the Serbs. On 29 July 1918 Austria declared war and attacked Serbia.

The Alliances

This regional conflict turned into a world war through a series of alliances. Prior to declining the Austrian ultimatum, the Serbian government contacted its ally Russia, which immediately called up its reserves and promised support. Russia, however, was being backed by the Triple Entente, a quasi-alliance between Britain, France, and Russia. Austria-Hungary, on the other hand, was in an alliance with Germany and Italy (and Romania). Germany shared a long border with the Russian Empire, and there had been substantial tensions between the two countries, but one of the major reasons behind World War I was the growing rivalry between Germany and Britain.

Up until 1871 Germany had not even been unified as a nation; the region consisted of a cluster of rather independent states. As late as 1848 these German states had elected Austrian Archduke Johann their Imperial Administrator, but when Austria insisted on bringing into the future Reich its entire realm, encompassing more than a dozen different peoples, the smaller Germany concept won the day, and Germany unified under Prussian leadership alongside the huge multilingual Austro-Hungarian monarchy. Within few decades the united Germany became a powerful industrialized nation with imperialistic overseas ambitions, while Britain, the principal Coal Age Power, rapidly lost its technological supremacy to Germany and the United States. Hence Britain felt threatened by this newly emerged power and was especially unhappy about Germany building up a battle fleet capable of challenging British naval preeminence. By 1914 the German navy was the second largest in the world, consisting of 17 dreadnoughts, 5 battle cruisers, 25 cruisers and 20 battleships (of pre-dreadnought design). Germany also had 30 petrol-powered and 10 diesel-powered submarines (and another 17 under construction).

When Russia declared that it would act to protect Serbian sovereignty and mobilized its troops, Germany viewed this as a declaration of war, because Germany had stated it would back Austria in the conflict. But Germany was highly vulnerable: It was situated right between Russia and Russia's ally France. Thus the Germans decided to declare war on both of these nations, attempting to neutralize France rapidly to free troops for the eastern front. In order to bypass the French defenses and quickly reach Paris from the north, the Germans took their troops through neutral Belgium. Britain, besides being in the Entente with France, was committed to defend Belgium (and Channel security) and declared war on Germany. Japan, an ally of Britain, declared war on Germany as well, while the Ottoman Empire (still consisting of Turkey and the entire Fertile Crescent), together with Bulgaria, entered the war on the side of the Central Powers (Germany and Austria-Hungary).

Italy was part of the Triple Alliance, a treaty by which Germany, Austria-Hungary, and Italy pledged to support each other militarily. But Italy first, in 1914, declared a policy of neutrality, and then, in May 1915, declared war on the Central Powers, after Britain had secretly promised Italy large territorial gains at Austria's expense. Similarly, Romania declared war on Austria-Hungary in August 1916, even though it had attached itself to the Triple Alliance. Romania, too, was hoping for territorial gains at Austria-Hungary's expense, but the immediate effect of this move was that the Central Powers had an excuse to occupy Romania rapidly and get access to the Ploesti (Ploiesti) oil fields. This was a welcome relief, or rather substitute, for oil from Austrian Galicia (in present-day Poland), where production had peaked in 1909 at 15 million barrels per year but was now in steep decline.

The War's Progression

Germany's attempt to immobilize France quickly failed, as the advance came to a halt 80 kilometers outside of Paris. In turn trenches and barbed-wire fences, separated by no-man's-land, were constructed on both sides of a front that stretched from Belgium all the way south to neutral Switzerland. Attempts to change this frontline cost many lives as the trenches were defended by heavy machine guns. (Neither light, portable machine guns nor tanks had yet entered the scene.) When the British started an offensive at the Somme, a river in northern France, they suffered over 20,000 casualties in single day. Altogether, the battle of the Somme cost the lives of 615,000 French/British and about 500,000 German soldiers. Similarly, more than 300,000 French and nearly as many German soldiers lost their life during the battle of Verdun that went on for several months. And on the Eastern front, which stretched from the Baltic Sea to the Black Sea, the death toll was enormous as well. When the Russians invaded East Prussia they were defeated by German and

Austrian troops in the battle of Tannenberg, which killed 30,000 Russians and saw another 90,000 being captured. The British attempt to get access to the Black Sea and link up with Russia ended in disaster as well: An estimated 36,000 Commonwealth troops, mainly from Australia and New Zealand, died in an unsuccessful nine-months campaign after Winston Churchill had instigated their landing at Gallipoli in Turkey in 1915.

In 1917 Russia surrendered (to the Central Powers) and immediately withdrew its troops from the Ukraine and southern Russia. Russia's army had suffered enormous losses and the population was ailing under the hardships of the war and famines. Russia's economy was ruined, a revolution broke out, and the Tsar was forced to abdicate. For Germany and Austria-Hungary this meant that their forces could now be concentrated on the western front. On the other hand, the United States entered the war on the side of the Entente in April 1917. The United States had initially declared itself neutral, but provided food, oil, and other supplies for Britain and France. (American spruce was critically important for French and British aircraft construction, for instance.) German vessels were cut off from the Atlantic by the British sea blockade, but German submarines passed through, or rather under, it and targeted British supply vessels arriving from Canada and the United States. These submarines had a policy of warning: when they found and targeted ships at sea, they allowed them to evacuate their passengers before they sank them. However, the only alert for the *Lusitania*, a liner under British registration carrying munitions for Britain and France, was a warning placed by Germany in international morning papers the day the vessel left from New York. Just as this warning said would happen, a German U-boat sank the *Lusitania* (off the Irish coast in May 1915).[116] Nevertheless the United States was outraged, because 128 Americans were among the 1,198 people who lost their lives in the incident. Germany then agreed to limit its submarine campaign against Britain, but in 1917 renewed unrestricted submarine warfare. In the light of pro-British propaganda in America, Germany was convinced that the United States would eventually fully enter the war, and attempted to keep it busy by promoting a North American conflict. Germany thus promised Mexico help in regaining the territories of Texas, New Mexico and Arizona, which the United States had conquered 60 years earlier. A telegram from the German foreign minister was intercepted, deciphered, and published in American newspapers. A month later, in April 1917, the United States formally declared war on Germany. This was important moral support for France and Britain, even though the first U.S. troops did not arrive in Europe until 1918. (In a November 1917 field trial, a U.S. battalion spent 10 days with a French division, which cost three American soldiers their lives.) Most importantly, the United States provided almost unlimited supplies, and from 1917 the U.S. oil companies cooperated with the Fuel Administration to deliver fuel to Europe.

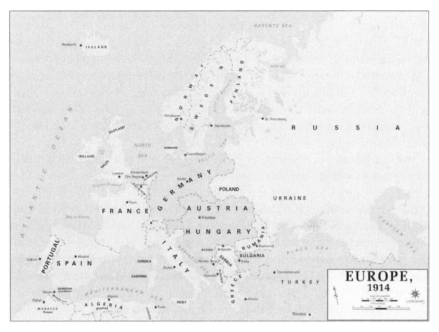

Europe before and after World War I The political map of Europe was a lot more fragmented after World War I. The Central Powers made substantial land gains during the conflict, especially after Russia surrendered in 1917. They obtained access to the Romanian Ploesti (Ploieşti) oil fields after Romania had declared war on Austria-Hungary in August 1916, but a lack of castor oil (for which a tropical plant was the source) eventually grounded their airplanes. After the war the Austrian-Hungarian monarchy was dissolved. Germany, too, faced substantial territorial losses, even though the Germans had not lost a single decisive battle during the war. (Maps published by the Department of History, United States Military Academy.)

The Central Powers, on the other hand, were starved for resources. One specific problem was lack of castor oil (Ricinus oil), which is obtained from the seeds of Ricinus communis, a tropical plant. Castor oil is still now a constituent of brake and hydraulic fluids, and was needed because the petroleum-based lubricants of the day tended to break down under the temperatures produced by high-performance aircraft engines: lack of castor oil in 1918 grounded German airplanes. But the Central Powers had more problems. The Ottoman Empire started to collapse under the continuous attacks. British forces, though they suffered heavy losses, finally captured Baghdad in March 1917, and Jerusalem in December 1917. Following the Russian withdrawal from the war, Britain also dispatched a small force to Baku to either destroy the Azerbaijani oil installations or to train local militias to defend them. This expedition failed, but by the time the Ottomans captured Baku, the war was just about over. The Austrian monarchy increasingly felt the internal pressure of different ethnic groups striving for national independence, and in Germany years of blockade had caused permanent shortage in food and fuel, while the population was suffering from the horrific influenza epidemic that killed over 400,000 civilians in the second half of 1918. The German military found it impossible to replace the sick and dying soldiers, the people and the army revolted, and Wilhelm II abdicated. In November 1918 the Germans concluded an armistice which ended the war that, according to the most conservative estimates, had witnessed some 8.6 million deaths, 3.5 million on the side of the Central Powers, and 5.1 million on the side of their opponents.[117] Another 21 million or so had been wounded, and over 7 million were missing or imprisoned. Toxic gas had been used, and bombs had been dropped from airplanes. Nearly 800,000 people had starved to death in Germany due to the sea blockade. The British Army had enlisted about 250,000 underage boys, of whom perhaps half were killed or wounded. In the Ottoman Empire, the Turks killed perhaps a million Armenians in an attempted genocide. (The approximately two million Christian Armenians living in the Ottoman Empire had begun agitating for territorial autonomy in 1894. Some 50,000 of them were then killed, and Russia could easily recruit Armenian troops to fight against the Ottomans during World War I. The Young Turk government then deported 1.75 million Armenians south to Syria and Mesopotamia, and at least 600,000 Armenians were either massacred or died of starvation during this operation.)

Shrinkage and Guilt

The First World War was not entirely global. (In fact, it might be argued that the Seven Years' War was more international.) Fighting in Africa remained relatively limited, and not a single battle was fought in the Americas. (Actually, no battle was fought on British soil either.) The peace treaties following World War I, on the other hand, had major global consequences. For starters, the map was radically redrawn in Europe and the Middle East, as the Austrian monarchy and the Ottoman Empire were dissolved. The Austrian emperor abdicated on November 11, 1918, and the former monarchy was turned into a series of new states. The new Austria, now a small, land-locked country comprising most of the former monarchy's German-speaking areas, was proclaimed a republic. In the south, Croats and Slovenes, won by Serbian propaganda, joined Serbia and Montenegro to found Yugoslavia, the state of the South Slavs. Poland, Czechoslovakia, and Hungary proclaimed their independence. Both Romania and Italy made substantial territorial gains. Greece extended its borders to the Black Sea, while the Ottoman Empire was reduced to Turkey. Azerbaijan briefly became independent, but in 1923 was re-integrated into the Russian empire, which was now called the Soviet Union.

Germany was stripped of its African and other overseas colonies. (Tanzania and German New Guinea were handed to Britain; Cameroon and Togo were both split between Britain and France; Namibia was ceded to South Africa; German Samoa to New Zealand; the Chinese province of Kiaochou (Shandong) as well as Germany's small islands in the Pacific Ocean were given to Japan.) In Europe, Germany had to surrender Alsace-Lorraine (Elsass-Lothringen) to France, made smaller cessions to Czechoslovakia, Lithuania, Belgium, and Denmark, and had to cede large areas, including West Prussia, most of the Posen province, plus part of Silesia, to Poland. Thus, Germany, though it still stretched from France to Lithuania, now saw the province of East Prussia cut off from the rest of Germany by the narrow Polish Corridor. No independent Polish state had existed for about 100 years, and the idea to provide the newly established Poland with access to the sea was only fair, but for Germany the new situation was difficult to accept, especially because the West Prussian capital of Danzig (Gdansk), an important Baltic seaport in which the overwhelming majority of citizens were German, was taken from it as well. The Free City of Danzig was put under the administration of the League of Nations (the predecessor of the United Nations) and ordered to provide services for the new Polish state. The Polish government should "undertake the conduct of the foreign relations of the Free City of Danzig as well as the diplomatic protection of citizens of that city when abroad," while the citizens lost their German nationality.[118] What is more, the Treaty of Versailles, which ended the state of war between Germany and the Allies, forced Germans to accept full responsibility for causing the war (the so-called War Guilt Clause).

Germany was neither crushed nor conciliated. The Germans had not lost a single decisive battle in World War I, and by the end of it occupied large territories from France to the Crimean Peninsula. Now Germany was being disarmed, and the Allies imposed a burden of huge reparations payments on the nation, which caused economic problems while a worldwide recession was looming. Hyperinflation and extreme unemployment helped Adolf Hitler to come to power in 1933 and to replace the constitutional republic by dictatorship. Hitler's Nazi regime immediately started to rearm Germany. (Nazi stands for Nationalsozialistische Deutsche Arbeiterpartei, i.e., National Socialist German Workers' Party.) In 1935 Hitler's troops marched into the Rhineland, Germany's western border region that was supposed to remain permanently demilitarized. Britain and France merely protested. In 1936 Hitler established an alliance with Italy, where fascist dictator Benito Mussolini had been in power since 1922. In 1938 German troops marched into Austria, and thereafter into Sudetenland, a German-speaking region now part of the new Czechoslovakia.

Then Danzig was on the agenda. With hindsight it is easy to say that the Treaty of Versailles was quite a blunder, but with respect to the Danzig/Polish Corridor situation France and Britain apparently felt right then that it was indefensible. Germany made various offers to the Allies and the Polish government to settle the situation, and Britain and France actually asked the Polish government to accept them, but Poland refused. (One suggestion was for Poland to return Danzig to Germany but to keep the corridor, another to let the people in the corridor vote for their preferred government.)[119] Nevertheless, Britain had pledged to defend Poland. In September 1939, after Hitler had demanded Danzig to be returned to Germany, and the Polish government had refused, German troops entered Poland. Britain thus declared war on Germany for the second time within 25 years. France declared war on Germany as well, but neither the Germans nor the Allies undertook any land operations during the following half year. Thereafter World War II, a fully fledged Oil Age war dominated by airplanes and tanks, was waged for five long years.

WORLD WAR II (1939–1945)

World War II had two main arenas, one in Europe, the other in the Pacific. In both of them attempts to get access to oil fields played a major role in the course of the events. Germany concentrated its resources to reach Baku, while Japan invaded the Dutch East Indies (Indonesia). Ultimately this war was decided by oil supplies. The United States poured 6 billion barrels of oil into the war, but emerged from it as a superpower. Similarly, the Soviet Union defended Baku and rapidly developed the rich oil fields of central Russia to keep Soviet tanks rolling and planes in the air.

Aircraft, more than anything, were an absolutely indispensable ingredient of victory on land and at sea. Trucks and tractors replaced horses to move field artillery and carry supplies behind tanks into the battlefield. Oil also fueled battleships (including aircraft carriers) as well as cars and motorbikes, while coal fueled trains and produced electricity. Coal also remained

The Very First Target of U.S. Bombers during World War II in Europe: Romanian Oil Installations, 1943 The Romanian oil fields were the first target bombed by American aircraft in Europe during World War II. The first U.S. raid on Ploesti, flown from North Africa on August 1, 1943, may have destroyed nearly 50 percent of the production capacity, but most of the damage was fixed within weeks, and the mission was a disaster. (Of 1,726 men who took off on the mission, 532 did not return. Of 177 planes 54 did not make it back to the base, and most of the remaining ones were too heavily damaged to fly again.) The year after, renewed attacks on Ploesti were launched from airbases captured in Italy, and in August 1944 Soviet troops seized the refineries. (Library of Congress image LC-USW33-036496-C DLC.)

indispensable for steel (and hence weapons) production, and was used to generate hydrogen for the Haber-Bosch synthesis (which provided ammonia for explosives).

German troops took Poland in just four weeks, occupied Denmark and Norway, and proceeded towards the Netherlands, Belgium, and into France. France surrendered after six weeks, and while the bombing of England continued, German forces took Yugoslavia and Greece. Germany is rich in coal, but has no oil at all. In 1940 Romania signed a mutual cooperation pact with the Axis Powers (Germany-Italy) that provided Germany with substantial amounts of Romanian oil. The Ploesti (Ploiesti) oil fields, with seven major refineries, then supplied about a third of Germany's wartime oil fuel needs, but the Romanian oil supplies were insufficient to meet the total oil demands of the Axis Powers. In fact, the German Army had to rely on horse traction for tactical logistics and field artillery mobility. What is more, a 1940 earthquake severely damaged some of the installations at Ploesti, and eventually British, Soviet and U.S. planes began bombing the oil complex.

In June 1941 German troops turned east to invade the Soviet Union. Italy, Romania, Hungary, Slovakia, and Finland immediately declared war on the Soviet Union as well. The Nazi regime speculated that if it could capture the vast farmland of the Ukraine combined with the rich oil fields of the Caucasus, Germany would be self-sufficient and invulnerable. However, the campaign turned into a disaster with heavy losses on both sides. Desperate to secure Baku, Hitler simply committed too many troops to his goal of reaching the Caucasus oil fields. The German advance in the north was halted just outside Moscow, and Leningrad (St. Petersburg) came under long-term siege (during which a third of the city's population died from frost and hunger). The Germans had severe supply problems, while the Russians cleared out the regions from which they retreated, and transferred all heavy industry to the east of the Ural Mountains.

The Run for Baku

The tractors of the Ukraine had already run dry when the Germans scheduled their attack on Azerbaijan for 25 September 1942. By August that year they had reached Maikop, the most westerly of the Caucasus oil centers, but the Russians had thoroughly destroyed all oil field installations. (Efforts to repair these installations yielded the Germans no more than 70 barrels of oil per day at Maikop as of January 1943.) The consequent commitment to cross the Caucasus mountains into Azerbaijan cost the Germans dearly. Hitler denied the request to transfer forces from the Caucasus to Stalingrad, a city northwest of the Caspian Sea. In the Battle of Stalingrad (Volgograd), fought from September 1942 to February 1943, some 100,000 Germans died and about 90,000 were taken prisoner. This battle was the beginning of the end for the German army, which in turn did not have the resources to reach Baku. Instead, Russia started a counter-offensive that was marked by the massive tank battle at Kursk. Thereafter the Soviet army advanced into the Ukraine, and began to push the German forces back westward, while German tanks in eastern Europe increasingly lacked fuel and stood still.

The Russians, on the other hand, had plenty of oil. Production in the Baku area was pushed to a record level of 172 million barrels a year in 1941, which accounted for over 70 percent of the entire Soviet oil extraction. (The Caucasus region as a whole accounted for 90 percent.) By the summer of 1942, about 76,000 workers were laboring 18 hour shifts in the Azerbaijani oil industries. (As increasingly more men had to fight at the front line, the percentage of women working in the oil industry rose to 33 percent by 1942, and to 60 percent by 1944.) However, by the late summer of 1942, when the Battle of Stalingrad began, Azerbaijan was under serious threat of being conquered by German troops. In response the Soviets started to seal wells in the Baku region and to transfer most movable equipment (and the work-force) to the oil regions of the Volga, the Ural Mountains, Kazakhstan, and Central Asia. In October 1942 more than 10,000 oil workers left for these areas from Azerbaijan. Half of the people and equipment was moved to the region near Kuybishev, at the Volga River, north of Kazakhstan, where five new oil and gas fields were in turn discovered, and huge oil refinery construction projects were undertaken. In consequence, Russia's Volga-Ural oil fields turned into the Soviet Union's most important post-World War II source of oil, allowing it to emerge as a superpower.

Meanwhile the United States joined the Allied Powers (Britain, France, the Soviet Union, the Commonwealth, and China), after Japan (during World War II an ally of Germany and Italy) had attacked Pearl Harbor, Hawaii, in December 1941. Thereafter the Allied Powers enjoyed a better supply of food, arms, and ammunition, and, above all, oil, as increasingly more American fuel was added to the oil supplies from Persia (Iran).

Germany's most important ally, Italy, collapsed under British and American attacks. Britain defeated Italy in North Africa and helped the United States invade Sicily and the Italian mainland in 1943. By 1944 Mussolini ruled merely over a small area in northern Italy, while Allied forces invaded Normandy (in northern France) and started to move towards fuel-starved Germany. Paris was freed in November 1944, and U.S. and British airplanes bombed German cities to the ground, deliberately killing hundreds of thousands of civilians. On February 13, 1945, when Germany had already been defeated, about 1,000 U.S. and British bombers in a single massive air raid dropped 3,900 tons of explosive and incendiary bombs over Dresden, a city packed with refugees fleeing the Soviets. Dresden had no industry of military significance, and had previously been spared of bombings. However, this raid destroyed 80 percent of the city and killed 135,000 people[120], many or most of them women, children and elderly.[121] In March 1945 Soviet troops conquered Berlin, and Hitler committed suicide.

The Holocaust

When ground troops entered the territories that had been occupied by Germany, the full brutality of the Nazi regime came to light. Starting in Dachau in 1933, the Nazi government had built concentration camps whose first inmates were primarily political prisoners (Communists and Social Democrats), habitual criminals, homosexuals, Jehovah's Witnesses, and so-called anti-social elements (beggars, vagrants, hawkers), plus others considered problematic, including writers and journalists, lawyers, and unpopular industrialists. Later on, with the first camp operating from mid-1942, the Nazi regime introduced another type of concentration camp that purposely served the Holocaust, the systematic extermination of Jews. In Germany, the Nazis killed 141,500 Jewish individuals, equivalent to 25 percent of the pre-Nazi Jewish population, while all six extermination camps, equipped with special apparatus designed for systematic murder, were located in Poland: Auschwitz-Birkenau, Belzec, Chelmno, Majdanek, Sobibor, and Treblinka. Three million Jews (90.9 percent of the Jewish population) were killed in Poland, 1.1 million (36.4 percent of the Jewish population) in the Soviet Union, and 143,000 (85.1 percent of the Jewish population) in Lithuania. To varying degrees, Jews living in the following countries were murdered as well: Austria, Belgium, Czechoslovakia, Denmark, Estonia, Finland, France, Greece, Hungary, Italy, Latvia, Luxembourg, the Netherlands, Norway, Romania, Slovakia, and Yugoslavia. The total number of Jewish victims amounted to approximately 5.86 million out of about 10 million Jews living in Europe before World War II. In addition, about 5 million non-Jewish civilians were murdered by the Nazis and their collaborators. Victims were variously starved, tortured, experimented on, and worked to death. Millions were executed in gas chambers, shot, or hanged. In almost every country they occupied, the German Nazis found plenty of locals who were willing to cooperate fully in the murder of the Jews. This was particularly true in eastern Europe, where a long standing tradition of virulent anti-Semitism existed, and where various national groups previously under Soviet (Russian) domination (including Latvians, Lithuanians and Ukrainians) fostered hopes that the Germans would restore their independence. (After 1945 many camp officials and others responsible were tried for war crimes and consequently executed or imprisoned.)[122]

War in the Pacific

In the Pacific, the fighting continued after Germany had capitulated. In this arena the war had already started in 1937, with Japan launching an attack on China. In the late 19th century Japan had emerged as the only non-Western Coal Age Power, and by the beginning of the 20th century had developed ample colonial ambitions. By 1905 Japan had acquired Port Arthur, the southern half of Sakhalin, Korea, and Formosa (Taiwan). As an ally of Britain and the Entente during World War I, Japan gained from Germany several Pacific islands north of the Equator plus China's critical Kiaochow (Jiaozhou) territory with the port of Qingdao (Tsingtao) in the Shandong province. Right after World War I Japan used its position to threaten China with war, attempting to make it a de facto protectorate. China in turn submitted to the extension of Japanese power in several Chinese regions, but refused to appoint Japanese political and financial advisers to the Chinese government. In 1922 China even managed to recover Kiaochow.

The global economic depression of 1929 hit Japan as hard as it hit Germany, and the civilian government lost ground to senior army generals who recommended solving the economic problems by winning new colonies abroad so that the industries there could be exploited for Japan. In 1931 Japan used an incident at the Manchurian railway (which the Japanese controlled since the Russians had left) as an excuse for locally based Japanese army units to take control over Manchuria, China's huge coal- and iron-rich province north of the Korean peninsula. The international community protested, but there was no action when Japan created the puppet state of Manchukuo, nominally led by a Chinese pretender to the throne. (In effect, the economy of Manchuria had been controlled by the Japanese South Manchuria Railway Company even before 1931.) During the following 14 years the Japanese developed Manchuria into a huge industrial agglomerate of coal, oil, metallurgical, and chemical industries that served the massive Japanese military complex. In 1933 Japan took Jehol, a Chinese territory bordering Manchuria, and in 1937 Japan fully attacked China, seizing Beijing, Tianjin, Shanghai, and much of the other coastal provinces and ports, while carefully avoiding Europeans in their enclaves. The Japanese annexation of internally fractured China involved tremendous brutalities, including the mass slaughter of civilians. (In the infamous Nanking massacre alone, the Japanese killed at least 200,000 Chinese civilians and raped 10s of thousands of women.[123])

When German troops rapidly defeated France in Europe, Japan in 1940 seized the moment to take control of French Indochina (Vietnam, Laos, Cambodia). This region was an important source of natural rubber, a resource of great strategic importance for Western countries. (The production of synthetic rubber was still immature and chiefly a German affair.) Since Japan now had essentially taken sides with one of the European war parties, the United States, maintaining its officially neutral position, stopped shipments of oil and other goods to Japan. These U.S. supplies were, however, essential to the Japanese economy and military complex, chiefly because Japan, like Germany, did not have oil deposits. This put the Japanese in a difficult situation. Either they had to return French Indochina, which was considered highly humiliating, or they were going to face severe economic difficulties and would have been unable to keep control of China.

The closest source of oil were the oil fields of Sumatra and Borneo in the Dutch East Indies (Indonesia). In January 1941 Japan demanded access to that oil, but the request was denied. Instead the Dutch and British, together with the Americans, in July 1941 organized oil (and general trade) sanctions against Japan, and by December 1941, the Japanese had decided to go to war against the Western powers and invaded the Dutch East Indies. Japanese national coal production had grown almost fourfold between 1910 and 1940 (to over 56 million tons annually), and now the Japanese military also had the oil it needed to continue its imperialistic goals in China. Or rather, these goals could now be expanded to much of Southeast Asia. While encouraging natives to get rid of the European imperialists, Japan managed to occupy virtually all the regions between Australia and China that had previously been held by the French, British, and Dutch, who were now busy fighting in Europe. This included Dutch East India, British Malaysia, and Burma, plus the U.S.-administered Philippine Islands.

Well aware that this territorial expansion might pull the United States into the war, Japan launched a preemptive strike against U.S. military installations on Hawaii and the Philippines on December 7, 1941. The United States had transferred its fleet to Pearl Harbor, Hawaii, 18 months earlier as a presumed deterrent to Japanese aggression, but the Japanese secretly sent an aircraft carrier force across the Pacific and without warning launched an aerial strike that rapidly sank five and damaged three of eight U.S. battleships at Pearl Harbor. In addition, several other ships and most U.S. combat planes based in Hawaii (and the Philippines) were destroyed. But if the Japanese had thought this demonstration of power would prevent American intervention in the Pacific, they got it all wrong. Their attack left about 2,400 Americans dead, and the United States declared war on Japan, entering World War II on the Allied side. By spring 1942, the United States was bombing Tokyo and blocking Japanese passage to Australia. In the Battle of Midway, in early June 1942, the United States destroyed a number of Japanese aircraft carriers and soon made steady gains. American, Australian, and New Zealand troops retook many of the Pacific islands in battles that resulted in heavy casualties, and by the end of it all, the United States, France, and Britain reclaimed much of their Southeast Asian Coal Age empires.

The battle of Okinawa, south of Japan, was the last, largest, and bloodiest campaign of the Pacific War. About 131,000 Japanese and Okinawan conscripts were killed, plus perhaps 130,000 Okinawan civilians. The United States suffered 12,000 killed and 36,000 wounded. After this three-month battle, in which the Japanese lost 7,830 aircraft and 16 combat ships, Japan was practically defeated. Nevertheless U.S. military experts warned that the invasion of Japan would cost the lives of many U.S. soldiers. The United States thus decided to drop two atomic bombs on Japanese cities in August 1945 to enforce unconditional capitulation six weeks after Okinawa had been taken. The Hiroshima bomb killed between 90,000 to 140,000 people, and the Nagasaki bomb between 60,000 to 80,000 people, immediately or within four months. This added to the enormous civilian casualties produced by the other U.S. raids on Japan, particularly the firebombing of Tokyo and other major cities in March 1945.

World War II was by far the single most deadly conflict the world had ever witnessed. Though estimates vary, it probably cost the lives of some 55 million people, including military, civilian, and Holocaust victims. The Soviet Union suffered over 20 million deaths (of which perhaps 7 to 10 million civilians), China 11 million (of which perhaps 8 million civilians), Germany some 7 million, and Poland between 5 and 6 million. France lost roughly 300,000 civilians and 200,000 soldiers, Britain nearly 70,000 civilians and some 350,000 soldiers. The United States had no significant civilian losses, but suffered the death of more than 400,000 soldiers.[124]

THE UNITED NATIONS AND BRETTON WOODS

By the end of World War II the United Nations (UN) was set up as the successor of the League of Nations, an international organization established after World War I to promote international peace and security. The original intention was for the UN to preserve the wartime alliance between the United States, the Soviet Union, China, Britain, and France, but this never materialized due to the immediate frictions between the Soviet Union and the United States. Nevertheless the UN developed into an organization that facilitated international cooperation (and employed peace-keeping forces with varying success), not least because it widened its base from originally 51 member states in 1945 to 154 members by 1980s (and 191 by 2005). In spite of that, the UN has been lacking adequate funds ever since its founding. Member states are supposed to contribute financially according to their resources, but by the early 1990s unpaid contributions brought the UN to the brink of insolvency: Only some 20 members had paid their annual dues in full, with the United States accounting for half the debt owed to the UN. (As of the year 2000, U.S. arrears had reached $1.3 billion.) But this did not necessarily harm the UN's authority. In fact, the Security Council, the United Nations' most powerful body, is arguably the world's most powerful international institution. The Security Council investigates international disputes and may call on all members to take economic or military measures to enforce its decisions. Practically, it decides whether or not a war is legal. The set-up of the Security Council put the power in the hands of the war parties that won World War II. The United States, the Soviet Union (from 1991 the Russian Federation), Britain, France, and China (until 1971 the Taiwan-based Republic of China) were declared permanent members of the Security Council, which can exercise a veto in that their support is requisite for all decisions. In addition, there are 10 temporary members who serve two-year terms and have no veto rights. The permanent members were also the only nations to develop or acquire nuclear weapons technology (during or) shortly after World War II. (By the early 1960s all five permanent members, including mainland China, not Taiwan, had nuclear arms capacity.) In turn the Nuclear Non-Proliferation Treaty of 1970 firmly established the de facto principle that permanent membership in the Security Council and membership in the nuclear club were one and the same.

By the end of the 20th century the world had changed. Both India and Pakistan had developed nuclear weapons without being awarded a permanent seat in the council. India's population had reached one billion and was set to pass that of China. Japan had become economically more powerful than Britain and France, and it was no longer obvious why these two nations, rather than Italy and Germany, for instance, should have permanent seats in the Security Council. However, the principal victors of World War II were still determined to cling to the existing power structure, which increasingly put the role, position, and meaning of the UN into question.

From Bretton Woods to Free Floating Exchange Rates

At the end of World War II the international community also set up a new exchange rate system to create a postwar economic environment that would promote worldwide peace and prosperity. In 1944 diplomats from 44 countries met at Bretton Woods, New Hampshire, and agreed to establish a fixed exchange rate system. In turn the currencies of nearly all members of the United Nations outside the Soviet sphere were fixed in terms of gold. However, only the United States pledged to redeem its currency for gold at the request of a foreign central bank, which made the U.S. dollar the keystone of the Bretton Woods exchange rate system. (A country's central bank, such as the Federal Reserve System of the United States, issues money and controls interest rates within a country.) All countries were obligated to keep their currency at the set value in comparison to the other currencies, and only under extraordinary circumstances should a readjustment of value be allowed. However, the Bretton Woods system was flawed, and ultimately failed. The international community allowed the U.S. dollar to take a central role, because it had faith in the economy of oil-rich America, the principal winner of World War II. (In essence, the dollar took over the role that the British Pound Sterling had during the Coal Age when Britain was the world's main power, and coal the principal fuel associated with prosperity and military might.) But the gold supply did not expand in the short run, and U.S. currency was soon the only source of liquidity to support the rapidly expanding international trade. During the 1950s and 1960s foreigners steadily held more U.S. dollars, which worked as long as the world believed in the integrity of the U.S. currency. However, despite its enormous oil wealth the United States had become a net oil importer from 1948, saw the volume of its proved oil reserves peak in 1959, and consequently was unable to avert a production peak: from 1970 annual U.S. oil production began to decline. (Even Alaskan oil, available from 1977, could not restore the production rate of the late 1960s.) Evidently, the United States was now going to be increasingly dependent on foreign oil, whose price was going to influence the general price level within the United States and hence the price of the U.S. dollar in terms of gold and foreign currency. The world lost its faith in the ability of the United States to redeem for gold the dollars held overseas, and a run on the U.S. central bank set in. In the first seven months of 1971, the United States sold one third of its gold reserves, and it became clear to the marketplace that it did not have sufficient gold on hand to meet the demands of those who still wanted to exchange their dollars for gold. In August 1971, President Nixon announced that the United States would no longer redeem gold at $35 per ounce, and the Bretton Woods system came to an end. Instead, a flexible (floating) exchange rate system was established, in which the value of currencies is determined by supply and demand. The U.S. dollar immediately fell relative to most of the world's major currencies, and countries in turn chose different systems, either allowing the value of their currency to truly float free or pegging their currency to one of the major currencies.

The World Bank, the principal international development institution, was set up as a specialized agency within the United Nations system at the Bretton Woods Conference as well. Its main component, the International Bank for Reconstruction and Development (IBRD), lends money to middle-income and creditworthy poorer countries. The World Bank also makes interest-free loans to its poorest member countries, offering insurance against noncommercial risks to encourage foreign direct investment in unstable countries. (Over 180 countries were members of the IBRD by the end of the 20th century.) But such help has never been unconditional: the world's richest countries provided most resources and in return had the most voting power in terms of influencing the guidelines on how borrowed money is to be invested and who is allowed to borrow. Nevertheless, it was a novelty of this Energy Era that the rich would help the poor (over which they did not rule) on a large scale and in an organized fashion. This trend continued as the United Nations Development Programme (UNDP) was formed in 1965 to promote general environmentally sustainable human development in low-income countries.

INDEPENDENT ASIA

But not all post–World War II development aid was granted for humanitarian reasons. Much of it was provided due to the immediate rivalry between the newly emerged superpowers, the United States and the Soviet Union. These two nations attempted to widen their influence in Asia, Africa, and Latin America, not least because the colonial empires that had been established or nurtured during the Coal Age fell apart after World War II, leaving newly independent countries on the outlook for new orientation. People living in the Asian areas temporarily occupied by the Japanese simply did not accept the return of the European colonial powers and demanded independence.

Dutch attempts to regain the oil-rich East Indies led to a bitter war that lasted several years, but finally, in 1949, the Netherlands accepted Indonesia's independence, and Achmed Sukarno was elected president of the new federal state. (Indonesia is the world's fourth largest country in terms of population: The country had 235 million inhabitants as of 2007.) In French Indochina semi-autonomy was achieved by Cambodia and Laos in 1949/1950, and full independence in 1953/1954. However, in Vietnam the French fought a bitter war for eight years against local forces before finally pulling out in 1954. The Japanese were also forced to get out of Korea, but the peninsula was divided between the two occupying forces, the Soviet Union north of the 38th parallel, and the United States south of it, which created a lasting division of the country as North Korea and South Korea.

China: The Dragon Regains Independence

China managed to get rid of both the Japanese occupying forces and European influence. Immediately after World War II, in 1946, a civil war broke out which had really begun in 1927 before being interrupted as China's various political groups pooled their military resources against the Japanese invaders. The civil war was waged between the Chinese National People's Party, which had been founded in 1894 and had overthrown the last Chinese emperor in 1911, and the Chinese Communist Party, which had been founded in 1921 and from 1935 was under the leadership of Mao Zedong. In 1949 the communists won and proclaimed the People's Republic of China, while the nationalists under Chiang Kai-shek (supported by the United States) took their remaining forces to Taiwan (Formosa). (Both Formosa and Manchuria had been returned to China after the defeat of Japan.) The two million or so nationalists who fled to Taiwan firmly established themselves on the island and maintained the permanent seat in the UN Security Council that had been awarded to China. Meanwhile the People's Republic of China expanded its territory in 1950 by conquering Tibet, which had declared itself independent from China in 1911.

In the following years (mainland) China's communist regime focused on economic reconstruction by nationalizing industries and introducing central planning and moderate land reform. (China is very rich in coal and had modest oil deposits.) Mao's policy of creating large self-sufficient agricultural and industrial communes suffered from bad leadership which, in combination with weather extremes, led to the death of over 20 million people during floods and famines between 1959 to 1961. (A figure of up to 27 million peasants who perished as a consequence of the forcible collectivization has often been cited. Several million were executed by Mao's regime as enemies of the people—mostly landlords and richer bourgeoisie as well as former Kuomintang officials and officers.[125]) In turn the communes were reduced in size, and private farming plots and markets reintroduced, but in 1966 Mao initiated the Cultural Revolution, a rectification campaign to expose enemies of socialism and oppose foreign influence as well as classical Chinese culture. This resulted in chaos and social unrest for several years, and eventually Mao had to deploy the army to maintain control. Estimates of violent deaths during the Cultural Revolution are in the order of one to two million.[126]

From 1970 Mao began conducting a more balanced policy again, and in 1971, the year the Bretton Woods system was abandoned, the People's Republic of China, the world's most populous country, was permanently admitted to the UN Security Council, while Taiwan was expelled from it. This reflected China's possession of nuclear weapons and ballistic missile delivery systems, which had its origin in technical assistance granted by the Soviet Union. The Soviets had provided China with three kinds of missiles (one of them being the Russian copy of the German V-2) in the 1950s, and though Chinese-Soviet relations had soured by 1960, China soon presented its first generation of nuclear weapons and short-range liquid-fueled ballistic missiles. China now had the capability to attack Taiwan (as well as Japan and the Philippines), and by the early 1980s had developed longer-range systems that could potentially target much of the Soviet Union (including Moscow), and even true intercontinental ballistic missiles that could reach the continental United States.[127]

Despite these achievements in terms of destructive energy technology, and a radical increase in constructive energy use, China remained a poor and overwhelmingly rural country. Mao died in 1976, when China consumed 20 times as much energy as in 1949, and his successor, Deng Xiaoping, and other leaders soon focused on market-oriented economic development. During the 1980s and 1990s China enjoyed extraordinary economic growth, and many were lifted out of the most extreme poverty. But political controls remained tight. At the beginning of the 21st century China was still the world's most populous country (with 1.3 billion people), and China's economic output was second only to that of the United States. (China's 2007 GDP (at PPP) was $7.0 trillion, compared to that of the United States, $13.9 trillion.)

Japan: An Economic Superpower

Japan (like Germany) recovered well from destruction during World War II. Japan regained full sovereignty in 1951, and U.S. military occupation ended the year after. A new constitution gave women the vote and restricted deployment of troops overseas. The emperor was forced to renounce his divinity and became a powerless figurehead. The United States initially provided aid to rebuild infrastructure and kick-start economic development, but soon Japan emerged as a major competitor for the United States economically (while remaining a strong ally politically and strategically). Japan's overall economic growth was truly spectacular for three decades. Output in real terms grew 10 percent annually on average in the 1960s, 5 percent in the 1970s, and 4 percent in the 1980s. From the mid-1960s Japan's GDP began to surpass that of Britain, France, and Germany, and by 1970 Japan had emerged as the world's second largest economic power after the United States, with Japanese electronics, machinery, and motor vehicles booming in the export markets of North America and western Europe.

Among several factors that presumably contributed to this extraordinary economic development are protection of domestic markets from foreign competition, and a low level of military spending (no more than one percent of GDP). However, the most unique feature of Japan's economy are the *keiretsu*, great family-controlled banking and industrial combines. (The largest ones are Mitsui, Mitsubishi, Dai Ichi Kangyo, Sumitomo, Sanwa, and Fuyo.) In 1937 the four leading *keiretsu*, or rather *zaibatsu*, as they were called prior to World War II, directly controlled one third of all bank deposits, one third of all foreign trade, one half of Japan's shipbuilding and maritime shipping, and most of the heavy industries. During the occupation the United States attempted to break up these family-based conglomerates, but they reemerged, maintaining close relations with the major political parties, and pooling their resources to foster economic growth. Their large corporations organized almost every aspects of Japanese life, guaranteeing lifetime employment to a substantial portion of the urban labor force, and maintaining a strong work ethic. (They also pioneered innovative management systems that later spread to the West.)

Japan's curious emergence as the world's largest car manufacturer has its early roots in technology-transfer from the West in the early Oil Age, combined with investments by the military.[128] Under the Military Vehicle Subsidy Law of March 1918 the military granted subsidies to seven Japanese automobile manufacturers, and in the 1920s Ford and General Motors established local production and began monopolizing the Japanese car and truck market. However, they pulled out (in 1939) after Japanese carmakers began mass producing from 1935, and business conditions were changed to favor domestic producers. The Automobile Manufacturing Industries Act of 1936 served to prepare for the invasion of China the year after. Eighteen companies, including Toyota and Nissan, were operating under this law. They were forced to produce almost exclusively trucks rather than cars, and after the war struggled due to steel and other supply problems. However, the government from 1952 supported them through tax advantages and low-interest loans, and most Japanese car makers (with the notable exception of Toyota and Fuji Seimitsu) in turn opted to collaborate closely with foreign partners (Nissan Austin, Isuzu Hillman, Hino Renault). Domestic Japanese automobile-manufacturing in turn surpassed that of Italy in 1963, France in 1964, Britain in 1966, Germany in 1967, and finally the United States in 1980. The United States took measures against Japanese competition (to protect domestic car manufacturers) after the first oil price surge of 1973, as it had become clear that consumers might increasingly opt for fuel-efficient Japanese cars. Threatening to impose ever higher tariffs, the United States forced Japan to "voluntarily" restrict its passenger car exports to the United States, and western European countries imposed similar limitations. However, Japanese cars in the 1980s and 1990s consistently ranked as the most reliable ones (in the comprehensive surveys compiled annually in western European countries such as Germany), and therefore retained a strong market position.

The Japanese rise to economic superpowerdom in the later Oil Age was especially remarkable because Japan does not have domestic oil resources. Had Japan maintained permanent control over the Indonesian oil fields, it might well have emerged as a true (military) superpower of the Oil Age. Instead, Japan had to base its economic growth on its rather meager domestic coal resources and hydropower capacities, and on imported oil and gas. In 1950 coal supplied half of Japan's energy needs, hydroelectricity one-third, and oil the rest. Thereafter Japan's reliance on imported oil steadily increased until the oil price shocks of the 1970s. In turn Japan emerged as one of the world's most efficient energy users and early on emphasized the import of natural gas in addition to the construction of nuclear power plants in order to spread energy price and supply risks. (In 1988 oil provided Japan with 57.3 percent of its energy needs, coal 18.1 percent, natural gas 10.1 percent, nuclear power 9.0 percent, and hydroelectric power 4.6 percent. The remainder came from other sources.) In fact, Japan's domestic oil consumption was slightly lower in 1990 (4.9 million barrels per day) than in the late 1970s (5.1 million barrels per day), while the consumption of nuclear power and natural gas had risen substantially. (About half of the liquefied natural gas imported in 1990 came from Indonesia, and roughly a fifth each from Malaysia and Brunei.)

Economic growth slowed markedly in the 1990s, averaging just 1.7 percent, and the influence of the *keiretsu* faded away. However, Japan remains a major global economic power, and the Japanese population of 127 million (2004) enjoys an extremely high life expectancy of 85 years for females and 78 years for males.

The Four Tigers

Except for Japan, hardly any Asian country industrialized during the 20th century, but four small nations, Hong Kong, Singapore, Taiwan, and South Korea, fared similarly well. Often referred to as the Four Tigers, they started to dominate much of the world's textile and electronics industries from the 1960s. With the exception of Hong Kong, the Tiger States were neither democratic nor free market economies during their industrialization. They were all supported by the West (struggling against the spread of communism in the region) and received ample foreign investment. Their industrialization was marked by strong government involvement and a focus on export markets. Generally there was a trend from lower-tech, labor-intensive production towards higher-tech, capital intensive manufacturing as industrialization progressed, and wages and the standard of living increased.

Hong Kong, a British dependency administered by a crown-appointed governor, had to absorb almost one million Chinese (predominantly Cantonese) refugees in the 1950s. Immigration then continued to boost the colony's population from one million in 1946 to 5 million in 1980, to 7.4 million in 2003. However, Hong Kong coped well with this population growth and emerged as one of Asia's major commercial, financial, and industrial centers. By 1987 Hong Kong's container port was the busiest in the world, in part because the city had traditionally served as a re-exporter of Chinese products, in part because Hong Kong had turned into a major producer of textiles, clothing, electronics, plastics, toys, watches, and clocks. On July 1, 1997 Hong Kong reverted to Chinese administration under an agreement by which China promised not to impose its socialist economic system on Hong Kong and that Hong Kong would enjoy a high degree of autonomy in all matters except foreign and defense affairs for the next 50 years.

Singapore, located on islands off the tip of the Malay Peninsula, was founded as a British trading colony in 1819 and gained independence in 1959. Between 1968 and 1980 dictatorial Prime Minister Lee Kuan Yew's People's Action Party maintained a monopoly of all parliamentary seats. During this time Singapore emerged as a major commercial and financial center with one of the world's busiest ports. Starting out as a center of trade, Singapore turned into a major producer of electronics, chemicals, mineral fuels, oil drilling equipment, rubber, and food products. Singapore's population increased from 2.8 million in 1993 to 4.6 million in 2003, of whom three quarters were Chinese.

Taiwan, then known as Formosa (literally, "the beautiful"), was ceded to Japan after China's military defeat in 1895. The Japanese further developed the subtropical island's agriculture, and by 1908 had established 50 new steam-powered sugar factories. The island reverted to Chinese control after World War II, and following Communist victory on the mainland in 1949, Taiwan became the refuge for the Chinese nationalist forces, who established a government using the constitution drawn up for all of China in 1947. The United States continued to recognize the nationalists as the legitimate government of China, and Taiwan thus held China's United Nations and Security Council seats until October 1971, when they were handed to the People's Republic of China, which until this day does not recognize Taiwan as an independent country. (Similarly, the government in Taiwan for a long time maintained an army of 600,000 in the hope of reconquering the Chinese mainland, over which it still claimed sovereignty.) Taiwan's ruling authorities only slowly incorporated the island's native population into the governing structure, and the first democratic presidential elections were not held until 1996. Government-owned banks and industrial firms invested according to central plans, and Taiwan quickly industrialized while the contribution of agriculture to the GDP sank from 32 percent in 1952 to 2 percent in 2002. Taiwan became known for its electronics, petroleum refining, chemicals, textiles, and machinery, and emerged as the world's largest manufacturer of notebook computers, at some stage producing 90 percent of all the world's laptops. (As of 2007 nearly 23 million people lived in Taiwan.)

Korea had been formally annexed by Japan in 1910, and after Japan's defeat in World War II the Korean peninsula was divided into a northern part, where a communist-style government was installed, and a southern part, where a republic with a U.S.-type constitution was set up. South Korea (the Republic of Korea) immediately had to cope with a massive influx of immigrants fleeing the communist regime in the north. The situation escalated when North Korean forces invaded the South on June 25, 1950, which started the Korean War (1950–1953). U.S. (and some other United Nations) forces intervened to defend South Korea, while North Korea was supported by troops from neighboring (also communist) China. North Korean forces overran almost all of the South, but U.S. troops regained the area, and in the end the original border along the 38th parallel was restored. Thereafter South Korea enjoyed stability and rapid economic growth, fostered by a system of close government-business ties. As in Taiwan, the land-owning class was dissolved, and the perceived threat of invasion (from the North) generally helped to justify a stern rule. South Korea's booming industries included electronics, chemicals, shipbuilding, steel, textiles, clothing, and footwear, and except for Japan, the country emerged as the largest shipbuilder in the world, producing oil supertankers and oil-drilling platforms. South Korea also turned into a formidable producer of steel and automobiles, delivering over one million motor vehicles per year by 1988. (South Korea had a population of 49 million as of 2007.)

India and Pakistan

British India achieved independence in 1947, but was divided into India and Pakistan, as the colony's Muslims were calling for their own state. This split-up was followed by complete chaos, as 12 million Hindus and Muslims were fleeing towards the new India and Pakistan, respectively. Bangladesh, previously known as East Pakistan, detached itself from Pakistan after the Indian-Pakistani war of 1971. Burma, formerly also part of British India, achieved full independence from Britain in 1948 and was renamed Myanmar in 1989. The borderline set between Pakistan and India by the British authorities before they took off was highly problematic. It placed two-thirds of the Kashmir province, where a Muslim majority lived, under Indian (Hindu) authority, partitioning a region that had been united since ancient times. Pakistan and India then fought wars over Kashmir in 1948, 1965 and 1971, while the skirmishing over the contested province went on almost continuously. The issue still remained unresolved by the beginning of the 21st century, and motivated both countries to develop nuclear weapons.

For India a conflict with China served as additional motivation to acquire nuclear weapons. China shares a long border with India and in 1962 invaded its southern neighbor because of territorial disputes that were also a legacy of British rule. China already had nuclear weapons by this stage, and India had no means to respond to China's massive attack. India had previously maintained friendly relationships with the Soviet Union for security reasons, but the Soviets did not react to the invasion. The United States, on the other hand, was happy to provide India with help against communist China, but the Chinese had partly withdrawn by the time U.S. troops arrived, keeping the Aksai Chin plateau of northern Kashmir. On the other hand, India managed to secure the North-East Frontier Agency (Arunachal Pradesh), which had been the actual reason for the invasion, and is still now claimed by China.

Yet another motive for India to develop nuclear weapons was to gain recognition and a permanent seat in the UN Security Council. India had historically been one of the world's most populous and powerful regions, but was refused a permanent seat because it happened to be a colony rather than an independent country in 1945, when the Security Council was established. (The Churchill administration vehemently opposed India's national aspirations at the time.) Just a short time later many would have agreed that independent India deserved a permanent seat, but by this stage it had become clear that permanent membership in the council was synonymous with membership in the nuclear club. It thus seemed necessary for India to develop nuclear weapons if it was to achieve the recognition it deserved.

India realized its nuclear ambition after Canada in 1955 sold India a 40 megawatt nuclear research reactor, and the United States delivered the heavy water (21 tons) required for its operation. The reactor was a design ideal for producing weapons-grade plutonium and was extraordinarily large (if intended for research purposes), capable of delivering enough plutonium for one to two bombs a year. (What is more, this sale set a precedent for similar technology transfers, which greatly assisted Israel in obtaining its own plutonium production reactor from France shortly thereafter.) In 1964, India commissioned a reprocessing facility to recover the plutonium produced by the Canadian reactor, and in turn used it for India's first nuclear test on May 18, 1974, described by the Indian government as a "peaceful nuclear explosion." The Canadian reactor also served as the prototype for a larger plutonium production "research" reactor built domestically. And even as late as the 1980s, India also began to develop a hydrogen (fusion) bomb.[129]

In March 1998 the Hindu nationalist Bharatiya Janata Party (BJP), which had declared it would make India an official nuclear power, achieved a strong electoral victory. Ending a 24-year period without testing, India in turn still set off five nuclear tests in 1998. In part, this was a reaction to Pakistan demonstrating a delivery system, a missile with a range of 1,500 km (900 miles) under a payload of 700 kg (1,540 lbs). Despite an international call for restraint, Pakistan responded to India's nuclear tests by conducting several nuclear tests of its own just two weeks later. And in April 1999 both India and Pakistan tested nuclear-capable ballistic missiles.

Pakistan's nuclear weapons program was established in 1972, right after the third war against India that ended with the loss of East Pakistan (Bangladesh). The program gained momentum following India's 1974 nuclear test, and with the arrival of Abdul Qadeer Khan. Qadeer Khan had been educated at universities in Germany and Belgium and then worked at a Dutch uranium enrichment plant that used highly classified centrifuge technology to separate fissionable uranium-235 from U-238. While working there he passed critical information to Pakistani intelligence agents, and in 1975 relocated to Pakistan, bringing with him stolen uranium enrichment technologies. Khan was immediately put in charge of the Pakistani nuclear program, began to build up a uranium enrichment facility, and developed an extensive clandestine network in order for Pakistan to obtain the necessary materials and technology it was still lacking. By 1986 Pakistan had produced enough fissile material for a nuclear weapon, and by the early 1990s operated an estimated 3,000 centrifuges to enrich uranium. However, Pakistan also began to pursue plutonium production capabilities, and with Chinese assistance built a 40 megawatt nuclear "research" reactor that was operational in April 1998.[130]

Soon thereafter the UN's International Atomic Energy Agency (IAEA) and various Western intelligence agencies forwarded evidence that Pakistan had passed its nuclear arms technology to other countries. In February 2004 Abdul Qadeer Khan, who was now a very wealthy man, admitted that he (and other Pakistani scientists) had sold nuclear weapons technology to several nations over more than a decade, from 1989 to 2000.[131] The countries were identified as Libya, North Korea, and Iran. While Libya renounced its atomic weapons program in 2004, North Korea did in fact develop a nuclear bomb. First, it used a British nuclear power plant for a plutonium-based program, and thereafter, in 1991, 1997, and until 2000, purchased uranium enrichment technology from Qadeer Khan. In 2003 North Korea announced it was developing a nuclear deterrent and in October 2006 conducted its first nuclear test, a relatively small explosion detected by several countries. Meanwhile the IAEA suspected that Iran's nuclear program was not exclusively peaceful either. In July 2006 the UN Security Council therefore demanded a suspension of Iran's nuclear enrichment and reprocessing activities, threatening sanctions for noncompliance. Abdul Qadeer Khan publicly apologized on Pakistani television in 2004, but was not prosecuted. He was pardoned by General Pervez Musharraf, who had come to power in October 1999 after a coup d'état ousting democratically elected Prime Minister Nawaz Sharif. The United States, and the West in general, had condemned this coup, but soon recognized Musharraf as the legal president of Pakistan, as they viewed and needed him as a strong ally in the region. In short, Abdul Qadeer Khan was off the hook, but given the chances that the countries who received nuclear weapons know-how from Pakistan will in turn also pass it on, Qadeer Khan may potentially go down in history as the man who killed the world.

Despite being nuclear powers both India and Pakistan remained quite poor countries at the beginning of the 21st century. India's 2006 per capita GDP was $3,700 (at PPP), and Pakistan's $2,600 (compared to the U.S. per capita GDP of $43,500). Both countries experienced extreme population growth in the later 20th century, and India was set to become the world's most populous country. (India's population was 1.1 billion in 2007, while Pakistan's was 165 million.) India enjoyed rapid economic growth from the mid-1980s, accompanied by a stronger integration into the global economy. In 2006 India's total economic output was the fifth largest in the world, well ahead of Germany. (India's 2006 GDP (at PPP) was $4.0 trillion, compared to the U.S. GDP of $13.0 trillion.)

AFRICA: INDEPENDENT BUT POOR

Africa also managed to shake off its colonial chains after World War II. Germany did not have any African (or other) overseas colonies left that could have been taken away from it after the war, but the Nazis' war-time ally Italy did. Italian forces had occupied Ethiopia in 1935, but the British in 1941 drove them out of the country and restored emperor Haile Selassie, who in turn ruled independently until 1974. Somalia, an Italian colony from 1927, came under British military rule from 1941 to 1950, but was then administered by Italy for 10 years, before achieving full independence in 1960. The Italian colony of Libya was also placed under British control after World War II, but the independent United Kingdom of Libya was already proclaimed in 1951.

Libya's western neighbor Algeria fought a bitter eight-year war against the French from 1954 to achieve independence in 1962. The French had even less intention to let go of Algeria after the Hassi-Messaoud oil field and the Hassi R'Mel gas field were discovered in the Algerian Sahara in 1956. (Besides, France began developing ballistic missiles in 1959 using a flight center in the Algerian Sahara.) Morocco, Tunisia, and Sudan all gained independence from France in 1956, and in 1960 the remaining French colonies of (West)

Africa were dissolved, which resulted in the founding of 15 new states. (The exception was Djibouti, which is of strategic importance for Suez Canal traffic and thus was not released into independence before 1977.) Also in 1960, the Belgians left the Congo, where an extensive civil war in turn broke out.

Britain slowly had to let go of its African possessions as well. Egypt had been granted full independence in 1936 and declared itself a republic in 1953, but British troops remained in the country to control the Suez Canal. Ghana gained independence in 1957, Nigeria in 1960, and virtually all other British colonies in Africa between 1961 and 1967. Some African tribes fought bloody rebellions to achieve their independence. In Kenya, for instance, tens of thousands of fighters were killed or imprisoned in camps during the Mau Mau insurgency, which contributed to Kenya's liberation in 1963. And finally, in 1980, Zimbabwe, too, achieved independence from Britain.

Portugal tried to hang on to its African colonies as well, but following a military coup and regime change in Lisbon in 1974–75, the long struggle of the Portuguese colonies came to an end, with Angola, Mozambique, and Guinea-Bissau quite suddenly gaining independence. Namibia, a German colony until World War I, achieved independence from South Africa in 1990, the same year Nelson Mandela was freed in South Africa, which was the beginning of the end of the apartheid policy. (South Africa itself had become entirely independent from Britain in 1961.) And in 1993, Eritrea (in northern Ethiopia) broke away from Ethiopia.

Though they were now independent, many African countries chose to maintain strong links to their former colonial powers, which often remained the main trading partner. Much of the former British Africa, including South Africa, Nigeria, Tanzania, Uganda, Kenya, Botswana, Namibia, and many more, joined Canada, Australia, New Zealand, India, Pakistan, and other former British dependencies in the Commonwealth of Nations, an association to promote (economic) cooperation between member states. (As of 2007, 16 out of 53 member states even maintained the British monarch as their head of state.) Nevertheless, Africa stayed rural and poor. The continent did not experience the transformations associated with entering the fossil Energy Eras, and the population grew rapidly (from 230 million people in 1950 to 800 million people in 2000), putting severe strains on the available resources. In many African countries the European colonial powers pulled out without leaving democratic structures, and the national borders rooted in the Berlin Conference of 1884 led to plenty of intra-national rivalry, as various hostile tribes were aspiring for power. This situation was often exploited by the two superpowers of the Oil Age, the United States and the Soviet Union, which attempted to expand their international spheres of influence. Both supported and maintained what were often totalitarian and brutal regimes through military and financial aid, not only in Africa, but also in parts of Asia and in Latin America that had achieved independence earlier on.

SOVIET UNION

The Russian Empire had grown to enormous size during the Coal Age, but, unlike the United States, did not enjoy a lot of inner cohesion. Much of the expansion had led into the sparsely settled areas, but the empire's more fertile, wealthy, populous, and developed regions were inhabited by people of various cultures, languages, and religions, many of whom had little reason to be loyal to the Tsar. The empire operated as absolute monarchy imposing strict rule and a coal-based industrialization onto an agricultural country in order to support the military complex. Most of Russia's industries were either state or foreign owned, and the empire never experienced the emergence of a strong factory-owning bourgeoisie as had been witnessed in western European countries. The struggle to improve the lot of exploited workers was therefore directed towards the central government rather than towards private factory owners. Russia's working class and peasants organized themselves in political parties early on, and during the 1890s and early 1900s, when abysmal working conditions, high taxes, and land hunger gave rise to more frequent strikes, socialists of different nationalities within Russia formed their own parties. The disastrous and unpopular Russo-Japanese War (1904–1905) led to the Russian Revolution of 1905, which forced Tsar Nicholas II to grant a constitution and to establish a parliament (duma), but the new democratic freedoms were soon curtailed again.

Russia entered World War I in alliance with England and France and suffered tremendous losses on its long front against Germany, Austria-Hungary, and the Ottoman Empire. Inflation, food shortages, and poor morale among the troops were all factors contributing to the outbreak of the February Revolution of 1917, which led to the execution of Tsar Nicholas in July 1918. A new central government was installed, but its authority was immediately challenged by the soviets (Russian for councils). At this stage there were about 900 of these elected local, municipal, and regional councils which represented workers, peasants, and soldiers. On November 7, 1917, the Bolsheviks, a wing of the Russian Social Democratic party led by Lenin Vladimir Ilyich, seized control and led the soviets to power. (This event became known as the October Revolution.) The Russian Social Democratic Labor Party was renamed the Communist Party, and Russia was proclaimed the Russian Soviet Federated Socialist Republic.

Marxism and Leninism

The idea behind the October Revolution was to establish a Marxist state. This goes back to the philosophical and economic system developed by 19th century German social theorists Karl Marx and Friedrich Engels, who promoted common ownership of the means of production. They conceived of communism as a society of abundance, equality, and free choice. Everyone should work according to his or her capacity and receive according to his or her needs, but they said little about how economic decisions would be made, other than that property would belong to society as a whole. Marx was especially unhappy with the fact that some members of capitalist societies (such as landowners) may become increasingly richer without doing any work (or taking any risks). In a sense Marxism overemphasized equality, the second of the three principles of the French Revolution, liberté—egalité—fraternité, while Western democracies tend to emphasize the first, freedom. There can never be truly equal opportunities in a society in which individuals are allowed to inherit or accumulate great wealth (money and other assets), but there can never be a truly free society if people are not allowed to make gains and distribute them however they choose, including leaving them for their children. What is more, communism, for all its noble ideas to remove social inequalities, will always be plagued by the "tragedy of common production:" Progress will be delayed as people will simply put in less effort if the reward is to be shared as a common good.

In the *Communist Manifesto* of 1848, Marx and Engels put forward the theory that human society, having passed through successive stages of slavery, feudalism and capitalism, must advance to communism. According to Marxism the final step towards a classless society is inevitable. Marx believed that capitalism had become a barrier to progress and had to be replaced by leadership of the proletariat (working class), which would build a socialist society. To fit conditions prevailing in Russia, Lenin modified Marxism by arguing that in a revolutionary situation the industrial proletariat needs strong central leadership. The responsibility for this was to be taken on by the Communist Party which was to act as the "vanguard of the proletariat," but the power was to be returned to the people as soon as the proletariat, in post-revolutionary times, achieved a full socialist awareness. In turn the power of the party, and ultimately the state itself, would wither away. But that did not quite fly.

Civil War and the Birth of the USSR

After the October Revolution of 1917 Russia immediately attempted to pull out of World War I. Lenin was so eager to conclude peace with Germany that he signed a humiliating treaty under which Russia lost much territory. (In fact, the area ceded to Germany contained about one third of Russia's population, one third of its cultivated land, and one half of its industry.) In 1918 World War I ended on all fronts, but a civil war broke out in Russia, partly because of the unpopular terms of the peace treaty with Germany. The civil war was soon complicated by foreign (including British, French, and U.S.) intervention, which was later going to be a source of anti-Western sentiment. The Soviet government under Lenin's leadership launched a brutal campaign aimed at eliminating political opponents, and by 1920 the Soviet regime emerged victorious. Poland, Finland, and the Baltic countries gained national independence in the aftermath of the civil war, while the Ukraine, Belarus, and the Caucasus countries (Azerbaijan, Georgia, Armenia) were occupied by Soviet armies despite proclaiming their independence. In 1922 they were united with Russia to form the Union of Soviet Socialist Republics (USSR). This Soviet state was going to last until 1991; however, Lenin's successors developed ideas that were quite different from his vision of returning the power to the people.

Industrialization under Stalin

In 1921 Lenin launched his liberal New Economic Policy (NEP), which allowed for private enterprise, and shortly before his death in 1924 he expressed regret at the direction the Soviet government had taken. He criticized its dictatorial manner and complex bureaucracy, and more than anything, denounced the aggressive behavior of his former disciple Joseph Stalin, then general secretary of the Communist Party. Nevertheless Stalin seized power after Lenin's death and emerged as the supreme leader of the Communist Party and the USSR. Contrary to Lenin, who was a lawyer, Stalin had no formal education. He was born in Georgia, the son of a shoemaker, studied for the priesthood, but was expelled from a theological school for insubordination. In turn he joined the Social Democratic party and was repeatedly exiled to Siberia between 1903 and 1913, the year he adopted the name Stalin (Russian for Man of Steel) at age 34. After the February Revolution of 1917 Stalin was released from exile by general amnesty, and after extensive bureaucratic maneuvering in 1922 became the General Secretary of the Communist party. In the later 1920s he turned into a ruthless dictator.

Stalin inherited from Lenin an efficiently operating machine for the mass destruction of political and social opponents, and he further improved on it.[132] He first eliminated all real and imagined enemies by having members of different factions of the Communist party arrested, deported, and executed. He then imposed a restructuring and industrialization program onto the Soviet Union that killed an unimaginable number of people, 20 million according to conservative estimates. Stalin's destruction of the peasant society in itself probably cost 7 million lives. Soviet agriculture nearly collapsed after Stalin in 1928 abolished Lenin's New Economic Policy (NEP) and forced peasants into state-owned collective farms. Millions died in artificially induced famines, chiefly in the Ukraine and Kazakhstan. A million or more Kazakh camel-and-sheep herders starved to death, for instance, after Stalin forcibly turned them into farmers in the 1930s. Some five million members of the (relatively wealthy peasant farmer) kulak class were being either killed or deported to Siberia. Deportation to Siberia or the Central Asian republics was also the fate of several potentially trouble-making ethnic groups, and more Jews lost their lives under Stalin than during the Holocaust. Perhaps 12 million people died in labor camps, and by 1936 some one million had been executed. Then another one million people were shot one by one in the years 1937–38 alone, when about 2 million died in labor camps.

The countryside was exploited to finance the country's industrialization as laid out in Five-Year plans starting from 1928. The urban population had doubled already within the first five years, and by the end of the second period, in 1939, heavy and light (state-owned) industries had been developed. The casualties and the social cost involved were dramatic, but Stalin did indeed achieve his goal: by World War II the USSR was fully industrialized.

World War II and the Warsaw Pact

In August 1939, after Britain had turned down his attempts to form an alliance, Stalin approached Germany, signing a non-aggression pact with Hitler. The two dictators promised not to attack one another, but Hitler later broke the pact, which Stalin probably expected. The idea was thus to win time to prepare for war, but the USSR was nonetheless ill-prepared for the German invasion due to Stalin's mass killings in 1937–38, which included the decimation of the military ranks. The Soviets then suffered tremendous losses with over 20 million casualties but, based on the newly-developed Volga-Ural oil fields, nevertheless emerged from the war as a superpower.

Soviet troops occupied all of Eastern Europe and half of Germany, and subsequently Stalin turned the entire region into a series of Soviet satellite states by installing national communist governments. If there had been time, the Soviet Union might possibly have entered the war in the Pacific as well to expand its sphere of influence in the east. (Arguably, the likelihood of this happening was partly responsible for the United States deciding to drop nuclear bombs onto Japan to end the conflict quickly.) However, the USSR soon started to provide economic and military support globally to countries that had achieved, or were hoping to achieve, independence from Western colonial powers. Relations with the West, and particularly the United States, soured fast, and in 1949 Germany was split into two separate countries, the western Federal Republic of Germany and the eastern German Democratic Republic. The same year the Soviets finalized their first nuclear bomb, and the United States, together with Canada and 10 Western European nations, founded the North Atlantic Treaty Organization (NATO) as a defense alliance against the perceived threat from the USSR.

Stalin died soon thereafter, in 1953, leaving a Soviet Union that used over 25 times the energy it had consumed in 1921, the year the country emerged from the civil war. Nevertheless it took no more than three years before

Stalin was officially denounced as a brutal despot, responsible for mass murders and deportations, by Nikita Khrushchev, new head of the Communist party of the USSR. When West Germany joined the NATO in 1955, the USSR and its Eastern European satellite states established the Warsaw Pact military alliance. The ideological, political, and economic tensions between these blocs were fought out in the Cold War that lasted until 1989, and the competitive forces unleashed in this conflict led to a space and arms race between the USSR and the United States, resulting in major technological advances on both sides.

Technological Supremacy and Economic Deficiency

The USSR emerged as a serious rival to the West in terms of weapons and space technology. After demonstrating a fusion bomb in 1949, and testing a hydrogen bomb in 1953 (just one year after the United States had fired one at Eniwetok Atoll in the Pacific Ocean), the USSR was the first country to launch an artificial satellite (Sputnik I) into space in October 1957. (The Russian rocket program, like the American, was initially based on German technology, but the Soviets did not directly involve scientists from Nazi Germany as America did.) Sputnik II, launched a month later, carried a dog into space. Vostok rockets then lifted the first man (1961) and the first woman (1963) into space. (Both were safely returned to Earth after up to five days in space.) In 1966 the USSR crash-landed an unmanned space probe on Venus, pioneering the landing of a spacecraft on another planet's surface. For comparison, the United States launched its first satellite in 1958, had the first man orbit Earth in space in 1962, and landed people on the moon in 1969. In 1970 the Russians landed an (unmanned) space probe on the moon that took photographs, drilled into the surface, and safely returned a lunar soil sample to Russia. In 1971 the USSR landed a space probe on planet Mars and established the first orbital space station. (The United States established their first station, Skylab, two years later.)

Down on Earth, the economic and political fate of the USSR was less favorable. Uprisings in East Germany (1953), Hungary (1956) and Czechoslovakia (1968) were suppressed by Warsaw Pact troops, and the arms race against the United States and NATO strained the USSR's economy and threatened world peace. In 1972 and 1979 the USSR and the United States agreed on strategic arms limitations (SALT I and SALT II), and the Helsinki Accord of 1975 brought Western recognition of the postwar division of Eastern Europe. However, by this stage it had become clear that the standard of living in the two blocs was more and more diverging in favor of the West. After 1985 political power was transferred to a new generation of USSR politicians led by Mikhail Gorbachev, who introduced a number of reforms and devised a new policy of glasnost (openness). He began to free farmers and factory managers from bureaucratic interference and to increase material incentives in a market socialist manner. His domestic, economic, and political program of restructuring (perestroika) faced opposition, but helped to end the Cold War officially in 1989. What is more, Gorbachev's glasnost policy nourished growing nationalist demands for secession among the republics of the Baltic and the Caucasus.

The Disintegration of the Warsaw Pact and the USSR

In the years following the collapse of international crude oil prices in 1986 (from about $22 to $6 a barrel) the strategic position of the USSR as a critical supplier of oil to its satellite states was severely weakened. More and more Eastern Germans wished to leave for the West, and in the summer of 1989 Hungary suddenly allowed them to cross its border with Austria freely. As thousands of Germans headed for the West via Hungary, the call for freedom within the eastern German Democratic Republic became louder. Following extensive demonstrations for free elections the government of Eastern Germany gave in, and the Berlin Wall, which had divided the historical German capital into an Eastern and Western part, fell on November 9, 1989. The German Democratic Republic ceased to exist the year after, on October 3, 1990, when the two Germanies reunified. The movement spilled over to other Eastern Bloc countries and even to the USSR itself. Hungary held its first multiparty elections in 1990, and Lithuania unilaterally declared its independence in March 1990, which the Soviet government finally recognized following a failed coup against Gorbachev in August 1991. Estonia and Latvia followed suit, and in turn Azerbaijan, Belarus, Uzbekistan, the Ukraine, Georgia, Moldova, Armenia, Kazakhstan, Turkmenistan, Kyrgyzstan, and Tajikistan all declared their independence. Gorbachev was unable to maintain control. On December 25, 1991, he resigned as president, and the USSR ceased to exist, splintering into the Russian Federation and 14 other independent republics.

To be sure, the Russian Federation easily remained the world's largest country even by itself. (The nation stretches from Finland to Japan, but is not all that large in terms of its population, which counted 141 million in 2007.) Russia took over the USSR's permanent seat and veto right in the UN Security Council. The USSR as a whole was formally succeeded by the Commonwealth of Independent States (CIS), a new, decentralized confederation with no real formal political institutions. The main objectives in founding the CIS were to ensure that some measure of cooperation continued in economic and military matters, but several member states rapidly initiated free market economies and even joined NATO. In 2004, Estonia, Latvia, Lithuania, the Czech Republic, Slovakia, Hungary, and Poland, together with Slovenia, Cyprus, and Malta, joined the European Union, and so did Bulgaria and

Romania in 2007. (The European Union, EU, is a supranational organization of countries across Europe, initially founded in 1951 between Belgium, France, West Germany, Italy, Luxembourg, and the Netherlands to establish a lasting peace through economic and political cooperation.) Russia itself struggled in its efforts to build a democratic political system and to overcome the strict social, political, and economic controls of the Communist period. Some progress was made, but under president Vladimir Putin, in office from 2000, power seemed to become recentralized and nascent democratic institutions eroded.

The fall of the Soviet Union is often portrayed as the victory of the market economy over centralized economic planning. The former creates more wealth, as the price mechanism indicates demand changes through rising or falling prices to direct resource allocation, while the latter does not receive immediate feedback as to where demand exists or changes. In addition, the Soviet communist system (in the pre-Gorbachev era) did not provide adequate incentives and rewards for people to work harder or take risks. But the political system was as problematic as the economic. The Soviet Union denied its people personal freedom and the right of self-determination. The centralized government censored authors and the media, and decided what was art. It also diverted immense resources to Cold War efforts. Nevertheless communism never spread much beyond the Soviet Union, the Warsaw Pact states, and China (which however combined accounted for about one third of the world's population). By the time the Soviet Union disintegrated, few communist states were left, including China, Cuba, North Korea, Laos, and Vietnam.

Soviet Oil

Perhaps the best explanation why the Soviet Union, despite its lack of internal coherence and flawed economic and political system, could manage to last as a superpower for several decades is its enormous energy resources. Owing to the Baku region, even the old Russian Empire had been the world's largest oil producer in the start of the 20th century. Annual production in Azerbaijan peaked in 1904 (at around 103 million barrels), but during the Russo-Japanese War of 1904–1905 the situation in the politically charged region escalated, and half the oil installations were destroyed during riots. At the end of World War I Azerbaijan briefly achieved independence, but the Russians were back in 1920, and the Baku region became part of the USSR. The Soviets ousted those foreign industrialists who had persisted through the years of tumults, and by 1921 Azerbaijan's oil production had fallen to the level of 1872. However, by 1930 the production level of 1904 had been restored to fuel Stalin's industrialization plans, and by 1941 war-time oil production reached 172 million barrels a year, a peak not to be reached

Soviet Union: Petroleum Deposits and Pipelines, 1982   Like the United States, the Soviet Union turned out to be very rich in oil. While Azerbaijan was initially the principal source of Soviet oil, the Urals-Volga and West Siberian Lowland oil basins took over from the end of World War II. (Map compiled by the U.S. Central Intelligence Agency. Courtesy of the University of Texas Libraries, The University of Texas at Austin.)

again for the remainder of the 20th century. Azerbaijani oil fueled the Soviet troops during World War II, but by late summer 1942 there was a good chance the Germans would actually reach and occupy Baku. Thus the Russians sealed off the wells in the region and transferred all movable equipment, plus over 10,000 oil workers, from Azerbaijan to the USSR's new oil regions. (Meanwhile Azerbaijan, due to lack of investment, never regained its key importance for the USSR's oil supplies. For this reason Azerbaijan, when it became an independent republic, was left with one of the largest known undeveloped offshore reserves in the world.)[133]

Central Russia and Western Siberia

The Soviets had discovered oil fields in the region of Timano-Petchorsk (in the Komi province situated in the extreme northeast of European Russia), and on Sakhalin (next to Japan), already in the 1920s. However, these areas were considered too remote, and production levels remained minimal (even during the war). But Soviet oil prospecting efforts continued, and in 1932 oil fields were found in central Russia in the Ishimbaev area as well as the Bashkiria province (in the region around Samara/Kuibyshev, Orenburg and Perm). It soon became clear that the Volga-Ural region might hold even larger oil deposits than Azerbaijan, but oil extraction was still very low there by the early 1940s. Then about half the people and equipment transported away from Azerbaijan arrived in the area around Kuibyshev (Samara), the city situated on the confluence of the Volga and Samara rivers, and a huge refining complex was set up to serve the five major oil and gas fields discovered in the region. The Romashkin oil field, discovered in 1948 in Tatarstan (the republic located at the confluence of the Volga and the Kama rivers), was the most significant find of the post-war years. During the second half of the 1950s about 75 percent of all Soviet oil was extracted in Volga and Urals basins, which also accounted for some 80 percent of the USSR's known oil reserves. This was of enormous strategic importance for Russia, because this fuel was situated in the heartland rather than in one of the politically challenging republics of the periphery. And when oil extraction in the Ural-Volga region finally peaked in 1975, oil had long since been discovered on the other side of the Urals, in western Siberia. The great Siberian oil fields are the largest in the world save those of the Persian Gulf. The first oil field was discovered in this region in 1960, in Trekhozerie, and prospecting during the following years was extremely rewarding. First, increasingly more oil fields of the so-called Shaimov Group were discovered, and finally the huge Samotlor field, with reserves of more than 15 billion barrels, was found in 1965.

Russia also emerged as a giant natural gas producer. The Soviet Union's European gas fields in the Volga-Ural region dominated production through the 1970s; thereafter production shifted to the giant fields of Siberia, among which the Urengoy (the largest gas field in the world) and Yamburg fields turned out to be most productive.

The World's Top Oil Producer (1973 to 1991)

The Soviet Union produced substantial oil surpluses from about 1955, and emerged as a major exporter with the 1964 opening of the Druzhba pipeline, passing through the Ukraine and Slovakia. (The first main gas pipeline from West Siberia was completed little later, in 1967.) In terms of total oil production the SR passed Venezuela as the second largest producer in the world in the early 1960s, while the United States maintained the top position. But there was a major difference between the two global superpowers. The United States, though the top producer, had to import large amounts of oil to meet its enormous domestic demand, while the USSR could afford a much more relaxed attitude in regards to its foreign policy when it came to the Middle East, where the world's largest oil fields are located. What is more, U.S. oil production peaked and began to decline in 1970, while Soviet oil production steadily and rapidly increased. Soviet oil production passed U.S. oil production in about 1973, and Russian production (not accounting for the oil produced in other Soviet republics) surpassed that of the United States in 1977. The USSR in turn remained the world's top oil producer until it disintegrated in 1991. (Soviet peak production was reached in 1988 at around 4.4 billion barrels per year.)

The Soviet Union's vast energy riches promoted inefficiencies. The Soviet government controlled and subsidized the energy sector, and set the price of oil far below the level of world market prices, which encouraged excessive consumption. Inefficient Soviet (heavy) industries in turned absorbed so much oil that relatively little was left to export. (Even at the 1988 peak production of 4.4 billion barrels per year, exports accounted for merely 0.73 billion barrels per year. In 1993 it took about 4.46 tons of oil equivalent (TOE) to produce US$1,000 of Russia's GDP, compared with an average of 0.23 TOE to produce US$1,000 of GDP for OECD member countries.) What is more, natural gas, as it was so abundant and so easily accessible, was consumed in such huge quantities that oil by the mid-1990s accounted for merely 20 percent of Russia's energy consumption, while gas accounted for more than half. (Besides, Russia also has abundant coal resources.)

The Post-Soviet Era

During the disintegration of the Soviet Union oil production declined, but remained above U.S. levels and picked up again in the mid-1990s. Production in the Russian Federation alone was 2.6 billion barrels in 2001, equivalent to that of the United States. But again, Russia produced 1.7 billion barrels of oil in excess of its own consumption, while the United States was the world's largest importer and saw its domestic production decline further. (Russia was hence the second largest oil exporter after Saudi Arabia.) By this stage roughly two-thirds of the Russian Federation's oil came from Siberia, mostly from the huge fields in West Siberia, but also from East Siberia and the Far East province. Nevertheless, the Volga-Ural region, the North Caucasus, and the Komi republic remained important sources of oil and gas for Russia east of the Ural mountains, accounting for one-third of Russian oil production.

During the Soviet period all oil operations and infrastructure had been organized by the government, while Russia in 1992 introduced legislation concerning private enterprise, privatization, and foreign investment that was to become the foundation for the reform of the energy sector. Within few years Russia's oil and gas sector was dominated by private and joint-stock companies, with Gazprom, Lukoil, Yukos, TNK, Surgutneftegas, Sibneft, and Slavneft among the largest. Lukoil (LangepasUraiKogalymneft), for instance, was set up in 1991 as a state-owned oil concern uniting three oil producers and three oil refiners, but from 1993 became an open joint stock company that by some measures was the largest oil firm in the world. In 2003 British Petroleum (BP) purchased 50 percent of TNK, then Russia's 3rd largest oil company (in terms of production and reserves). And when Yukos, Russia's second largest oil company, merged with smaller rival Sibneft in October 2003, it became Russia's largest oil and gas firm and the fifth largest private oil company in the world.

However, it also became clear that the Russian government would not entirely let go of the control over the energy sector. True, oil production in Russia was now mostly in the hands of the private sector, but Transneft, the company controlling the pipeline network, remained state-owned. Similarly, Russia's gas production and transmission was controlled by the State Natural Gas Company (Gazprom), a company that was reorganized into a joint-stock company in which 40 percent of the shares remained under government control. Thus, the government controlled all exports of oil and gas, which gave it considerable foreign policy power in regards to receivers of Russian energy in eastern as well as western Europe. (Some 85 percent of Russian gas exports are delivered to western Europe via the Ukraine, which itself receives some two thirds of its natural gas needs from, or through, Russia.)

Though virtually all large international oil corporations have attempted to invest in Russia (either through joint-ventures or by purchasing stakes in local companies), such undertakings remained risky. Russia's president Vladimir Putin, a former KGB official, on the one hand promoted liberal reformers to key posts, but on the other hand squashed Russia's independent media and manipulated the 2004 elections so that he won by a land-slide victory. Putin also oversaw the arrest of Mikhail Khodorkovsky, the former chief executive of Yukos, who had publicly criticized the president. (The government practically hijacked Yukos.) Putin also sacked the prime minister, Khodorkovsky's chief governmental ally, along with his government.

Nevertheless, substantial foreign investments (by Western firms) in the Russian oil and gas infrastructure helped Russia to pass Saudi Arabia as the largest oil producer in the world. (In 2001 Saudi Arabia had still produced well above Russian and United States levels, extracting 3.1 billion barrels per year.) By 2004 Russia produced at the level of 3.3 billion barrels a year, targeting European and Asian export markets. However, with just about five percent of the world's proven oil reserves Russia was expected to exhaust its reserve base relatively fast, within few decades. At the beginning of the 21st century it remained unclear whether the USSR's successor, the Russian Federation, would reemerge as a superpower. It certainly had the energy resources (and nuclear weapons) to do so, but it remained doubtful whether Russia would be able to challenge the position of the United States, the last remaining superpower at the early 21st century.

UNITED STATES OF AMERICA

The United States is now the third largest country in the world, well behind Russia, but only slightly smaller than Canada, and ahead of China. The nation grew large during the Coal Age mainly at the expense of its southern neighbor Mexico and Native Americans, and in the later 19th century the United States established itself as an international Coal Power with strong overseas imperialistic ambitions, chiefly in South America and in Asia-Pacific. In addition to huge coal stocks and vast areas of fertile land in temperate climate, America also turned out to be endowed with tremendous oil riches. The nation's first big gusher was struck at Spindletop, Texas, in January 1901, and for the remainder of the 20th century no other country was able to rival cumulative U.S. oil production. (From the very beginning of oil extraction to 1995 the United States produced a cumulative total 165.8 billion barrels of crude oil, compared to Russia with 92.6 billion barrels of crude oil.) Annual production was challenged by the Soviet Union and Saudi Arabia towards the end of the 20th century, but the United States remained a top-three producer and by far the world's largest consumer of oil.

U.S. rise to international power is closely associated with high energy use. In the 1850s the United States was still an overwhelmingly rural country of only marginal global importance, but a century later, after more than tripling its per capita consumption of energy, it was both an economic and military superpower. Much of the initial oil technology, including Otto engines, Diesel engines, jet engines, rocket engines, automobiles, oil tankers, and oil refining technology, was imported from western Europe, but many of these innovations were further developed in the United States.

Despite both nations' enormous oil wealth, the U.S. rise to superpowerdom was quite different from that of the Soviet Union. No war was fought on U.S. soil during the entire 20th century. (The attack on Pearl Harbor, Hawaii, far away from American mainland, was the one exception.) No U.S. factories have ever been bombed, no U.S. infrastructure destroyed. America attempted to stay out of both World Wars and was pulled into both of them at a late stage. The Europeans lost millions of adults in both wars, while the U.S. workforce remained practically unharmed. (In World War II the United States suffered 295,000 casualties, while the Soviet Union suffered over 20 million.) America enjoyed an extraordinarily safe geographic position: It is separated from Eurasia by the Atlantic and Pacific Oceans; it has been at peace with its sole northern neighbor Canada since 1812; and it never had to view its southern neighbor Mexico as a threat. While Europe was drawn into political turmoil and war over border disputes, the United States did not have to fear outside aggression and maintained a huge internal domestic market that allowed the economy to prosper even when foreign markets were going up in flames. U.S. stability also resulted in stable democracy: the nation was spared dictators of the kind of Hitler, Stalin, Mussolini, or Franco. In turn, the large degree of personal (and religious) freedom combined with economic prosperity kept on attracting large numbers of immigrants, who were being filtered for their skills and potential contributions to the nation's wealth. (Already the Immigration Act of 1924 further reduced quotas for immigrants considered less desirable. Immigrants from Great Britain, Germany, and Ireland were assigned generous quotas, while those for southern and eastern Europeans were cut back, and Asians were practically barred from entering the United States. The latter had principally begun with the Chinese Exclusion Acts of 1882.)

Between the Wars

Nevertheless, United States economic development has not always been smooth. Rapid economic growth in the 1920s ended abruptly with the stock market crash of 1929, which marked the start of a deep economic depression that severely affected not only America, but most of the world. The Soviet Union was not affected at all, which influenced various countries' response to the crises: Central economic planning and state ownership appeared to provide a structure that would avoid economic downturns. Britain adopted far-reaching measures in the development of a planned national economy, Nazi Germany pursued rearmament, conscription, and public works programs, and in Italy fascist dictator Mussolini tightened the economic controls of his corporate state. In the United States, where output fell by 30 percent between 1929 and 1933, while unemployment soared from three percent to 25 percent, Democratic nominee for the presidency Franklin Roosevelt in 1932 promised "a new deal for the American people." Following the elections Roosevelt's New Deal increased government intervention to stabilize the economy. The Agricultural Adjustment Act (AAA), introducing subsidies and production controls, fostered a long-lasting federal role in the planning of the agricultural sector. Industrial production gained momentum again when the government began directing enormous funds towards the military-industrial complex that was soon growing rapidly in preparation for World War II. Oil and gas increased from 32 percent in 1929 to 45 percent in 1939 in terms of their contribution to overall U.S. energy consumption, while huge hydroeclectric schemes were being constructed under the New Deal's Tennessee Valley Authority (TVA) to generate electricity for the production of aluminum needed to build airplanes and bombs. By 1942 the TVA employed 28,000 people and the U.S. economy had achieved full recovery and full employment (and the electrification of the countryside). 1942 was also the year in which the United States entered World War II. After investing 6 billion barrels of oil into the war, and dropping two nuclear bombs on Japan, the United States emerged from the conflict as a superpower.

The Golden Era (1945 to 1970)

As it had been the war preparations that had pulled the United States out of the Great Depression, many feared that the economy would contract again once the conflict was over. But these concerns were unfounded: between 1950 and 1970, when U.S. oil production peaked, GDP more than doubled (in real terms). This extraordinary economic growth was accompanied by extreme oil production and consumption levels, and lack of foreign competition on national and international markets. Western Europe, Eastern Europe, the Soviet Union, Japan, and China had all been devastated during the war, and U.S. companies could easily capture all global markets that would accept American goods and services. Domestic markets were booming, too. Americans had saved much of their disposable income during the war and were eager to spend it once peace was restored and the long troubling period since the 1929 Wall Street crash was finally over. The 1950s saw the advent of professional advertising campaigns, many of which were targeted at the rapidly swelling middle class. An unprecedented house construction boom pulled people from city apartments into rapidly growing suburbs, while the rural poor, pushed by increased mechanization in farming, moved into the cities. (The mechanical cotton picker, for instance, pushed black sharecroppers off the land and towards such cities as Chicago, Cleveland, Detroit, and New York that whites were abandoning.) Eighty-five percent of new homes were built in suburbs during the 1950s, and by 1960 one-third of all Americans lived in suburban homes, relying on oil-powered cars for their mobility. Eisenhower, inspired by Hitler's Autobahn, in 1954 signed the act that created the Interstate Highway System (at the expense of the railroad). Government intervention remained generally high, and the U.S. Golden Era economy deserves being called a mixed rather than a free market economy. The government financed the housing boom, for instance (by offering mortgages for new home owners and providing suburban home constructing businesses with production advances). It also financed public education. Most post-war jobs were assembly line jobs that required little formal training, but after the Sputnik shock of 1957, the National Defense Education Act of 1958 provided a huge infusion of money into education (in order to keep pace with the Soviets). From 1965 the federal government committed itself to large-scale aid to elementary and high schools, and it generally financed military-related research as well as much of private industry's research and development, which helped the United States to emerge as a global technological leader. (Government funding during this period also included the research into what was to become the Internet.) Meanwhile such assistance programs as Medicare and Medicaid, begun during President Johnson's (1963–1969) War on Poverty, directly helped individuals and families. In turn the percentage of Americans living in poverty declined from 21 percent in 1960 to 11 percent in 1970.[134]

To be sure, much of government spending kept on flowing into the military-industrial complex. This created a lot of employment and economic growth, especially when the Cold War was warming. The government was also responsible for the import of foreign experts in the years after World War II. This included hundreds of German military experts, many of whom were ardent Nazis. (U.S. authorities cleansed their files of Nazi references, organized the transfer of their families from Germany, and within short time handed them U.S. passports.) Kurt Blome, who had conducted experiments on humans in concentration camps, was hired by the U.S. Army Chemical Corps to work on chemical warfare. Walter Schreiber, who had organized such experiments, was employed at the Air Force School of Medicine at Randolph Field, Texas. (Schreiber was flown out to Argentina in 1952 when evidence of his previous activities reached the American public.) Reinhard Gehlen, Hitler's chief of intelligence responsible for the Eastern Front, was hired by the U.S. Central Intelligence Agency (CIA). Wernher von Braun, who had reported to Hitler personally and had formally been an SS member, was imported to America together with 126 scientists of his staff to develop missiles. One of von Braun's rockets lifted the first American satellite into space for the military, and eventually he was put in charge of the NASA Apollo project to fly people to the moon in 1969. Arthur Rudolph, an early (1931) member of the NSDAP and operations director at the Mittelwerk missile factory, was the project director of the Saturn V rocket program.

After the Oil Peak (The Post-1970 Era)

Due to resource depletion the United States was unable to increase its oil production in the 1970s even when international supply interruptions drove up the oil price. Economic growth rates declined, and America faced increasingly more economic competition from western Europe and Japan. With the boom years over, economic inequality increased sharply in the United States from 1970. U.S. workers earned less in 2004 in real terms than in 1969. Middle income households (defined in 2001 as households earning between $33,315 and $53,000 a year) were also worse off than three decades earlier. Meanwhile the share of national income earned by the top five percent of America's wealthiest households increased steadily to reach 22.4 percent of the total by 2001. Nearly all household gains since 1975 went to the top fifth of U.S. households, comprised by people having the professional skills to be employed by American firms that remained technological leaders in such industries as computing (software), medical, aerospace, and military equipment.[135]

U.S. Imperialism in the Name of the Cold War

Though the United States and the USSR had been allies during World War II, the newly emerged superpowers soon thereafter opposed one another in a conflict that is commonly known as the Cold War and lasted nearly until the collapse of the Soviet Union around 1990. To be sure, the bonds between the two nations had never been especially tight. The Soviets had every reason to be reserved, as the United States, alongside Britain, France, and Japan, had intervened militarily in Russia at the end of World War I. (The Western governments wanted to overthrow Russia's new communist leaders, and to restore the collapsed Eastern Front.) The United States then did not recognize the Soviet government until 1933. Britain in 1939 turned down the Russians' offer to cooperate (which prompted Stalin to make a flimsy pact with Hitler), and the United States dragged its heels for two and a half years during World War II before opening a second front in Europe to take some pressure off of the Soviets (who lost 20 million people in the conflict).

After the Russians had chased the Germans out of all of eastern Europe in 1945, they attempted to install friendly governments along their western border: After all they had been invaded from the west twice within just one generation. They also considered access to (non-Russian) eastern European resources vital to achieve economic recovery. The United States agreed that the Soviet Union would become the predominant power in eastern Europe (and especially Poland and Romania), while the Soviets agreed to Western influence in Italy and Greece.

The United States was in a tremendously strong position. While most of the industrial world was devastated, America had come out of the war largely unharmed, and had just demonstrated to the world that it was (the only country) able to produce (and willing to use) nuclear bombs. The western European powers were about to lose the colonial empires they had built up in the Coal Age, and the United States, which had traditionally limited its sphere of influence to the Americas and the Pacific, was in a prime position to expand its imperialistic ambitions by taking over the remnants of the European colonial empires (including spheres of influence in Africa and Asia), and to dominate western Europe itself. As an additional presentation of its power, the United States in 1946 carried out the first of 23 well-publicized nuclear bomb tests in the South Pacific, on Bikini Atoll (Marshall Islands).

In the midst of this situation communist rebels in the southern European country of Greece started to fight the monarchical regime. Britain supported the existing government, but in 1947 informed the United States that it did not have the resources to maintain these efforts. U.S. president Truman in turn portrayed this uprising as not just an act by rebellious local communists, but rather by Soviet Communism, whose aim it was to control the Middle East (which was already known to be rich in oil), South Asia, and Africa. The American public did not buy it at first, and Congress initially turned down Truman's demand for hundreds of millions of dollars to give military and economic aid to Greece and Turkey. However, anti-Soviet propaganda continued, and Congress approved what became known as the

Truman Doctrine (1947) to extend full aid to Greece and Turkey, and to abandon the traditional U.S. commitment not to intervene in Europe during peacetime. The Doctrine established a public consensus for Americans to be willing to fight the Cold War for the purpose of containing the spread of Communism. This was soon used as justification to apply American troops overseas whenever U.S. interests were at steak. What is more, the National Security Act of 1947 established the Central Intelligence Agency (CIA) and the National Security Council (NSC). Henceforth, U.S. presidents could commit federal funds through national security directives without informing Congress or the public. (Several U.S. presidents in turn used this tool to finance covert operations to interfere in the affairs of foreign countries.) And soon the United States was going to support corrupt regimes around the world, to assist in the murder of thousands of people, and to help the overthrow of democratically elected governments, all in the name of stopping the spread of communism. In effect, these activities created a new kind of global empire that included strongholds on all continents. (Covert operations were usually carried out by the CIA, acting on the U.S. president's orders. Many CIA documents remained classified even after 30 years, while others reached the public through Freedom of Information Act lawsuits.)

Obviously the Soviets were not pleased about this polarization in American politics. From their standpoint they were being encircled by Western influence zones, but their country was devastated and there was no way they would be able to match American overseas economic or military aid. The United States began providing western Europe with substantial financial aid within the Marshall Plan, named for the World War II general who was Truman's Secretary of State from 1947 to 1949. Under this scheme (known officially as the European Recovery Program) the United States spent more than $13 billion ($88.2 billion in terms of 1997-$) over a four-year period (1948–51) to promote quick economic recovery and U.S. political influence in western Europe. (The perceived danger of communist takeover in postwar Europe was the main reason for the aid effort.) The USSR had no means to undertake similarly large efforts in eastern Europe, which had historically been poorer and less industrialized than western Europe anyway. Hence the differential between the two blocs was bound to increase drastically, which clouded the West-East relations even more. When the United States in 1948 decided to unite Germany's American and French military zones and bring them under the Marshall Plan scheme, the Cold War began to heat up. The Soviets opposed this idea, as they preferred a divided and economically weak Germany (at least until eastern Europe could fully recover from the war), and blocked all road traffic into Berlin, Germany's largest city and historical capital, situated in the country's east. The United States responded with an air-lift, which after some 11 months persuaded the Soviets to reopen the corridor to Berlin. However, before the crisis ended Truman had threatened nuclear war and transferred B-29 bombers carrying atomic bombs to England. In 1949 the Soviets presented a nuclear bomb of their own, and the same year the United States founded NATO, a military alliance of Western countries. Germany was split into two separate nations, West Germany being the one, and East Germany, consisting of the Soviet military occupation zone, being the other. In 1955, when West Germany joined NATO, the Soviets established the Warsaw Pact military alliance with their Eastern European satellite states to counter NATO.

U.S. Imperialism in East and South Asia

In 1949, Mao's communists won out in China's civil war against the Chiang Kai-shek's nationalist forces. The nationalists had been supported by the United States, but when some 80 percent of the American equipment given to them had fallen into the hands of the communists, the Truman administration decided to terminate aid, arguing that the communist take-over of China was beyond the control of the United States. With the Russians in possession of a nuclear bomb, and China fallen to communism, America was fully alert. And right then, in 1950, North Koreans (acting independently) crossed the border and overran almost all of South Korea. The international community (with passive support of the Soviet Union) decided to oppose them, but communist North Korea was supported by China. U.S. troops were landed in Korea and worked their way through the entire Korean peninsula, but then the Chinese (not yet in possession of nuclear weapons) entered the war and pushed the U.S. forces back into South Korea. In 1953 this first armed conflict of the Cold War period ended with the restoration of the original boundary between North and South Korea, along the 38th parallel. (The Korean War cost the lives of perhaps 4 million Koreans (North and South, two-thirds of them civilians; up to 1 million Chinese soldiers; 36,934 Americans (plus 103,284 wounded); and 3,322 from other UN nations.)[136]

In addition to South Korea, the United States maintained Japan and Taiwan as its strongholds in East Asia. All three countries were under the protection of the U.S. military and received economic aid. In both Japan and South Korea the Allies introduced radical political reforms, and both countries essentially became Western-style democracies. Taiwan in 1949 became the refuge for the defeated Chinese nationalist government forces of Chiang Kai-shek. The nationalists continued to claim sovereignty over mainland China and were recognized by the United States as the legitimate government of China until 1971, when Taiwan's permanent seat in the UN Security Council was finally transferred to the People's Republic of China. Thereafter the United States kept protecting Taiwan, but China never recognized the island as an independent country.

THE PHILIPPINES In the Philippines the United States supported corrupt president Ferdinand Marcos. The Philippines had been under U.S. rule from 1898, were occupied by Japan from 1942 to 1945, and became an independent republic in 1946. Marcos was elected in 1965 and 1969, even though he had been convicted of murdering a political opponent of his father in 1939. In 1972, some months before his second term was completed, he declared martial law, suspended the constitution, and began to rule by decree, while pro-Marcos groups terrorized and executed his opponents. Marcos fully lifted martial law only in 1981, and in turn validated his power through a fraudulent presidential election. When opposition leader Benigno Aquino returned from exile in 1983, he was assassinated within minutes after his arrival at Manila airport. This triggered the massive non-violent protests (led by Benigno's widow Corazon Aquino) that forced Marcos into exile in the United States (Hawaii) in 1986. Following an investigation, a U.S. grand jury alleged that Marcos and his wife had embezzled over $100 million from the government of the Philippines, but Marcos was too ill to stand trial and died in 1989. U.S. economic and military aid to the Philippines continued after Marcos's abdication (about $1.5 billion between 1985 and 1989 alone), but the Philippine senate in 1991 rejected a renewal of the U.S. lease for its Subic Bay naval base.

INDOCHINA (VIETNAM, CAMBODIA, AND LAOS) In Indochina the French attempted to regain their lost colony when the Japanese were gone, but were drawn into a war against local independence fighters (supported by China) from 1946. By 1954 the French were finally defeated, and the permanent members of the UN Security Council met with representatives from Vietnam, Laos, and Cambodia, agreeing that Vietnam would be temporarily partitioned into a northern and a southern part before being reunited in 1956 following national elections. The Viet Minh, a Communist group that was concentrated in North Vietnam and had led the fight for Vietnamese independence, was expected to win these elections. Hence South Vietnam, where the French and their Vietnamese supporters were based, refused to hold the elections. In turn North Vietnam attempted to overthrow the South Vietnamese government. The French left, but the United States backed South Vietnam, dispatching more and more military aid and advisers to the region, while North Vietnam was fighting a guerilla war against the South. By 1960 there were 900 American military advisers in South Vietnam, and from 1965 the United States sent a total of over one million combat troops to fight a long and bloody conflict alongside the South Vietnamese. The draft, the high war casualties, and the undeclared nature of the war all made the conflict unpopular in America, but newly-elected President Nixon nevertheless expanded it to Laos and Cambodia before phasing out U.S. involvement in the region between 1973 and 1975. By 1976 South Vietnam was annexed by North Vietnam, and the country was finally reunited under the name Socialist Republic of Vietnam.

In short, America lost this conflict, despite pouring an astounding five billion barrels of oil into the Vietnam War (much of it extracted in the Middle East), compared to six billion barrels the United States had invested into World War II. Between 1961 and 1975, one million North Vietnamese soldiers, 200,000 South Vietnamese soldiers, 56,555 U.S. soldiers, and 500,000 Vietnamese civilians were killed. In the end of 1967 there were half a million American soldiers present in Vietnam, some of them committing war crimes, including murder, rape, and torture on a large scale. In the My Lai incident alone (March 16, 1968) U.S. soldiers massacred as many as 500 unarmed civilians, including old men, women and children. (The massacre was in turn covered up. Twenty-six men were eventually charged, but only a single soldier was temporarily imprisoned. Colin Powell, later Secretary of State under the second U.S. President Bush, was involved in the investigation. He was later accused of contributing to the cover-up.)[137]

The war also spilled over to neighboring Cambodia and Laos. Cambodia had achieved independence in 1954, but Prince Sihanouk, head of state from 1960, steered a neutral course rather than supporting South Vietnam. In 1969–1970, President Nixon's national security adviser Henry Kissinger sanctioned the (secret) illegal carpet-bombing of neutral Cambodia that killed thousands of civilians. The target was suspected communist base camps that supported North Vietnam. The United States then helped local military leader Lon Nol to organize a coup and overthrow Sihanouk in 1970, and American and South Vietnamese troops invaded Cambodia to eradicate North Vietnamese forces. Lon Nol assumed total power in Cambodia and fully supported the United States in the war, but when the Americans left Vietnam in 1975 Lon Nol fled to the United States as well. In the chaos created by the United States in Cambodia, one of the Oil Age's most brutal dictators, Pol Pot, came to power. Pol Pot was the Chinese-backed leader of the Khmer Rouge, a radical communist guerilla movement that found plenty of sympathizers after U.S. forces had started to operate in the region. Pot's regime ruled Cambodia between 1975 to 1979. During this time it executed hundreds of thousands of people, and over a million died from starvation and hardship. Currency was abolished, workers were turned into slave laborers, and every sign of Western culture or technology (such as speaking English or wearing glasses) was punished by execution. In 1979 the Soviet Union and Vietnam together invaded Cambodia and removed Pot's regime, but the Khmer Rouge established itself in the hinterlands. UN-sanctioned elections were finally held in 1993, and Sihanouk was restored to the monarchy. Pol Pot died in April 1998 from an apparent heart attack. He was never brought to trial.

Laos during the Vietnam War earned the unenviable distinction of being the most heavily-bombed place on earth, as the United States dropped more bombs on the country than fell on all of Europe during World War II.[138] Laos fought hard to gain independence from France in 1953, but the country was thereafter de facto split into a communist north ruled by the Pathet Lao (Lao State) movement, and the southern part ruled by the internationally acknowledged royal Lao government. Laos was then marked by civil war and instability, and its neutrality was violated by both the United States and North Vietnam during the Vietnam War. America recruited, equipped, and trained a mercenary force of Hmong tribesmen to fight the Pathet Lao, and carpet-bombed northern Laos. (As about 30 percent of the unimaginable large number of American bombs dropped during 580,344 missions did not detonate, Laos even 30 years later remained a dangerous place: It seemed impossible to clear the ground from leftover-bombs found just about everywhere.) However, with the United States and South Vietnam defeated, the Pathet Lao in 1975 with support of their North Vietnamese allies made its move, overthrew the royal government, and proclaimed the Lao People's Democratic Republic, a communist state that lasted into the 21st century.

INDONESIA In Indonesia, the world's fourth most populous country, the United States supported brutal army general Raden Suharto. After the Japanese had surrendered in 1945, the leader of the Indonesian Nationalist Party (PNI), Achmed Sukarno, proclaimed Indonesia's independence. However, Dutch attempts to retake the former colony had to be defeated before Sukarno in 1949 was elected president of the new federal state. His plans to establish a centrally-ruled democracy failed, however, as conflicts between communists, Muslims, and regional groups resulted in a series of coups and violent confrontations. From 1955 the United States attempted to undermine and discredit the regime of President Sukarno, who allied himself with the communist party PKI and the army to achieve relative stability in Indonesia. However, by the 1960s inflation soared, a huge foreign debt accumulated, and Sukarno's increasingly corrupt rule strained international relations. Indonesia then tightened its relations to the Soviet Union, which supplied arms during the conflict with the Dutch over the recovery of the western part of New Guinea in 1960–62, and the conflict with Malaysia over Borneo in 1963.

After Sukarno had nationalized several Dutch-owned industries and (in 1965) Indonesia's oil reserves, an attempted coup, purportedly by communists, was defeated in October 1965 by Raden Suharto, a hitherto unknown army general. Suharto in turn seized power and during the following months oversaw the mass-murder of between 700,000 and a million people. Targeted was anyone remotely suspected of having communist sympathies, including large numbers of Indonesian Chinese claimed to have links to communist

China. The United States (CIA) assisted the killings, providing Suharto with lists of communist suspects. (From what is known, U.S. agents supplied some 5,000 names to the Indonesian military.) Ten years later, in December 1975, U.S. President Gerald Ford and his Secretary of State, Henry Kissinger, met with Suharto and approved Indonesia's immediate invasion of East Timor, a former Portuguese colony that had just achieved independence. This led to the massacre of 200,000 Timorese. (Suharto's troops received military training and arms from the United States and Britain.) By 1989 over one-third of East Timor's population of about 650,000 had lost their lives, and independence was only re-achieved in 2002. Suharto also brutally suppressed separatist movements in the Moluccas Islands, Western New Guinea, and other parts of Indonesia. He reacted by relocating over six million people, mainly from populous Java, Madura, and Bali into outer, separatist islands.

Suharto phased out support of China and the Soviet Union, encouraged foreign investment, and during his three decades of uninterrupted rule delivered oil to world markets. Hence the West was satisfied. Oil revenues enabled the government to finance numerous development programs during the 1970s, but the oil wealth also gave rise to corruption on an unprecedented scale. The public focused on this issue following a severe economic downturn, and Suharto finally resigned in 1998. It had been known all along that much of the oil revenues went into the pockets of Suharto's close family, which enjoyed various monopolies and amassed a fortune estimated at US$ 15 billion in 1998. Suharto was placed under house arrest in 2000, but was declared unfit to stand in trial. (In January 2008 Suharto died at age 86, after suffering multiple organ failure.) His son Tommy Suharto was sentenced to 15 years in jail for arranging the murder of a judge who sentenced him for his role in a land scam in September 2000, but Tommy serves this sentence in a comfortable open environment. In 2001 Mrs. Megawati, the daughter of former President Sukarno, became President of the Republic of Indonesia.

U.S. Imperialism in Latin America

In Latin America the United States had a long tradition of directly intervening in the affairs of various countries. Latin America had been declared a U.S. sphere of influence in the Coal Age, and interests in the region were to be protected against those of western European powers. During the Cold War the first U.S. (CIA) covert operation in Latin America targeted Guatemala, where democratically elected president Jacobo Arbenz Guzmán continued reforms to curb the power of the army and install political freedoms. However, Arbenz's agrarian reforms in 1954 included the nationalization of the plantations owned by the U.S. United Fruit Company, which caused the United States to sponsor a revolution led by army general Carlos Castillo

Armas. But President Truman had authorized an effort to overthrow Arbenz even earlier, in 1952. The CIA then produced lists of Guatemalans to be eliminated "through Executive Action." According to its own files, the CIA considered assassinating Arbenz, but he and his top aides fled the country.[139] Others were less lucky. The new U.S.-backed military dictator killed hundreds of Guatemalans right away, and successive military regimes tortured and murdered more than 100,000 civilians in Guatemala between 1954 and 1990. The Guatemala coup spread anti-U.S. sentiments all over Latin America, but the activities continued. In the small country of Guyana, neighboring Venezuela, the government of Cheddi Jagan's left-wing People's Progressive Party was repeatedly destabilized, but after winning the 1953 elections, Jagan was reelected in 1957, 1961 and 1992.

CUBA U.S. relations with Cuba turned out especially strained. Cuba had been ceded from Spain to the United States in 1898, and became an independent republic in 1901. However, the United States retained a naval base on Cuba (at Guantánamo Bay) in perpetuity and asserted a right to intervene in internal affairs until 1934. Fulgencio Batista, an army sergeant, ruled Cuba (with support from the United States) from 1933 to 1944, and, following an army revolt, from 1952 to 1959. Fidel Castro, a young lawyer and son of a sugar planter, unsuccessfully tried to overthrow Batista's corrupt and brutal regime in 1953 and 1956, and finally deposed Batista in 1959 with the help of his 5,000 or so guerrilla fighters. Castro nationalized all foreign businesses without compensation, and the United States immediately enacted an economic embargo. America then organized several failed assassination plots against Castro, and sponsored an invasion by 1,500 Cuban exiles at the Bay of Pigs in April 1961. The invasion failed and motivated Castro to seek protection from the Soviet Union. In December 1961 Castro proclaimed a communist state whose economy would develop along Marxist-Leninist lines. The U.S. government then had Operation Northwoods drafted, which intended to win public support for an invasion of Cuba by staging attacks on various U.S. targets and making it look as if Cuba was behind them. In 1962 Castro allowed the Soviet Union to station missiles with nuclear warheads in Cuba. The Cuban missile crisis brought the United States and the USSR to the brink of nuclear war, but conflict was averted when the USSR gave in and agreed to dismantle the missiles. In the 1960s Cuba provided military services in different Latin American (and African) countries, notably sponsoring guerrilla forces in Bolivia and Venezuela. Castro remained prime minister until 1976 and became president thereafter. He raised the standard of living for most Cubans, but hundreds of thousands of the middle class fled the country. He also dealt harshly with dissenters. Castro's administration introduced a centrally planned economy based on the export of sugar, tobacco, and nickel, with aid for development being provided by the Soviets. Castro was still in power at the beginning of the 21st century.

BRAZIL   In Brazil João Goulart became president in 1961 and won more than 80 percent of the votes cast at a plebiscite in 1963. He expropriated oil refineries and uncultivated land owned by foreign companies, passed a law limiting the share of profits multinational corporations could transfer out of the country, and adopted an independent foreign policy opposing sanctions on Cuba. A U.S.-backed army coup overthrew him in 1964 and installed General Castelo Branco to dictatorial power. Brazil then went through 15 years of military dictatorship, with all the usual features, including disappearances, death squads, and torture. Meanwhile Brazil broke relations with Cuba and became one of the United States' most reliable allies in Latin America.[140]

CHILE   In Chile Marxist Salvador Allende was elected president in 1970, and in turn nationalized the banking and (U.S.-owned) copper industries. The United States had already sabotaged Allende's electoral endeavor in 1964, and now attempted to destabilize Chile. U.S. President Nixon ordered maximum CIA covert operations to "prevent Allende from coming to power or unseat him." Following Allende's electoral victory, Nixon ordered that "No impression should be permitted in Latin America that they can get away with this" and further stated: "I want to work on this and on the military relations—put in more money. On the economic side we want to give him cold turkey." The United States then installed a full-scale economic embargo, cut off international loans to Chile, and supported strikes. By 1973 Chile's economy took a major downturn. The price of copper, Chile's main export, had radically declined, and the country was plagued by hyperinflation. The military received vast payments from the United States and on September 11, 1973, staged a coup during which the presidential palace was bombed. Allende died during this coup d'état, possibly through suicide. (As documents concerning the coup remain classified, the full extent of U.S. involvement remains unknown.)[141]

Ruthless General Pinochet took power and democracy would not return to Chile until his retirement in 1990. Pinochet brutally crushed all opposition, and thousands were killed in mass executions or disappeared. Tens of thousands were imprisoned and tortured. The United States actively supported the military junta and put some of Pinochet's officers on CIA payroll. Pinochet agents were then active in various countries, including even the United States. (They killed Orlando Letelier, former member of the Chilean Allende government, for instance, through a car bomb in Washington, D.C., in 1976.) In 2005, when he was in his 90th year, Pinochet was finally stripped of immunity in Chile and ordered to stand trial on charges related to human-rights abuses and embezzlement, but he died in 2006.

ARGENTINA   In Argentina the United States backed the 1976 coup that installed a military government which was going to kill some 30,000 people. Argentina had a civilian government between 1932 and 1943, and thereafter military officer Juan Domingo Perón came to power. Strengthened by the popularity of his wife Evita Perón, he was overthrown only after her death in 1952. Civilian rule was restored between 1955 and 1966, then another military regime took over. Perón came back to power in 1973, and though he died the year after, the Peronist government ruled until the next military coup in 1976. The United States had provided $10.6 million worth of training to the Argentine military between 1962 to 1976, and the new three-person junta led by Jorge Videla immediately amended the constitution, banned political and labor-union activity, and waged what became known as the Dirty War against left-wing elements. U.S. Secretary of State Henry Kissinger told the Argentine military men, "If there are things that have to be done, you should do them quickly," and while the dictatorship committed massive human rights abuses in 1976, Kissinger advised "If you can finish before Congress gets back, the better."[142] The Argentine junta forces ruthlessly killed those viewed as opponents, with tens of thousands disappearing between 1976 and 1983. (Political prisoners were, naked and heavily sedated, thrown out of airplanes into the Atlantic Ocean, for instance.)

NICARAGUA   In Nicaragua the United States had established military bases at the request of the local government in 1912 and withdrew its forces in 1933 after setting up and training a National Guard, commanded by General Anastasio Somoza. Backed by the United States, the Somoza family began a near-dictatorial rule that was to last for over 40 years. The leader of a guerrilla group that opposed U.S. military presence in the country, Augusto César Sandino, was assassinated in 1934, but his followers continued their guerrilla activity and in 1979 finally ousted the Somoza right-wing dictatorship. The left-wing Sandinistas then appointed a council of state and scheduled elections for 1984 (in which they were endorsed), but relations with the United States deteriorated quickly. After Reagan was elected U.S. president, he claimed that the Sandinistas were promoting the overthrow of the government of nearby El Salvador, and his administration in turn supported the Contras, who were forces mainly consisting of ex-Somoza soldiers. In March 1982 the Nicaraguan government declared a state of emergency in the wake of attacks on bridges and petroleum installations, and in 1984 the CIA mined Nicaraguan harbors as part of a series of covert operations. The Contras bombed villages, burned schools and hospitals, and killed and tortured civilians. (High-ranked officials of the Reagan administration were eventually found guilty for their implication in illegal arms sales to the Contras, mainly because they also sold weapons illegally to Iran during the Iran-Iraq war. The arms sales of the Iran-Contra affair were worth some $30 million.[143])

The International Court of Justice (ICJ) ruled in 1986 that the United States was in breach of international law and ordered it to pay $17 billion in reparations to the Nicaraguan government, but the payment remained outstanding.[144] Instead the U.S. Congress approved $100 million in overt military aid to the Contras in June 1986. (Total U.S. aid to the Contras was $300 million.) It also came to light that U.S. government officials had sanctioned that cocaine originating from the Contras was sold in the United States to finance the Contras' terror activities. (The 1987 investigation of the Senate Subcommittee on Narcotics, Terrorism, and International Operations, led by Senator John Kerry, concluded that "senior U.S. policy makers were not immune to the idea that drug money was a perfect solution to the Contras' funding problems.") In February 1990 the (first) Bush administration spent $9 million on the election campaign of the National Opposition Union, which in turn defeated the Sandinista government. The Bush administration then lifted its economic embargo in exchange for Nicaragua dropping its claim to the damages of $17 billion awarded to it by the ICJ.

PANAMA In Panama the United States operated a large army base in the Canal Zone. Manuel Noriega, chief of intelligence from 1970, effectively ruled Panama from 1982 as head of the National Guard. U.S. officials worked closely with Noriega, though it was known as early as 1972 that he was involved in drug trafficking. The Torrijos-Carter Canal Treaties of 1977 specified that U.S. forces in Panama were present only to defend the canal, but Noriega allowed the United States to use Panama as an intelligence, training, supply, and weapons base for the Reagan administration's campaigns in Nicaragua and El Salvador. The Panama Canal Zone also hosted a U.S. communications installation that served Operation Condor, a covert campaign coordinated between the rulers of six Latin American countries (Argentina, Chile, Brazil, Uruguay, Paraguay, and Ecuador) to eliminate their opponents. Henry Kissinger was actively involved in the establishment of Operation Condor, which from 1976 and through the 1980s assassinated hundreds, perhaps thousands, of political rivals. (The murder of Orlando Letelier was a Condor operation, for instance, but teams were also sent to European countries.)[145]

In the late 1980s, toward the end of Reagan's second term, U.S.-Panamanian relations soured. The scheduled transfer of the Canal Zone to Panama was coming close, and the United States suspended financial and military aid.

Noriega, not least because of his knowledge of U.S. activities in the region, was now considered a problem. The United States publicly accused him of corruption, cocaine trafficking, the murder of a political opponent, and of rigging the 1989 elections. In December 1989 U.S. President Bush (senior) ordered the bombardment and invasion of Panama, during which perhaps 3,000 civilians were killed. This was the largest United States overseas operation since Vietnam, involving 24,000 U.S. troops. Noriega was captured for trial in the United States and sentenced to 40 years in prison. (His jail term was later reduced.)

U.S. Imperialism in Africa

In Africa the United States was not as active as in Latin America, or at least not as successful. America's western European NATO allies insisted on maintaining some influence over their former colonies, and those newly independent African countries that attempted to fully wean themselves from Western influence turned to the Soviet Union rather than the United States. The Soviets offered generous economic and military assistance, but in the end only few countries made concerted efforts to introduce a Soviet-style regime. Moscow achieved an important strategic breakthrough in November 1963, when the Republic of Somalia chose Soviet over Western military aid.[146] Somalia's neighbor Ethiopia, ruled by pro-Western emperor Haile Selassie, strongly protested the deal, but a decade later, in 1974, Selassie was overthrown in a military coup. Ethiopia was declared a socialist state with a one-party system, but sank into chaos following various coups and uprisings. In 1977 ethnic Somalis living in eastern Ethiopia's Ogaden region (with the help of Somalia) began a war to gain self-determination, but the USSR and Cuba helped Ethiopia to recover the region the year after. Millions of refugees left Ogaden for Somalia, where the United States provided military and humanitarian aid, and was in turn allowed to use the Berbera naval base, which had previously been under Soviet control. (In Ethiopia severe droughts condemned millions to hunger and starvation in the 1980s, and in 1991 the military government was finally forced from power.)

The Republic of the Congo, the small neighbor of the former Belgian Congo, gained independence from France in 1960 and proclaimed itself a country along Marxist-Leninist lines in 1964. The government obtained foreign aid from the Soviet Union and China, but a coup organized by the army and more militant leftists installed Marien Ngouabi as head of state in 1968. He was assassinated in 1977, and his successors kept the country's pro-Soviet orientation. After the disintegration of the Soviet Union, multiparty elections were allowed in 1992.

A military coup and regime change in Portugal in 1975 finally led to Portugal's African colonies becoming independent. The two principal ones,

Angola and Mozambique, opted for communist-type governments. In Angola the USSR and Cuba helped to establish a one-party regime, but a nationalist group received assistance from the United States and South Africa. A civil war broke out immediately, and Cuba came to the rescue of the government party, deploying 50,000 troops as well as construction workers, doctors, and teachers in Angola. The Cuban troops withdrew only after South Africa, Angola, and Cuba had announced a cease-fire in August 1988. In 1991 the Russians and Americans brokered a peace accord signed by the two opposed Angolan parties, but the fighting eventually continued until the civil war finally ended in April 2002.

In Mozambique a Marxist-Leninist group, led by Samora Moises Machel, assumed power and immediately set out to collectivize agriculture, eradicate nomadic practices, and to curb the power of village elites and the Catholic Church. Mozambique soon became embroiled in a vicious civil war between the central government and a national rural guerrilla movement financed initially by Rhodesia (Zimbabwe) and later by South Africa. Machel negotiated with the West to cut off support for the guerrilla movement and by the mid-1980s had given up most Marxist economic policies. Foreign investment was encouraged, and a multi-party constitution and other political reforms were adopted in the early 1990s. (In short, even the last of Africa's few communist regimes disappeared with the disintegration of the USSR.)

THE CONGO The former Belgian Congo turned into a principal stronghold of the United States in Africa. This enormous country achieved full independence in June 1960 under the name Republic of the Congo and was renamed Democratic Republic of Congo in 1964. The nation was supposed to be centrally governed by Prime Minister Patrice Lumumba, but the rich mining province of Katanga declared itself independent. The Belgians supported this secession, as they figured it would serve their future influence over the mines. Lumumba thus asked and received an international peacekeeping force from the United Nations to restore order and recover Katanga, but was disappointed by its work. He therefore asked for additional help from the Soviets. America was not at all pleased, and U.S. President Eisenhower ordered the assassination of Lumumba. (Robert H. Johnson, an executive member of the National Security Council, later recalled his own "sense of that moment quite clearly because the President's statement came as a great shock" to him.) Lumumba was overthrown in late 1960 and flown to Katanga, where he was executed immediately. Possibly the CIA, the Belgians, and the army chief of staff Joseph Désiré Mobutu, were all implicated in the murder. Mobutu declared himself president in 1965 and ruled until 1997. The country, which was called Zaire between 1971 and 1997, suffered and declined under his repressive regime, while he enriched himself to become one of the most wealthy people in the world. The West kept him in power, providing Mobutu with all the military assistance he needed to put down rebellions, even though the dictator banned European names and dress, and seized some 2,000 foreign-owned businesses. In exchange, Mobutu supplied critical minerals to the West and allowed the United States to use Zaire as a basis for covert operations in other African countries. (This included the CIA's support for anti-Communist Angolan rebels operating out of southern Zaire.)[147]

Mobutu's overthrow in 1997 was engineered by Zaire's tiny neighbor Rwanda. The story began in 1994, when Rwanda's Hutu tribe tried to exterminate Rwanda's Tutsi, a prosperous minority, killing 800,000 within three months. While the international community remained inactive, an army of exiled Tutsi invaded Rwanda from Uganda to stop the slaughter. When the killers fled to Zaire, the new Tutsi-dominated government of Rwanda feared the Hutu might regroup within Mobutu's country. They therefore organized a rebellion that ousted Mobutu and put Laurent Kabila in power. Kabila, however, rearmed the Hutu, and a bloody civil war unfolded in the Congo that cost the lives of over three million (and perhaps even five million) people. The government of Rwanda, with assistance from Uganda and Burundi, nearly managed to topple Kaliba, but he was saved by troops from five friendly nations, Angola, Zimbabwe, Namibia, Chad, and Sudan. The war eventually reached a stalemate, but the various armies began or kept plundering the Congo. Angola joined the Congolese government in an oil venture, Zimbabwe helped itself to diamonds in the south, and Rwanda and Uganda harvested timber and ivory, and mined diamonds and coltan. Coltan, short for columbite-tantalite, has been especially blamed for financing the conflict. Spewed out by the local volcanoes, this mineral is globally scarce but scattered widely in the Congo. After refining it yields tantalum oxide, a material critically important to make mobile phones (and other electronics and gadgetry). Demand for coltan soared to a peak price of $840 a kilo in 2000, and the Rwandan army purportedly made 100s of millions of dollars from coltan sales. In January 2001 Laurent Kabila was shot, and succeeded by his son Joseph, who attempted to end the war in the Congo.[148]

Post–Cold War Global U.S. Militarism

After the end of the Cold War the United States kept on providing military (and economic) assistance to many countries to ensure political influence. The International Military Education and Training (IMET), created in 1976, is the traditional military training program offered by the United States to its allies.[149] IMET is overseen by the State Department and implemented by the Defense Department. It is designed to influence and strengthen foreign militaries and their governments. IMET has schooled foreigners in some 150 military training institutions throughout the United States, offering courses from counter-intelligence to helicopter repair. IMET also keeps providing much of the funding to run the School of the Americas (SOA), dubbed School of the Assassins,[150] originally located in the Panama Canal Zone. (This center trained military personnel from various Latin American countries and used manuals that recommended interrogation techniques such as torture, execution, blackmail, and arresting the relatives of those being questioned. Several hundred SOA graduates have been accused of serious human rights violations.) IMET also finances American teams to teach in foreign countries. In 2001 funding for IMET worldwide was $55 million; the budget request for 2003 was $80 million. Meanwhile assistance to "problematic" countries is typically provided through less transparent programs. Under the Joint Combined Exercises and Training (JCET) program, for instance, Special Operations Forces trained Indonesian troops even when Indonesia's poor human rights records were openly discussed in the 1990s. Under the African Crisis Response Initiative, launched in 1996, the United States provided military training in various African countries. In 1997 41 of Africa's 53 armed forces received United States military training, among them most the countries that were participating in the war in the Congo. (Zimbabwe's Robert Mugabe in 2001 still received IMET assistance, though his torture and killing habits had long been publicized.) Worldwide U.S. military support for foreign armed forces reached $21.3 billion in 1997. Recipients were 168 of the world's 193 nations. $15.6 billion went to 123 nations in the developing world, of which 52 had non-democratic regimes. Arms sales are also a major component of U.S. efforts to create international allies and broaden its influence. (In the mid-1990s the United States accounted for half of international arms sales, NATO as a whole for over three quarters.) The international proliferation of U.S. arms in turn serves as an incentive for the continuous development of new and better arms by U.S. companies that are major employers and have significant influence on the government.

When the International Criminal Court (ICC) was established in 2002 to try persons accused of committing war crimes, the United States refused to ratify the treaty to avert the possibility that American troops could be held accountable in front of this international court. Signatory countries (in total over one hundred) in which United States soldiers were active were asked to sign "bilateral immunity agreements." Some 35 countries chose not to sign such agreement and hence had their United States military aid canceled.

U.S. Overseas Troops and Military Bases

By the end of the 20th century the United States kept some 200,000 troops stationed in 144 countries, and usually had another 20,000 deployed on ships at any given time. Though the Cold War had now been over for more than a decade, some 100,000 American troops were stationed in Europe (mainly in Germany, Italy, Britain, Turkey, Spain, Iceland, Belgium, and Portugal) and some 75,000 in East Asia (in Japan and South Korea). In Panama the United States kept 6,000 troops; at the Guantánamo base, Cuba, nearly 2,000; and another 4,000 in the Persian Gulf region (Kuwait and Saudi Arabia). Among the more recent United States military bases are those established in former Soviet republics (Kyrgyzstan, Uzbekistan), former eastern bloc countries (Bulgaria, Romania), and in Djibouti at the Horn of Africa (ca. 1,800 troops). In 2004 the United States announced that it would gradually reposition, and cut by one third, its overseas troops over a 7 to 10 year period to adjust them to the changed geopolitical environment. The changes were to affect mainly Europe (and here chiefly the 68,000 troops present in Germany) and Asia (Japan and South Korea).

The United States has been able to expand its worldwide political and economic influence ever since World War II. U.S. foreign policy was to contain communism and to ensure access to international markets and raw materials. Maintaining or promoting democracy and the personal freedom of people internationally was a goal sidelined whenever it clashed with the former objectives. Substantial investments in armed forces and weapons technology, with the setup of hundreds of overseas military bases and the buildup of a large fleet of destroyers and aircraft carriers, all served the realization of foreign policy goals. Since 1945 the United States has carried out serious interventions into more than 70 nations. And all these activities were fueled by oil.

U.S. Oil Production

At the beginning of the 21st century the United States had cumulatively produced more crude oil than any other nation in the world. From Drake's find in Pennsylvania in 1859 until the early 1970s the United States remained the world's top producing country, with a short interruption at the beginning of the 20th century, when Russia (Azerbaijan) was the leader. And even though U.S. production began to sink from 1970, the United States remained a top-three producer for the remainder of the 20th century. Even in 2001 only Saudi Arabia (producing 3,100 million barrels per year) and Russia (producing 2,552 million barrels per year) were able to rival U.S. oil production of 2,578 million barrels per year.

It was clear from the start that America was generally rich in oil, with vast deposits located in several different states. The same year Drake found oil in Pennsylvania, a settler drilling for water in the area that was to become Oklahoma struck oil. The Indian Territory was quickly invaded by oil prospectors, but the first commercial oil well was not established here before 1897. (The Indian Territory and Oklahoma Territory were united in single statehood in 1907, shortly after a true oil-boom had set in in the area.) The Oklahoma oil fields were part of the huge Mid-Continent Oil Region that stretched from central Texas across Oklahoma to eastern Kansas. Between 1918 and 1935 the Mid-Continent Oil Region accounted for more than half of total U.S. production, with over 8.8 billion barrels of crude being pumped between 1900 and 1935.

The Lucas Gusher at Spindletop, Texas

Texas emerged as the most important oil state in this area. The first economically significant Texan find was in 1894 in Navarro County near Corsicana. This field was gradually developed to produce a record of 839,000 barrels of oil per year in 1900. However, the true birth of the Texan Oil Industry, and arguably the birth of the modern oil industry as such, is ascribed to the Lucas gusher, which blew out on January 10, 1901, at Spindletop near Beaumont in East Texas when Captain Anthony F. Lucas drilled a well to 1,020 feet. Captain Lucas, unlike his Pennsylvania counterpart "Colonel" Edwin Drake, really held a military rank. Lucas was born in the Austrian monarchy in present-day Croatia, and from the age of 19 studied engineering at the

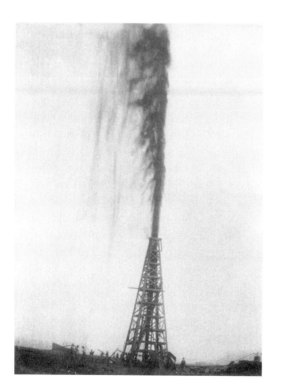

The Lucas Gusher at Spindletop, Texas, 1901 The Lucas gusher is widely regarded the most important oil find in American history. Only after eight days did it drop off enough to be capped and controlled. Like Nikola Tesla, Anthony Lucas (Anton Luchich) had studied engineering at the Technical University of Graz, Austria. (Photograph by Francis John Trost.)

Technical University of Graz, Austria. (This was shortly before Nikola Tesla studied there. Tesla was one year younger than Lucas.) At the age of 24, when he served in the Austrian navy as second lieutenant, he took a leave-of-absence to visit his uncle in the United States and decided to stay. Changing his name from Anton Luchich to Anthony Lucas, he worked as a salt miner in Louisiana, and soon understood that oil was to be found at underground salt domes. The Lucas gusher produced an astounding 800,000 barrels of oil in just 8 days before it dropped off enough to be capped and controlled. Within a year, over 200 wells surrounded Lucas' drill site, all competing for space on top of Spindletop, which produced over 17 million barrels of oil in 1902. Thereafter production at Spindletop declined quite rapidly, but lots of oil was discovered across the nearby border, in Louisiana, the state that emerged as a major center of petroleum refining.[151]

The Lucas gusher has been dubbed the most important oil find in (American) history, because it changed the general attitude towards oil, creating the opinion that oil resources were in fact plentiful and could replace coal in many applications. (Notably, coal-fired ships and trains were then increasingly converted to oil.) The Spindletop arena experienced a second boom in 1925, when oil was discovered on the flanks of the dome. Production was pushed to an all-time high of 27 million barrels per year in 1927, but even this was minimal compared to the extraction volumes associated with the Texas oil boom of 1930, when Columbus Marion ("Dad") Joiner near Tyler struck what was perhaps the largest oil pool ever found in America. It stretched over 140,000 acres of East Texas flatland and contained as much as 5 billion barrels. However, the Joiner strike came at a somewhat inopportune time, at the onset of the Great Depression. The price of oil plummeted to ten cents a barrel in 1931, creating chaos in the industry for several years.

Californian Oil

In California oil was produced early on as well. In 1850 petroleum from hand-dug Los Angeles pits was distilled to produce lamp oil, and in 1854 natural gas from a water well was used to light the Stockton courthouse. The first productive Californian oil well was drilled in California's Central Valley in 1865, just six years after Drake's monumental find in Pennsylvania. This area, east of San Francisco, became the scene of much of the Californian drilling activity for the remainder of the 19th century, but none of these wells were considered major strikes. Nevertheless they provided enough oil to meet the demand in San Francisco, by far the largest population center in California at the time. Los Angeles (LA) was still a sleepy seaside village, but this was going to change rapidly as soon as southern California's major oil fields were discovered. The find at Pico Canyon, north of LA, in 1875, was followed by a real boom when the Los Angeles Field was discovered in 1892.

Within two years some 80 wells were producing oil in the area bounded by Figueroa, First, Union and Temple Streets, and by 1897 the number of wells had increased to 2,500. The population of LA doubled between 1890 and 1900, then tripled again between 1900 and 1910. Offshore wells were drilled from piers in Summerland (south of Santa Barbara) from 1896, and in 1899 oil was discovered at the Kern River close to Bakersfield (north of Los Angeles). In 1900 the state of California produced 4 million barrels of crude a year, and by 1903 California was the top oil-producing state, with the Kern River field by itself producing 17 million barrels per year. In 1910 Californian production reached 77 million barrels a year, accounting for 22 percent of world production and exceeding that of any foreign country. Soon the oil fields at Huntington Beach (1920), Santa Fe Springs (1921), and Signal Hill (1921) added to Californian production, which accounted for a quarter of the world's oil output in 1923. (Signal Hill Field by itself produced 614.5 million barrels of crude cumulatively by 1938, 750 million barrels by 1950, and over 900 million barrels by 1980, making it one of the most productive fields per acre the world has ever known.)[152]

The Rockefeller Standard Oil Monopoly

The abundance of oil soon fueled much of America's economic growth, but the oil industry also exposed a fundamental weakness in the free market system, which relies on price levels to indicate whether supply meets demand. (Rising prices indicate supply shortages and tell companies where to allocate resources to satisfy demand and make profit.) The price signaling mechanism depends on a sufficient number of independently competing market players, but as it turns out, an unregulated free market economy will eventually evolve into an economy with features similar to those of centrally planned Communist countries (in which prices are fixed and production volumes predetermined). Firms that are doing well will acquire more and more market share and will eventually force smaller competitors out of business (possibly by buying them out). The remaining market players may in turn merge to eliminate the remaining competition, monopolize markets, raise prices, and reap high profits. To keep free market economies functioning, legislators have introduced anti-trust legislation, and in the United States much of it was initially instituted to keep John D. Rockefeller's Standard Oil Company in check.

Rockefeller was 20 years old when he heard about Edwin Drake's 1859 find. He then operated a commission firm in Cleveland together with a partner, but they sold it to build a small oil refinery that delivered kerosene from 1863. Three years later Rockefeller had bought out his partner, and in 1867 he created the firm that was to become the Standard Oil Company (of Ohio). By 1870 Standard Oil was the dominant oil refiner in Pennsylvania, and had begun to acquire much of the pipeline system that transported oil (kerosene) to the nearest railroad. Two years later, in 1872, Rockefeller's reputation became permanently tainted by his involvement in the scandal surrounding the South Improvement Company (SIC), a secret alliance between selected major refiners and the railroads. Launched by the president of the powerful Pennsylvania Railroad, the scheme was intended to raise freight charges. According to the pact, the railroads would raise their rates, but would pay rebates to Rockefeller and other large refiners, while the smaller market players were going to face higher transportation costs and hence would not be able to get their product to market at the same price. The deal was leaked to newspapers, but many (non-member) refiners felt threatened and decided to sell out. The Pennsylvania legislature was going to quickly repeal the SIC's charter, but Rockefeller was able to buy up 22 of his 26 Cleveland competitors within less than six weeks: Standard Oil had thus taken its most important step towards monopolizing America's oil refining industry. The firm had now become big enough to threaten competitors with price wars, and during the following years Standard Oil marched on to acquire most refineries in Pittsburgh, Philadelphia, Baltimore, New York, and other oil processing centers. By 1877 the firm controlled 90 percent, and a few years later 95 percent, of U.S. oil refining operations. In 1882 the Standard Oil Trust was established (as something of a holding company), and the firm emerged as one of the first and largest modern multinational corporations, evolving into a complex global enterprise that delivered its products (kerosene, lubricants, and fuel oil) to the far corners of the world. (By 1885 some 70 percent of the business was overseas.) Rockefeller conceived of, and implemented, the principle of vertical integration, whereby Standard Oil centrally controlled every step in the oil industry value chain, from exploration to production to transport to refining to distribution.[153]

John D. Rockefeller, who emerged as America's most powerful industrialist and the world's richest person, lived on until 1937, but retired in 1895, a year before Ford built his first motor car. With a monopoly on oil refining in place right at the time when oil began to define mobility and military power, Standard Oil was set to grow even more rapidly and to become a corporate giant powerful enough to threaten even the federal government. In 1900, when Californian oil production began to boom, Standard Oil purchased the Pacific Coast Oil Company (which had its roots in the Pico Canyon) and started to operate on the West Coast. Standard Oil used various tactics to influence legislatures and Congress, but resistance by competitors and the public grew. President Theodore Roosevelt called Standard Oil's executives the biggest criminals in the country and was determined to enforce the initially quite ineffective Sherman Antitrust Act of 1890. He, and his successors Taft and Wilson, attempted to create federal regulatory machinery to control the behavior of U.S corporations, but they failed to institute federal chartering for large corporations, which were able to avoid accountability by fleeing to states such as New Jersey and Delaware (which instituted lenient chartering regulations). Nevertheless, the law suit filed by the federal government against Standard Oil under the Antitrust Act in 1906 ended with the Supreme Court declaring that the Standard Trust had operated to monopolize and restrain trade, and ordering the break up Standard Oil Trust in 1911. The largest of the resulting separate entities were Standard Oil of New Jersey, which from 1972 was called Exxon; Standard Oil Company of New York, which in 1966 became Mobil; Standard Oil of California (Socal), which together with Standard Oil of Kentucky in 1984 became Chevron; Standard Oil of Ohio (Sohio), which in 1987 became the American arm of British Petroleum (BP); Standard Oil of Indiana, which in 1954 became Amoco; Continental Oil, which became Conoco; and Atlantic Oil, which in 1966 merged with Richfield Oil to become ARCO. In short, nearly all sizeable U.S. oil corporations of the 20th century had their origins in the Standard Oil break-up. Among the notable exceptions were Gulf Oil and Texaco, firms that have their roots in the Texan Spindletop oil boom, and Union Oil Company (from 1983 called Unocal), which was already large when Standard Oil entered California. (Towards the end of the 20th century there was a trend towards consolidation in the industry. In 1998 BP merged with Amoco, and the year after Exxon merged with Mobil. In 2000, BP-Amoco took over ARCO as well as Burmah Castrol. Chevron, after buying Gulf Oil in 1984, merged with Texaco in 2001.)

Close Ties between Government and Oil Companies

The progeny of Standard Oil soon grew into true corporate giants as well, but the government no longer interfered. Quite on the contrary, U.S. energy firms soon enjoyed all kinds of government subsidies, open or hidden, direct or indirect, including tax cuts and special accounting rules. (A study conducted for the Energy Information Administration at the request of Congress delineated an annual cost of between $5 billion and $10 billion in Federal energy subsidies in 1990.[154] A report released by the Congressional Joint Committee on Taxation titled "Estimates of Federal Tax Expenditures for Fiscal Years 1999–2003" showed that the oil and gas industry accounted for 62 percent of tax breaks, reaching a level of $17.8 billion by 2003.[155] Other estimates put federal subsidies at $21 billion to fossil fuels industries and $11 billion to the nuclear industry in 1989.)[156] Energy prices are critically important for economies, because they influence virtually all other prices (and thus inflation). Hence many Oil Age governments kept tight control of domestic oil prices or set up state-owned energy companies. In the United States, oil firms were not state-owned, but maintained close ties with the government. It was appreciated that oil corporations had to be large, because their operations are very capital intense, and that internationally active American oil corporations serve the purpose of ensuring U.S. access to (and control of) overseas oil, which was especially important in the second half of the 20th century, when domestic American oil production was declining. Access to international oil was critical, as preservation of domestic oil resources was sidelined and automobile production emerged as America's leading industry by many measures. Some 22,000 cars were manufactured in the United States in 1904, but a decade later car registration totaled more than a million. U.S. car manufacturer General Motors (GM) arose as the largest corporation in the world, and one of America's most important employers. In 1936, when U.S. car sales had slowed, GM teamed up with Standard Oil of California (Chevron) and Firestone Tire to purchase public electric transit systems in 16 states to convert them into GM bus operations. GM was convicted in 1949 of criminally conspiring to replace electric transportation, as well as monopolizing the sale of buses, but the firm continued to grow, selling buses and cars to a post-war American society that became totally dependent on oil-fired vehicles. (The U.S. railway system declined, but in 1930 still accounted for 75 percent of freight and 68 percent of passenger traffic. After World War II practically all trains were converted to diesel. By 1987 railroads accounted for 36 percent of intercity freight traffic and 3 percent of passenger service.) The close ties between the U.S. oil and automobile industries promoted the trend towards large, fuel-inefficient cars. Even when the Corporate Average Fuel Economy (CAFE) law was introduced (which boosted the average fuel efficiency of new American-made cars by over two-fifths from 1978 to 1987), the industry achieved a loophole that exempted trucks and sport-utility vehicles (SUVs). In result, the motor vehicles driving on American streets used more gasoline per mile on average in 2001 than they did at any given time during the previous 20 years.

Oil Production Decline since 1970

In light of dwindling domestic U.S. oil resources, this was a major problem. The United States poured an astounding six billion barrels of oil into World War II (a quantity equivalent to the total capacity of the great East Texas oil field), but became a net oil importer from 1948. Domestic oil production remained high, and was still growing, but huge amounts of fuel were absorbed by rapid economic growth and the soaring fleet of motor vehicles. Thus, it did not take long before America's vast oil resources showed signs of depletion, and the rate of new oil field discoveries fell behind the rate of oil field exhaustion. The level of known (proved) U.S. oil reserves peaked as early as 1959, and in 1970 U.S. oil production reached its all-time high at about 3.5 billion barrels of oil per year. (U.S. natural gas production peaked shortly thereafter, in 1973.) Ten years later, in 1980, oil production (in the lower 48 states) had fallen by 30 percent (despite a quadrupling in oil well completions), and at the start of the 21st century production had decreased to just 2.1 billion barrels per year. Alaskan oil extraction slowed the decline, but it was unable to restore U.S. production to the 1970 level.[157] Oil was first discovered in Alaska in 1957, and the supergiant Prudhoe Bay oil field (holding about 10 billion barrels) was found at Alaska's Arctic North Slope in 1968. To access this oil a pipeline was built north-to-south across the entire state of Alaska. Following its completion in 1977, the U.S. oil production decline came to a temporary halt between 1980 and 1986, before Alaskan production peaked at 0.73 billion barrels per year in 1988. Proved U.S. oil reserves amounted to 39 billion barrels in 1971, but only 23 billion barrels in 1999. The only substantial undiscovered U.S. reserves are now expected in northern Alaska's Arctic National Wildlife Refuge (ANWR), which may hold between 4 billion and 12 billion barrels of oil. However, the Senate in 2003 rejected a proposal to allow oil exploration there. (In addition, some 0.6 billion barrels have been left in the ground at the U.S. Gulf Coast as Strategic Petroleum Reserve.)

Of all the oil extracted in the United States between 1859 and the end of the 20th century (roughly 180 billion barrels), about one third came from Texas, 14 percent each from California and Louisiana, and some 8 percent each from Oklahoma and Alaska. In 2001 total U.S. oil production (including crude oil, shale oil, oil sands, and natural gas liquids, that is, the liquid content of natural gas where this is recovered separately) was about 2.8 billion barrels (with almost a fifth coming from Alaska). That same year U.S. proven oil reserves (including natural gas liquids) amounted to merely 30.4 billion barrels, indicating that domestic oil production would come to a standstill in less than 11 years at constant production and reserves levels. Total U.S. oil consumption was 6.6 billion barrels per year in 2001, with 3.9 billion barrels per year being imported (net). (U.S. natural gas reserves have been similarly depleted. Proved U.S. natural gas reserves in 2000 were two thirds of the 1970 level.)

The combination of rising oil consumption and falling domestic oil production made the United States ever more dependent on foreign oil. Imports as a percentage of U.S. oil consumption grew from 28 percent in 1973 to 35 percent in 1983 to 44 percent in 1993, to 60 percent in 2001. Crude oil imports amounted to 483 million barrels in 1970, but 3,320 million barrels in 2000.[158] With some four percent of the world's population, the United States in 2001 accounted for 26 percent of the world's oil consumption, 10 percent of global production, and 3 percent of the world's proven oil reserves. Obviously, foreign policy had to focus on securing access to overseas oil. Without imports, the U.S. economy would stand still within less than five years (based on 2001 consumption and reserves levels). And it will depend on much larger imports in the future to make up for the further decline in domestic

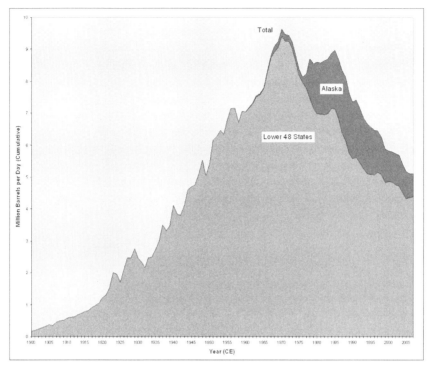

U.S. Oil Production   U.S. oil production peaked in 1970, despite substantial efforts to restore higher annual production volumes during and after the two Oil Price Shock episodes. Alaskan oil production only slowed U.S. oil production decline, but could not stop it. (Based on data in Energy Information Administration, "U.S. Crude Oil Field Production," http://tonto.eia.doe.gov/dnav/pet/hist/mcrfpus1A.htm, and "Crude Oil Production and Crude Oil Well Productivity, 1954–2007," http://www.eia.doe.gov/emeu/aer/txt/ptb0502.html.)

production. But securing access to oil is not the only issue. The U.S. economy is critically dependent on modest international oil prices. Oil imports accounted for about one-quarter of America's enormous $550 billion yearly trade deficit in 2004, for instance. The United States will therefore always do what it can to avoid disruptions of oil deliveries from the main exporting nations to world markets.

Mexico and Venezuela: Two Major (and Early) Sources of Oil for the United States

The oil of Latin America has at times been viewed as U.S. oil as well. The United States had made clear since the Monroe Doctrine of 1823 that it would have some sort of supremacy over Latin America in comparison to the European Coal Age Powers, and throughout the 20th century Latin American oil was chiefly consumed in the United States. The two major suppliers were Mexico and Venezuela, both of which temporarily held the position as the world's second largest oil producer (after the United States). Oil production in both countries started early on, and in the 1920s, when U.S. production accounted for 60 percent of world oil production, Latin America accounted for 25 percent. What is more, both countries were not industrialized themselves, had therefore little domestic demand, and produced mainly for export. (In fact, Venezuela was the world's largest oil exporter between 1929 and 1970.) But their low level of industrialization also meant that these countries depended on foreign investment and expertise to find, access, and sell their oil.

Mexico

In Mexico, an exploration boom started after the Mexican Eagle Oil Company hired Captain Anthony F. Lucas in 1902, a year after the Spindletop gusher had made him famous. The company was set up by Englishman Weetman Dickinson Pearson, later to be titled Lord Cowdray, who had been sent to Mexico in 1889 by S. Pearson & Son, Ltd., a civil engineering firm which, after the collapse of the 1860s railway construction boom in Britain, operated mainly internationally and was contracted by the Mexican government to drain swamps, build railways, and construct new ports. Pearson was a personal friend of Porfirio Díaz, the dictator who ruled Mexico between 1876 and 1911, and he began buying oil drilling concessions in 1901. (The earliest attempts to drill for oil in Mexico date to 1869.)[159]

Once Lucas had located two oil fields near Coatzacoalcos, gradually more fields were discovered, and in 1910 international oil companies moved into Mexico to widen the prospecting efforts. However, that year the Mexican Revolution (1910 to 1920) broke out and claimed over one million lives during a long and bloody period of repeated regime change. With Díaz ousted, Mexican Eagle had to pay off various revolutionary leaders, but Mexico managed to export oil from 1911. In 1918 Pearson sold out to Royal Dutch/Shell. (Pearson thereupon turned into one of the richest men on the planet. The Pearson company later evolved into a media empire. Purchasing print media from the start of the century, it eventually acquired the *Financial Times* (1957), the *Economist*, Longman Books (1968) and Penguin Books (1971). It also became a major investor in the Lazard Brothers merchant bank.) Mexican oil production grew rapidly during and after the revolutionary years due to foreign investments made during and right after World War I, when it became clear that wars were from now on going to be won through oil. Mexican oil production peaked in 1921 at 193 million barrels a year (some

25 percent of world production), while output 10 years later had fallen to just 20 percent of the 1921 level due to political instability; increased taxation; decreased demand as a consequence of worldwide economic depression; lack of new oil discoveries; and Venezuela's emergence as a more attractive source of oil in the region. However, production picked up again once the Poza Rica field near Veracruz was discovered in 1932. (This was Mexico's principal source of oil until the late 1950s.)

The constitution of 1917 gave the Mexican government a permanent and inalienable right to all subsoil resources, which immediately led to a lengthy dispute with foreign oil companies. In 1938 Mexico finally took a quite radical step and nationalized the 17 foreign-owned oil companies operating in Mexico, placing their assets under the control of PEMEX, the newly established national oil company. Although Mexico offered compensation, U.S. oil companies pressured their government to embargo all imports from Mexico in order to discourage similar nationalizations in other countries. However, the boycott was extremely short due to U.S. security interests in World War II: in 1943 the oil corporations agreed to a final settlement under which Mexico paid them $24 million, a fraction of the book value of the expropriated facilities.

The nationalization of its oil industry deprived Mexico of foreign capital and expertise for some 20 years. Production increased from 44 million barrels per year in 1938 to 78 million barrels per year in 1951, but Venezuela's output passed that of Mexico in 1939, and from 1957 Mexico was unable to meet even domestic demand, remaining a net oil importer until 1975. Following new oil discoveries in the early 1970s, Mexico in 1973 finally surpassed the peak production of 1921 (193 million barrels per year). International loans (for which Mexico's oil reserves served as collateral) in turn financed the investments necessary to increase output to a record 1,000 million barrels in 1984.[160] Mexico hence had once again emerged as a top oil exporter, but production quickly leveled off again. The government then decided to reform the oil sector radically. It had previously tolerated PEMEX's inefficiencies because the company produced nearly all public revenues, but from the early 1990s the poor administration, low productivity, overstaffing, and corruption was eliminated. Within four years, PEMEX's 1989 workforce of 210,000 was halved, and foreign involvement in Mexico's oil sector was once again encouraged. In result, Mexico was able to restore the 1984 output by 1996, and to keep annual oil production stable at 1.1 billion barrels per year in the late 1990s.

Venezuela

Venezuela was one of three countries that emerged from the collapse of Gran Colombia in 1830 (the others being Colombia and Ecuador). During the first half of the 20th century, Venezuela was mostly ruled by corrupt military dictators, who enriched themselves by working closely with international oil corporations, offering them low nominal taxes. The first of them was Juan Vicente Gómez, who was in power from 1908 until his death in 1935, when he was the wealthiest man of Latin America. Under his rule international oil corporations moved in (from 1910) and oil was found in 1914, right at the start of World War I. Commercial production at the Lago de Maracaibo oil fields began in 1922, and seven years later Venezuela had become the world's largest exporter of oil, a position the country would maintain for nearly half a century. Most Venezuelan oil has always been shipped to the United States, but Venezuela also served as Britain's principal supplier of oil. Compared to its size Venezuela had a very small population (which even in 2007 had grown to no more than 26 million people), and the level of industrialization remained low. The vast majority of Venezuelans did not profit from the oil boom, unemployment rose (as the oil industry is capital-intensive rather than labor-intensive), and farmers had problems to compete against imported food, paid for by oil revenues.[161]

The rulers succeeding Gómez transferred some political power from the military to civilians, and in the midst of World War II, in 1943, Venezuela decided to follow Mexico's example and nationalize its oil industry. (This was when Mexico "got away with it," achieving the end of the U.S. embargo and a bargain settlement with the oil corporations.) However, Venezuela chose a moderate stance by agreeing to split oil profits fifty-fifty with the oil companies. The country then reaped substantially greater benefits from its oil endowment, but the military once again overthrew the government in 1948, and in 1952 Pérez Jiménez declared himself president. During the next five years, Pérez brutally tortured and murdered hundreds, if not thousands. He controlled the press, harassed labor unions, and shut down the Central University of Venezuela (when it became a center of opposition to the regime). But the United States openly supported the dictator, mainly because of his liberal policies toward the foreign oil companies and his staunch anti-communism. (President Eisenhower even awarded him the Legion of Merit in 1954.) The people of Venezuela did not share these sympathies. A general uprising, supported by parts of the military, in 1958 ousted Pérez, who took what remained of the national treasury and fled to the United States.

Finally a democracy, Venezuela held elections in December 1958, which were won by the left-wing (but anti-communist) party of Rómulo Betancourt, who led the country into a long civilian period. At the beginning of Betancourt's rule Venezuela was the world's second largest oil producer, though by 1960 its annual oil production (1.04 billion barrels) was passed by that of the USSR (1.06 billion barrels) and well behind that of the United States (2.57 billion barrels). However, Venezuelan production kept expanding until a peak was reached in 1970 (at 1.35 billion barrels a year). Betancourt retired in 1964, but supported Carlos Pérez, who was elected president in 1973. Carlos Pérez fully nationalized the oil industry in 1975, but managed to retain experienced foreign personnel. He slowed production to conserve oil resources, and channeled petroleum income into hydroelectric projects and steel mills. However, it soon became clear that some of his ambitious industrial projects were designed to maximize bribes and commissions. Nevertheless Carlos Pérez was reelected in 1989, survived two attempted coups, and was imprisoned on charges of embezzlement and misuse of public funds only in 1993. In the mid-1990s Venezuela reopened its nationalized oil sector to foreign investment, which pushed output to a post-1970 record of 1.20 billion barrels per year in 1997. (By 2001 production had sunk to 1.10 billion barrels again.)

Hugo Chávez, a military man who had helped staging the unsuccessful 1992 coup against Carlos Pérez, was democratically elected president of Venezuela in 1998, 2000 and 2006. Under his rule Venezuela seemed to remain internally unstable and dangerously dependent on international oil prices, with the oil sector accounting for roughly one-third of GDP and 80 percent of export earnings. (A two-month national oil strike starting December 2002 completely halted economic activity, for instance.) This instability, combined with Chávez' ardent criticism of U.S. foreign policy (and his eventual claims that the U.S. government was attempting to assassinate him), was increasingly worrying for the United States, which kept receiving much of its oil imports from Venezuela.

At the start of the 21st century the principal countries delivering oil (and oil products) to the United States were Saudi Arabia, Canada, Mexico, and Venezuela, with all supplying roughly fifteen percent of total imports. Among the geographically closer suppliers of the Western hemisphere Canada (traditionally a much smaller producer than Mexico and Venezuela) had the lowest proven reserves levels at a meager 6.6 billion barrels (2001), compared to Mexico, 26.9 billion barrels, and Venezuela, 77.7 billion barrels. (The United States itself had 30.4 billion barrels in proven reserves, while consuming 6.6 billion barrels per year.) Venezuela's 2001 excess production (total production minus domestic consumption) was 1.276 billion barrels, topped only by Russia (1.655 billion barrels) and Saudi Arabia (2.640 billion barrels). Any major supply disruptions of Venezuelan oil to world markets are therefore bound to impact international oil prices.

PERSIAN GULF OIL

Of all the regions in the world, none holds as much oil as the one centered around the Persian Gulf. In the beginning of the 21st century (in 2001) only Saudi Arabia (261.8 billion barrels) and Iraq (112.5 billion barrels) had more than 100 billion barrels in proved oil reserves, and only five other countries had over 31 billion barrels: the United Arab Emirates (97.8 billion barrels), Kuwait (96.5 billion barrels), Iran (89.7 billion barrels), Venezuela (77.7 billion barrels), and the Russian Federation (48.6 billion barrels). (The global total was 1,050.0 billion barrels.) Due to this concentration of oil wealth in the Middle East, a region that was hardly industrialized and thus would export most of its oil, the Western powers attempted to control the Persian Gulf countries throughout the 20th century. Western Europe had practically no domestic oil until the North Sea reserves were discovered, and remained a major oil importer even thereafter. The United States was the world's largest oil producer, but consumed such large quantities that it turned into a net importer in 1948, and soon emerged as the world's largest oil importer. (By 2001 the United States accounted for 27 percent of all the world's oil imports, Europe, including eastern Europe, accounted for 26 percent, Japan for 12 percent, and the rest of the world for 32 percent.) The Soviet Union, on the other hand, had much less incentive to control Middle Eastern oil, simply because it kept producing large surpluses of oil domestically, enough even to supply its allies.

America, or rather the West, seemed to have no limits as to the measures applied to ensure domination of the Persian Gulf. This included anything from financing political propaganda to targeted assassinations to large-scale covert operations to outright warfare in order to install or empower so-called friendly regimes, which was usually preferred over actual occupation. As in South America, Africa, or elsewhere in Asia, the West did not seem to care much whether or not these regimes were oppressive dictatorships, as long as they served the interests of the United States and western Europe. (On the other hand, such regimes were quickly removed, usually under the dictum of promoting freedom and democracy, if they did not conform with Western wishes.) This strategy worked quite well. America received about a quarter of its oil imports from the Middle East, and Europe a third, towards the end of the 20th century.[162] And yet even these large shares understate the importance of Middle Eastern oil for the world: if it does not reach international markets, the oil price rises globally and slows economic growth around the world. This happened twice, during the two oil price shocks of 1974 and 1979–80, but the United States nevertheless decided to maintain its support for Israel (which angered many Muslim governments in the region) and to rely on its economic and military power to ensure access to Middle Eastern oil. This was increasingly more important as U.S. oil production declined, while U.S. oil consumption and imports were rising. As percentage of world production, U.S. oil production decreased from 52 percent in 1950 to 16 percent in 1980, while Persian Gulf oil increased from 17 percent in 1950 to 37 percent in 1973. Western oil corporations played an important role in producing Middle Eastern oil. Globally, the five large U.S. oil corporations eventually produced two thirds of all oil, and the two British ones (BP, Shell) produced one third, but the power of these firms (and thus of the West) was bound to decline, as Arab nations with the world's largest oil reserves demanded more control over their oil resources and began nationalizing their oil industries. (OPEC, the Organization of the Petroleum Exporting Countries, was initially founded in 1960 by Iran, Iraq, Kuwait, Saudi Arabia, and Venezuela to negotiate a better share of the immense profits the giant Western oil corporations were making from their oil, but then developed into a cartel that controlled the international oil price.)

Dissecting the Ottoman Empire

The Western Coal Age powers began their struggle to control the Persian Gulf oil fields even before World War I had broken out. After a British initiative had discovered oil in southern Persia (Iran) in 1908—the first major oil find in the Middle East—the Anglo-Persian Oil Company (APOC) was incorporated as a subsidiary of Burmah Oil and became the new owner of the oil concession that applied to all of Persia except the five most northern provinces. In 1912 the British government decided to switch its navy from coal to oil despite Britain's lack of domestic oil resources, and henceforth access to foreign oil was a top national security priority. Thus, the British government in 1914 bought a majority share of APOC, which was later to be renamed BP, British Petroleum. In neighboring Mesopotamia, which was soon to become Iraq, things were a bit more complicated for the British. This area was also suspected to be rich in oil, but it was part of the Ottoman Empire, which still encompassed much of the Near and Middle East, and had friendly relations with Germany, then Britain's principal economic and political rival. However, in 1914 APOC managed to acquire half of Turkish Petroleum Company, the firm that had in 1912 been promised oil concessions in Mesopotamia. (The other half was owned by Deutsche Bank and Royal Dutch/Shell, except for a five percent share owned by a private mediator, Armenian millionaire Calouste Gulbenkian.)[163]

When World War I started, and the Ottomans sided with Germany and the Austrian monarchy, Britain sent 1.4 million troops to conquer the Ottoman Empire, viewing the conflict as an opportunity to take control of the region's potential oil areas. In 1916, in the midst of the war, Britain and France secretly worked out an arrangement that became known as the Sykes-Picot Agreement. It documented how the Ottoman Empire was to be dissected into British and French colonies (or zones of influence), practically creating the mostly arbitrary boundaries of the future countries of Jordan, Syria, Lebanon, and Iraq. For the British it was most important to gain control of oil-promising Iraq, but they were also going to administer Transjordan (later called Jordan) and an area around Haifa in Palestine, while the French were awarded (control over) south-eastern Turkey, a part

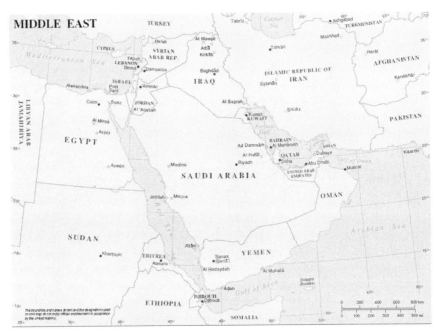

The Middle East (2007 borders)   The borders of several countries in the Middle East, the region with the world's largest oil deposits, were created somewhat arbitrarily by Western Powers after World War I. Various aspects that became relevant in geopolitical terms can be exemplified by looking at this map. Israel is relatively small and surrounded by Muslim nations. Saudi Arabia, with a comparatively small population but more oil reserves than any other nation, covers wide parts of the desert interior of the Arabian peninsula. Qatar, where U.S. troops have recently been transferred to move them out of Saudi Arabia, is in close proximity to Saudi capital Riyadh. Sudan, where Osama bin-Laden had been residing after being expelled from his native country Saudi Arabia, is situated very close to the Saudi Islamic center of Mecca. Kuwait, all the way inside the Persian Gulf, is very small in comparison to its northern neighbor Iraq. Afghanistan borders Turkmenistan, which until 1991 was part of the Soviet Union, as well as Pakistan and Iran, and stretches relatively close to the Gulf of Oman. Turkey, a NATO member and contestant for EU membership, borders Iran and Iraq and is close to the Caspian oil region. (Map compiled by the United Nations' Department of Field Support.)

of northern Iraq, Syria, and Lebanon. Palestine, except for those parts already allocated to Britain, would come under international administration. The Russians and Italians were later included in the agreement. Italy was promised some Mediterranean islands and a sphere of influence around Izmir in Anatolia. Russia was to receive Armenia and parts of Kurdistan. Potential aspirations by the Kurds to create their own state were disregarded in the agreement, which sealed the tragic fate of the Kurds for the entire 20th century. (The Kurds are Sunni Muslims of neither Arabic nor

Turkic descent. They were variously persecuted in the successor states of the Ottoman Empire and still did not have their own country at the beginning of the 21st century. Some 8 million lived in Turkey, 5 million in western Iran, and 4 million in northern Iraq.) To the great embarrassment of the western European nations, the Russians decided to release a copy of the confidential Sykes-Picot Agreement as well as other treaties. As it revealed that the Coal Age powers had made decisions about western Asia as they pleased, without consulting regional representatives, it spurred Arab distrust against Europe. Nevertheless, the content of the Sykes-Picot Agreement became reality.

Britain maintained over a million troops in the region of the Ottoman Empire by the end of World War I, and in 1920 took over the administration of Palestine, Jordan, and Iraq. In 1921 the British appointed a king in Iraq and insisted that the oil concessions granted to the Turkish Petroleum Company were still valid. As the Germans were to be excluded, the one quarter previously held by Deutsche Bank had to find a new owner: it was handed to French investors, which was the beginning of Compagnie Française des Pétroles (CFP), eventually renamed TOTAL. (TOTAL then emerged as the largest of all French companies in the later Oil Age.) The Turkish Petroleum Company struck oil north of Kirkuk on October 15, 1927, and half a year later the United States pressured Britain to hand a 23.75 percent share of the firm to the Near East Development Corporation (NEDC), a consortium of five U.S. companies. (NEDC was subsequently equally divided between Standard Oil of New Jersey, later called Exxon, and Standard Oil Company of New York, later called Mobil). Another 23.75 percent each were maintained by Anglo-Persian Oil (BP), Shell, and CFP (TOTAL), while Gulbenkian kept 5 percent. (These five percent reputedly made him the single most wealthy person in the world for many years.) Besides, the Anglo-Persian Oil Company retained exclusive rights to the oil found in 1923 in the "Transferred Territories," which were now located in Iraq but fell under the concession granted by the Shah of Persia in 1901. In 1929, a year after its ownership change, the Turkish Petroleum Company was renamed the Iraq Petroleum Company, and in the late 1930s oil exports from Iraq reached significant quantities.

The Arabian Peninsula

Britain also kept controlling the oil production in Persia (Iran) through Anglo-Persian Oil, but the new Iranian government of Reza Shah Pahlavi in 1932 temporarily cancelled the company's concession. The British therefore had a strong interest in spreading their oil supply risk, and began looking towards the opposite side of the Persian Gulf, the area of the present-day United Arab Emirates (UAE) and Oman. This region was also suspected to hold a lot of oil, and Britain was already well positioned to control it. The

Coal Age superpower had initially dealt with this coast line to eliminate pirates, who were threatening British trade between India and Europe. Britain launched attacks from 1809, and controlled this coast from 1819, but had little interest in making the desolate region a colony. Instead, Britain concluded a series of treaties with tribal leaders who had not been involved with, and promised to suppress all, piracy. As a result of this continuing policy, the Arab side of the Persian Gulf ended up with various, small, independent states that all had signed contracts with Britain during the Coal Age. Abu Dhabi became a British protectorate in 1835, for instance, while Qatar signed an arrangement with Britain as late as 1916. The various local rulers were kept in power, but had to hand control over all foreign policy to Britain. Notably, most members of the present-day UAE in 1882 signed the Exclusive Agreement, which specified that they were not allowed to make any international agreements or host any foreign agent without British consent. To be sure, the completion of the Suez Canal in 1869 had dried up trade through the Persian Gulf, and the region's earnings from the slave trade disappeared by 1900 (due to international, mostly British, pressure), but Britain held on to the region to keep other powers (such as the Ottomans and the French) out. Thus, nearly the entire Arab side of the gulf, from Oman to Kuwait, was under British protection by the end of the Coal Age.

Kuwait

Kuwait, located at the head of the Persian Gulf, immediately south of Iraq, was nominally under Ottoman rule, but nevertheless became a British protectorate in 1899. The sheikhdom's rulers had traditionally maintained friendly relations with the Ottomans, their Arab neighbors, and the British, but Abd Allah Al Sabah II developed close ties to the Ottomans, and in 1871 even took on the formal Ottoman title of provincial governor (qaimaqam). His successor was murdered after four years in power, with the murderer, the half-brother of the legitimate ruler, becoming known as Mubarak the Great. Britain is suspected to have assisted this coup, because Mubarak, who ruled Kuwait from 1896 to 1915, completely reversed the sheikhdom's policy of the previous three decades, turning it over to the British. The Ottomans, who had supported the legitimate murdered ruler, protested the 1899 cession of Kuwait, but Britain was quick to maneuver coal-fired gunboats off the Kuwaiti coast. Britain's policy was influenced by its concerns about the growing influence of other Western nations in the region. Worst of all, the Germans had earlier that year negotiated a deal with the Ottomans to construct a railroad from Constantinople to Baghdad to Basra, the city that gave name to Iraq's most southern province, which had included Kuwait. The German-Ottoman agreement came with concessions to the mineral (oil) resources on both sides of the tracks, and it was proposed to extend the railroad to

Kuwait, where the Germans intended to build a coaling station to support their ships operating in Asian and African waters. (Germany had colonies in nearby East Africa at that time.)

The British, in exchange for supporting Mubarak, received control over Kuwait's foreign policy for the following six decades. They learned about the region's oil seepages that had indicated the presence of oil below the desert surface for centuries, and in 1913 Britain signed an agreement with Mubarak, according to which any oil concessions granted to companies would have to be approved by the British government.[164]

Saudi Arabia

The Arabian Peninsula's west coast had been the birthplace of Islam in the 7th century, but quickly lost importance within the expanding Islamic Empire, which prospered in more fertile and more densely populated regions around such cities as Damascus, Baghdad, and Cairo. Much of the peninsula's coastal areas eventually became part of the Ottoman Empire, while the inner parts were mainly covered by desert and populated by tribal nomads. (Europeans knew nearly nothing about these interiors still at the beginning of the Oil Age.) In the 18th century an Islamic revivalist movement, founded by Muhammad Ibn Abdul-Wahhab, sprang up in the Arabian peninsula. It claimed that the practice of Islam had lost its original purity, in luxurious living, the veneration of saints, and ostentation in worship. (Wahhabi mosques are therefore simple and without minarets, and adherents of Wahhabism dress plainly and do not smoke tobacco or hashish.) The founder of the Wahhabi sect was driven from Medina and relocated to the interior desert area, where he managed to convert the Saud tribe. In 1744 his daughter married the son of the Saudi sheik, who decided to wage a holy war (jihad) against all other forms of Islam, and began subduing his neighbors from around 1763. As of 1811, the Wahhabi founder's Saudi grandson ruled over much of Arabia (with the notable exception of Yemen) from his capital Riyadh. Thus it was time for the Ottomans to react. They sent troops (from Egypt) and by 1820 had driven the Wahhabis back into the interior desert where they had come from. Nevertheless, the tribe rose again some 30 years later, this time threatening the Persian Gulf coast of Arabia. But this region was now protected by the British, and the Saud family in turn declined entirely, losing even its capital Riyadh in 1884. The clan therefore retreated to Kuwait, but this still was not the end of the story. A short two decades later, in 1902, the new leader of the Saud dynasty, then 22-year-old Ibn Saud (Abd al-Aziz bin Abd al-Rahman Al Saud) left Kuwait with a party of some 20 relatives and servants to recapture Riyadh. They assassinated the governor of the city and within 10 years consolidated control over the clan's traditional desert region. Two years later World War I broke out, and the British, chiefly through

T. E. Lawrence (of Arabia), encouraged the Arabian tribes to revolt against Ottoman leadership. In 1915 the House of Saud concluded a treaty with Britain. The Saudi lands were made a British protectorate, and Britain was to supply weapons and cash. However, Ibn Saud remained inactive during World War I, and instead attacked a neighboring dynasty in 1920. He doubled his territory within two years, while British subsidies continued until 1924. (In 1923 Ibn Saud actually granted a British company a concession to explore for oil in his territory, but it was never exploited.)

The other strongman in the neighborhood was Husayn ibn Ali (Hussein ibn Ali), the former sherif of Mecca (Makkah), who had proclaimed himself king of the newly independent Hejaz region in 1916. Husayn ibn Ali was a member of the clan of Hashim that traces its ancestry from the great-grandfather of Muhammad, founder of Islam. (The Hashemite dynasty traditionally acted as guardians of the holy cities of Mecca and Medina.) He had been a leading figure in the Arab revolt against Ottoman rule and initially had strong support from the British, who had made two of his sons kings of Iraq and Jordan in 1921. However, Ibn Saud defeated Husayn ibn Ali in war in 1925. In turn rule over the Hejaz gave the House of Saud a greatly enhanced status as guardian of the great shrines of Mecca and Medina, the most holy places of Islam. And thereafter Ibn Saud's ability to keep the pilgrimage routes open won him acknowledgement and respect from other Muslim leaders. In consequence, the British in 1932 accepted the founding of the Kingdom of Saudi Arabia, and acknowledged the newly proclaimed king, Ibn Saud.

British-American Rivalry for Oil Concessions

The time between the World Wars was marked by an increased rivalry between Britain and the United States for oil concessions in the Middle East. Britain still maintained a strong global position, though its Coal Age superpowerdom was clearly on the decline, and America had not yet risen to true superpowerdom, though it was already, and by far, the world's largest oil producer. Compared to the United States, Britain's urge to control Middle Eastern oil was a lot more imminent in the early Oil Age, simply because it did not have any domestic (and hardly any colonial) supplies. (Britain's North Sea oil was first discovered in 1969 and produced from 1975.) Hence Britain argued that the United States, controlling 82 percent of the world's oil production (63 percent domestically and 19 percent in Latin America), had no need or reason to get a share of Middle Eastern oil. However, U.S. Secretary of State Charles Evans Hughes and Secretary of Commerce Herbert Hoover pressured American companies to look for oil overseas immediately after World War I, and the United States was outraged when it was initially left out of the Iraqi oil deal (that is, the Turkish Petroleum Company), and was later excluded from the newly negotiated contract between the Shah of Persia and the Anglo-Persian Oil Company. On the other hand, the United States managed to bring its corporations into the Turkish Petroleum Company in 1928, and Standard Oil of California (later Chevron) in 1929 received a concession in Bahrain, a tiny island sheikdom in the Persian Gulf, even though it was a British protectorate. (Standard Oil of California chartered the Bahrain Petroleum Company in Canada, a self-governing dominion of the British Empire, in order to meet the requirement that oil concessions were to be granted to British companies.) It was here, in Bahrain, that in 1932 oil was struck for the first time in substantial quantities on the Arab side of the Persian Gulf. In return for an immediate payment of $50,000 in gold, Standard Oil of California was the following year (1933) also granted a concession to drill for oil in eastern Saudi Arabia. To handle these substantial operations Standard Oil of California in 1936 formed a partnership with Texaco. Together these two American corporations owned the Arabian American Oil Company (Aramco) and the Bahrain Petroleum Company, and they established a joint venture, the California-Texas company (Caltex), as outlet for future oil production in Bahrain and Saudi Arabia. Oil was discovered in Saudi Arabia in commercial quantities in 1938, but oil operations were disrupted by World War II.

The United States also pressured the British government to approve (in 1932) that the U.S. Gulf Oil Corporation would negotiate a concession in Kuwait together with the Anglo-Persian Oil Company through the Kuwait Oil Company (KOC), established as a 50–50 joint-venture between the two firms. The sheik granted this concession in 1934 (for a time period of 75 years), but Britain, through a subsequent agreement, established control over the Kuwait Oil Company. Surveying began in 1935, drilling in 1936, and the first gusher was struck in 1938 in an area subsequently called the Al Burqan field, one of the largest and most productive fields in the world. (Thereafter KOC discovered another seven oil fields.)

Post-World War II Development

World War II entirely changed the power balance between Britain and the United States. Britain had depended heavily on American support during the war, and the nation's decline since the Coal Age was nearly complete. The United States, on the other hand, emerged from the conflict as a true superpower with global influence, and had more reason than ever to seek control of Middle Eastern oil. America had poured an astounding 6 billion barrels of oil into World War II, and its economy now grew very fast (as it produced free of competition for both domestic and international markets, while the rest of the world lay largely in ruins), thus absorbing a lot of fuel. In 1948 the United States turned into a net oil importer for the first time, and it soon became obvious that U.S. domestic oil production would peak and start to decline within a matter of decades. What is more, Mexico (1938) and Venezuela (1943), the traditional oil suppliers to the United States, had just nationalized their oil industries to achieve more economic independence.

The years after the war showed that the Persian Gulf oil reservoirs were the largest in the world, and that Middle Eastern oil was extremely cheap to produce. Hence, access to this region's oil was once and for all made a top U.S. priority. The major American oil corporations were instituted to serve this foreign policy goal, but for a moment after World War II the alliance between government and industry was jeopardized by a renewed anti-trust movement. It had become clear that five U.S. corporations, (later called) Exxon, Mobil, Chevron, Texaco, and Gulf, together with Shell and BP (Anglo-Iranian Oil), had between 1928 and 1948 engaged into illegal cartel building activities, and that these "Seven Sisters" had divided the international oil market among them. All the new shareholders of the Turkish Petroleum Company had in 1928 signed the Red Line Agreement, for instance, which prohibited the signatories from independently seeking oil interests in the territories that had of 1914 been part of the Ottoman Empire, including the Arabian Peninsula. Also in 1928, Shell, Exxon, and BP signed the Achnacarry Agreement, which Mobil, Gulf, and Texaco had joined by 1932. This agreement set out the rules for the major oil corporations to keep other market players out, and to keep prices high and stable on international markets. (It was followed by various local agreements.) To be sure, Rockefeller interests were still the largest shareholders in Mobil, Exxon, Chevron, and Amoco, which were all descendents of the Standard Oil Company, the firm that had prompted legislators to introduce strict antitrust laws in the United States in the first place. In 1952 the Senate investigated the Seven Sister's cartel-building activities, but then Eisenhower was elected president and shifted the policy focus towards anti-communism: the large oil corporations were off the hook and once again viewed as an indispensable means to achieve future energy security and control over Persian Gulf oil.[165]

To be sure, the Middle Eastern countries depended on those oil corporations as well. Much of the region was thinly settled and entirely underdeveloped by Western standards, or even compared to international averages. (Herding and pearling were typically the main source of income.) Getting the oil out of the ground and to international markets thus required foreign investment and expertise. The bargaining position of the various local Arab rulers was weak, many of them were interested in making a quick buck, and nobody knew how big those Middle Eastern oil reserves really were. The early oil concessions sold in the 1930s did therefore not make the Arab rulers rich overnight. Quite the contrary. At first they sold rights to explore for oil, but the prospecting activities were interrupted by World War II. Then they collected export fees, but it often took many years before substantial oil exports became reality. Oil from Qatar's relatively small fields was exported from the 1950s, Abu Dhabi's offshore oil from 1962, and the rulers of Dubai and Oman had to wait for export revenues until the late 1960s. Kuwait enjoyed a bit of a head-start. This British protectorate exported commercial quantities of oil from 1946, and by 1953 emerged as the largest Persian Gulf oil producer (maintaining this position until 1965). Kuwait achieved independence from Britain in 1961, but Iraq immediately claimed sovereignty over the emirate on the grounds that it had been part of Iraq's most southern province during the time of the Ottoman Empire. (This was actually the renewal of the claim Iraq had put forth in 1938, when Kuwaiti oil had initially been discovered.) Thus, Britain quickly dispatched a brigade to Kuwait to deter Iraq.

Saudi Arabia, on the other hand, was an entirely American affair. Aramco, the Arabian American Oil Company (owned by Chevron and Texaco), had received an oil concession in 1936, merely four years after the kingdom of Saudi Arabia had been founded. In turn the United States and Aramco helped build much of the newly created desert nation's infrastructure (such as roads, housing, schools, and wells) from scratch. In 1948 Exxon and Mobil joined Chevron and Texaco to own a share in Aramco, which held a monopoly in Saudi oil operations. (Chevron, Texaco, and Exxon henceforth owned 30 percent each, and Mobil 10 percent. In short, Aramco was dominated by Rockefeller interests.) This was the year (1948) that Saudi Arabia's Ghawar Field, the largest oil field in the world, was discovered. (At the end of 2004 it still held 70 billion of its original 80 billion barrels.) Three years later, in 1951, the world's largest offshore oil field, the Safaniya field, was discovered. Thus, Saudi Arabia emerged as the country with the largest oil reserves in the world, and from 1966 produced even more oil than Kuwait. (That year, 1966, Saudi Arabia produced 949 million barrels of oil, Kuwait 905 million barrels, Iran 777 million barrels, and Iraq 507 million barrels, compared to Venezuela, 1,230 million barrels, the Soviet Union, 1,909 million barrels, and the United States, 3,030 million barrels.)

The United States certainly hit the jackpot by securing a monopoly oil concession in Saudi Arabia. At the start of World War I, U.S. corporations controlled less than 13 percent of Middle Eastern oil supplies, while it was 60 percent by the end of the 1950s. However, the Saudi Arabian enterprise also required investments of hundreds of millions of American (tax and corporate) dollars into oil infrastructure such as pipelines, refineries, and terminals. The investments also included the U.S. Air Force base at Dhahran, built close to Saudi Arabia's first oil fields. Construction of this base began during World War II, and was completed in 1946. The United States was then allowed to operate the installation until 1962 in exchange for technical aid and for granting Saudi Arabia permission to purchase arms under the Mutual Defense Assistance Act. Ibn Saud died in 1953, but his family became fabulously rich through cooperation with the United States and its corporations. Saudi

Arabia's oil revenues soared from $5 million in 1945 to $334 million in 1960 (comprising about 80 percent of the country's total income) to $1,945 million in 1970. This was possible because the share of Persian Gulf oil profits was gradually shifted further towards the host countries.

The Founding of OPEC

Saudi Arabia in December 1950 negotiated a new agreement with Aramco under which Saudi Arabia was to receive 50 percent of the company's net earnings. Then Mohammad Mossadegh assumed premiership in Iran in April 1951 and nationalized (expropriated) the entire Iranian oil industry, which gave other countries in the Persian Gulf region a chance to improve their terms as well. A 70–30 split in profits between oil corporation and host country generally changed to a 50–50 split (in addition to royalty fees), which dramatically increased the oil income of local rulers. (Kuwait, for instance, shared the earnings equally with the Kuwait Oil Company from 1951 and in exchange extended the firm's concession for another 17 years in addition to the original 75.) Nevertheless, the Seven Sisters still dominated the international oil markets, as they controlled oil operations in the Middle East (through their subsidiaries) as well as the international distribution networks. This led five nations, Kuwait, Iran, Iraq, Saudi Arabia, and Venezuela, to set up the Organization of the Petroleum Exporting Countries (OPEC) in 1960. Initially the idea was to negotiate a larger share of the enormous profits the major oil corporations were reaping from their international marketing and sales activities, but soon the producing countries realized that their cooperation would allow them to control the supply, and hence the price, of oil. By the early 1970s another seven countries had joined OPEC, and together they now controlled over half the world oil production outside the Soviet Union, and an even larger share in exported oil. (The later members were Qatar (1961), Indonesia (1962), Libya (1962), the United Arab Emirates (1967), Algeria (1969), Nigeria (1971), Ecuador (1973) and Gabon (1975). The latter two withdrew membership in 1992 and 1994, respectively, but Azerbaijan and Angola joined OPEC in 2007.) With British and Norwegian North Sea oil coming on line to add to Russian, Canadian, and Mexican production, OPEC saw its share in world production plummet to 31 percent as of 1985, but it was back at about 40 percent in 2000.

OPEC is a cartel that regulates the production volume of each member state in order to control the international oil price and to maximize profits for its members. Obviously, OPEC immediately exerted enormous political power. The oil price influences prices of all other goods and services in Oil Age economies and OPEC soon had the economic well-being of the world at its finger tips. On the other hand, oil is the main, and in some cases the only, source of income for OPEC members. Hence, OPEC never had an interest in pushing the oil price up too far for a long period of time, as this would stifle international economic growth, which would in turn decrease the demand for oil and promote investment in energy alternatives. OPEC therefore set price goals at which it believed the overall profit would be maximized, and long-term prosperity for the member states ensured. And sometimes OPEC has engaged in negotiations with key non-members such as Russia, Norway, Mexico, and Oman, usually to persuade them to decrease exports during times OPEC considers the oil price too low. The strategic position of OPEC has gradually become stronger, simply because its members are holding the largest reserves and thus control gradually more of the world's remaining oil deposits. (In 2001, OPEC countries accounted for roughly 80 percent of the known oil reserves.) However, the West was to feel the organization's full might for the first time as early as 1973.

Israel and the First Oil Price Shock (1973)

Even though the Soviet Union had plenty of oil domestically, America was apparently concerned the Soviets might have plans to invade the Middle East. It was well known that securing oil resources was a main objective for the West after World War II (codified, for instance, in the Truman administration's National Security Council Resolution 138/1), but President Truman went as far as to approve a detailed plan to store explosives near Persian Gulf oil fields to blow up oil installations and refineries, and plug the oil reserves, in the event of an imminent Soviet invasion.[166] These explosives were placed near Middle Eastern oil fields in coordination with the British government and the Seven Sisters, but without the knowledge of local Arab governments. Later on, Truman's successor Eisenhower made clear that his statement about the United States being prepared to use armed forces to assist any country "requesting assistance against armed aggression from any country controlled by international communism" applied to the Middle East. And in 1955 the United States instigated the creation of CENTO, the Central Treaty Organization (CENTO), to counter the perceived threat of a Soviet expansion into the Middle East. Otherwise known as the Baghdad Pact, this mutual-security organization was established among Turkey, Iran, Iraq, Pakistan, and Britain, while the United States did at first not officially participate in order to avoid alienating Arab states with whom it was still attempting to cultivate friendly relations. (In reality the United States used a facility in one member state, Pakistan, for illegal reconnaissance flights over Soviet airspace.) To be sure, this pact achieved little to prevent the expansion of Soviet influence in the area. Quite on the contrary, states such as Egypt and Syria felt excluded from CENTO and thus turned to the Soviets (who supplied them with weapons). In short, it was quite difficult for the United States to maintain friendly relations with all Persian Gulf countries in this

Cold War atmosphere, especially because many of them were angered about the West's support for Israel, the state founded in 1948 in the relatively small region known as Palestine.

Early Roots of the State of Israel

Situated between the eastern shore of the Mediterranean Sea and the Jordan River valley, Palestine was at the very heart of the world's first agricultural center, but remained politically unimportant, because the region's major population centers arose along the large rivers in nearby Mesopotamia and Egypt. Empires that centered in these two powerful neighboring regions intermittently ruled over Palestine, but following the Egyptians, Assyrians, and Babylonians, the area was absorbed into the empires of the Macedonians, Ptolemaics, Seleucids, Romans, Byzantines, Arabs, and finally Ottomans.

Early settlers in Palestine included the Canaanites, Hebrews, and Philistines. Shortly after 600 B.C.E. the Babylonians put down an uprising of the Hebrews and deported the majority of them to Babylon. In exile, the Hebrews, who were known as Jews, managed to maintain their identity as a people by adhering to their religion, Judaism, and it was probably during this time (the 500s B.C.E.) or shortly thereafter (around 440 B.C.E.) that most of the books of the Torah were written down in their currently known form. (For Christians the Torah became the main part of the Old Testament, the older part of the bible.) These religious writings referred to legendary Hebrew forefather Abraham, who supposedly lived some 1,500 years earlier and was promised heirs and land in Canaan (the land of the Canaanites) for his people by the god of the Hebrews. The writings also describe mythical King David, who supposedly lived several centuries earlier, making Jerusalem his capital. (To be sure, present-day Israeli archaeologists have expressed doubt that a King David ever existed as described.[167]) In 539 B.C.E. the Persians defeated the Babylonians and allowed the Jews to return home (though many decided to stay in Mesopotamia), and when the Romans entered the scene, they named the region and province around Jerusalem Palestine, because it was inhabited by the Philistines, though plenty of Jews were living in the area as well. (By this stage Judaism had split into various divisions, sects, and orders, and during Roman times Jesus Christ offered yet another interpretation of Judaism, which subsequently emerged as a religion in its own right.) Following various rebellions against the Romans, increasingly more Jews were forced into exile, and after the revolt of 73 C.E. the Romans destroyed the temple of Jerusalem. By 300 C.E. only few Jews were left in Palestine. Instead, they were spread out over the whole Middle East and around the Mediterranean Sea, where they settled down along the important trading routes and at busy seaports.[168] However, prejudice and discrimination against

Jews in the Roman Empire increased in the 4th century, when Christianity was adopted as the official religion, and later continued in Christian Medieval Europe as well as Islamic regions. As the Jews hung on to their own religion and customs, they did not fully integrate into local societies, were considered suspicious, and often housed in separate quarters, as was normal for foreign merchants in the Middle Ages. (The first Jewish ghettos appeared in Spain, Portugal, and Germany in the 13th century, but the term ghetto actually derives from the 14th century ghetto of Venice, Italy.) Jews were excluded from craft guilds, and in turn excelled in trade, including slave trade, often serving as intermediaries between the Christian and the Islamic World. These trading activities were often facilitated by vast information networks spanning Jewish communities living widely dispersed over western Eurasia. However, the Jews were expelled from various countries, including England (1290), where they were readmitted only in 1655; the Netherlands (mid-13th century) with the decision being reversed in later centuries; and eventually Spain and Portugal (15th century).

The fact that the Jewish masters of trade often did especially well, added Christian jealousy to the cocktail of resentment. However, Jews were also appreciated as moneylenders by Christian business communities and nobles. Several of Europe's royal families began using Jewish financial advisers in the Middle Ages, and by the early 17th century the employment of Court Jews had practically become institutionalized in most of the principalities of the (German) Holy Roman Empire, plus some of the adjoining states, such as Poland and Denmark.[160] This was especially awkward as most provinces had expelled their Jews, but Court Jews were able to finance wars or other undertakings, and provided valuable international intelligence during conflicts such as Thirty Years' War (1618–1648). Court Jews provided princes and their courts with merchandise and money, metal for the mint, and provisions for the army. They also undertook commercial and diplomatic missions, and investigated proposals for the promotion of trade and industry. With time, several especially powerful Jewish banking families arose in Europe. Perhaps the best known is the Rothschild family, which originally came from Germany, where their house in Frankfurt on Main (in the days before houses had numbers) was aptly marked by a red shield. (Their true surname was actually Bauer, German for farmer.) The Rothschilds had a few decades of banking and merchant experience when they earned a fortune during the Napoleonic Wars. Nathan Mayer Rothschild was sent to Manchester in 1798 to deal in the thriving cotton industry, but then raised money for British troops and their allies, and profited from the family's information network that facilitated trading and banking operations despite the turmoil in continental Europe. In 1816 Nathan's four brothers, though all living in different countries of continental Europe, were elevated to nobility in Austria, and so was Nathan two years later. Lionel Rothschild, of the London branch of the family, became the first Jewish Member of Parliament in 1858, and his son the first Jewish peer, carrying the title Baron from 1885.[170]

During the late 18th and early 19th century the liberal thought of the Enlightenment period improved the position of Jews in European society. In the Austrian monarchy Jews were allowed to own land, for instance, and after the French Revolution the rights of man were extended to French Jews in 1790. However, the rise of nationalism and unscientific theories of race during the Coal Age promoted new resentments. The term anti-Semitism was coined in 1860, and in Russia waves of anti-Jewish riots (mainly between 1881–1884 and 1903–1906) were at least in part organized by Tsarist authorities. The pogroms (Russian for devastation) spilled over to Poland and other eastern European regions, and many Jews thus decided to emigrate to western Europe and America. (In Russia, Jews were also victims of later pogroms that were targeted towards wealthy Russians in general during the 1917 revolution. In consequence a total of some two million Jews emigrated to the United States between 1880 and 1920.) However, anti-Semitism was also on the rise in France, Germany, Austria, and the United States. (Jews were, for instance, routinely excluded from academic posts at American universities up until World War II.)

In Vienna, Austria, Theodor Herzl, a successful Jewish playwright and journalist, launched political Zionism in 1896 by publishing a book titled *Jewish State*. The year after he became the first president of the newly-founded World Zionist Organization. Herzl was convinced that the only solution to the problem of international anti-Semitism was the resettlement of the Jews in a state of their own, and Zionism was defined as the movement aiming to have Jews return to their ancient homeland in Palestine, with Jerusalem, the city of Zion, as its capital. As Palestine was part of the Ottoman Empire, Herzl outlined a scheme for setting up an autonomous Jewish commonwealth under Ottoman suzerainty. (To be sure, Zionism was not solely a religious concept, but heavily influenced by the rise of nationalism, a major trend in 19th century European politics. Germany, for instance, had resisted unification as a national state up until the Coal Age.) As many members of the international Jewish community were wealthy and influential, the Zionist movement had significant support from the very beginning. Nevertheless, this was going to be difficult to implement as long as the Ottoman Empire was intact. Notably, the area was now native to Arabs, while no significant number of Jews had been living there for one and a half millennia. Quite obviously, the people now indigenous to the region would not be willing to forego their land to satisfy someone else's religious or nationalistic concept.

The Balfour Declaration

Nevertheless, there was hope for the Zionists, because Britain, the Coal Age's principal superpower, was open to the idea of a Jewish state in Palestine.

Frankly, Britain welcomed just about everything that would weaken the Ottoman Empire, which had close ties to Britain's rival Germany. However, when World War I broke out, the British made promises to Arab leaders in order to promote an armed rebellion within the Ottoman Empire. Britain essentially promised to establish an independent united Arab state covering most of the Arab Middle East, once the Ottoman Empire would have been defeated. But at the same time, the British made contradicting promises to the international Jewish community, which supposedly helped pull the United States, where many wealthy Jews were now living, into World War I. Perhaps another reason to support the Zionists was for Britain to reward Chaim Weizmann, a Swiss-educated Jewish chemist. Weizmann had come to England in 1904 and famously developed a method to synthesize acetone, which the British urgently needed to produce explosives during World War I. Weizmann helped British foreign secretary A.J. Balfour to write a letter sent on November 2, 1917, to Baron Lionel Walter Rothschild (grandson of the first Jewish member of the British parliament and eldest son of the 1st Baron Rothschild), who held the chair of the British Zionist Federation. This letter expressed Britain's support for "the establishment in Palestine of a national home for the Jewish people," though without prejudicing its non-Jewish communities. Due to this statement, known as the Balfour Declaration, together with the content of the Sykes-Picot Agreement (in which Britain and France divided the southern parts of the Ottoman Empire between them), the Arabs felt they had been double-crossed by the British.

After World War I Palestine was placed under British administration, with Jews wishing to settle in the region and indigenous Arabs opposing them. The 1918 pogrom in Poland promoted an influx of Jewish settlers, but thereafter Jews still accounted for only about ten percent of Palestine's total population. However, another 100,000 Jewish immigrants arrived during the 1920s, and, with anti-Semitism on the rise in Europe, the rate of Jewish emigration increased. The British restricted Jewish immigration into Palestine, but tensions grew and led to violent clashes. Arabs attacked Jews, while organized militant Jewish groups, such as Etzel (Irgun) and Lehi (Stern Gang), attacked both Arab and British targets. In 1937 Britain proposed separate Arab and Jewish communities, but the idea was rejected by the Arabs, who felt they were being marginalized in their own country, and even by parts of the Jewish community. In turn Arab and Jewish terrorist groups spread increasingly more violence, and the British put down a general Arab uprising.[171]

During World War II the British closed Palestine for Jewish refugees, supposedly to win the full support of the Arab countries against Nazi Germany and its allies, while taking support from the Jewish community against Hitler's Germany for granted. However, Jewish terrorist groups attacked British installations in Palestine for this policy. The best known incident was the 1946 bomb attack on the King David Hotel, which housed the British administrative and military headquarters in Jerusalem. The attack killed 92 people. It was conducted by the Irgun group, which was then led by later Israeli prime minister and Nobel Peace Prize winner Menachem Begin.

By the end of World War II the atrocities experienced by Jews during the Holocaust had won a lot more people to the idea of a Jewish state. These included international Jews (many of whom had been anti-Zionist for various reasons) and large parts of the international Christian community, while the Arabs still objected to the proposal. Arabs felt they had nothing to do with the Holocaust, and that the international community, if it insisted to create a new state for all those Jews who could not, or did not want to, return to their original home countries, should establish such a state elsewhere. However, the United States, the principal winner of World War II, welcomed the idea of a Middle Eastern state of Israel, not just to satisfy the demands of American and international Jews. It was obvious that such state would be a sure ally and convenient base in the midst of the Arab world, close to the rich Persian Gulf oil fields and not too far from the Soviet Union. In 1947 the newly founded United Nations (UN) announced it would support and execute the original British plan of dividing Palestine into a Jewish and an Arab state, with Jerusalem being administered as international city. Violence erupted immediately, as the Arabs were extremely unhappy with this decision, but many Jews were unsatisfied as well, chiefly because they wanted Jerusalem as their capital.

Proclamation of the State of Israel

The British Army initially intervened in the general uprising, but in May 1948, when the British mandate in Palestine expired, Britain withdrew and an independent State of Israel was proclaimed. (Chaim Weizmann, who had helped formulate the Balfour Declaration, was made the first president of Israel.) Since the Arabs rejected both the partition of Palestine and the existence of Israel, troops from Iraq, Syria, Lebanon, (Trans-)Jordan, and Egypt immediately crossed Israel's proclaimed borders. However, they were actually outnumbered by Israeli defense forces, who had obtained weapons illicitly from Czechoslovakia. When a cease-fire agreement ended this first Arab-Israeli War in 1949, Israel controlled much more land than had been originally allocated to it by the UN. (In fact, the Israelis by this stage controlled three quarters of what had been Palestine under British mandate.) Perhaps 800,000 Arab Palestinians had to flee the area during the conflict and were not allowed to return to their homes (which were expropriated or destroyed) despite international pressure on the Israeli government. The luckier refugees were moving to different Arabic countries, while most of them found themselves permanently in camps, mainly in three Arab states: Jordan, Syria, and Lebanon. (These refugee camps were home to over two million

Palestinians at the beginning of the 21st century.) Meanwhile hundreds of thousands of Jews who were no longer welcome in neighboring Arab countries and Iran settled in Israel. In addition, many more Jews arrived from Europe. Thus, Israel's population doubled within one year of its existence.

The only remaining parts of Palestine that were not under Israeli control were the West Bank (of the Jordan River), which was occupied by Jordan from 1949, plus the small Gaza Strip, which was controlled by Egypt. Under President Nasser, Egypt in the 1950s emerged as a leader in the Arab world, and in 1956 Nasser nationalized the Suez Canal and closed off access to the Israeli Red Sea port of Eilat. Israel immediately sent troops towards Egypt, and Britain and France, without consulting the United States, joined the attack in an attempt to reassert international control over the canal zone. However, these European Coal Age powers were now long past their zenith, and as soon as the United States and the Soviet Union disapproved of this military intervention, Britain and France were forced to withdraw their forces. The United Nations declared the canal Egyptian property, but Israel had used the conflict to occupy the Gaza Strip and withdrew only after considerable U.S. pressure in 1957. The true reason why the West could afford giving up international control of the Suez Canal (which was closed by Egypt once again between 1967 and 1975) was that transport economics of the Oil Age's most important product suggested the construction of larger tankers and have them circumnavigate Africa. The Suez Canal, for long the main gateway of Middle Eastern oil to the West, had limited the size of tankers to about 30,000 tons gross, while the Suez Crisis of 1956 promoted the construction of supertankers, which by 1976 had reached the enormous size of around 480,000 tons deadweight, and by 1981 around 560,000 tons deadweight.[172]

The Arab-Israeli War of 1967 (Six-Day War)

After the Suez Crisis the Israeli government kept on encouraging Jewish immigration, which helped the total number of Jews arrived since 1948 to soar to two million by 1962, while more Arab residents moved from Israel to neighboring countries. In 1964 a number of Palestinian Arabs in exile founded the Palestine Liberation Organization (PLO) with the goal of overthrowing Israel and campaigning for an independent Palestinian state. But this was not going to be easy. Israel was now heavily armed with U.S. weapons systems, and by the mid-1960s very likely had become a nuclear power, after purchasing a plutonium-yielding reactor from France. In a strange arrangement the UN Security Council allowed Israel to conduct a policy of ambiguity in regards to nuclear weapons, meaning that Israel is allowed to neither acknowledge nor deny the possession of nuclear weapons without having to face inspections by the UN's International Atomic Energy Agency.

Following Palestinian guerrilla attacks on Israel from bases in Syria, the Syrian government feared an invasion by Israel and turned to Egypt for support. In 1967 Egypt, Syria, and Jordan concentrated troops on Israel's borders, seemingly in preparation for an attack, but Israel launched what it felt was a pre-emptive (non-nuclear) strike, targeting Arab air fields and destroying the Egyptian air force on the ground. (By this stage Israel was probably in possession of two nuclear bombs.) This started a short war that killed some 700 Israeli and 25,000 Arab troops within less than a week. Israel gained the whole of Jerusalem (now including the old city), the West Bank area of Jordan, the Sinai peninsula of Egypt, and the Golan Heights of Syria. And Israel showed little interest in returning these areas. It erected massive lines of fortification in both the Sinai (along the Suez Canal) and the Golan Heights. To be sure, UN Security Council Resolution 242 (November 22, 1967) affirmed "the inadmissibility of the acquisition of territory by war" and called upon Israel to withdraw "from territories occupied in the recent conflict." More specifically, it called for Israel to withdraw to pre-1967 borders in return for Arab recognition of Israel's right to exist peacefully within them. But this resolution was not going to be implemented for the remainder of the 20th century.

The Arab-Israeli War of 1973 and the First Oil Price Shock

Egypt was increasingly upset when Israel for years refused to return the Sinai peninsula. The Egyptians thus declared on several occasions that they would attack, but nevertheless took Israel by surprise when they sent troops across the Suez Canal into Sinai in October 1973, on the holiest day of the Jewish year. Syria attacked simultaneously at the Golan Heights and the Israeli army suffered heavy losses. However, Israeli forces, with massive ammunitions supply provided by the United States, managed to push back the Egyptian, Syrian, and other Arab forces beyond the original lines. After three weeks of fighting, with some 19,000 casualties, U.S. and Soviet pressure achieved a ceasefire.

During this conflict, OPEC demonstrated its considerable political power for the first time. OPEC now controlled over half the world oil production outside the USSR, and the contribution of Middle Eastern oil to total world production had increased from 17 percent in 1950 to 37 percent in 1973, while the U.S. share had sunk from 52 percent to 22 percent. Middle Eastern oil was cheap to produce and had flown out to the world to fuel astonishing economic growth during the 1960s. (It currently costs some 20 cents per barrel to produce oil in the Middle East, compared to 80 cents in Venezuela and 90 cents in Texas.[173]) However, when many Western nations showed strong support for Israel in October 1973, the Arab oil countries imposed an embargo on them. This had disastrous effects on the world economy as prices per barrel of oil quadrupled from around $3 to around $12 by the end of the year, and stayed high after the embargo was lifted in March 1974. (Expressed in 2004 dollars this would be a jump from $10 to around $40.)

The Western World experienced a serious combination of sudden inflation (the rise of general price levels) and economic recession, but in part this crisis had already been in the making. The dollar-based Bretton Woods system had collapsed two years earlier (after U.S. domestic oil production had began to decline), and the dollar had been greatly devalued since. International oil prices (and other commodities delivered by developing countries) had, however, been fixed in U.S. dollars, which made energy extraordinarily cheap, especially in America. Now the United States, with only six percent of the world's population consuming a third of the world's energy, was especially hard hit.[174] Imports of Arab oil dropped from 1.2 million barrels a day to 19,000 barrels a day, while national average retail prices per gallon of gasoline increased from 38.5 cents in May 1973 to 55.1 cents in June 1974. Gasoline was rationed, and long queues of cars lined up at the gas stations. A national speed limit of 55 miles per hour and a year-round Daylight Saving Time was introduced to help reduce energy consumption. Schools and offices closed down to save on heating oil, but the central banks of the industrialized nations were largely helpless. Their attempt to raise interest rates to curb inflation slowed down the economy even more (as it became more expensive to borrow money for investments), but general price levels remained high due to high energy prices. The real standard of living decreased for several years in a row for just about everyone in the West, because inflation was raising the cost of living far faster than salaries increased. Decreased consumption then forced factories to cut production and lay off workers. In short, unemployment rose and GDPs sank even in such countries as Britain and France, which had been exempted from the oil embargo (as they refused to allow U.S. airplanes to use their air bases to supply Israel) but were fully affected by the rise in international energy prices and the contraction of global trade. The only major industrialized country that somewhat benefited from the crisis, at least in the longer run, was Japan, whose automakers had specialized in more compact and fuel-efficient cars. (The crisis also prompted Japan to shift focus away from oil-intensive industries and to invest heavily in the electronics sector.) Much of the developing world stepped into a debt trap during this time: The OPEC countries enjoyed record earnings that were deposited into international banks which could not find borrowers in the West due to the recession. Hence, these banks pushed less industrialized countries into borrowing at bargain prices, but the short- and medium-term loans later had to be renewed at much higher rates (as interest rates quickly rose to counter inflation).

The United States seriously contemplated an invasion of Saudi Arabia and Kuwait in 1973, but refrained from military action, probably because the embargo ended quite fast.[175] The oil price, on the other hand, stayed high throughout the 1970s and early 1980s. Several OPEC countries, including

Libya (1971), Iraq (1972), Iran (1973), and Venezuela (1975), nationalized oil assets during the early 1970s, others began purchasing major shares of the subsidiary oil companies operating in their countries. The Kuwaiti government acquired a 60 percent ownership in KOC in 1974, the Bahraini government a 60 percent interest in BAPCO in 1974, and 100 percent in 1980. Saudi Arabia's oil revenues climbed to an astonishing $4.35 billion in 1973, but reached as much as $36 billion in 1978, and an astronomic $116 billion in 1981. The Saudi government bought a 25 percent interest in Aramco in 1973, and the remainder in 1980. (In 1984 Aramco had its first Saudi president, and in 1988 the company was renamed Saudi Aramco, the Saudi Arabian Oil Company. In the beginning of the 21st century, Saudi Aramco was easily the world's largest oil firm. The best outside guess of 2004 put the company's oil production at 10 million barrels a day, one-eighth of the world's consumption, with at least 260 billion barrels of proved oil reserves, 20 times the size of Exxon Mobil Corporation, the largest private-sector oil firm. Saudi Aramco's 2003 revenues amounted to some $93 billion.[176]) This change in ownership of oil assets further shifted the balance of oil power to OPEC countries, but during the second oil price shock something even worse happened from the standpoint of the Western nations: they lost control of a long-standing partner at the Persian Gulf.

Iran and the Second Oil Price Shock (1979)

U.S. (or Western) policy in terms of controlling the oil-rich Middle East rested on three pillars. One was to maintain democratic, non-Muslim, pro-Western Israel as a base and partner. Another was to sustain close and friendly relations with Saudi Arabia, and to keep the House of Saud in power. The third was to control Muslim, but non-Arab Iran: but here things did not work out as planned.

The state of Iran, until 1935 called Persia, is centered around one of the Middle East's principal and ancient power and population centers. (Iran had some 65 million inhabitants in 2007, of whom 51 percent were Persian, 24 percent Azerbaijani, and the remainder various ethnic groups.) After the revolutions of 1905 and 1909 a parliamentary regime was established, and the Anglo-Persian Oil Company (APOC), which held a concession over the Persian oil found by British efforts from 1908, operated freely during subsequent years. (The British government was the majority shareholder of APOC from 1914.) After World War I both Britain and Soviet Russia had a strong influence in Persia until Reza Khan, an army officer, in 1921 staged a coup and seized control. Shortly thereafter, in 1925, he was crowned shah (king) and assumed the name Reza Shah Pahlavi. The shah improved the country's infrastructure, opened several hospitals and a university, but faced increasingly more resistance due to his repressive methods. The British were upset when he revoked the APOC's oil concession in 1932. Though Pahlavi and APOC soon agreed on new terms, the British no longer regarded Persia as a reliable supplier of oil and began looking for other sources. During World War II Pahlavi sympathized with Hitler, and when German troops advanced into Russia in 1941, British and Soviet troops invaded and occupied Iran to secure the country's rich oil fields. Reza Shah Pahlavi was immediately forced to abdicate in favor of his son Mohammed Reza Shah Pahlavi, and was exiled to South Africa, where he died in 1944. After the war, the United States insisted that Iran allowed two American oil companies to operate in the country next to the Anglo-Iranian Oil Company (renamed from APOC in 1935): both Exxon and Mobil signed a 20-year contract with the Iranian government in 1947.

Ousting Mossadegh

At first the new, young shah seemed to let the parliament rule the country, but in 1951 Dr. Mohammed Mossadegh, a seasoned politician, was elected prime minister and obtained legislative approval to nationalize Iran's petroleum industry. The shah opposed this, but Mossadegh had popular support, and the shah (temporarily) left the country. Though Exxon and Mobil were also affected, Britain was the hardest hit by Mossadegh's move, as the vast assets of the Anglo-Iranian Oil Company were expropriated. Britain organized a sea blockade of Iran's principal refining city of Abadan and considered a full-scale war, but Mossadegh had sympathizers in the United States, where Iran was viewed as a bulwark against communism. Nevertheless the American oil corporations joined the boycott that made it impossible for Iran to sell its oil on world markets. The country thus suffered extreme economic hardship, but Mossadegh was nevertheless able to maintain popular support and was approved by parliament for a second term in 1952. However, when President Dwight Eisenhower entered office in 1953, the United States decided to cooperate with the British government under Winston Churchill to remove Mossadegh and to restore the shah to absolute power. CIA director Allen W. Dulles approved $1 million to be used "in any way that would bring about the fall of Mossadegh," and Kermit Roosevelt, grandson of former U.S. President Theodore Roosevelt and chief of the CIA's Near East and Africa division, directed Operation Ajax, through which Mossadegh was ousted. The money was mainly used to fund Mossadegh's opponents and to discredit him. Undercover CIA agents especially targeted the religious community, stirring anti-Mossadegh sentiments by delivering false information. Within a short time Mossadegh's popularity was eroding, and nearly 300 people died when anti- and pro-monarchy protestors began violently clashing in the streets. In turn pro-Shah tank regiments stormed the capital Tehran and bombarded the prime minister's official residence. Mossadegh was arrested on August 19, 1953; sentenced to three years in prison for treason; and after his release put under house arrest until his death in 1967. (Much later, in 2000, the United States under the Clinton administration expressed regret over the overthrow of Mossadegh, admitting that "the coup was clearly a setback for Iran's political development and it is easy to see now why many Iranians continue to resent this intervention by America.")

Restoring the Shah and Arming Iran

The shah returned from his exile in Italy and was reinstalled to power. He immediately reprivatized (internationalized) the Iranian oil industry, granting oil concessions to a consortium of eight Western oil companies in 1954: the Seven Sisters plus the French CFP. (CFP was the predecessor of TOTAL, the company that in 1999 merged with French Elf Aquitaine and Belgian PetroFina into TotalFinaElf. CFP had its roots in Iraqi, and PetroFina in Romanian oil prospecting. The TOTAL group was the world's fourth largest publicly-traded oil and gas firm based on market capitalization as of 2006.) It was actually at this stage that Anglo-Iranian Oil, the former APOC, was renamed British Petroleum, BP, which was initially a small British subsidiary of Europäische Petroleum Union, EPU. (EPU was a company set up in 1906 by Deutsche Bank, Berlin, together with the Nobels and the Rothschilds of Paris, who both held vast interests in the Baku oil industry. The primary objective of EPU was to deliver oil products from Azerbaijan to western Europe. EPU's branch in England, called British Petroleum, Ltd., was during World War I seized by the British government, which in 1918 sold it to the Anglo-Persian Oil Company.[177]) The British government remained the majority shareholder of BP after 1954 (until 1987), but the reprivatization under the shah was yet another step in the decline from former British superpowerdom at the expense of America: Britain had previously dominated the Iranian oil industry through Anglo-Iranian Oil, but in the new set-up the five American "**S**isters" each got 8 percent of the shares of the new oil consortium, which combined equaled the 40 percent assigned to BP. Shell received 14 percent, and CFP 6 percent. Iran had no control or share in the oil consortium for the next 20 years. This was a tremendous success for the Western oil corporations and a clear warning to all countries who may have considered expropriating their international assets. (Nevertheless Mossadegh's initially successful nationalization of the Iranian oil industry had quite an effect on the entire Middle East, as it helped to shift more share of oil profits towards the host countries.)

With the shah restored to full power, America and its allies turned Iran's army into the largest and best equipped in the region, counting some 200,000 troops. Iran was to be the West's watchdog in the region, as it had the infrastructure and population size that Saudi Arabia was lacking. Altogether the West sent well over four thousand military advisors to Iran and provided the country with literally billions of dollars worth of weaponry, including hundreds of airplanes such as M-60A1s, F-4s, and F-14s. Iran prospered economically throughout the 1960s, and the Western media presented the shah as an able, modern, and modest leader. In reality, Iran's new wealth benefited a rather small part of the population, and the shah's autocratic rule included murder, systematic torture, and other human rights violations. The security and intelligence service SAVAK was founded in 1957 with CIA assistance. It was installed to protect the shah from political opposition and had practically unlimited powers to arrest and detain people.

Interestingly, the West let the shah get away with nationalizing Iran's oil industry in mid-1973, but this occurred shortly before war between Israel and its Arab neighbors started in October that year: this allowed the shah to show immediately that he remained a reliable partner of the West. Iran did not join the Arab oil embargo, but supplied international markets with urgently needed fuel. Iran then profited handsomely from the high oil prices, and used the earnings for a further modernization of industry and even more defense spending. The United States exported over $20 billion worth of arms to Iran between 1970 and 1978, which led to what was perhaps the most rapid peacetime buildup of military power in history.

Meanwhile the shah turned increasingly more autocratic, and resistance against him grew, initially mainly from the (fairly secular) urban middle class. In 1975 the shah abolished the multiparty system of government, which allowed him to rule as absolute dictator, and in 1977 even the United States reacted. President Carter threatened to freeze arms shipments if the human rights situation in Iran did not improve. In turn more than 300 political prisoners were released and censorship was somewhat relaxed, which led to campaigns and demonstrations demanding personal freedom. What is more, the shah's land reform failed, as it did not help the farmers while depriving the landed elites of their traditional powers. From 1978 the protests developed a much more religious cast: Islamic leaders managed to rally the largely illiterate rural population against the shah, portraying his reforms as evil westernization.

The Iranian Revolution and the Second Oil Price Shock (1979)

Following a year of extreme turmoil in which the shah used full military force against demonstrators and revolutionaries, he fled the country in 1979 together with much of the pro-Western elite and some 20,000 U.S. advisors. The shah died soon thereafter, in July 1980, while the Iranian Revolution replaced the monarchy with an anti-Western Islamic Republic run by conservative Muslims under the leadership of Ayatollah Khomeini. The 25-year period during which Iran had been the mainstay of Western defense structures in the Middle East thus came to an end, and oil prices jumped from little under $15 per barrel to around $35 per barrel. (In 2004 dollars that's a jump from $40 to over $80.)

The U.S. policy of installing a non-elected pro-Western ruler to power were now backfiring. Iran introduced a new constitution which created the post of Supreme Leader for Khomeini, who in turn controlled the military and security services and could veto candidates running for office. The new theocratic regime was extremely conservative, cut women's rights, introduced the strict Islamic dress code (hijab), and curtailed freedom of speech and the press. It was about as repressive as the shah's regime and (especially in the direct aftermath of the revolution) murdered political critics as well as adherents of non-Islamic religions such as the Bahá'í. (The Bahá'í faith is a monotheistic religion that combines the teachings of several of the world's major religions.) What is more, a mob of some 500 Iranian students took 63 staff members of the U.S. embassy in Tehran hostage during the revolution, while a crowd of thousands gathered around the embassy. Fifty-two of the hostages were held for more than a year. (It was during this hostage crises that oil prices in 1980–81 in real terms reached their highest level to date.)

The United States may have considered invading Iran, to free the hostages and to overthrow the regime of the ayatollahs. But Iran was now heavily armed, and anti-American sentiments had spread throughout the Muslim World during the previous Arab-Israeli War. Some of the region's Arab states may thus have considered supporting Iran or imposing another oil embargo on the West. Besides, it would have been difficult to win support for an invasion by the U.S. public a short six years after the Vietnam War had finally ended. Hence, the United States merely imposed economic sanctions on Iran and engaged in negotiations to free the hostages. A secret military rescue mission, Operation Eagle Claw, ended in disaster when two helicopters broke down in a sandstorm and a third one crashed, killing eight U.S. soldiers. In turn the bodies of some of these soldiers were paraded through the streets of Tehran in front of international television cameras. The 52 hostages were only released after 444 days in captivity, a few minutes after Reagan's inauguration as U.S. president on January 20, 1981.[178]

To be sure, even many of those who had supported a revolution against the shah eventually reconsidered their support for Khomeini's regime once it had become clear how the situation would develop. The Iranian Revolution did indeed achieve removal of foreign influence, but it left Iran economically isolated by much of the international community, including the Soviet Union.[179]

Iraq and Its Wars against Iran and Kuwait

The United States responded to the new situation in Iran in various ways. As the pro-Western Iranian troops were now history, President Jimmy Carter announced on October 1, 1979, that he had ordered the formation of a Rapid Deployment Force (RDF), which were U.S. troops to be employed overseas, especially to protect U.S. interests in the Persian Gulf region. The United States also began to support Iran's aggressive but secular neighbor Iraq. Iraq's leader Saddam Hussein had long been cooperating with the United States, and was more than willing to exploit Iran's military and political chaos during the revolution to gain control of the oil-rich Khuzestan region, which borders Iraq at the head of the Persian Gulf and was mainly populated by Arabs, not Persians.

Iraq as a nation was created in the region of Mesopotamia by Britain (and its allies) when the Ottoman Empire was dissected after World War I. In 1921 Britain installed a pro-Western Sunni Muslim monarch, who was to guarantee access to the Iraqi oil fields and to maintain order in the new country whose population was divided religiously into a Shia Muslim majority and a Sunni Muslim minority, and ethnically into an Arab majority and a Kurdish minority living in the north. (Kurds tend to be Sunni Muslims. In 2004 Iraq's population was 25.4 million with about 62% Shia and 34% Sunni, and about 77% Arabs and 17% Kurds.) In 1932 Iraq became the first Arab country to formally gain independence. To be sure, the kingdom was then not ruled by the monarch, but by pro-Western general Nuri-el-Said, who was prime minister from 1930 to 1958 and made Iraq the leading nation within CENTO, the anti-Soviet Baghdad Pact founded in 1955. However, the Egyptian president Nasser, who led the movement of Arab nationalism at the time, opposed the Baghdad Pact and challenged the legitimacy of the Iraqi monarchy, calling on the officer corps to overthrow it. In 1958 a swift coup led by Brigadier Abdel Karim Kassem (Abdul Karim Qasim) did indeed overthrow the monarchy. General Nuri and the royal family were executed, a republic was proclaimed, and Kassem seized power. He immediately declared the existing oil concessions null and void, but was willing to allow Western corporations to continue operations under contracts that provided higher revenues for Iraq. Kassem pulled out of the Baghdad Pact and included Iraq's large Communist Party in his coalition cabinet. The United States was quite concerned about this development and apparently sponsored young Saddam Hussein to assassinate Kassem.

Saddam, born into a poor farming family near Tikrit in 1937, had come to Baghdad in 1955 and joined the Baath party, an Arab nationalist movement. Saddam rose quickly within the party, developed a reputation for his brutality, and did in fact attempt to murder Kassem. The assassination was set for October 7, 1959, but was completely botched. Saddam killed Kassem's driver, while Kassem was only injured, and so was Saddam himself. While CIA documents regarding this episode remained undisclosed, rumors gathered that Saddam was guided by the CIA before, during, and after the assassination attempt.[180] The source of these rumors seem to be retired CIA agents. As the story goes, Saddam's CIA handler was an Iraqi dentist working for both the CIA and Egyptian intelligence. Agents of these two organizations helped Saddam to escape to Tikrit after the shooting and then moved him to Syria from where Egyptian agents transferred him to Beirut (Lebanon). While in Beirut, the CIA allegedly paid for Saddam's apartment, put him through a brief training course, and helped him get to Cairo, where Saddam briefly attended law school and continued Baath party-affiliated activities. (Meanwhile Saddam was sentenced to death in Iraq in absence.)

Saddam Hussein Comes to Power

In early 1963 Kassem was overthrown and executed in a coup led by Colonel Abdul Salam Arif, who initially supported the Baath Party. (Hence, Saddam Hussein immediately returned to Iraq from Cairo.) Noting that the Baath Party was hunting down Iraq's communists, the CIA (allegedly) provided lists of suspected communist sympathizers who were then jailed, interrogated, and summarily gunned down. The number of suspected communists killed at this time is in the thousands. In October 1963 Arif withdrew his support for the Baath Party and began suppressing it. (Saddam Hussein, like other Baath Party members, was jailed, but was able to escape.) However, Arif died in a helicopter crash in 1966, and two years later the Baath party seized power under Ahmad Hassan al-Bakr, who made his relative Saddam Hussein his deputy and head of security. Throughout much of the 1970s Saddam was de facto the man ruling Iraq, though he first assumed the presidency in 1979, when al-Bakr formally retired.

Saddam presided over the killings of various political opponents including communists, Kurds, and former Baath party members, while taking a leading role in addressing the country's major domestic problems. In 1972 he led the nationalization of the Iraqi oil industry, but immediately flew to Paris to assure French president Georges Pompidou that the French oil interests would be upheld. (Iraq in turn became France's second largest oil supplier, while France delivered weapons systems to Saddam.) As oil prices soared in 1973, right after Saddam had nationalized the oil industry, he soon had plenty of money to enrich himself and secure his position, but also to pursue an ambitious economic development program that included new schools, universities, hospitals, and factories. Saddam also promoted education for women, and turned out a very secular leader who strongly opposed Islamic fundamentalism.

In 1970 Saddam granted autonomy to the separatist Kurdish leaders of Iraq's north, which is also rich in oil. However, this agreement soon broke down, and in 1974 the Kurds, with support from Iran, waged a bloody war against the Baath regime. In March 1975 Iraq reached a preliminary agreement with Iran (then still under the shah) which settled a border dispute between the two countries and called for a stop of Iranian support for the Iraqi Kurds. The improvement of Iran-Iraq relations at this stage equaled an improvement of U.S.-Iraq relations. In December 1975 U.S. foreign minister Henry Kissinger met the foreign minister of Iraq for the first time since relations between the two nations had deteriorated following the 1967 Arab-Israeli War. In the meeting Kissinger clearly stated that Israel's existence was not up for discussion, but painted a picture of diminishing Israeli sway on U.S. policy, partly because of "our new electoral law" which will mean "the influence of some who financed the elections before isn't so great."

Following the revolution in Iran in 1979, both the United States and Iraq feared that the wave of Islamic fundamentalism would spread to Iraq. Shia Muslims account for the majority of the population in Iran as well as Iraq, while Saddam represented a Sunni Muslim government. Hence, it was a concern that Iran's new leaders might successfully encourage Iraq's Shia majority to revolt against Saddam's secular regime. (Worldwide, about 80 percent of all Muslims are Sunni. Shia are concentrated in Iran, Lebanon, and Pakistan, but also in Iraq and Bahrain.)

Start of the Iran-Iraq War

In mid-September 1980 Iraq abandoned the 1975 border agreement with Iran and invaded its neighbor in the mistaken belief that Iranian political disarray would guarantee a quick victory. In reality the war against Iran was going to be waged for eight years, cost the lives of hundreds of thousands of soldiers[181], and end with the borders being set exactly where they were before the war. As Iraq maintained relatively good relations with both the United States and the USSR, it did not have to fear objections by either one of the Oil Age's principal superpowers. And the Israelis in 1981 exploited Iraq's preoccupation with Iran to execute Operation Opera, in which they destroyed a nuclear reactor Iraq had acquired from France. (The UN, supported by the United States and most the world's countries, condemned the attack, but prime minister Begin pointed at Saddam's anti-Semitic threats and stated, "On no account shall we permit an enemy to develop weapons of mass destruction against the people of Israel." The United States had doubled aid to Israel in 1980 due to the revolution in Iran, and needed Israel as an ally in the region more than ever.)

Between 1980 and 1982 Saddam's military campaign was successful, but thereafter Iraq began to lose ground and sought to negotiate peace. In Iran, on the other hand, Khomeini enjoyed a nationalistic wave of solidarity directed against the new enemy, which strengthened the government. Besides, Iran was heavily armed due to U.S. support for the shah, and Khomeini thus refused the peace offer, which turned the war into a bloody stalemate. Iraq also had the problem that the Soviet Union opposed the war and cut off arms exports to both Iraq and Iran. Some Western nations, most notably France, also objected to arms sales to both war parties, but America eagerly filled the gap. Prior to mid-1982, U.S. weapons reached Iraq only via third countries (including

Israel); thereafter the United States delivered weapons and supplies directly to Saddam. The Soviets and the French in turn resumed arms deliveries as well. (France supplied Iraq with fighter jets and Exocet missiles, for instance.) In the end weapons companies from various Western countries made true fortunes by supplying both war parties. The West also delivered lots of dual-use technology, such as computers, helicopters, chemicals, trucks, foundries, and the like, with potential civilian as well as military uses. These exports were not declared as weapons, but nevertheless kept the Iraqi war machinery going.

The United States also funneled vast financial support to Iraq, mainly through loans. The Reagan administration pressured the Export-Import Bank to provide Iraq with financing, to enhance its credit standing, and to enable it to obtain loans from other international financial institutions. The obscure Atlanta branch of Italy's largest bank, the state-owned Banca Nazionale del Lavoro (BNL), funneled an astronomical $5 billion to Iraq from 1985 to 1989. (Henry Kissinger, the former secretary of state, was a paid member of BNL's advisory board for international policy at the time, and so was David Rockefeller, the chairman of the Rockefeller Group and a director of Chase Manhattan Bank.) Over $2 billion in BNL loans to Iraq went to Iraqi government entities involved in running a secret Iraqi military technology procurement network. The U.S. Agriculture Department provided loans for purchases of American grain (which benefited the U.S. farm sector), but both the Federal Reserve and the Agriculture Department warned of suspected abuses of these loans by Iraq to free other funds for munitions purchases. The United States also coordinated the 1981 founding of the Gulf Cooperation Council, which would foster a collaboration between Saudi Arabia and the smaller states in its neighborhood: Kuwait, Bahrain, Qatar, the United Arab Emirates, and Oman. The Gulf Cooperation Council, backed by the high oil price at the time, then financed much of Iraq's efforts in the war against Iran. The debt Iraq in turn accumulated with Kuwait would later be of special relevance. (Meanwhile Iran had nearly no support in the Arab world, with the notable exceptions of Libya and Syria.)[182]

The Iran-Iraq War also provided the background for the United States to broaden its direct military presence in the Gulf region. By 1983 the Rapid Deployment Forces had expanded into CENTCOM (the U.S. Central Command), the main and permanent American presence in the Middle East, which included the 5th Fleet of the United States Navy, headquartered in Bahrain. This was but part of a string of bases set up by the United States in the Gulf states.

Iran-Contra Affair

To be sure, the United States double-crossed Iraq by secretly selling weapons to Iran as well. Iran had been equipped with American weapons systems during the rule of the shah, and the new Muslim government, though anti-Western, depended on U.S. spare parts and ammunition supplies if it was to stay in the war. Ironically, some of the these supplies were delivered by Israel, a principal enemy of Iran's Islamic fundamentalist regime. However, the Israelis viewed Iraq as the greater threat and had an interest in Iran and Iraq weakening one another. The United States officially called for a worldwide arms embargo on Iran, but secretly opened channels to deliver weapons to Khomeini's regime. Reagan administration officials sold about $30 million worth of weapons to Iran in 1985. The arms deals, which included anti-aircraft missiles, were in part designed to induce pro-Iranian groups in Lebanon to release Americans held captive. (Lebanon was the one area outside Iran where Khomeini's Islamic Revolution had a substantial impact as the Hezbollah, a major party in the Lebanese Civil War, allied itself with Iran.) President Reagan's officials then used proceeds from weapons sales to Iran to supply anti-government, right-wing Contra guerrillas in Nicaragua, even though U.S. law also specifically prohibited military assistance to the Contras. Thus, the whole matter eventually became known as the Iran-Contra affair. (Following Congressional hearings in 1986–87, high-ranking members of the Reagan administration were convicted on several felony counts of lying to Congress, obstruction of justice, conspiracy, and altering and destroying documents pertinent to the investigation. In December 1993 independent prosecutor Lawrence Walsh published his final report, which asserted that President Reagan and Vice President Bush, who in turn became U.S. president, were aware of attempts to free U.S. hostages in Lebanon in 1985–86 by means of unsanctioned arms sales to Iran and that Reagan may have participated in, or known about, a cover-up.)[183]

Chemical Weapons

The Reagan administration also maintained friendly relations with Saddam after it had become clear that Iraq used chemical weapons, a major violation of the Geneva Protocol and the first incident of chemical warfare since World War I. Iran reported that it had been a victim of chemical weapons attacks from late-1980, and officially asked for a UN Security Council investigation in November 1983.[184] However, anti-Western and anti-Communist Iran was diplomatically isolated. The United States was well aware that Iran's claims were correct, but the Reagan administration issued a National Security Decision Directive (NSDD 114; dated November 26, 1983), defining U.S. policy priorities in respect to the Iran-Iraq war. It called for heightened regional military cooperation to defend oil facilities and measures to improve U.S. military capabilities in the Persian Gulf: "Because of the real and psychological impact of a curtailment in the flow of oil from the Persian Gulf on the international economic system, we must assure our readiness to deal promptly with actions aimed at disrupting that traffic." However, the claims were now investigated by the International Committee of the Red Cross, and two days before the Red Cross confirmed the Iranian allegations, the U.S. State Department on March 5, 1984, issued a public statement criticizing Iraq for using chemical weapons in violation of the 1925 Geneva Conventions. Saddam was upset, but the Reagan administration quickly told Iraq that the statement was not "a pro-Iranian/anti-Iraqi gesture," and that the "The U.S. will continue its efforts to help prevent an Iranian victory, and earnestly wishes to continue the progress in its relations with Iraq." It was during this time that Donald Rumsfeld, secretary of defense under President Ford (1975 to 1977) as well as under the second President Bush (2001 to 2006), made two trips to Iraq as President Reagan's special Middle East envoy to personally charm Saddam Hussein, once in December 1983, the second time in late March 1984, right after the State Department's chemical weapons statement. During the second visit Rumsfeld told Saddam that U.S. interests in improving U.S.-Iraqi ties "remain undiminished," and in November 1984 the United States restored formal relations with Iraq (which had been suspended since the 1967 Arab-Israel War).[185]

One of the main objectives of the Reagan administration was the construction an oil pipeline from northern Iraq to the (Red Sea) Gulf of Aqaba in Jordan, because Iran disrupted Iraqi oil shipments through the Persian Gulf, while Syria (supporting Iran) had closed the pipeline that had previously delivered Iraqi oil to the Mediterranean.[186] (The billion-dollar pipeline project was proposed by the Bechtel Group of San Francisco that had given Reagan his secretary of state George Shultz, formerly president of Bechtel, and his secretary of defense, Caspar Weinberger, who had been Bechtel's general counsel.) Saddam was eventually going to turn the pipeline project down, but governmental and business ties between the United States and Iraq intensified in the years following the Rumsfeld visits, even though Iraq kept on using chemical weapons. The United States also provided direct operational assistance, supplying battlefield intelligence obtained from surveillance aircraft stationed in Saudi Arabia. Notably, the CIA and DIA (Defense Intelligence Agency) helped Saddam in his fierce attack on Iranian positions on the al-Fao peninsula in February 1988, during which America blinded Iranian radar for three days.

The Iran-Iraq War also saw a resurgence of the war that Saddam had fought in the mid-1970s against northern Iraq's Kurds, who had long been dreaming of their independent state. Iran supported the Kurdish insurrection against the Baath regime, but Saddam reacted harshly: the 1987–88 Anfal campaign probably cost the lives of between 50,000 and 100,000 Kurds, and included chemical weapons attacks, particularly in the Kurdish town of Halabja, where perhaps several thousand civilians were killed on March 16, 1988. In total Iraqi chemical weapons are estimated to have killed between 3,000 and 5,000 Kurds and 45,000 Iranians.

Following the Halabja attack, the U.S. Senate, led by Democrat senators, on September 8 unanimously passed a bill called the Prevention of Genocide Act of 1988. It aimed at cutting off U.S. loans and military (and non military) aid for Iraq, and banned the import of any oil or petroleum products from Iraq. It also requested the secretary of state to bring before the UN Security Council the matter of Iraq's use of poison gas against its own nationals. However, the Reagan administration immediately launched a campaign to prevent the passage of this bill in the House of Representatives (the larger of two houses that make up the U.S. Congress), and was ultimately able to defeat the Prevention of Genocide Act. (National security adviser Colin Powell, later secretary of state under the second President Bush, coordinated the opposition to the bill. Dick Cheney, from 1989 secretary of defense under the first President Bush, and vice president from 2001 under the second President Bush, was part of the team that opposed efforts to revive the sanctions bill in 1989 and 1990.)

End of the War and Continued U.S. Support for Saddam Hussein

When Iraqi troops gained ground again, Iran finally agreed to a cease fire in August 1988. There was no clear winner, and both countries had been equally devastated by bombing raids and missile attacks during eight years of warfare. Both had used irregular military units and attacked civilian populations, and both played down their own losses. There were at least one million casualties (though some estimates talk about two million), with Iran suffering at least twice as many as Iraq.

The United States kept on financing Iraq even though it was clear that Saddam Hussein was still pursuing the production of weapons of mass destruction, including chemical weapons, biological weapons, and even nuclear weapons. U.S. intelligence had already learned in November 1986 that Iraq's Saad 16 research center at Mosel was attempting to develop ballistic missiles, capable of delivering weapons of mass destruction over long distances. In January 1989 a German news magazine made this information public, and later in the year it became known that the projects at Saad 16 included a nuclear weapons program. But nothing changed when Reagan's vice president and former CIA director George H. W. Bush (the first President Bush) started his term as President in January 1989. In March 1989 the current CIA director acknowledged to Congress that Iraq was now the largest producer of chemical weapons in the world, but in October Bush signed National Security Directive 26, stating that "Access to Persian Gulf oil and the security of key friendly states in the area are vital to U.S. national security," and that "Normal relations between the United States and Iraq would serve our longer term interests and promote stability in both the Gulf and the Middle East." Bush also approved another $1 billion in agricultural loan guarantees which allowed the Iraqi government to continue the development of its weapons. America then delivered additional financial aid as well as dual use technology, including modern computers, to Iraq.

Iraq's Annexation of Kuwait

Nevertheless Iraq had problems. It had suffered economic losses of at least $100 billion because of the war against Iran, and it had been borrowing heavily, especially from its allies on the Arabian Peninsula, thus accumulating a debt of about $75 billion. What is more, much of Iraq's oil infrastructure had been damaged during the war. Iraq's oil output was thus relatively small, and on top of that the international oil price was at this stage very low. In short, it seemed impossible for Iraq to achieve economic recovery and serve its international debt.

Iraq's small southern neighbor Kuwait was involved in Iraq's problems. Both countries were OPEC members, and Iraq felt that Kuwait's high production quota within the cartel made it impossible for the oil price and Iraq's income to rise. OPEC had assigned a production ceiling of 1.5 million barrels per day to Kuwait in the early 1980s, which was reduced to 1.25 million barrels per day in 1986, after the oil prices had collapsed. (This did not account for output from the so-called Divided Zone between Kuwait and Saudi Arabia, as oil from this area was earmarked as aid for Iraq during the war.) After the fighting between Iran and Iraq had ended, and more oil from these countries reached world markets, OPEC in 1989 once again reduced production quotas. However, Kuwait refused to adjust its oil production to the assigned volume of just under 1.1 million barrels per day, and instead produced nearly 2 million barrels per day in early 1990. What is more, a major share of Iraq's debt, some $14 billion out of $75 billion, was owed to Kuwait.

Saddam Hussein asked Kuwait to reduce its oil production (in accordance with the wish of OPEC) and to forgive (some of) Iraq's debt. When Kuwait refused, Iraq brought up a long-standing border dispute with its neighbor, and questioned the legitimacy of the Kuwaiti monarchy. To be sure, Iraq had claimed supremacy over Kuwait long before Saddam Hussein had come to power. The first official claim to the whole of Kuwait was aired in 1938, when oil was first found in Kuwait. After all, the territory that was now Kuwait had during Ottoman times been a part of the Basra province, which was centered around Iraq's second largest city. As soon as Kuwait achieved independence from Britain in 1961, then-Premier Kassem renewed these claims. However, Kuwait had by this stage become the Persian Gulf's largest oil producer, and the Kuwait Oil Company (KOC), a British-American joint venture, had an oil license valid until 2026. Hence, Britain reacted rigorously to Kassem's claim and immediately dispatched a brigade to the sheikdom to deter Iraq.[187]

It was quite clear that by the early 1990s the West's attitude had not changed much since the early 1960s. To combine the oil reserves of Iraq (11 percent of the world's proven reserves in 2001) and Kuwait (9 percent of the world's proven reserves in 2001) under one government which would then control one fifth of the world's proven oil reserves would have increased the long-term oil supply risk even if that government was to be pro-Western. But to let go of well-controlled Kuwait and hand such vast combined oil reserves to Saddam Hussein was entirely unacceptable, not least because he was also perceived as a threat to neighboring Saudi Arabia, now by far the Middle East's largest oil producer. Nevertheless, Saddam considered the annexation of Kuwait a viable option to rid Iraq of its serious economic problems. And he certainly had the means to go through with his plan: a large, highly experienced army inherited from the Iran-Iraq War, with super-modern weaponry supplied by the West.

On August 2, 1990, Saddam Hussein sent 150,000 troops backed by 300 tanks to Kuwait and declared it a part of Iraq. On the day of the invasion the UN Security Council put out Resolution 660[188], which condemned the Iraqi invasion of Kuwait and demanded that Iraq withdrew immediately and unconditionally. A few days later, on August 6, the UN Security Council imposed a full trade embargo on Iraq/Kuwait, meaning that Iraq could not export any oil.

U.S.-Led War against Iraq (January 1991)

As it soon became clear that Saddam Hussein was unwilling to withdraw from Kuwait, Western governments began to prepare the public for war. This was quite a problem for the United States. The current Bush administration looked back on three Republican administrations that had fostered increasingly friendly relations with Saddam Hussein, even after he had begun using illegal chemical weapons. How should they suddenly portray their cherished long-term ally as an aggressive invader? The issue was complicated by the fact that Kuwait was not exactly a showcase for democracy and freedom either. The country was run by the al-Sabah family, and most work in the country was done by immigrant guest laborers with very limited rights. (According to the CIA, Kuwait is still a destination for South and Southeast Asians immigrating legally in search of domestic and low-skilled labor, but in turn being subjected to conditions of involuntary servitude by Kuwaiti employers, including conditions of physical and sexual abuse, non-payment of wages, confinement to the home, and withholding of passports to restrict their freedom of movement.) In 1915, when Mubarak the Great died, Kuwait had merely 35,000 inhabitants. Now it was 1.5 million, of which 850,000 (or 57 percent) were non-nationals. Iraq claimed that it was supporting a popular uprising against the ruling al-Sabah family, and that elections would eventually be organized. The Iraqi troops would retreat as soon as a free Kuwaiti government asked for it. (Besides, Saddam argued that the UN should not force Iraq to leave Kuwait as it had not forced the Israelis to leave Arab territories they occupied during and after the Six-Day War of 1967.)

But then wealthy Kuwaitis intimate with the al-Sabah family teamed up with Americans who shared their interests, and paid well over 10 million dollars to well-established public relation firms and newly founded initiatives (Hill & Knowlton, Citizens for a Free Kuwait, etc.) to win public support for a war against Iraq. A principal strategy was to circulate false reports of Iraqi human rights abuses in Kuwait on American television networks, but the most ambitious stunt was targeted directly at Congress. In October 1990 a 15-year-old Kuwaiti girl testified eloquently and effectively before Congress about Iraqi atrocities involving newborn infants. Her account of Iraqi soldiers flinging babies out of incubators purportedly tipped Congress (and the public) towards a war. (The Senate supported military actions in a narrow 52–47 vote.) Even though journalists and human-rights groups challenged the girl's account as exaggerated, President Bush, facing a broad "No Blood For Oil" opposition, referred to the story six times in the next five weeks, and it was reported widely in the media. Eventually the story was shown to be a fraud: the girl was the daughter of the Kuwaiti Ambassador to Washington, Sheikh Saud Nasir al-Sabah, who was later on named Minister of Information in Kuwait. But for now the Bush administration had won support for the war that was marketed as a war over principle rather than energy security.[189]

In November 1990 the United States doubled its troop strength in Saudi Arabia to 400,000, and on November 29, 1990, the UN Security Council in Resolution 678 authorized Member States "to use all necessary means" (which is UN lingo to allow a military intervention) if Iraq did not withdraw before January 15, 1991.[190] Iraq ignored the deadline, and the United States led a coalition of about 30 nations (including several Arab countries) to war against Iraq. The campaign involved over 500,000 U.S. troops and about 160,000 non-U.S. coalition forces. Saudi Arabia contributed about $16 billion to the campaign's direct cost and indirectly paid 10s of billions of dollars more. (Notably, Saudi Arabia accepted the Kuwaiti royal family and 400,000 refugees.) The war started on January 16–17, 1991, with a large-scale air offensive in which the United States bombed Iraq for some six weeks. These attacks destroyed most of Iraq's infrastructure, including bridges and communication, electricity, and sanitation facilities. It also killed thousands of Iraqi civilians. Western media coverage was largely censored and manipulated. It focused on precision bombs (so-called smart bombs), which however accounted for no more than seven percent of the bomb load dropped over Iraq. In total, the campaign against Iraq is said to have killed some 56,000 Iraqi soldiers and 3,500 civilians, but the number of civilians left dead is unknown. Baghdad maintained that U.S. bombs killed more than 35,000 civilians; some estimates talk as many as 200,000 civilians. In comparison, coalition losses were extremely light (240 were killed, 148 of whom were American).[191] Most of the ballistic missiles launched by Iraq against Saudi Arabia and Israel (in an attempt to break the Arab alliance against Iraq) were intercepted. They were all loaded with conventional (not nuclear) explosives, and the United States asked Israel not to retaliate.

Actual fighting on the ground was limited. Iraqi forces in Kuwait were easily targeted from the air, and the ground offensive, from February 24 to 28, quickly ousted what was left of them. On their way out, the Iraqis set Kuwaiti oil fields on fire, supposedly to cause damage as well as to obstruct visibility, hoping that the enormous smoke clouds would impede the air attacks on them in the open desert. (American oil companies would in turn earn a fortune in rebuilding Kuwaiti oil infrastructure.) However, Saddam Hussein's troops, though they were in possession of them, did not use chemical and biological weapons. (Exposure to chemical weapons from Iraqi weapons depots destroyed by U.S. bombs has been suggested as one of the possible reasons behind the Gulf War Syndrome, from which thousands of U.S. troops were suffering. This included diarrhea, insomnia, short-term memory loss, rashes, headaches, and blurred vision.)

U.S. troops did not advance very far into Iraq, but retreated quickly into Kuwait. Obviously, the Bush administration would have liked to depose Saddam, but it maintained that it had no UN mandate to actually invade Iraq. Another explanation for the quick retreat was that Saddam's well-equipped and war-seasoned army would have inflicted too many casualties on coalition troops if it actually had been given a chance to fight them on the ground. Others have argued that fear of photo journalism was behind the retreat, as scores of independent Western journalists would eventually have followed U.S. troops to document the devastating result of the six-week bomb campaign.

Promoting an Uprising against Saddam

Thus America instead promoted an uprising against Saddam from within Iraq. This was supposed to be easy, because the now weakened president represented the country's Sunni minority, while the Shia majority, and the Kurds of northern Iraq, were longing for a leadership change or outright independence. On February 15, 1991, President Bush publicly urged "the Iraqi military and the Iraqi people to take matters into their own hands and force Saddam Hussein, the dictator, to step aside." These remarks were translated into Arabic and broadcast into Iraq by the CIA. However, the United States and most Western governments in turn decided that if Saddam was eliminated, Iraq might disintegrate and the whole region would destabilize. Hence, considerations to support the Shia and Kurdish uprisings that emerged in the aftermath of the war were put aside, even though the United States had initially promoted them. Instead, Saddam's military was permitted (after the ceasefire with the coalition forces) to keep and operate its fleet of helicopters. The world was now standing by and watching how Saddam was brutally suppressing the rebellion against him, literally slaughtering tens of thousands. As 35,000 victims of this rebellion (though it may have been twice as many), and 111,000 victims of "postwar adverse health effects" were added to the initial death toll, Iraq easily lost over 200,000 people in the aftermath of the U.S.-led campaign.

Continued Trade Embargo

The comprehensive UN trade embargo remained in effect after the end of the campaign, pending Iraq's compliance with the terms of the armistice. Iraq was asked to accept liability for damages and to destroy its weapons of mass destruction programs. Iraq also had to accept international inspections to ensure these conditions were met. Iraq agreed to get rid of its illegal weapons and allowed inspections, but refused to pay for damages, claiming that its withdrawal from Kuwait was sufficient compliance. (Following several weeks of aerial bombardment, Kuwait had to spend more than $5 billion to repair oil infrastructure damaged in 1990–91.) The UN sanctions thus remained in place for more than 10 years and, according to some estimates, resulted in 90,000 Iraqi deaths per year, mainly because the trade embargo included pharmaceuticals.[192] As this catastrophic effect became clear, the UN amended the sanctions to allow for Iraq to sell limited amounts of oil for food and medicine, if it also designated some of the revenue to pay for damages caused by the war. From March 1997, the UN's oil-for-food program helped improve living conditions in Iraq, while half of the proceeds from Iraqi oil sales were used for reparation payments and to cover the costs of UN inspections. In December 1999 the UN Security Council authorized Iraq to export under the program as much oil as was required to meet humanitarian needs, and in 2001–02 per capita food imports increased significantly, while medical supplies and health care services steadily improved. The UN's oil-for-food program in total sold some $65 billion worth of Iraqi oil, but some funds were appropriated, especially in the later years. What is more, Iraq sold some oil outside the program throughout the 1990s, mainly via Turkey and Jordan, despite the embargo. This was known (and even condoned) by UN Security Council members, earned Iraq's neighbors a great deal of money, and helped the Iraqi economy to keep going.

Continued Attacks on Iraq

Iraqi relations with the United States and Britain remained highly strained. As Iraq allegedly obstructed the work of UN weapons inspectors and infringed the no-fly zone in the country's south, the United States and Britain intermittently carried out air strikes on Iraqi military and other targets between 1991 and 1993. In April 1993 former President Bush, who had retired three months earlier, as he had lost the elections to Bill Clinton, visited Kuwait. During this visit a car bomb was recovered from a Kuwaiti farm, allegedly built in Iraq to assassinate Bush. It was later concluded that the Iraqi government had almost certainly nothing to do with the plot, but President Bill Clinton ordered 23 Tomahawk guided missiles to be fired at the headquarters of the Iraqi intelligence service in downtown Baghdad. Three of these missiles missed their target and landed on nearby homes, killing eight civilians. (Clinton's official statement read: "Regrettably, there were some collateral civilian casualties.")

In 1995 the United States attempted again to overthrow Saddam Hussein, this time by advising and financing a Kurdish rebel group of northern Iraq through CIA agents. (The Kurds had achieved a UN-guaranteed area in the north of Iraq, but no Kurdish self-rule.) However, on the eve of the battle the U.S. government backed down, probably in fear of a negative public relations effect if the attack failed. And so it did. A year later, while President Clinton was campaigning for reelection, Saddam launched his counter-attack and executed over 100 rebel leaders. The Kurds had begged for American air cover, but none came. With lack of outside support, the largest Kurdish group struck a deal with Saddam. In exchange for some autonomy, they opened their border checkpoints to Turkey to smuggling and helped Baghdad beat the economic sanctions.[193]

Saddam Hussein insisted that the sanctions against Iraq should have been lifted as Iraq had complied with UN resolutions. He accused the United States of seeking not to disarm Iraq, but to overthrow his regime. Following a series of disputes over interferences with UN weapons inspectors, the Iraqi government agreed that the inspections would continue in February and November of 1998, but blocked them again in December. In turn, the United States and Britain launched air strikes on Iraqi military and industrial targets for four days, the most intense bombardment since the 1991 campaign. In response, Saddam Hussein declared that Iraq would allow no further UN inspections. Only four years later, in November 2002, did he submit to a UN resolution ordering the immediate return of weapons inspectors to Iraq.

Saudi Arabian Oil Power and the Low-Oil-Price Period from 1986 to 1999

The whole conflict between Iraq and Kuwait, and later the oil export embargo imposed on Iraq, had the potential to drive the international price of oil to record levels. However, the price of crude stayed very low throughout (nearly) the entire 1990s. The oil price had actually collapsed as early as January 1986, even though Iran and Iraq were at this stage still at war. Behind this collapse was Saudi Arabia, the country that also managed to stabilize international oil prices quickly after they briefly soared to $42 following the annexation of Kuwait by Iraq.

The 1980s were generally a time when the industrialized world began to use energy more efficiently (more fuel efficient cars and industrial processes, better insulation in houses), and substituted oil where it was possible (shifting more electricity production to coal and nuclear). Another effect of the oil price climbing to $35 per barrel in 1981 was that a new search for oil outside OPEC countries set in. As non-OPEC production between 1980 and 1986 increased by 10 million barrels a day, while demand for oil decreased due to a global recession, the oil price was bound to fall, and OPEC reacted by reducing production quotas. However, several OPEC members lacked discipline and produced above their assigned quotas between 1982 and 1985, while Saudi Arabia attempted to counter the falling prices and cut its production. But eventually the Saudis were fed up, and radically changed their oil strategy. In early 1986 Saudi Arabia more then doubled its oil production, and by mid-1986 the international oil price had fallen by more than half to below $10 a barrel (just over $20 in 2004 dollars). This ended the 1973 to 1985 period of high oil prices that many people in the West, even if they were children at the time, remember as a period when their standard of living was low in comparison to the years before or after.

Saudi Arabia practically began to wage a price war to recover market share it had lost in previous years, especially to non-OPEC producers. The kingdom had an enormous spare production capacity that had been built up due to miscalculations in terms of future demand for oil. Saudi Arabia's spare production capacity in the early 1980s was not only twice as high as its actual production, it was also larger than the total exports of all other oil-exporting countries except Russia. This large spare capacity was backed by Saudi Arabia's huge oil reserves and the low cost of developing and producing this reserves base. (Discovery and development costs in 2004 amounted to some 50 cents a barrel, less than one-tenth of what private-sector rivals pay in Russia, the North Sea, or the Gulf of Mexico.) Thus Saudi Arabia was easily capable of disciplining other oil producers who would act contrary to Saudi Arabia's strategic goals. (Typically, this would be another producer attempting to free-ride on increasing oil prices that Saudi Arabia wants to achieve by curbing its production. If such a producer would increase production, Saudi Arabia would threaten to trigger its spare capacity and lower the oil price for an extended period of time, which tended to hurt other oil exporters a lot more than Saudi Arabia.) Other oil producers suffered tremendous income reductions in 1986, but had to give in and cut their production enough for Saudi Arabia to regain market share and maintain the minimum production level it targeted. This was problematic for oil producers in northern Europe and the Persian Gulf region, but truly catastrophic for such countries as

Mexico, Nigeria, Algeria, and Libya, whose economies depended heavily on oil exports that had greatly increased in previous years. Mexico, Nigeria, and Venezuela were actually pushed to the fringe of bankruptcy.[194]

Obviously, Nigeria, Venezuela, and several other OPEC members were not pleased, and concerted OPEC actions were more difficult in subsequent years. By 1994 the oil price reached a historic low when it came down to the level of early 1973 in real terms. Thereafter the global economy experienced a boom period of enormous growth for several more years, which was widely associated with the spread of IT technology and the Internet, but really reflected low energy prices. In the later 1990s Saudi Arabia disciplined Venezuela, after Caracas attempted to triple its production capacity despite its already high quota within OPEC. To achieve this, Venezuela had reopened its oil sector to international investment, and by the winter of 1996–1997 was producing enough to knock out Saudi Arabia from its position as number one supplier to the United States. In addition, demand for oil was suddenly reduced in 1997 due to a major economic crisis in Asia. Hence the Saudis tried to reason with Venezuela, but diplomacy failed. Saudi Arabia then raised its production and induced the oil price collapse of 1998. The kingdom suffered economically, but reestablished itself as the prime supplier of oil to the United States; reasserted its OPEC leadership; and prompted non-OPEC producers Mexico and Norway to support OPEC's revenue-maximizing goals. Saudi Arabia was thus a key player in coordinating the successful campaign in which OPEC reduced its production quotas to allow crude oil prices to rise to $25 per barrel in 1999. This ended nearly two and a half decades of (almost uninterrupted) extraordinarily cheap oil. (At this time, the IT bubble or dot-com bubble burst in the spring of 2000.)

Saudi Oil for the United States

With the world's largest oil exports, reserves, and production capacities, Saudi Arabia emerged as an increasingly important partner for the West. Together with the United States and Russia the kingdom was now a top-three producer, but U.S. production was on the decline, and Russia's reserves comparatively meager. Saudi Arabia produced mainly for export, and thus had the power to influence the standard of living in the entire (industrialized) world, by manipulating world petroleum prices. (By 1995 Saudi cumulative crude oil production had reached 71.5 billion barrels, compared to the U.S. 165.8 billion barrels.) On the other hand, Saudi Arabia, like many other OPEC countries, had practically nothing else to offer than oil and oil products. Saudi Arabia therefore depended on friendly relations with its principal customers, the West, Japan, and the emerging Asian economies, for its own well-being. In principle, it is not that important who actually ends up buying Saudi oil, as long as it reaches world markets to meet demand and keep the oil price down. But U.S.-Saudi relations were nevertheless special in terms of business as well as politics. America made sure that the House of Saud would stay in power, but it was also the fastest growing export market for oil in the 1990s. (This was due to a combination of fast economic growth and the continuing decline of domestic U.S. oil production.) The increase in U.S. oil imports during the 1990s was in itself larger than the total oil consumption in any other country, save Japan and China. Hence, America emerged as the most important future market for the Saudis. At the beginning of the 21st century, Saudi Arabia sold oil to the United States at a price of roughly $1 per barrel below the price it was selling to Europe and Asia, which amounted to a discount of about $620 million per year. Saudi Arabia absorbed this expense, which stems from the higher transportation costs compared to selling to Eurasia, in order to compete with oil suppliers that are geographically closer to the United States (Canada, Mexico, and Venezuela). The amount is actually small, considering that $1 per barrel is perhaps 5 percent of the total. But if oil trading had been left to market forces, Saudi Arabia's oil exports to the United States would have fallen by about half, despite favorable supertanker transport economics. For the United States, this translated into a risk-reducing diversification in oil supplies, which was especially welcome, because Venezuela, America's second largest oil supplier after Saudi Arabia, was troubled by political instability and corruption. What is more, President Hugo Chavez, elected in 1998, took a pronounced anti-American stance. He had the constitution changed so it would be illegal to reprivatize Venezuela's oil industry, and he accused the United States of having been involved in a failed coup against him in 2002. In 2005 he repeated the accusation that the United States had plans to assassinate him, and in 2007 he announced plans to change the constitution to allow him to stay in office indefinitely. (Nevertheless, Venezuela remained a major oil supplier to the United States. In 2004 it ranked fourth, after Canada, Mexico, and Saudi Arabia, but all four suppliers delivered similar amounts of oil to the United States.)

Western Arms for the House of Saud

Part of the U.S.-Saudi relationship was to recycle petrodollars back to America through construction projects and arms purchases. Saudi Arabia spent astronomical amounts of money on Western arms, especially after the Iranian Revolution of 1979. Just in the decade 1978–87, Saudi Arabia concluded arms agreements worth about $31 billion with the United States alone. (These included F-15 fighters, an air defense system, AWACS surveillance aircraft, KC-707 tanker aircrafts, sidewinder missiles, and more.) France sold Saudi Arabia warships, missiles, artillery, and other weapons systems worth some $10 billion during the same period. And Britain completed the enormous Al Yamamah weapons deals with Saudi Arabia, which included Tornado fighter planes. Signed under Prime Minister Margaret Thatcher in 1985, the *Financial Times* called this largest weapons deal in history, worth £43 billion ($85 billion), Britain's "biggest sale ever of anything to anyone." During the next two decades allegations into bribery were repeatedly aired, but the British government eventually closed the investigations in order not to harm relations with its strategic partner Saudi Arabia, even though the BBC and the *Guardian* alleged that the British government was directly involved in payments of more than £1 billion to Saudi Prince Bandar, the 20-year Saudi ambassador to the United States who had brokered the deal and used some of the money to pay the running costs of his private Airbus. (After Prime Minister Tony Blair had stopped the investigations, British defense contractor BAE Systems in 2007 announced that it had completed another £20 billion deal to supply 72 Eurofighter Typhoon jets to Saudi Arabia.)[195] So it happened that Saudi Arabia, due to its enormous military spending, managed to run a budget deficit even in times when record oil prices had provided it with exorbitant revenues. For Western arms industries and economies it could not have been any better: the money paid for imported oil flew back into the country and created employment, while the Western arms systems made sure that Saudi Arabia was safe from outside aggressors and the House of Saud would stay in power to keep the oil coming.

The Palestine Problem

Nevertheless, the relations between Saudi Arabia and the United States, or more generally the West, were not without problems. The major issue was the fate of the Palestinians and U.S. support for Israel. Saudi forces did not participate in the Arab-Israeli (Six-Day) War of 1967, but played a central role in organizing the oil boycott towards the West during the Arab-Israeli War of 1973, as the kingdom disapproved of the massive U.S. arms lift to Israel. However, between 1979 to 1982, Israel finally returned the Sinai Peninsula to Egypt under the terms of the Camp David Accords brokered by President Jimmy Carter. Egypt then fully recognized the state of Israel, but unfortunately the second part of the Camp David Agreements, which proposed to grant autonomy to Palestinians in the West Bank and the Gaza Strip, was not implemented. Israel continued its occupation of these regions, while the Palestinians and several Arab states rejected the autonomy concept: it did not guarantee full Israeli withdrawal from the areas captured in 1967, and it did not guarantee that an independent Palestinian state was to be established. The problem of hundreds of thousands of Palestinians living in refugee camps therefore remained unresolved, and it turned especially acute in Lebanon, Israel's neighbor to the north.[196]

As of 2004 Lebanon had 3.8 million inhabitants, of which 60 percent were Muslims and 40 percent were Christians. Over 200,000 Palestinian refugees had come to southern Lebanon immediately after Israel was founded in 1948, and by 1975 over 300,000 Palestinians lived in the country's refugee camps.[197] In 1970 the Palestine Liberation Organization (PLO) had moved its headquarters to the Lebanese capital Beirut, but the Lebanese government was actually dominated by Christians, who are concentrated in the north of the country. In turn Lebanese Muslims, in alliance with the PLO, started a fierce fight against the government, which effectively led to the 1976 partitioning of Lebanon in a Christian-dominated north and a Muslim-dominated south. Syrian and UN troops were present to maintain a cease-fire between 1976 and 1982, but Palestinian guerrillas kept attacking targets in northern Israel out of southern Lebanon. Thus Israel in 1982 retaliated with a full-scale invasion of southern Lebanon that cost the lives of some 19,000 people (90 percent civilians) and forced the PLO to relocate to Tunisia. Israeli forces under Defense Minister Ariel Sharon allowed Lebanese Christian militiamen (the fierce civil war enemies of PLO and the Lebanese Muslims) to enter two refugee camps, ostensibly to root out further PLO fighters. These Christians in turn murdered some 2,000 Palestinian civilians, including children, women, and elderly, within a few days in September 1982. (An Israeli court of inquiry later found Sharon indirectly responsible for the massacre, and he was forced to resign in 1983. However, Sharon held several cabinet positions throughout the 1980s and 1990s and was elected prime minister of Israel in 2001.) Both Israeli and international UN troops stayed in Lebanon, but on April 18, 1983, a pick-up truck loaded with explosives was driven into the lobby of the U.S. embassy in Beirut, flattening the seven-story building in an explosion that killed 63 people, including 17 Americans, of which six were CIA officers. (This bombing was never officially solved, but it is assumed that the PLO was behind it.) Then, in October 1983 more than 300 American and French troops, 241 of which were U.S. marines, were killed by a truck bomb in their barracks in Beirut. Shortly thereafter, the United States and other countries withdrew their troops from Lebanon. (The Reagan administration was later criticized for failing to retaliate against these bombings which allegedly fueled the worldwide perception that America was vulnerable to such attacks.) However, the Israeli troops stayed, even though they were fiercely attacked by Hezbollah, a Lebanese militia group (and political party) founded in southern Lebanon in 1982 as a response to Israel's invasion. In consequence, Israel in 1985 withdrew its troops from all but a narrow buffer zone in southern Lebanon.

In this atmosphere Americans actually had reservations about weapons delivered to Saudi Arabia: Congress refused to allow large shipments of U.S. arms to the Saudis in the mid-1980s due to concerns that such weapons might be sold (or simply passed on) to the Palestinians, or Yemen, for instance. (At this point Britain under Prime Minister Margaret Thatcher gladly stepped in and signed the enormous Al Yamamah weapons deals with Saudi Arabia.) However, support for Israel became an even larger liability for the United States in the following years. In 1987 Palestinians launched the first intifada (struggle or shaking-off) against Israeli occupation in the West Bank and the Gaza Strip. Palestinians used slingshots and homemade (gasoline) explosives to attack the Israeli army, which used tear gas and rubber bullets, and destroyed Palestinian homes to quell the popular resistance. The clashes continued for several years and cost the lives of around 1,300 Palestinians and 80 Israelis by 1991. The United States opened direct peace talks with the PLO in 1988, right after the PLO issued a statement renouncing terrorism and recognizing Israel's right to exist. However, Israel's population grew by nearly 10 percent in just three years between 1989 and 1992, due to immigration from the collapsing Soviet Union. This promoted the development of Jewish settlements in the Israeli-occupied territories. Nevertheless Israel began peace talks with PLO officials in Oslo (Norway) in 1993, after the Labor party had won the elections. The Oslo Accords proposed an interim Palestinian Authority, conducting limited self-government in part of the occupied territories, and a phased plan leading to a permanent peace settlement. In 1995, the Oslo II agreement was signed, under which Israel handed over security responsibility to the Palestinian Authority in parts of the occupied territories. Although the implementation of the Oslo accords was delayed, and Palestinian suicide bombers kept on attacking Israelis, the last West Bank city still under Israeli occupation was handed over to Palestinian control in 1997. The year after, Israel agreed to withdraw from additional West Bank territory, while the Palestinian Authority pledged to take stronger measures to fight terrorism.

Many Saudis felt very strongly about the fate of the Palestinians, but the House of Saud nevertheless did not take a clear stance against America, which remained Israel's principal financial, military, and political backer. (The UN's Security Council, whose recommendations are to be executed, has virtually never condemned Israeli action simply because the United States would veto all such resolutions. One exception was the April 1992 Security Council statement condemning Israel for allowing "the continued deterioration of the situation in the Gaza Strip," after clashes between Israeli troops and Palestinian demonstrators left five Palestinians dead and more than 60 wounded.) The issue strained U.S.-Saudi relations every now and then, but the House of Saud remained a loyal partner, which was especially important as the West was losing its other partners in the Middle East, first Iran, and then Iraq. Notably, Saudi Arabia increased its oil production during the Gulf War of 1990, when about five million barrels of oil per day (in Iraqi and Kuwaiti supplies) were suddenly removed from the market, and it allowed U.S. and other troops to operate from its territory. Nevertheless, keeping the House of Saud in power was in itself a liability.

Maintaining the House of Saud—Rejecting Democracy

The United States claimed for decades that it aimed to promote the international spread of democracy and personal freedom, but supporting and maintaining the House of Saud was quite the opposite. Saudi Arabia is an absolute monarchy with no written constitution, no legislature, and no political parties. The king rules, in accordance with Islamic law, by decree. According to their Wahhabi roots, the Saudis imposed strict segregation of the sexes, an absolute prohibition of the sale and consumption of alcohol, a ban on women driving cars, and many other social restrictions. The rules are enforced by the mutawwain (mutawwiyya), the religious police, officially known as the Commission for the Protection of Virtue and the Prevention of Vice, who patrol streets and shopping centers on the look-out for anyone breaking the rules. The press is strictly controlled, and there is no free access to the Internet. (No Saudi newspaper reported the 1991 Iraqi invasion of Kuwait until five days after the event, for instance.) In the early 21st century cinemas were still banned, and so was music in public places. Amnesty International, a human rights organization, recorded over 1,400 executions between 1980 and 2000.[198] Some 40 people a year are still publicly beheaded on market squares for crimes including drug offences and robbery. Saudis and foreigners are being flogged for such crimes as selling illicit alcohol. Saudi Arabia simply has a history of brutal repression and a "persistent pattern of gross human rights abuses," according to a 1999 Amnesty International Report. Prisoners are tortured until they confess. In mid-2002, a fire in a girls' school in Mecca left 14 dead, when members of the mutawwain forced the victims back into the burning building because they were insufficiently covered.[199]

As of the beginning of the 21st century all kings of Saudi Arabia had been sons of Ibn Saud, who died in 1953. His eldest son abdicated in 1964 in favor of his brother Faisal. In 1975 Faisal was assassinated (by a nephew), and his half brother, Khalid, succeeded him. Khalid died 1982 and was succeeded by King Fahd, the fourth of Ibn Saud's sons to rule. After King Fahd suffered a series of strokes in 1995, his half-brother Crown Prince Abdullah in effect took control. It is not exactly known how power is allotted within the phenomenally rich royal family. Ibn Saud had some 44 recognized sons, and estimates of the total number of Saud princes vary widely, from around 5,000 to 10,000. The extended family numbers perhaps between 20,000 and 27,000, but there are no published accounts of how much of Saudi Arabia's wealth these individuals receive. Saudi Arabia still has no personal income tax, either for princes or general citizens, but the latter have nearly no influence on the country's politics. In 2005 Saudi Arabia held elections, but these were for only half the seats on toothless town councils (the other half was appointed by the king), and women were not allowed to vote at all.

Saudi Arabia's oil wealth arrived quite suddenly and failed to trigger a quick social or technological transition to turn the country into an Oil Power. The desert nation did not have a settled agricultural base, and never entered the Coal Age. Slavery was not prohibited until 1962. Up until the 1960s most of the population, which was organized in clans, was nomadic or seminomadic. Saudi Arabia was hence thinly settled, but the government forced the nomadic tribes of camel herders to settle down, and soon Saudi Arabia experienced a population explosion in concert with the breakdown of traditional lifestyles. The population grew from 3.2 million in 1950 to 5.7 million in 1970 to 16.6 million in 1990, to 22.1 million in 2000, and in 2002 Saudi women still bore 6.21 children on average. The rapid population growth was especially worrying in light of the low oil price of the 1990s. While the population more than doubled from 9 million to over 19 million between 1980 and 1995, Saudi Arabia's GDP dropped by 25 percent from $157 billion to $126 billion. This translated into a 62 percent drop in terms of per capita GDP, threatening the standard of living that Saudis had become used to.

By 2007 the population counted 27.6 million, including 5.6 million non-nationals. These foreigners play an important role in the Saudi economy, as they comprise over a third of Saudi Arabia's total workforce. Many are highly educated foreigners in key positions of the oil and service sectors. Others are low-skilled workers from South and Southeast Asia employed under the same scandalous conditions as in Kuwait. Meanwhile many young Saudi Arabians are unemployed. The government's more recent economic five-year plans have therefore encouraged private sector growth (and diversification into non-oil sectors, such as manufacturing and re-exports) to increase employment opportunities for the swelling Saudi population.[200]

Nevertheless the House of Saud, despite or because of their long-standing success in blocking attempts to introduce democracy and modernity, was facing serious problems at the beginning of the 21st century. Increasingly more Saudi citizens begun to resent the absolute power of the Saud clan that kept on limiting their influence in public affairs while absorbing most of the country's oil income. Some also objected the power of the clerics to control their social life, while the conservative Wahhabi clergy criticized the House of Saud for its worldliness and for its close ties to America, which was allowed to keep troops in Saudi Arabia. (In 2003 the United States thus withdrew nearly all its troops to neighboring Qatar, located only 500 km away from the Saudi capital Riyadh.) In a survey conducted in spring 2001, 63 percent of Saudis characterized the Palestinian situation as "the single most important issue to them personally," and another 20 percent ranked it among the top three. This translates into straight anti-Americanism, as Muslims all over the world are well aware that the United States is Israel's principal backer and thus viewed responsible for the fate of the Palestinians. For the West, this makes clear that if Saudi Arabia was a democracy with an elected parliament representing the opinion and will of the nation's people, support for Israel would immediately led to a conflict with Saudi Arabia, with all its consequences for the oil market and international economic well-being. The West has therefore preferred keeping the House of Saud in power rather than promoting democracy. After all, Saudi Arabia holds about a quarter of all the world's remaining proven oil reserves. Asia, too, has been interested in a regime that reliably delivered oil to the world. (China, Japan, South Korea, India, and other Asian countries combined receive about 60 percent of Saudi Arabia's crude oil exports, as well as the majority of its refined petroleum products.) However, much of the discontent of Saudi Arabian citizens translated into anger directed at the United States. And strangely, the Saudi government added to this trend by financing religious schools that produced Muslim fundamentalists with strong anti-Western ideas.

September 11, 2001

In the morning of September 11, 2001, a group of 15 Saudi Arabian nationals and four or five individuals of other nationalities boarded four commercial domestic U.S. flights in groups of five and took control of the airliners soon after takeoff.[201] Some of the hijackers had received commercial flight training in the United States and piloted the four planes towards different targets to commit suicide attacks. At 8:45 a.m. the first of the planes crashed straight into the North Tower of the World Trade Center, a 1973 New York City landmark also known as the Twin Towers. The second plane struck the South Tower some 18 minutes later at around the 80th floor. Although the buildings had in fact been designed to withstand an airplane impact, both of them soon collapsed due to the immense heat released by the airplanes' burning fuel inside the buildings.[202] Around 2,750 people died at the site. (A strike just a little later, during true office hours, would have killed a lot more people. The two 110-story buildings, with about one acre on each floor of each tower provided office space for some 50,000 people. About 55,000 had permanent access badges and 5,000 more entered the building each day on visitor IDs.) The third plane crashed into the Pentagon, the headquarters of the U.S. Department of Defense near Washington, D.C., at 9:40 a.m.. About 20,000 people were at work in the complex consisting of five concentric three-story rings that is considered the largest office building in the world. Fortunately, the death toll here was comparatively low: 125 Pentagon employees were killed along with 64 passengers of the ill-fated airliner. Within the next hour the fourth plane crashed into the ground in Pennsylvania, killing its 40 passengers and crew. The plane went down after four passengers, aware of the other plane crashes via cellular phone calls, attacked the hijackers in an attempt to gain control of the airplane. (This is the official story. Suspicions have been aired that the government may have ordered the military to take the plane down. Some claimed on the grounds of widely scattered debris that the plane was shot down. Airplane parts have been found up to eight miles away from the crash site.)

America was in shock. There had hardly been any attacks on the United States in its entire history. The Japanese attack on Pearl Harbor, Hawaii, in December 1941 was still in living memory, and the Japanese submarine attack on the Californian Ellwood oil production facilities in February 1942 was quoted as the only previous attack on mainland United States by another nation. However, psychologically this was closer to the War of 1812, when the British captured Washington and burned down the White House and the Capitol. But despite the fact that most of the attackers were from Saudi Arabia, the U.S. government managed to prevent public anger from being directed towards its critical Persian Gulf ally. Instead, it was quickly concluded that the well-planned and synchronized operation was the work of terrorist network al-Qaeda (Arabic for the Base), an Islamic extremist group that opposed U.S. presence in Persian Gulf countries. This group had its roots in Muslim resistance against Soviet involvement in Afghanistan.

Soviet Involvement in Afghanistan

During the Coal Age Afghanistan had been a buffer region between British India (present-day Pakistan and India) and the expanding Russian Empire. Both Britain and Russia attempted to increase their influence in the region, but it was the British who actually invaded the country and were in turn ousted from Afghanistan in three consecutive wars. The peace treaty that ended the third and last Anglo-Afghan War in 1919 recognized Afghanistan as an independent state. In 1926 Amanullah took the title of king (shah), and his family ruled this rural Muslim country until Mohammed Zahir Shah, the last shah of Afghanistan, was ousted in 1973. Mohammed Zahir had installed his cousin and brother-in-law, Muhammad Daud, as prime minister in 1953. Daud had plans to recover what was now Pakistan's North-West Frontier Province, but was rebuffed by the United States for arms sales and loans. Thus Daud in 1956 turned to the Soviet Union, Afghanistan's northern neighbor. When Pakistan closed its border to Afghanistan in 1961, Soviet-Afghan relations became very close in terms of trade. The Soviets sold Afghanistan tanks, striker jets, and artillery at a bargain price, but in 1963 the king removed Daud as prime minister in an attempt to improve Afghanistan's relations to Pakistan (and the West). However, 10 years later Muhammad Daud was back, overthrowing the monarchy in 1973 to make Afghanistan a republic and himself president.[203]

Daud's dictatorial style was in turn opposed by the leftist People's Democratic Party of Afghanistan (PDPA), and in April 1978 he was ousted and assassinated by military officers. The PDPA formed a new government and renamed the country Democratic Republic of Afghanistan (DRA), announcing an extensive program that included land reforms, measures to emancipate women, and a campaign against illiteracy. But these reforms were opposed by Islamic traditionalists and ethnic leaders who objected to rapid social change. An armed revolt began in late 1978, and half a year later the insurgents controlled much of the countryside. As the guerrillas seemed to seriously threaten the Afghan government, the Soviet Union decided to intervene, sending troops into Afghanistan on December 25, 1979. This was the beginning of a disastrous 10 year involvement.

To be sure, mountainous and rural Afghanistan was not worth invading for its resources. The Soviets valued the country for its geographic position. Afghanistan's southern border was just 450 km away from the Arabian Sea, separated from potential warm-water ports (and the mouth of the oil-important Persian Gulf) only by Pakistan's sparsely populated Baluchi province. The Soviets also had lots of investments to protect in Afghanistan. These investments into Afghan infrastructure, military, and other projects had been made over years to avoid a potential expansion of Iran, Afghanistan's neighbor to the east that had been heavily armed by the United States. Earlier in 1979 the U.S.-sponsored shah of Iran had finally been overthrown, but his radical Islamist successors had inherited Iran's U.S. weaponry, a position that may have tempted them to invade Afghanistan to help the cause of the Muslim insurgents.

U.S. Support for Islamic Fundamentalists

The United States supported Iraq during the 1980–88 Iran-Iraq War, but shared with now radical Islamist Iran the goal of supporting Afghanistan's Muslim rebels to weaken America's Cold War archenemy, the Soviet Union. In response to the Soviet invasion of Afghanistan, President Jimmy Carter in his State of the Union Address on January 23, 1980, declared that the United States would use military force if necessary to defend its national interests in the Persian Gulf region:

> The region which is now threatened by Soviet troops in Afghanistan is of great strategic importance: It contains more than two-thirds of the world's exportable oil. The Soviet effort to dominate Afghanistan has brought Soviet military forces to within 300 miles of the Indian Ocean and close to the Straits of Hormuz, a waterway through which most of the world's oil must flow. The Soviet Union is now attempting to consolidate a strategic position, therefore, that poses a grave threat to the free movement of Middle East oil. This situation demands careful thought, steady nerves, and resolute action, not only for this year but for many years to come. It demands collective efforts to meet this new threat to security in the Persian Gulf and in Southwest Asia. It demands the participation of all those

who rely on oil from the Middle East and who are concerned with global peace and stability. And it demands consultation and close cooperation with countries in the area which might be threatened... Let our position be absolutely clear: An attempt by any outside force to gain control of the Persian Gulf region will be regarded as an assault on the vital interests of the United States of America, and such an assault will be repelled by any means necessary, including military force.[204]

However, the United States had begun supporting Islamic fundamentalists in Afghanistan through covert CIA operations long before the Soviet army entered the country.[205] Arms shipments to the insurgents began in July 1979, and the day Soviet forces crossed the Afghan border President Carter's national security adviser reported: "We now have the opportunity to give the USSR their Vietnam war." In turn the United States spent millions of dollars to supply Afghan schoolchildren with textbooks that were filled with talk of jihad (that is, a holy war waged on behalf of Islam as a religious duty) and, to teach children to count, featured drawings of guns, tanks, bullets, soldiers, and mines. (These books full of violent images and militant Islamic teachings served as the Afghan school system's core curriculum for the rest of the 20th century.)[206] Afghanistan in turn bred an army of mujahideen (literally, persons who wage jihad), as the Muslim guerrillas called themselves. In the West they were stylized as freedom fighters, and the United States spent hundreds of millions of dollars a year to arm and train them. Most of them were based in Pakistan, though some were operating from Iran. U.S. support continued for more than 10 years, and notably America from 1986 began supplying the mujahideen with Stinger missiles capable of shooting down Soviet armored helicopters. In exchange for being able to support the mujahideen out of Pakistan, the United States tolerated the spread of nuclear arms technology. The Reagan administration downplayed Pakistan's nuclear program throughout the 1980s, but Pakistan had produced enough fissile material for a nuclear bomb by 1986, and operated some 3,000 centrifuges to enrich uranium by the early 1990s.

Saudi Arabia spent about as much money to support the mujahideen as the United States did. Saudi Arabia, together with the United States, supported Saddam Hussein's secular regime in the Iran-Iraq war, and viewed the conflict in Afghanistan as an opportunity to demonstrate that the House of Saud nevertheless also supported the cause of Islam. Various religious groups had accused Saudi Arabia of being too pro-Western and too moderate in terms of upholding and spreading Islam. In November 1979, just one month before the Soviets entered Afghanistan (and the same month 63 staff members of the U.S. embassy in Tehran were taken hostage), a group of some 200 militant Muslims, many of whom had studied Islam at Medina, shocked Saudi Arabia by occupying the Great Mosque in Mecca, Islam's holiest site. Only after a two-week siege Saudi troops (allegedly with support by Western military) drove out the insurgents, killing some 100 of them in the heavy fighting, and publicly beheading 63 of their ringleaders in selected town squares all over the country. The Saudi regime then turned even more repressive to smother the seeds of rebellion, and the war between the Soviets and Afghanistan's mujahideen subsequently served as outlet for Saudi Islamic fanaticism. In short, Saudi Arabia provided more than money to support Afghanistan's Muslim insurgence: It dispatched militants to the base camps in Pakistan to join the rebels.[207]

Years and Years of Civil War

Afghanistan's central government and various mujahideen groups fought one another for 14 years, from 1978 to 1992. Soviet troops were involved in the Afghan Civil War for nearly 10 years, between 1979 and 1989. They quickly won control over Kabul and other important centers and installed as president a new PDPA leader who promised to combine social and economic reform with respect for Islam and Afghan traditions. However, he remained unpopular and the brutal fight continued. By 1986 some 118,000 Soviet and 50,000 Afghan government troops were facing perhaps 130,000 mujahideen guerrillas. The Soviets used tanks and bombers, but the mujahideen were trained and armed by the United States and enjoyed local support in familiar mountainous terrain. In the end, the war left some one million government troops, mujahideen and civilians dead. (The Soviets lost around 15,000 soldiers, but possibly many more, and had 54,000 wounded.) Some three million Afghanis fled to Pakistan and about 1.5 million to Iran. (For comparison, Afghanistan's total population was 28.5 million in 2004.) Many more were displaced inside the country. Beginning in 1985, the Soviets (under Mikhail Gorbachev) tried to pull out of the costly war, and in May 1988 Afghanistan, Pakistan, the Soviet Union, and the United States finally signed agreements providing for an end to foreign intervention in Afghanistan. The Soviets then began the withdrawal of their forces, which was completed in February 1989.

However, the civil war continued after the Soviets had pulled out, and the United States kept on supplying arms to the mujahideen who kept on fighting the central government. The Afghan government offered peace, which was rejected, and temporarily gained the upper hand as the Soviets delivered food, fuel, and weapons. Meanwhile, the mujahideen were still terrorizing the civilian population by bombarding the cities with rockets. In 1991, the year the Soviet Union disintegrated, the United States and the Russians finally signed an agreement to end military aid to the region. The PDPA government was overthrown the year after, and soon the country was ruled by different local or regional warlords, while the Russians and Americans turned their back to Afghanistan. In the capital Kabul a new government was formed by various rebel factions, but rival mujahideen groups immediately opposed it and besieged Kabul. Among them were the well-equipped Taliban (Arabic for students).

The Taliban

The Taliban were by all standards extremely radical Islamists. They emerged in 1994 from the Pashtun tribe (whose homeland is divided between Afghanistan and Pakistan's North-West Frontier Province), and many of their leaders had been trained in madrasas, Islamic religious schools, in Pakistan. They were heavily armed through the channels that the CIA had built up to support the mujahideen from northern Pakistan, and soon the Taliban took control of Afghanistan's Kandahar province, where they were joined by 10,000 Afghani and Pakistani Pashtuns, mostly between 14 and 24 years of age, by December 1994. (Most had been studying in madrassas and have been described as the displaced youth of the war, grown up in refugee camps with no other education than Islamic studies.) In September 1996 the Taliban seized Kabul, and by the late 1990s they ruled over nearly all of Afghanistan (with the notable exception of a small northern region ruled by the so-called Northern Alliance).

The Taliban regime immediately enforced its fundamentalist rules that particularly affected women. Females were prohibited from attending school or working outside their homes, were ordered to cover themselves from head to toe in burkas (long, tent-like veils), and were publicly beaten if they were improperly dressed or escorted by men not related to them. The Taliban also made theft punishable by amputation of the hand, and publicly executed persons convicted for adultery or drug dealing, for instance. Various human-rights groups were alarmed, but the United States tolerated the Taliban, and Saudi Arabia, Pakistan, and the United Arab Emirates even recognized these black-turbaned clerics as the legitimate government of Afghanistan. (Pakistan valued the Taliban for removing the tolls imposed by various warlords on Pakistani commerce and smuggling along the long Afghan-Pakistani border.)

The oil companies appreciated a new, strong central government that would allow them to pursue pipeline projects in Afghanistan. Argentine oil firm Bridas met the Taliban as early as August 1995, as it had obtained concessions to produce oil and gas in Turkmenistan, which it attempted to deliver to Pakistan. Although the fighting in Afghanistan continued, Bridas in 1996 signed an agreement with the Taliban to build a pipeline through the country. However, the deal was challenged by Unocal of California, which spent millions to secure the project, and welcomed Taliban representatives in the United States in February 1997. Yet in the end Unocal had to pull out of the project, and all American Unocal staff was asked to leave Afghanistan in August 1998, as the United States fired missiles onto Afghanistan in an attempt to kill Osama bin Laden.

What Goes around Comes around: Osama bin Laden

Osama bin Laden was born in Jidda, Saudi Arabia, in about 1955 as one of some 52 children of fabulously rich Yemen-born Saudi construction magnate Muhammad bin Laden, who made his fortune building palaces and mosques for the Saudi royal family. (Muhammad had 11 acknowledged wives during his lifetime, but, in compliance with Islamic law, he never had more than four wives at the same time.) Hence Osama had a secure upbringing, traveled to Western countries, and enjoyed a quality education that earned him degrees in civil engineering and public administration. When his father died in a plane crash in Texas in 1968, Osama inherited between $30 million and $300 million, while his brothers took over the family business, the Binladin Group, which grew into an international multibillion dollar conglomerate with tens of thousands of employees.

While studying in Jidda, Osama got acquainted with the radical Muslim Brotherhood, and in 1979 he joined the Saudi contingent that was sent to Pakistan to support the mujahideen of Afghanistan. Osama was quite different from the many foreign fanatics gathering in Pakistan. Most were poor and had marginal other opportunities, while Osama was a member of the family known as the richest in Saudi Arabia except for the royal family. Thus the Saudi intelligence chief, Prince Turki al Faisal, recruited Osama to raise funds for the mujahideen, and to use his family connections to deliver construction equipment (such as bulldozers) to build tunnels and caves for the mujahideen. Osama also financed the so-called Services Office, which recruited and trained militant Muslims from over 50 countries. And he fought personally on the side of Afghan warlord Gulbuddin Hekmatyar (Hikmatyar), whose Hezb-e-Islami party became the largest recipient of U.S. and Saudi funding. Hekmatyar developed a reputation of being especially brutal and having extreme religious views. He is considered responsible for the death of thousands of Afghan civilians killed in the period before the Taliban came to power. However, Osama bin Laden, after fighting in the Afghan Civil War for 10 years, in 1989 returned to Saudi Arabia.[208]

Back home he was by no means happy. He publicly accused the Saudi monarchy of being corrupt, cruel, and un-Islamic, and he was outraged when American troops were stationed in Saudi Arabia during the January 1991 war against Saddam Hussein's Iraq. Even worse, these troops were allowed to remain in the country of Islam's two holiest sites (Mecca and Medina) even after that war was over. To be sure, Osama fought on America's side in Afghanistan, but he would later state that during Israel's 1982 invasion of Lebanon he had already decided to hit the United States, as he detested

American support for Israel. The Saudi government reacted by placing Osama under house arrest in 1991, but later that year he left Saudi Arabia for Sudan, where he established his own businesses and set up terrorist training camps. (Sudan is less than 200 km [125 miles] away from Saudi Arabia, just across the Red Sea.) Osama's plan was to carry the jihad far beyond Afghanistan, to what he viewed as the corrupt kingdoms of the Persian Gulf, and even to Western countries. He had forged alliances with radical Islamist groups in Egypt and elsewhere already while in Afghanistan, and perhaps as early as 1988 had begun organizing a network of terrorists that grew out of the Services Office.

1993 World Trade Center Bombing

During the early 1990s, Muslim fundamentalists committed a series of terror attacks, but Osama's involvement in those is not entirely clear. In December 1992, for instance, bombs exploded outside two hotels in the Yemeni city of Aden, south of Saudi Arabia. (In this incident the intended targets, U.S. soldiers on their way to Somalia, had already left.) U.S.-Islamic relations soured even more when a U.S. missile in January 1993 hit the Al Rashid Hotel in Baghdad, where 200 delegates from 51 countries had gathered for an Islamic conference. America apologized for the attack, saying it was an accident, and that all other missiles fired into Iraq that day hit their intended targets, factories suspected of producing weapons components. However, little later, on February 26, 1993 (when president Clinton had been in office for just one month), a bomb exploded in the underground garage of the World Trade Center's north tower in New York City, creating a 22-foot-wide, five-story-deep crater. The bomb detonated at lunch time, when many employees had left the building, but it still took five hours to evacuate the complex and two hours to extinguish the flames in the heavily damaged skyscraper. Six people were killed and more than 1,000 were injured. (The towers were repaired and reopened in less than one month.) Five men of various nationalities (including Pakistani, Palestinian, and Iraqi) were eventually arrested and jailed. (They were tracked down, because one of them returned to the car rental, where he had rented the van in which the bomb was exploded.) No connection to Osama bin Laden was established, but the CIA would later claim that Osama was the prime financier behind the bombing.

Osama bin Laden was based in Sudan until 1996. In 1994 the Saudi government stripped him of his citizenship and officially froze his assets, though he allegedly continued to have access to substantial funding through his family network (though his law-abiding brothers disowned him). Then things began to heat up. Al-Qaeda, as the terrorist network built up by Osama bin Laden was going to be known, assumed responsibility for a 1995 car bomb attack that killed seven people, among them five Americans, right outside the

Saudi National Guard building in Riyadh. And in June 1996 a truck bomb exploded at Saudi Arabia's U.S. Air Force base at Dhahran, killing 19 American servicemen and wounding about 400. In 1996 the United States and Saudi Arabia pressured Sudan to expel Osama. Apparently Sudan offered to hand Osama over to either the United States or Saudi Arabia, but neither wanted him at this stage. So it happened that Osama returned to Afghanistan, where the Taliban in September 1996 took power in Kabul. Allegedly, Osama then maintained terrorist training camps in Afghanistan, but al-Qaeda was later described as a decentralized organization with autonomous underground cells in many countries and ties to other terrorist organizations that shared al-Qaeda's Sunni Muslim fundamentalist views.[209]

1998 Bombing of the U.S. Embassies in Kenya and Tanzania

In February 1998 Osama bin Laden and Ayman al-Zawahiri, leader of the Egyptian Islamic Jihad, issued a statement that "to kill Americans and their allies, civilians and military, is an individual duty of every Muslim who is able." (The Egyptian Islamic Jihad was a terrorist organization seeking to overthrow the Egyptian government and replace it with an Islamic state. The two men merged their networks.) This was the starting point for al-Qaeda to begin militant operations against civilians. A little later, on August 7, 1998, the U.S. embassies in Kenya and Tanzania were (truck)bombed, which was the first major terrorist act reliably attributed to al-Qaeda. The attacks cost the life of about 224 people, all but twelve non-Americans, and injured some 4,500 locals.

The FBI immediately put Osama on the agency's Ten Most Wanted List, and America unsuccessfully tried to persuade the Taliban to surrender him. On August 20, 1998, President Clinton ordered Tomahawk cruise missiles to be fired on alleged terrorist facilities in Afghanistan, but also to target and destroy a Sudanese pharmaceuticals factory which the United States claimed was helping al-Qaeda to produce chemical weapons. About 20 people died in the Afghan strikes, while the nighttime attack on the factory in Sudan killed the watchman and his family, and injured several more people. This retaliation was a fiasco. For one, Osama escaped the U.S. attack in Afghanistan, and while America failed to delivered evidence to the contrary, Muslims around the world doubted that the destroyed pharmaceutical factory in Khartoum was in fact producing the ingredients for nerve gas. What is more, it was soon revealed that the factory had been open to Western visitors all along, and that the facility was Sudan's primary source of pharmaceuticals, many of which were difficult to replace through imports during the following months. This made the United States indirectly responsible for the death of perhaps tens of thousands in this impoverished African nation.[210] Massive protests and demonstrations were staged in Muslim countries around the world, and

Osama enjoyed a sudden publicity push, as everyone wanted to know who he was. Thus, the exiled Saudi millionaire became somewhat of a cult hero to many of his countrymen (and Muslims in general) who shared his sentiments about America's role in the Arab-Israeli conflict and U.S. troop presence in Saudi Arabia.[211]

The other major al-Qaeda plot against the United States prior to 9/11/2001 was the suicide bombing attack against the USS *Cole* on October 12, 2000. It did not involve civilians, but killed 17 American soldiers and injured 39. It was conducted by two men in a small bomb-laden boat blasting a 40-foot hole in the side of the destroyer as it was refueled in Aden, Yemen.[212]

As far as the September 11, 2001, attacks were concerned, Osama let the public know that he had nothing to do with it, while the U.S. national security adviser stated as soon as September 23 that "the United States has evidence of bin Laden's role in terrorist acts that it will present in due time." However, such evidence actually remained thin as far as the current attack was concerned, though U.S. authorities claimed that "investigators have found financial records, communications among al-Qaeda members, and other evidence linking bin Laden to September 11." One of the September 11 bombers was said to have been an "associate of Cole bomber," referring to the attack on the USS *Cole* in Yemen, but even evidence that al-Qaeda was responsible did not immediately certify involvement of Osama, given the alleged decentralized nature of the organization. However, three years after the attack, Osama, who had become something of the most wanted person in the world, changed his mind and claimed he was behind 9/11 anyway. And a few days before the fifth anniversary of the September 11 attacks, Arab television channel Al-Jazeera aired an old videotape showing Osama bin Laden with two men identified as being among the 19 hijackers who had carried out the 9/11 attacks.[213]

Ousting the Taliban

Immediately after the September 11 attack, the United States, backed by nearly the entire world, demanded that the Afghan government extradite Osama bin Laden and other al-Qaeda members. When the Taliban refused, the United States and allied forces aerial bombed and invaded Afghanistan. (Perhaps a few thousand civilian Afghans lost their lives during these operations.) A pro-Western interim government was installed in Kabul, and thousands of international peacekeeping soldiers maintained a measure of law and order in the capital. However, Osama bin Laden as well as other al-Qaeda and Taliban leaders had gone into hiding. Thus, even in the years after the Afghan government in 2004 had pledged to curb the power of regional warlords and to build an effective national security force, international troops stationed in Afghanistan kept being attacked by resistance forces, increasingly through suicide bombings. And the civilian population continued to suffer as well. In June 2007, for instance, an aerial raid by foreign troops killed 65 civilians together with 35 Taliban fighters.

As scores of Afghans fled to refugee camps in border areas of Pakistan (and Iran) to escape U.S. bombings during the initial invasion, the cooperation of Pakistan was indispensable from the very start of the campaign. However, the army had carried out a (bloodless) coup in Pakistan on October 12, 1999, and General Pervez Musharraf had taken power and suspended Pakistan's constitution. The United States had condemned the overthrow of the democratically elected government, imposed economic sanctions on Pakistan, and was even more outraged when Musharraf declared himself president in June 2001. But all this changed immediately after September 11. Now the United States called Musharraf an important ally, paid his regime well for supporting the war against the Taliban and anti-American terrorism, and even kept quiet when Musharraf in February 2004 pardoned Abdul Qadeer Khan, Pakistan's top nuclear scientist who had been covertly aiding Libya, North Korea, and Iran in developing their own nuclear programs. Musharraf also refused to honor his promise to quit as army chief at the end of 2004, but he pleased the United States by conducting a military operation along Pakistan's Afghan border in March 2004 in an attempt to flush out insurgents, and by cracking down on al-Qaeda remnants, which resulted in several critical captures and kills. However, parts of the Taliban and its al-Qaeda allies kept on waging sporadic guerrilla campaigns against U.S. forces, which tried to rout them out from the rugged Tora Bora cave region and the Shah-i-Kot Valley, both in eastern Afghanistan. But the whereabouts of Osama bin Laden; his most important aide Ayman al-Zawahiri; and principal Taliban leader Mohammad Omar, remained unknown.

The Problematic "War on Terrorism"

The campaign in Afghanistan was part of a larger effort that the U.S. government called the "War on Terrorism," designed to curb the spread of terrorism through various military, political, and legal actions. While the invasion of Afghanistan had the nature of a conventional war, the war on terrorism was criticized from the start on various grounds. One issue was its effects on civil liberties and the violation of constitutional rights of citizens. Passed by Congress almost immediately after September 11, the so called Patriot Act radically expanded the authority of U.S. law enforcement agencies. (These would, for instance, now be able to quite freely search medical, financial, and other records, plus telephone and e-mail communications, of anyone they declared suspicious.) Another problem was that the United States, operating internationally with various allies, did not abide by the rules and regulations regarding warfare (including the rights of prisoners of war and unlawful combatants) as set out by the Geneva Conventions. The United States used the Predator, an unmanned aerial vehicle, to target suspected terrorists, for instance. (One fired a Hellfire missile at a car in Yemen in November 2002, killing all six passengers, including an al-Qaeda suspect allegedly involved in the attack on the USS *Cole*. Others killed al-Qaeda suspects in Afghanistan and Pakistan.) Such attacks mirrored Israel's assassination policy that has often been condemned by the United States. Eventually it became known that the CIA set up a network of secret prisons in up to eight countries (including several democracies in eastern Europe), where thousands of suspects were held without trial, and many were tortured to extract information. However, the public soon focused in on the 660 or so individuals held at Guantánamo Bay, a naval base the United States maintained in Cuba ever since ousting the Spanish from the island in 1898. (It is based on a perpetual lease which Cuba cannot terminate without U.S. consent.) The U.S. government argued that Guantánamo Bay detainees (who were mainly captured in Afghanistan) had no rights under the American constitution as they were held in territory that was not under U.S. sovereignty. The public was outraged when it became clear that these people from 42 countries (including Britain and Australia) included minors below the age of 16; that they were held for years without time limit, charge, or access to legal counsel; and that they were subjected to ongoing torture, sexual degradation, forced drugging, and religious persecution.[214]

U.S. authorities maintained that such methods were necessary to extract information that helped in breaking up al-Qaeda cells in the United States, Britain, Italy, France, Spain, Germany, Albania, and Uganda, for instance. But decentralized terror groups, whether truly part of the al-Qaeda network or not, remained active and kept striking in such countries as Turkey, Morocco, Indonesia, Spain, and Britain. And notably, Osama bin Laden himself remained free and sporadically addressed the public through audio or video tapes (though there was a long gap between his video appearances in 2004 and August 2007).

Losing the Grip on Saudi Arabia?

One of the West's major concerns after September 11 was the situation in Saudi Arabia, the country whose stability was most critical for the international oil price and economic well-being. Just two days after the September 11 attacks, when U.S. air traffic was severely restricted, 140 Saudi nationals were allowed to leave America without even being interrogated. (Some of them were close relatives of Osama bin Laden.) This conduct has been ascribed to the close (business) ties between the presidential Bush family, the House of Saud, and the bin Laden family. (Both Presidents Bush were especially close to Prince Bandar, the 20-year Saudi ambassador to the United States who had brokered the Yamamah weapons deal with Britain, which probably earned him over $2 billion in British bribes. Bandar used parts of this money to maintain a private Airbus: it was he who arranged the flight that transported Saudis out of the United States after September 11.)[215]

On the other hand, the U.S. Department of State had denounced Saudi Arabia's human rights problems, including abuse of prisoners and incommunicado detention; prohibitions or severe restrictions on the freedoms of speech, press, peaceful assembly and association, and religion; denial of the right of citizens to change their government; systematic discrimination against women and ethnic and religious minorities; and suppression of workers' rights. By the time of the September 11 attack the United States had supported non-democratic, repressive Saudi governance for some 70 years, and now the Saudi regime was turning even more repressive in order to deal with terrorists.[216] But the West was less concerned about House of Saud dictatorship, and more about whether the Saudi government would actually be able to curb Islamic fundamentalism in their country. The vast majority of September 11 hijackers had been Saudis, and during the following years Saudi Arabia experienced a series of attacks that just did not seem to end. In May 2003 suicide bombers killed 35 people at housing compounds for Westerners in Riyadh, for instance, and in November 2003 terrorists attacked a compound housing for foreign workers from mainly Arab countries, killing at least 18. Several attacks followed between April and May 2004, and in December 2004 an assault on the U.S. consulate in Jeddah killed five staff members. Notably, this was long after Osama bin Laden and al-Qaeda had achieved one of their major goals: The United States in 2003 had removed nearly all their troops from Saudi Arabia to neighboring Qatar.

Many Saudis remained unhappy about the close link between the House of Saud and the United States. For one thing, the Saudi government under Western pressure curbed or stopped its payments to various Islamic schools and groups outside the country. (These were mainly payments to promote the Wahhabi movement, or similarly radical views of Sunni Islam in such countries as Lebanon, Egypt, Pakistan, and Afghanistan. Several of these circles have been shown to generate terrorists.) But it was considered even more upsetting that well over 100 Saudi nationals were detained at Guantánamo, and that U.S. support for Israel remained untarnished. Prince Nayef bin Abdul-Aziz, the powerful minister of the interior, even blamed the September 11 attack on Jewish agents. (The rationale behind such allegation is that Israel benefited from September 11 in that it could act more freely against suspected Palestine terrorists within a global War on Terror. Right in the time period before 9/11 Israel had been vehemently criticized by Western countries for its harsh reprisals that harmed Palestine civilians, and the targeted killing of suspects.) All this appeared yet more concerning as Osama bin Laden rose in the region to be the most admired Saudi national in history, with esteem for him growing on university campuses as much as among the rising number of unemployed.[217]

The West was especially worried about the security of Saudi oil installations (as well as the 3,000 Western oil workers in the kingdom).[218] A tape from Osama bin Laden surfacing in December 2004 threatened that al-Qaeda would specifically target Saudi oil infrastructure. A successful attack on Ras Tanura, the world's largest oil-export terminal, or Abqaiq, a vital oil-processing centre, would undoubtedly cause an immediate oil price surge, but Aramco, the state-owned oil company, declared that it employed 5,000 security guards to protect oil facilities, while military and other state forces protected Saudi Arabia's oil fields, pipelines, ports, refineries, and other oil facilities. The Saudis spent $5.5 billion in 2003, and over $8 billion in 2004 on oil security, and assured the world that the situation was under control. But not everyone was so sure. With opposition against the House of Saud rising, and a new, restless generation of Saudis rapidly growing up, the United States decided to take control of Saudi Arabia's neighbor Iraq, land of the second largest known oil reserves, and the only country besides Saudi Arabia to have reserves above 100 billion barrels.

U.S. Invasion of Iraq (2003)

Barely five hours after American Airlines Flight 77 on September 11, 2001, plowed into the Pentagon, Defense Secretary Donald H. Rumsfeld was telling his aides to come up with plans for striking Iraq. Rumsfeld wanted "best info fast. Judge whether good enough hit Saddam Hussein."[219] A little later Richard Clarke, the first National Coordinator for Security, Infrastructure Protection, and Counter-terrorism, told Rumsfeld that the United States had to bomb Afghanistan, as al-Qaeda is in Afghanistan. But Rumsfeld replied that there aren't any good targets in Afghanistan, and that there are lots of good targets in Iraq. According to Clarke, President George W. Bush also immediately sought to link the attacks in New York and Washington to Iraq. And so it happened, despite all evidence indicating that there was no link between Saddam Hussein and al-Qaeda, that the Bush administration began to advertise an invasion of Iraq in the context of the War on Terrorism.[220] Bush equated 9/11 terrorism with Saddam Hussein again and again, and top administration officials repeatedly asserted that there were extensive ties between the Iraqi government and Osama bin Laden's terrorist network. Vice President Cheney went as far as saying that evidence of a link was "overwhelming," and Bush in his State of the Union Address on January 28, 2003, said: "Evidence from intelligence sources, secret communications and statements by people now in custody reveal that Saddam Hussein aids and protects terrorists, including members of al-Qaeda." Hence, half America's population eventually believed that Saddam Hussein was actually behind 9/11. For many of these Americans it came as a shock when President Bush finally said it loud and clear: "We have no evidence at all that Saddam Hussein had anything to do with the 9/11 attacks." But this was long after the United

States had invaded Iraq, and after the September 11 Commission (that is, the National Commission on Terrorist Attacks), after studying files previously accessible to the government, but not the public, had reported that there was no collaborative relationship between Iraq and al-Qaeda.[221] Al-Qaeda did in fact contact the Iraqi government several times, but Saddam Hussein had no interest at all in collaborating with these Islamic fundamentalists.

Weapons of Mass Destruction—or Not?

The Bush administration's other justification for going to war in Iraq was fabricated as well: both the American and British government maintained they had evidence that Saddam Hussein had an active weapons-of-mass-destruction program, and that Iraq was therefore a threat to U.S. and global security. After the invasion, even the most comprehensive search could not reveal any signs of such weapons in Iraq, and U.S. weapons inspectors (in October 2004) concluded that Saddam Hussein was neither in possession of weapons of mass destruction, nor had Iraq maintained an active program to develop weapons of mass destruction after the war in 1991, nor had he a formal plan to revive such weapons programs.[222] Thus the world was wondering how the American and British allegations before the war could have been so specific. Britain's prime minister Blair had published an intelligence dossier that claimed some of Iraq's weapons of mass destruction could be ready within 45 minutes of an order to use them. And President Bush had said in January 2003 in his State of the Union address: "The British government has learned that Saddam Hussein recently sought significant quantities of uranium from Africa." (Both the CIA and the U.S. State Department had discredited these allegations many months earlier, as they were based on false information and forged documents, but this did not keep the Bush administration from using them for its purpose.) In February 2003 secretary of state Colin Powell told the U.N. Security Council that biological warheads had been distributed across western Iraq, and on March 16, 2003, vice president Dick Cheney said that Saddam Hussein "has been absolutely devoted to trying to acquire nuclear weapons. And we believe he has, in fact, reconstituted nuclear weapons." The latter statement was made about a week after the International Atomic Energy Agency (IAEA) had delivered its March 7 report to the UN Security Council in which it reaffirmed that "The IAEA had found no evidence or plausible indication of the revival of a nuclear weapons program in Iraq."

When the Bush and Blair administrations were later accused of lying, they blamed inaccurate intelligence for their own false prewar allegations and statements. However, the best intelligence was actually officially available: Hans Blix, head of the UN weapons inspectors, was present in Iraq between December 2002 and March 2003, reporting that Iraq was cooperating

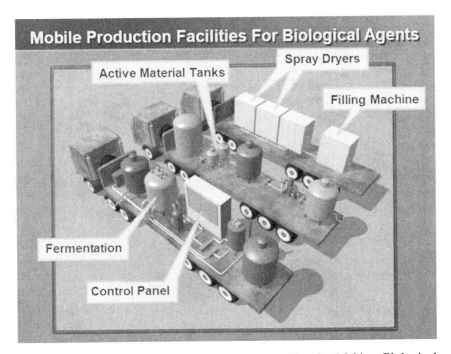

Misleading the World: Alleged Iraqi Mobile Facility for Making Biological Weapons President Bush had a problem. The Pentagon's war plans were completed and March 2003 had been secretly set as the deadline for the Iraq invasion, but there was little foreign support and an uncertain American public. In February 2003 Bush's Secretary of State Colin Powell addressed the UN Security Council, presenting highly doubtful evidence with great confidence to mislead the world. Powell said he was convinced that Saddam Hussein was attempting to build nuclear weapons ("There is no doubt in my mind"), and went on to explain that Iraq was in possession of as many as 18 trucks used as mobile facilities for making biological weapons. Displaying slides illustrating what the United States believed these mobile labs might look like, he explained, "Ladies and gentlemen, these are sophisticated facilities. For example, they can produce anthrax and botulinum toxin. In fact, they can produce enough dry, biological agent in a single month to kill thousands upon thousands of people. A dry agent of this type is the most lethal form for human beings." (Karen DeYoung, "Falling on His Sword: Colin Powell's Most Significant Moment Turned Out to Be His Lowest," *Washington Post*, October 1, 2006, Page W12, http://www.washingtonpost.com/wp-dyn/content/article/2006/09/27/AR2006092700106.html). In the end, it all turned out to be fake. Iraq had no active weapons of mass destruction program. (U.S. Department of State, CMS images, Slide 21_600.) (Secretary Colin L. Powell, "Remarks to the United Nations Security Council," New York City, February 5, 2003, http://www.state.gov/secretary/former/powell/remarks/2003/17300.htm).

"rather well" and that his staff was admitted to all the places they visited, almost always promptly. Blix reported he had found no signs that weapons of mass destruction in fact existed in Iraq and said he needed some more time to conclude his work: "not weeks, not years, but a few more months." However, the United States asked the weapons inspectors to leave right before America began bombing Iraq on March 20, 2003.

Illegal War

There was plenty of resistance to the U.S. invasion of Iraq within and outside America. Most notably permanent UN Security Council members France, Russia, and China, as well as Germany and many other nonpermanent members, opposed this war, arguing that it was too soon to give up on the weapons inspections. The United Nations General Secretary called this war illegal.[223] But the United States (initially 200,000 troops) and Britain (45,000 troops) went ahead and invaded Iraq with support by Australian (2,000) and Polish (200) forces. This was a very minor coalition when compared to the liberation of Kuwait twelve years earlier and for a good reason. Back then, UN Security Council Resolution 660, which had demanded Iraq withdraw from Kuwait, had authorized an invasion by threatening Iraq with the "use of all necessary means," which is UN lingo for a military intervention. The current UN Security Council Resolution 1441, issued in November 2002 due to British and American pressure, only warned that Iraq would face "serious consequences" if it failed to fully cooperate with UN weapons inspectors.[224] But Bush and Blair nevertheless used Saddam's alleged noncompliance with Resolution 1441 to justify to the world that they were starting a war. And even though it was the United States who asked the UN inspectors to leave Iraq before the attack, Bush would in a press conference on July 14, 2003, say about Saddam: "We gave him a chance to allow the inspectors in, and he wouldn't let them in." Iraq had been under a severe international embargo ever since the UN-sanctioned liberation of Kuwait in 1991, but it was supposed to be removed once the Iraqi government agreed to pay war reparations to Kuwait and prove sufficiently to UN inspectors that all its chemical weapons (or more generally programs to produce weapons of mass destruction) had been destroyed. Baghdad allowed the weapons inspections, but from the mid-1990s claimed that the true purpose of the sanctions was to worsen the situation in Iraq to the degree that the regime would be overthrown. In turn, Saddam Hussein began interfering with the work of UN weapons inspectors, agreed to let them continue in February and November of 1998, but blocked them again in December. U.S. and British airplanes then bombed Iraqi military and industrial targets for four days, and Baghdad responded by declaring that Iraq would allow no further UN inspections. Saddam Hussein did, however, agree to the resumption of UN weapons inspections in

September 2002, even before Resolution 1441 was issued in November that year. Permanent Security Council members China, Russia, and France made clear that Resolution 1441 was not intended to authorize war, but the United States, the sole remaining Oil Age superpower, was powerful enough not to care. The United States and its ally Britain clearly undermined UN authority and international law. Clare Short, Britain's secretary of state for international development at the time, would eventually reveal that Britain bugged the office of UN secretary-general Kofi Annan in 2003 during the run-up to the invasion of Iraq, and two chief UN weapons inspectors, Hans Blix and Richard Butler, also said they believed their private conversations had been recorded.

Getting back at Saddam?

Some were concerned that the Republican Bush administration simply wanted to get back at Saddam. George W. Bush won the 2000 presidential elections by the tiniest margin.[225] The democratic candidate Al Gore actually received more votes, winning the national popular vote, but Bush gained the presidency when the U.S. Supreme Court reversed a recount order by the Florida Supreme Court, enabling him to secure a narrow majority in the electoral college. (Bush had a majority of no more than 537 votes in Florida, where his brother was governor, while thousands of ballots were disputed due to a faulty voting technology.) Bush then seemed to continue where his father had left off after losing the 1992 presidential election to Democrat Bill Clinton. He picked John Ashcroft, a controversial religious conservative, as his attorney general, and otherwise chose (or was told by more senior Republican party members to choose) seasoned former members of his father's and President Ford's cabinet for his government. (Ford was made vice president under Nixon in 1973, and became U.S. president when Nixon resigned in 1974. He was actually never elected.) The new vice president, Dick Cheney, had been secretary of defense, the new secretary of state, Colin Powell, had been chairman of the Joint Chiefs of Staff (which is the President's principal military advisor), and the new national security advisor, Condoleezza Rice, had been a member of the national security team under the elder Bush. The new secretary of defense, Donald Rumsfeld, had filled the same position under President Ford and was responsible for the elder Bush's promotion to director of the CIA in 1976. (Dick Cheney and the new treasury secretary Paul O'Neill had also held positions under Ford.) In light of this experienced Republican crew, many viewed Bush as some sort of puppet president who had done well in using his popular name to win the elections, but lacked high-level rhetoric skills and foreign policy experience. (The White House then employed hordes of script writers; paid handsome sums to conservative print, radio, and TV commentators; and admitted rightwing journalists to

White House press briefings to "throw the President softball questions," that is, questions that are easy to answer.)

It was Donald Rumsfeld who had visited Saddam Hussein as President Reagan's special Middle East envoy, telling Saddam that U.S. interests in improving US-Iraq ties "remain undiminished" after it had become known that Iraq had used illegal chemical weapons in the Iran-Iraq War. And it was Bush's father who was President when Iraq was bombed during the liberation of Kuwait, who promoted a Shia revolution in Iraq against Saddam, but then did not back it up (which led to the death of tens of thousands when Saddam was allowed to suppress it); and who had instigated a massive international embargo on Iraq, which failed to oust Saddam, but (according to United Nations estimates) cost the lives of 100,000 Iraqi children.[226] And eventually George W., the second President Bush, was going to advertise the invasion of Iraq to oust Saddam Hussein by saying, "After all, this is the guy who tried to kill my dad," though analysts had long concluded that Iraq was not behind the alleged assassination attempt on the first President Bush during a visit of Kuwait in April 1993.[227] Hence, it appeared as if this Republican crew had personal motives, fueled by guilt or outrage, to finish the job and remove Saddam from office. But whatever the motive, it was eventually revealed that the U.S. State Department actually made detailed plans for an invasion of Iraq *before* 9/11, shortly *after* George W. Bush's inauguration in 2001. (This was initially brought to light by Bush's former treasury secretary Paul O'Neill.) Richard Clarke, the first National Coordinator for Security and Counter-terrorism, claimed that Bush was so pre-occupied with Iraq before and after 9/11 that he failed to effectively confront threats from the al-Qaeda terror network. What is more, Clarke asserted that Bush "launched an unnecessary and costly war in Iraq that strengthened the fundamentalist, radical Islamic terrorist movement worldwide." (To be sure, right when Clarke published and presented his book *Against All Enemies*,[228] Israel assassinated Palestinian spiritual leader Sheikh Ahmed Yassin in a missile strike, which drew tens of thousands of Palestinians to the streets and took most of the media coverage off Clarke's revelations and accusations.)

It has also been suggested that Bush needed the war against Iraq in order to win a second term. He had taken office during an economic recession and did the obvious after the 9/11 attacks by declaring a full-scale campaign against terrorism. However, Osama bin Laden was not expected to be caught until the next elections (though the bounty on his head was doubled from $25 million to $50 million in 2005), and the decentralized al-Qaeda terrorist network remained active. The ousting of the Taliban from Afghanistan was not likely to be viewed as a satisfactory retaliation for the 9/11 attack either (nor was it to be entirely completed). Thus any Democratic presidential candidate would have easily challenged Bush for not having done enough to reinstall America's security. Saddam Hussein, on the other hand, was a well-located target, and, following a 10-year international embargo, Iraq would not have the means to defend itself at all. (The initial invasion did indeed last for no more than three weeks. Many of Saddam's war-seasoned soldiers of the Iran-Iraq War and the annexation of Kuwait had by this stage retired.) War-time presidents tend to have high approval ratings (as they benefit from a wave of nationalistic unity), and America has an impressive record of pulling out of recessions by funneling investments towards the military-industrial complex in preparation for war. (Besides, a war tends to distract from domestic problems or may excuse them.)

No Blood for Oil

To be sure, many of the millions demonstrating worldwide against the Iraq invasion were motivated by the "No blood for oil" spirit. (The demonstrations on February 15/16, 2003, involved some 10 million people, which made it the largest global demonstration for or against anything in world history. Over 3 million people marched in Rome, over one million in both London and Barcelona, some 500,000 in Berlin, 200,000 in Damascus, and some 100,000 or more each in New York, Los Angeles, and San Francisco.) The "No blood for oil" motive had already been a problem for the first President Bush during the liberation of Kuwait, but now it was even more obvious to the general public. Former South African president and Nobel Peace laureate Nelson Mandela in January 2003 called Mr. Bush "a president who can't think properly and wants to plunge the world into holocaust. They just want the oil. We must expose this as much as possible."[229]

U.S. domestic oil production had peaked long before, while British North Sea oil production had started to decline a short four years earlier, in 1999. Venezuela, a principal oil supplier to the United States, was plagued by political turmoil and organized strikes in the oil industry in the months prior to the U.S. invasion of Iraq. (In December 2002 a national strike entirely shut down Venezuelan oil production.) Russia, a major oil supplier to western Europe, had increased its exports, but had a relatively small reserves base and struggled to become a true democracy. And last not least, the future stability of Saudi Arabia was now somewhat doubtful, because the West's support for House of Saud totalitarianism was increasingly questioned. Iraq, on the other hand, had enormous proven oil reserves that were second only to those of Saudi Arabia, and was ranked first, at least according to some estimates, in terms of undiscovered (expected, but not yet found) oil reserves. But the door between Iraqi oil and the world had been closed by the UN embargo imposed on Saddam Hussein's regime after the invasion of Kuwait over 10 years before. If a pro-Western government was to be installed in Iraq, the country's pre-1990 oil production of 1.3 billion barrels per year was expected to double quickly, though it was clear that this would require substantial investments. At this point a commercial motive came into play. The Bush administration had close ties to the U.S. oil industry, in which Bush himself had been working (though not very successfully[230]). Bush was closely affiliated with energy firms when he was governor of Texas, and Vice President Dick Cheney until 2000 was chairman and CEO of Halliburton Company, a major energy services company with a long history of service to the government. (In 1998 Halliburton had merged with its competitor Dresser Industries, for which the first president Bush had worked before he founded Zapata Oil.) Under Saddam Hussein, Iraqi oil rights had been contracted to the French (TotalFinaElf), the Russians (Lukoil), the Italians, and the Chinese. (Some argued that the reason why France and Russia had declared to veto any U.N. Security Council resolution which would directly allow for military action against Iraq was to protect their oil interests in Iraq. Lukoil had signed a $20 billion deal to develop the West Qurna field.) The ideas was that these contracts would be annihilated following a U.S.-led invasion. The Iraqi National Oil Company would be privatized, and the Iraqi oil rights handed to U.S. oil corporations. The Iraq invasion plan drawn up by the Bush administration prior to 9/11 actually included plans to sell off the Iraqi oil industry, which would then deliver enough oil to destroy the OPEC cartel. But American oil industry representatives interestingly pushed for a change of this plan after the actual invasion. They opposed the total privatization of the Iraqi National Oil Company and preferred a U.S.-friendly state oil company. This was to calm insurgents by signaling that Iraqi oil belonged to Iraqi people (and not to shareholders of Western oil corporations), and would avoid the risk that a future Iraqi government would bar U.S. oil companies from keeping their investments or bidding for oil reserves.[231] Besides, the oil companies had no interest in the destruction of OPEC and a fall in oil prices. After the invasion the Bush administration gave Iraq $2.3 billion to be invested directly into the Iraqi oil sector, and hired various oil service companies (including Halliburton) to fix Iraq's desolate oil infrastructure, which was now being protected by a massive force of U.S. soldiers and private contractors. In March 2004, Iraq's daily production reached the pre-war level of 2.5 million barrels, but pipelines and refineries kept being targeted by saboteurs, and no one dared to say when the production level would reach the fabled 6 million barrels per day that U.S. officials had proclaimed for the end of the decade. The Iraqi National Oil Company (INOC) was now in the hands of American-installed technocrats, and large oil corporations were lining up to get a share of the Iraqi oil. However, in order for them to invest, political stability had to be guaranteed; the future taxation level known,[232] and the status of contracts that oil corporations from France, Russia, China, and Italy had signed with Saddam Hussein's regime declared void.[233]

Say Nothing Good about a Bad Man

The public was quite aware of it all: the old bills the Bush administration had open with Saddam; the oil motive; the issue that it would be easier to withdraw U.S. troops from Saudi Arabia as soon as Saddam was ousted; and that a U.S.-friendly regime in Baghdad would improve Israel's strategic position in the region. And yet the Bush administration pretended it was necessary to invade Iraq because Saddam was supporting al-Qaeda and producing weapons of mass destruction.

On the other hand, many opponents of the war had a hard time speaking up, simply because they did not want to say anything to defend a bad man. Saddam was indeed a brutal dictator, he had indeed used chemical weapons during the Iran-Iraq war, he had indeed at some stage sought to produce a nuclear bomb, and he had indeed suppressed both the Shia majority and the Kurd minority in his country. But it would have been difficult to communicate through the media (which largely supported the views of the U.S. government) that Saddam's use of chemical weapons, for instance, happened in the days when Iraq was still an ally of the United States. The brutal use of such weapons against the Kurds occurred during the Iran-Iraq War, when the Kurds, after having waged a war against Saddam's regime in the mid-1970s, renewed this fight with support by Iran. Nevertheless, much of the U.S. media simplified the case into "Saddam even used chemical weapons against his own people." Similarly, Saddam was later to be hanged for crimes committed against residents of Dujail in 1982, while Donald Rumsfeld, sent by U.S. President Reagan, visited Saddam in 1983 and 1984 to cordially shake his hand. (Residents of the Shia town of Dujail unsuccessfully attempted to assassinate Saddam during a visit, and he retaliated by having 148 of the town's men killed during an attack and subsequent executions.)

Creating Chaos

The internal frictions of Iraq were the main reason to expect chaos in Iraq after Saddam was ousted. Iraq was a state with arbitrary borders created by Britain and the West after World War I. The region's Shia majority had been ruled, or suppressed, by a Sunni government for decades (or even centuries, if the Ottoman Empire is taken into account), and the Kurds living in the north had always wanted their own state. (The British had initially handed power in Iraq to the Sunni minority by installing to the throne the son of (Sunni Arab) Husayn ibn Ali, the former sherif of Mecca and king of the Hejaz region who had been defeated by the House of Saud.) With all the open wounds inflicted in the recent past to Iraq's Shia (ca. 62%) and Kurd (ca. 17%) population, a democratically elected central government tolerant to the country's Sunni Arabs was going to be difficult to realize. This implied severe Sunni resistance to such model. Civil war and the disintegration of Iraq was looming. (The Kurds had already achieved wide autonomy under Saddam and were likely going to ask for their own state.) Opponents of the war hence warned that the chaos created by an invasion would lead to a major humanitarian crisis. Human rights group Amnesty International, for instance, stated that military action could easily precipitate a huge disaster, and that it believed that sacrificing the human rights of people for the sake of geopolitics is unacceptable.[234] And that's what happened. The U.S.-led invasion caused, directly or indirectly, the death of hundreds of thousands of people. In November 2004, one and a half years after the initial invasion, the British medical journal *The Lancet* published a statistically based study according to which most likely about 140,000 people had perished due to the conflict since the beginning of the invasion. This study, conducted by researchers at Johns Hopkins University, was later revised and updated to show that an estimated 654,965 additional people died in Iraq between March 2003 and July 2006.[235] The U.S. military, famously and arrogantly, refuses to publish data on how many people it kills ("We don't do body counts"[236]), but iraqbodycount.org, by analyzing official press releases, by mid-2007 registered over 90,000 documented civilian deaths by violence in Iraq as a result of the U.S.-led invasion.[237] As far as foreign soldiers are concerned, some 182 American soldiers had died as a direct result of the invasion when President Bush prematurely declared an end to major combat operations on May 1, 2003. By September 8, 2004, the number had reached 1,000, and by February 7, 2005, it was 1,448 Americans, in addition to 86 Britons, 20 Italians, 17 Ukrainians, 16 Poles, 11 Spaniards, and 21 citizens of other countries. By May 2007 some 3,600 of US armed forces were dead, and 25,830 wounded in action. (This excludes contractors and Western workers, many of whom were kidnapped and killed.)[238]

But the Civil War that followed the invasion continued, some two million refugees were eventually on the move, and the occupying Western troops were being attacked by guerrilla resistance. Not even Iraq's Shia population greeted the U.S. troops as liberators. The first U.S. advance into Baghdad killed about 1,000 Iraqis (according to American estimates), and when a U.S. tank tore down a massive statue of Saddam Hussein on April 9, 2003, American soldiers had to organize a cheering crowd to stage effective pictures to be broadcast live by Western television stations.[239] Opinion polls indicated that the majority of Iraq's population opposed the U.S. occupation and wanted the coalition troops to leave. In February 2004 Iraqis were still evenly divided on whether the invasion had humiliated or liberated Iraq, and whether life was better without Saddam.[240] Abuse of Iraqi prisoners by U.S. troops at Baghdad's Abu Ghraib prison, documented in January 2004, did not help the situation. (U.S. military personnel had abused Iraqi civilian detainees by subjecting them to sexual humiliation and other acts of degradation, including beatings, threats of rape and electrocution, stripping detainees naked, pouring phosphoric liquid on detainees, and forcing them to simulate sex acts in public. The United States held about 7,000 people at the site.) British soldiers were involved in a similar scandal. Besides, there were also less spectacular cases of unlawful killings by U.S. forces in Iraq. (In one incident, six members of one marine battalion were accused of a role in the killing of 24 civilians in the town of Haditha, north-west of Baghdad, in late 2005.)[241] Iraqis saw their country descend into chaos, as resistance fighters attacked power plants, bridges, and oil infrastructure that the U.S. bombardment had deliberately spared. Hence, when BBC reporters interviewed Iraqis in June 2005, they said, "The situation went from bad to worst. Today we are missing security and electricity, probably the most important things." And in April 2005 tens of thousands of Shia Muslims had taken the streets, demonstrating against U.S. occupation and demanding that American troops leave Iraq.

The violence escalated. Shia and Sunni militia targeted each other and civilians, while various Iraqi and non-Iraqi groups, including Saddam loyalists and imported terrorists, attacked the occupying Western troops (through suicide bombings and sniper fire) as well as Iraqi police forces and their families.[242] Nevertheless general elections were held on January 30, 2005 and December 15, 2005.[243] The Sunni minority opposed these elections as well as a newly written constitution, because it knew this would end its ruling position in Iraq. Still, a new Iraqi government of Shia Muslims and Kurds was sworn in in April 2006. Saddam Hussein, who had been captured in December 2003, was sentenced to death by an Iraqi court established by the U.S.-appointed Iraqi Governing Council.[244] He was less than three months short of his 70th birthday when he was executed on December 30, 2006.

Total Failure?

By this stage support for the Iraq War by the U.S. public had deteriorated. Americans were tired of the continuous stream of pictures documenting the ongoing violence in Iraq. They had seen ordinary Iraqis dancing over American corpses and had come to understand that all groups of the Iraqi population bitterly resented the U.S. occupation. Spain had withdrawn its 1,300 troops from Iraq as early as April 2004, Britain kept 9,000 troops in the somewhat safer areas of southern Iraq, but the United States had over 150,000 soldiers in Iraq. The list of U.S. soldiers dying in Iraq was getting longer almost by the day, and the war was also outrageously expensive in financial terms. The Bush administration had initially estimated the total cost of the engagement to be $50 billion, but towards the end of 2006 the costs had risen to nearly $400 billion, with some economists estimating the ultimate cost at one trillion US dollars.[245] And if U.S. troops would stay in Iraq through 2015, it may exceed two trillion US dollars. This was a likely scenario, given that it had been revealed to the press as early as April 2003 that the U.S. Army expected to stay at four key military bases in Iraq for a long time.[246] Republican Senator McCain, as presidential candidate in 2008, maintained that U.S. troops should remain in Iraq for 100 years if necessary, while the commitment to a war by this stage drained billions of dollars from the U.S. Treasury every month.

What is more, the Bush administration (in light of the developments in Iraq) announced in July 2007 that arms sales and military aid to the region would be stepped up. Over a period of 10 years Egypt was to receive a package of military aid worth $13 billion, and Israel one worth $30 billion, while an estimated $20 billion worth of sophisticated weaponry would be sold to Saudi Arabia over an unspecified period of time. (This included advanced air systems that would greatly enhance the striking ability of Saudi warplanes.) A renewed and increased arming of the Middle East was generally questionable, but the announced weapons deals with Saudi Arabia were especially criticized, because American military officials claimed that Saudi Arabia played a counter-productive role in the Iraq War; that nearly half of an estimated 70 foreign fighters who entered Iraq each month were coming from Saudi Arabia; that Saudi Arabia attempted to finance Sunni insurgents in Iraq; and that the Saudi Arabian government viewed Iraq's elected (Shia) prime minister as an agent of Iran.

By now nearly all experts agreed that the U.S. engagement in Iraq had made America and the world (or at least the Western World) a more unsafe place, because it increased anti-U.S. sentiments in the Middle East, did nothing to reduce the activity of al-Qaeda and similar terrorist networks, and had turned Iraq into something of an anarchic launch pad for terror. (In June 2007 a new record was reached, as attacks on civilians, infrastructure, and U.S. and Iraqi forces averaged 178 per day! And though there was much talk during the following year about the violence winding down, this was far from reality. Nearly 100 people were killed by two suicide bombers in Baghdad in February 2008, for instance, and 68 people killed, and 120 injured, in two Baghdad bomb attacks on March 6, 2008.) Besides, the engagement in Iraq required so many resources that the United States became weakened in terms of its potential responses to other threats, prominent among them the nuclear programs of North Korea and Iraq's neighbor Iran, where president Mahmoud Ahmadinejad publicly challenged the right of Israel to exist.[247] In the Israeli-Palestine conflict things did not go well either. In 2000 peace talks at Camp David failed and Palestinian militants launched a second intifada, which this time had the character of a guerrilla war, marked by suicide bomb attacks. Israel then began reoccupying areas controlled by the Palestinian Authority, and in 2002 closed down and sealed the borders to the West Bank and the Gaza Strip. In 2004 Israel unveiled a proposal to evacuate all Jewish settlements in the Gaza Strip (for security reasons), while expanding some (and abandoning a few) settlements in the West Bank. (By this stage Israel had 6.2 million inhabitants, including some 187,000 Israeli settlers in the West Bank, about 20,000 in the Golan Heights, more than 5,000 in the Gaza Strip, and some 175,000 in East Jerusalem, while 2.3 million Palestinian

Arabs lived in the West Bank, and 1.3 million in the Gaza Strip.) The militant Islamic group Hamas kept on attacking Israel out of the Gaza Strip (and Israel kept on retaliating harshly), and won a surprise victory in the Palestinian parliamentary elections of January 2006. Israel said right away that it would not negotiate with a Palestinian government that included Hamas, which is listed as terrorist organization in most Western countries, but Hamas in June 2007 took control of the Gaza Strip by force, and was in turn ousted from its positions in the Palestinian National Authority government residing in the West Bank. What is more, Israel in mid-2006 waged a short war in southern Lebanon against Shia Muslim political and paramilitary organization Hezbollah, which fired rockets into Israeli border villages, and crossed the border to capture Israeli soldiers. Hezbollah has close links to Iran, and senior leadership in Iran was also suspected of having brought Hezbollah operatives into Iraq to help train and organize militants.

Soaring Oil Price

U.S. policy with respect to Iran had been to impose gradually more severe sanctions ever since the revolution in 1979. The Iran and Libya Sanctions Act (ILSA) of 1996 even threatened non-U.S. companies and their countries if they were to attempt making investments over $20 million for the development of petroleum resources in Iran.[248] This measure, intended to deprive Iran of its major source of income by impeding the modernization of its oil industry, may have been useful to punish Iran for its attempts "to acquire weapons of mass destruction" and its "support of acts of international terrorism," but it also reduced the amount of oil arriving on world markets. (Iran has the third largest proven oil reserves in the world, and the second largest natural gas reserves.) However, this was not considered a major problem, as Iran is a member of OPEC, and OPEC had enough spare production capacity (some 20 percent of combined full production capacity at the beginning of the 21st century) to realize its price goals even if a gradually (but not abruptly) declining Iranian production decreased world oil supplies enough for oil prices to rise above the price band targeted by OPEC.

This price band was defined as $22 to $28 per barrel in 2001. To be sure, the "price of oil" may refer to different internationally traded types of oil, or baskets of oils, or market price indications. The price of West Texas Intermediate (WTI) crude oil is reflected by the NYMEX futures price for crude oil, while the price of Brent refers to a combination of crude oil from 15 different oil fields in the North Sea, and the Imported Refiner Acquisition Cost (IRAC) to an average price of the many different types of crude oil imported into the United States. The IRAC is usually similar to the OPEC Basket price, and both are typically some $5 to $6 per barrel below the Brent price, and $6 to $8 per barrel below the WTI spot price.[249] Oil prices were down as far as $10 per barrel (WTI) at the end of 1998, when Iraqi oil reached world markets through the UN-controlled oil-for-food program, and demand was reduced following the East Asian Financial Crisis of 1997. Thereafter, the oil price rose sharply to peak at nearly $40 per barrel (WTI) in mid-2000, and then quickly stabilized somewhere around $28 per barrel (WTI). Following the September 11, 2001, attack the oil price crashed below OPEC's target price band. OPEC thus cut its production, but in turn launched a price war to force Russia to join it and reduce production as well.[250] This sent the oil price well below $20 per barrel (WTI), but subsequent years showed that the low oil price levels of the 1990s were definitely history.

As the oil price began to increase, non-OPEC members restored their production cuts in mid-2002, and U.S. inventories reached a 20-year low later that year. Then Venezuelan production plummeted due to oil industry strikes, and in early 2003 OPEC increased quotas by 2.8 million barrels per day. However, immediately thereafter the invasion of Iraq and the consequent chaos removed Iraqi production from the international market, and oil producers accessed more and more of their spare capacity, which eroded from over 6 million barrels per day in mid-2002 to below 2 million in mid-2003 and below 1 million during much of 2004 and 2005. Such low levels of excess oil production capacity, compared to actual world consumption of 80 million barrels per day, added a significant risk premium to crude oil prices. (Supply risks were enhanced due to political instability in the Middle East and Venezuela, hurricanes damaging oil infrastructure in the Gulf of Mexico, the Russian government hijacking YUKOS, and civil unrest in West African oil producer Nigeria.) Thus oil prices increased to above $50 per barrel towards the end of 2004, and then soared to above $65 a barrel in August 2005 and to above $77 a barrel in July/August of 2006. A year later, in September 2007, the oil price moved above $80 a barrel for the first time ($82.39 on September 19, 2007; West Texas Intermediate). However, by mid-2006 OPEC countries produced about 1 million barrels per day less than a year earlier, and had a recovered spare production capacity of 3 million barrels per day, though global demand had risen to 84 million barrels per day. (Saudi Arabia's crude oil production capacity alone reached 10.8 million barrels per day, and was scheduled to reach 12.5 million barrels per day by 2009.) Hence OPEC, though producing some 26.8 million barrels of oil per day in July 2007, now had capacity reserves to raise output substantially, if it felt like doing so.[251]

Demand for oil remained strong as the U.S. economy was picking up, but most importantly because of strong economic growth in China (and other populous Asian nations, including India). Thus it looked as if the West and Japan would henceforth have to share much of their Middle Eastern oil imports with the emerging markets of Asia. China was about to become a major producer of, and the largest market for, motor vehicles. In 2003 China had passed Japan as the second largest oil-consuming nation after the United

States, and with a population of 1.3 billion (compared to 290 million in the United States) China was bound to take the first place eventually. On its way to become a significant global economic power, China sold more and more goods to the world, while the United States, due to imports of oil and (Chinese) manufactures, saw its trade balance worsen. America was thus interested in allowing the dollar to devaluate, which makes imported goods more expensive in the United States, and American goods cheaper overseas. Oil, however, is traded internationally in terms of U.S. dollars, and oil exporters needed to adjust the dollar-based price for crude upward to reflect what a barrel was worth in their own currencies. Notably, the oil price at the August 2006 peak was in real terms still well below the level of the second oil price shock in 1979–1980, which was around $90 a barrel in terms of 2006 dollars. What is more, gasoline prices in most Western countries (but notably not the United States) now included very high taxes, which in a true crisis could be removed to relieve the critical cost for transportation.[252] However, in light of the fiasco in Iraq, the problems with Iran, and the worsening situation in Saudi Arabia, the West reacted to the soaring oil prices by promoting alternative energy sources, a campaign designed to wean the West of Middle Eastern (and Russian and Venezuelan) oil, but was justified largely as a climate protection measure (to decrease carbon dioxide emissions). By July 2008 the oil price reached $147 and some began calling this a kind of slow-motion oil price shock. However, the word shock may have been inadequate altogether, as it had to be expected that meeting future demand for oil at prices previously considered reasonable would be difficult. After all, the global oil discovery rate (of cheaply producible oil) had not kept up with global production rates for a while. Thus, higher oil prices were likely to become a permanent condition, unless or until being automatically adjusted downward as soon as they would lead to a global economic crisis that would in turn curb the demand for oil. Such scenario had at least to be expected if the world was indeed getting close to a global oil production peak, and the great Age of Oil was bound to come to an end.

NOTES

116. Colin Simpson, *The Lusitania* (Boston: Little, Brown and Company, 1972).
117. All these figures are uncertain. Of the 65 million men who were mobilized, more than 10 million were killed and more than 20 million wounded. William R. Keylor, "World War I," Microsoft Encarta Online Encyclopedia, http://encarta.msn.com/encyclopedia_761569981/World_War_I.html#s1; Michael Duffy, ed., "Feature Articles: Military Casualties of World War One," First World War.com, http://www.firstworldwar.com/features/casualties.htm.
118. Yale Law School, "The Versailles Treaty June 28, 1919: Part III," The Avalon Project- Documents in Law, History and Diplomacy, http://avalon.law.yale.edu/imt/partiii.asp.

119. John V. Denson, "Reassessing the Presidency," Ludwig von Mises Institute, Auburn, Alabama, 2001, p.480, http://books.google.com/books?id=hJGpAT7IWhwC&printsec=frontcover&dq=Reassessing+the+presidency.

120. "Dresden," Microsoft Encarta Online Encyclopedia, http://encarta.msn.com/encyclopedia_761552600/Dresden.html.

121. The bombing of Dresden is a highly controversial issue that has only been discussed publicly in more recent years. Various intellectuals have called it a war crime. Winston Churchill was ultimately responsible for the raid on Dresden, but Britain's Prime Minister in turn distanced himself from it. Detlef Siebert, "British Bombing Strategy in World War Two," August 1, 2001, http://www.bbc.co.uk/history/worldwars/wwtwo/area_bombing_05.shtml.

122. All figures are from the website of the Simon Wiesenthal Center at http://motlc.wiesenthal.com/site/pp.asp?c=gvKVLcMVIuG&b=394663.

123. "Commemorating The 60th Anniversary of The Nanking Massacre, NANKING 1937," Princeton University, http://www.princeton.edu/~nanking/.

124. Earl F. Ziemke, "World War II," Microsoft Encarta Online Encyclopedia, http://encarta.msn.com/encyclopedia_761563737/World_War_II.html.

125. Zbigniew Brzezinski, *Out of Control—Global Turmoil on the Eve of the 21st Century* (New York: Touchstone, 1995), Introductory Chapter: "The Century of Megadeath."

126. Ibid.

127. "Country Profiles: China, Ballistic Missile Defense Organization," Department of Defense, April 1995, http://www.fas.org/nuke/guide/china/bmdo1995.pdf.

128. "Japan's Auto Industry," Japan Automobile Manufacturers Association (JAMA), http://www.jama.org/about/industry.htm.

129. "India's Nuclear Weapons Program—The Beginning: 1944–1960," The Nuclear Weapon Archive, http://nuclearweaponarchive.org/India/IndiaOrigin.html; The Federation of American Scientists (FAS), "India-Nuclear Weapons," http://www.fas.org/nuke/guide/india/nuke/index.html.

130. The Federation of American Scientists (FAS), "Pakistan-Nuclear Weapons," http://www.fas.org/nuke/guide/pakistan/nuke/.

131. David E. Sanger, "Pakistan Leader Confirms Nuclear Exports," *New York Times*, September 13, 2005, A10.

132. Zbigniew Brzezinski, *Out of Control*. In his introductory chapter, "The Century of Megadeath," Brzezinski further writes: "Because of Lenin—through mass executions during and after civil war, through massive deaths in the Gulag initiated under Lenin's direction (and powerfully documented in Solzhenitsyn's Gulag Archipelago), and through mass famines induced by ruthless indifference (with Lenin callously dismissing as unimportant the deaths of "the half-savage, stupid, difficult people of the Russian villages")—it can be estimated that between 6–8,000,000 people perished. That number subsequently was more or less tripled by Stalin, who caused, it has been conservatively estimated, the deaths of no less than 20,000,000 people, and perhaps even upward of 25,000,000."

133. Vagif Agayev et al., "World War II and Azerbaijan," *Azerbaijan International* (3.2), Summer 1995, http://www.azer.com/aiweb/categories/magazine/32_folder/32_articles/32_ww22.html; "History of the Oil Industry in the Soviet Union," TNK, http://www.tnk.com/company/history/industry.html.

134. Rosalind Rosenberg, "Lecture Notes: American Civilization Since the Civil War," http://www.columbia.edu/~rr91/1052_2002/class_page/schedule_of_lectures_and_reading.htm.
135. "United States," CIA—The World Factbook https://www.cia.gov/library/publications/the-world-factbook/geos/us.html.
136. Bruce Cumings, "Korean War," Microsoft Encarta Online Encyclopedia 2008, http://encarta.msn.com/encyclopedia_761559607_3/Korean_War.html#s14.
137. Doug Linder, "An Introduction to the My Lai Courts-Martial," http://www.law.umkc.edu/faculty/projects/ftrials/mylai/Myl_intro.html.
138. "Laos: 'Most-heavily Bombed Place'," *BBC News*, January 5, 2001, http://news.bbc.co.uk/2/hi/asia-pacific/1100842.stm.
139. "CIA and Assassinations: The Guatemala 1954 Documents," National Security Archive Electronic Briefing Book No. 4, http://www.gwu.edu/~nsarchiv/NSAEBB/NSAEBB4/index.html; The National Security Archive of George Washington University, http://www.gwu.edu/~nsarchiv/NSAEBB/, served as a general source for information on CIA activities during the Cold War. Find information on such activities in Latin America at "Electronic Briefing Books-Latin America," http://www.gwu.edu/~nsarchiv/NSAEBB/index.html#Latin%20America.
140. "Brazil Marks 40th Anniversary Of Military Coup—Declassified Documents Shed Light On U.S. Role," http://www.gwu.edu/~nsarchiv/NSAEBB/NSAEBB118/index.htm.
141. "Chile and the United States: Declassified Documents relating to the Military Coup, 1970–1976," National Security Meeting on Chile Memorandum of Conversation, November 6, 1970, http://www.gwu.edu/~nsarchiv/NSAEBB/NSAEBB8/nsaebb8.htm; "The Kissinger Telcons," National Security Archive Electronic Briefing Book No. 123, http://www.gwu.edu/~nsarchiv/NSAEBB/NSAEBB123/index.htm#chile; "CIA Acknowledges Ties to Pinochet's Repression," http://www.gwu.edu/~nsarchiv/news/20000919/.
142. "Kissinger To Argentines On Dirty War: 'The Quicker You Succeed The Better'," http://www.gwu.edu/~nsarchiv/NSAEBB/NSAEBB104/index.htm.
143. "The Oliver North File: His Diaries, E-Mail, and Memos on the Kerry Report, Contras and Drugs," http://www.gwu.edu/~nsarchiv/NSAEBB/NSAEBB113/index.htm#doc4.
144. "Military and Paramilitary Activities in and against Nicaragua, Nicaragua v. United States, decision of 27 June 1986," International Court of Justice (The Hague), Collection of decisions, advisory opinions and rulings, page 14, quoted in Monique Chemillier-Gendreau, "The International Court of Justice. Between Politics and Law," *Le Monde diplomatique*, November 1996, Global Policy Forum, http://www.globalpolicy.org/wldcourt/icj.htm#fn1.
145. "Operation Condor: Cable Suggests US Role," http://www.gwu.edu/~nsarchiv/news/20010306/.
146. Jeffrey A. Lefebvre, "The United States, Ethiopia and the 1963 Somali-Soviet Arms Deal: Containment and the Balance of Power Dilemma in the Horn of Africa," *The Journal of Modern African Studies* 36 (04) (1998): 611–643, http://journals.cambridge.org/bin/bladerunner?REQUNIQ=1078150879&REQSESS=48303&118000REQEVENT=&REQINT1=17499&REQAUTH=0.

147. Roger T. Housen, "Why Did The US Want To Kill Prime Minister Lumumba Of The Congo?," National Defense University, http://www.ndu.edu/library/n2/n025603K.pdf; "The Story of Africa—Independence, Case Study: Congo," *BBC World Service*, http://www.bbc.co.uk/worldservice/africa/features/storyofafrica/14chapter7.shtml.

148. "A report from Congo: Africa's Great War," *The Economist*, July 4, 2002, http://www.economist.com/displaystory.cfm?story_id=1213296; "War, Coltan and Conservation: Digging a Grave for King Kong?," *The Economist*, July 31, 2003, http://www.economist.com/displaystory.cfm?story_id=1956903.

149. Federation of American Scientists, "International Military Education and Training (IMET)," http://www.fas.org/asmp/campaigns/training/IMET2.html; Center for International Policy, "IMET: International Military Education and Training," http://www.ciponline.org/facts/imet.htm.

150. "Human Rights Advocate Examines U.S. 'School of the Assassins' in Lawrence Lecture," May 8, 2000, quoting a 1996 New York Times article on the School of the Americas, Lawrence University (Appleton, WI), http://www.lawrence.edu/dept/public_affairs/media/release/9900/bourgeois.html.

151. "Spindletop Gusher," The History of the Oil Industry, San Joaquin Geological Society, http://www.sjgs.com/history.html; The Paleontological Research Institution, Ithaca, NY, "Spindletop, Texas," History of Oil, http://www.priweb.org/ed/pgws/history/spindletop/spindletop.html.

152. "The History of the Oil Industry; The Paleontological Research Institution, Ithaca, NY, "The Story of Oil in California," History of Oil, http://www.priweb.org/ed/pgws/history/signal_hill/signal_hill.html.

153. "The Rockefellers: People & Events: South Improvement Company / Cleveland Massacre, 1871–1872," and "The Rockefellers: The Ludlow Massacre," Public Broadcasting Service (PBS) Online, http://www.pbs.org/wgbh/amex/rockefellers/peopleevents/e_south.html, http://www.pbs.org/wgbh/amex/rockefellers/sfeature/sf_8.html.

154. The authors argue that federal energy subsidies are not large compared to the total value of energy production. The annual cost of between $5 billion and $10 billion from Federal energy subsidies for 1990 compared to the total value of production in all energy industries of close to $475 billion. Thus, Federal subsidies were approximately 1 to 2 percent of the value of sales. EIA Service Report: "Federal Energy Subsidies: Direct and Indirect Interventions in Energy Markets," DOE/EIA/SR/EMEU-92-02, November 20, 1992, Energy Information Administration, Office of Energy Markets and End Use, http://tonto.eia.doe.gov/ftproot/service/emeu9202.pdf.

155. Joint Committee on Taxation, "Estimates of Federal Tax Expenditures for Fiscal Years 1999–2003 (JCS-7–98)," House of Representatives, 105th Congress, 2nd Session, December 14, 1998, http://www.house.gov/jct/s-7-98.pdf. Find more "Publications on Tax Expenditures" by the Joint Committee Staff at this site: http://www.house.gov/jct/pubs_taxexpend.html.

156. Doug Koplow and John Dernbach, "Federal Fossil Fuel Subsidies and Greenhouse Gas Emissions: A Case Study of Increasing Transparency for Fiscal Policy," *Annual Review of Energy and the Environment* 26 (2001): 361–89, http://arjournals.annualreviews.org/doi/full/10.1146/annurev.energy.26.1.361; Roland Hwang, "Money

Down the Pipeline: The Hidden Subsidies to the Oil Industry," Union of Concerned Scientists, http://www.ucsusa.org/publication.cfm?publicationID=149. Find more estimates in the following Greenpeace book: Doug Koplow and John Dernbach, *Fueling Global Warming—Federal Subsidies to Oil in the United States*, http://archive.greenpeace.org/~climate/oil/fdsub.html; Friends of the Earth, "Analysis of the Bush Energy Plan," http://www.foe.org/camps/leg/bushwatch/energyplan.pdf; Greenpeace, "The Subsidy Scandal—The European Clash between Environmental Rhetoric and Public Spending," http://archive.greenpeace.org/~comms/97/climate/eusub.html.

157. Energy Information Administration, "US Crude Oil Field Production," http://tonto.eia.doe.gov/dnav/pet/hist/mcrfpus1A.htm.

158. Energy Information Administration, "U.S. Crude Oil Imports from All Countries," http://tonto.eia.doe.gov/dnav/pet/hist/mcrimus1A.htm.

159. Library of Congress Country Studies, "Mexico-Oil," http://countrystudies.us/mexico/78.htm.

160. Find historic oil production data by country in "World Crude Oil Production, 1960–2007," Energy Information Agency, http://www.eia.doe.gov/aer/txt/ptb1105.html.

161. Library of Congress Country Studies, "Venezuela—Energy—Petroleum," http://countrystudies.us/venezuela/30.htm

162. To be sure, this was also in the interest of the Middle Eastern countries. Many of them have little else to sell than oil and petroleum products.

163. Find a comprehensive history of the oil industry, covering all regions, in: Daniel Yergin, *The Prize: The Epic Quest for Oil, Money & Power* (New York: Free Press, 1991).

164. Library of Congress Country Studies, "Persian Gulf States/Kuwait: Treaties With The British," http://countrystudies.us/persian-gulf-states/13.htm or http://lcweb2.loc.gov/frd/cs/kwtoc.html#kw0015; Library of Congress Country Studies, "Kuwait-Oil Industry," http://countrystudies.us/persian-gulf-states/21.htm; *Persian Gulf States: A Country Study*, ed. Helen Chapin Metz (Washington: GPO for the Library of Congress, 1993), http://countrystudies.us/persian-gulf-states/.

165. Anthony Sampson, *The Seven Sisters* (New York: Viking Press, 1975). "Memorandum for the Attorney General Relative to a Request for Grand Jury Authorization to Investigate the International Oil Cartel, June 24, 1952," http://www.mtholyoke.edu/acad/intrel/Petroleum/jury.htm. Find more information about the Achnacarry Agreement, the Red Line Agreement, and the Draft Memorandum of Principles of 1934 at the "Documents on the International Energy System" page of Vincent Ferraro at Mount Holyoke College, MA, http://www.mtholyoke.edu/acad/intrel/energy.htm; David Osterfeld, "Voluntary and Coercive Cartels: The Case of Oil," *The Freeman: Ideas on Liberty* 37 (1987), http://www.fee.org/publications/the-freeman/article.asp?aid=2333; Eric V. Thompson, "A Brief History Of Major Oil Companies In The Gulf Region," Petroleum Archives Project, Arabian Peninsula and Gulf Studies Program, University of Virginia, http://www.virginia.edu/igpr/apagoilhistory.html.

166. President Truman's plan to destroy the Middle Eastern oil installations in case of a Soviet invasion is described in a National Security Council directive known

as NSC 26/2 and later supplemented by a series of additional NSC orders. Shibley Telhami and Fiona Hill, "Does Saudi Arabia Still Matter? Differing Perspectives on the Kingdom and Its Oil," *Foreign Affairs* 81 (6) (November/December 2002), http://www.foreignaffairs.org/20021101faresponse10002/shibley-telhami-fiona-hill/does-saudi-arabia-still-matter-differing-perspectives-on-the-kingdom-and-its-oil.html.

167. This view is promoted by Israel Finkelstein and Ze'ev Herzog of Tel Aviv University. Herzog writes, "Perhaps even harder to swallow is the fact that the united monarchy of David and Solomon, which is described by the Bible as a regional power, was at most a small tribal kingdom." Ze'ev Herzog, "Deconstructing the walls of Jericho," October 29, 1999, http://mideastfacts.org/facts/index.php?option=com_content&task=view&id=32&Itemid=34.

168. John Haywood, *World Atlas of the Past, Vol. 1–4* (Oxford: Andromeda Oxford/Oxford University Press, 1999). I used the Danish edition: John Haywood, *Historisk Verdensatlas* (Köln: Könemann, 2000), 33.

169. "The Court Jews," Gates to Jewish Heritage, http://www.jewishgates.com/file.asp?File_ID=53.

170. Glyn Davies, *A History of Money: From Ancient Times to the Present Day* (Cardiff: University of Wales Press, 1994).

171. "Palestine, Israel and the Arab-Israeli Conflict-A Primer," The Middle East Research and Information Project (MERIP), http://www.merip.org/palestine-israel_primer/toc-pal-isr-primer.html.

172. Alan B. Mountjoy, "The Suez Canal at Mid-Century," *Economic Geography* 34 (2) (1958): 155–67, http://www.jstor.org/pss/142300; "Tanker History," GlobalSecurity.org, http://www.globalsecurity.org/military/systems/ship/tanker-history.htm.

173. Leonardo Maugeri, "Not in Oil's Name," *Foreign Affairs* 82 (4) (2003), http://www.foreignaffairs.org/20030701faessay15412/leonardo-maugeri/not-in-oil-s-name.html.

174. Misguided energy policies in the years before the embargo apparently contributed significantly to the shortages in America. In 1971 the Nixon administration imposed price controls on the energy industry that caused oil companies to reduce their imports, and Congress in September 1973 attempted to allocate oil to various sectors of industry and different parts of the country through bureaucratic fiat. "Special Report: OPEC, Still Holding Customers Over a Barrel," *The Economist*, October 23, 2003, http://www.economist.com/displaystory.cfm?story_id=2155405.

175. Paul Reynolds, "US Ready to Seize Gulf Oil in 1973," *BBC News Online*, January 2, 2004, http://news.bbc.co.uk/1/hi/world/middle_east/3333995.stm.

176. "Big Oil's Biggest Monster," *The Economist*, January 6, 2005, http://www.economist.com/displaystory.cfm?story_id=3545061.

177. Find more information on EPU in Wolfgang Sartor, "International and Multinational and Multilateral Companies in the Russian Empire before 1914: The Integration of Russia into the World Economy," XIII Economic History Congress, Buenos Aires, July 2002, The International Economic History Association, http://eh.net/XIIICongress/cd/papers/43Sartor202.pdf.

178. Iran is said to have in exchange been promised that its frozen assets in the United States would be returned. Allegedly, representatives of the Reagan presidential campaign, including ex-CIA director and upcoming vice president George H.W. Bush, made a deal with the Iranians to delay the release of the hostages until after

the November 1980 presidential elections to make sure Reagan's opponent, President Jimmy Carter, would not gain a popularity boost before election day. The allegations center around a purported Bush trip to Paris to meet representatives of Iran. Gary Sick, *October Surprise* (New York: Three Rivers Press, 1992).

179. Shortly before his death in 1989 Khomeini declared that the murder of Indian Cambridge graduate Salman Rushdie was a religious necessity for Muslims due to the contents of Rushdie's novel *The Satanic Verses*. Rushdie remained unharmed as he was guarded by British security police, but lived in constant fear for his life for years. In 1991 Rushdie's Japanese translator was killed, the Italian translator was beaten and stabbed, and Rushdie's Norwegian publisher was shot and severely injured in 1993.

180. Find information on the rumors of Saddam Hussein's early CIA connections in Richard Sale, "Saddam Key in Early CIA Plot," *United Press International*, April 10, 2003, http://www.upi.com/print.cfm?StoryID=20030410-070214-6557r.

181. Total casualties (lives lost and people wounded) are typically given at one million, but sometimes at two million. Reliable figures do not exist. Iraq claimed to have killed 800,000 Iranians, while neutral estimates, which are admittedly uncertain, are closer to Iran's assertion that 200,000 or fewer of its citizens were killed. Nathan J. Brown, "Iran-Iraq War," Microsoft Encarta Online Encyclopedia, http://encarta.msn.com/encyclopedia_761580640_2/Iran-Iraq_War.html.

182. Russ W. Baker, "Iraqgate-The Big One That (Almost) Got Away, Who Chased it—and Who Didn't," *Columbia Journalism Review* (March/April 1993), http://backissues.cjrarchives.org/year/93/2/iraqgate.asp; Henry B. Gonzalez, "Kissinger Associates, Scowcroft, Eagleburger, Stoga, Iraq, and Bnl," Congressional Record, House of Representatives, April 28, 1992, http://www.fas.org/spp/starwars/congress/1992/h920428g.htm.

183. "The Oliver North File," The National Security Archive, http://www.gwu.edu/~nsarchiv/NSAEBB/NSAEBB113.

184. The Nuclear Threat Initiative, "Iran—Chemical Chronology 1929–1983," http://www.nti.org/e_research/profiles/Iran/Chemical/2340.html.

185. "Saddam Hussein: More Secret History," The National Security Archive, http://www.gwu.edu/~nsarchiv/NSAEBB/NSAEBB107/; "Shaking Hands with Saddam Hussein: The U.S. Tilts toward Iraq, 1980–1984," The National Security Archive, *Electronic Briefing Book No. 82*, ed. Joyce Battle, February 25, 2003, http://www.gwu.edu/~nsarchiv/NSAEBB/NSAEBB82/.

186. Bob Herbert, "Ultimate Insiders," *New York Times*, April 14, 2003, http://www.commondreams.org/views03/0414-02.htm.

187. Library of Congress Country Studies, "Kuwait—Oil Industry," http://countrystudies.us/persian-gulf-states/21.htm, in *Persian Gulf States: A Country Study*, ed. Helen Chapin Metz (Washington: GPO for the Library of Congress, 1993), http://countrystudies.us/persian-gulf-states/.

188. Find a list of all Security Council Resolutions on Iraq at Federation of American Scientists, "Security Council Resolutions on Iraq," http://www.fas.org/news/un/iraq/sres/.

189. Seymour M. Hersh, "A Case Not Closed," *The New Yorker*, November 1, 1993, http://www.newyorker.com/archive/content/?020930fr_archive02.

190. Federation of American Scientists, "Security Council Resolutions on Iraq," http://www.fas.org/news/un/iraq/sres/; RESOLUTION 678 (1990), Adopted

by the Security Council at its 2963rd meeting on November 29, 1990, "Authorizes Member States co-operating with the Government of Kuwait. . . to use all necessary means to uphold and implement resolution 660 (1990) and all subsequent relevant resolutions and to restore international peace and security in the area." Federation of American Scientists, "Resolution 678 (1990)," http://www.fas.org/news/un/iraq/sres/sres0678.htm.

191. Jack Kelly, "Estimates of Deaths in First War Still in Dispute," *Post-Gazette*, February 16, 2003, http://www.post-gazette.com/nation/20030216casualty0216p5.asp.

192. Ramsey Clark, *The Impact of Sanctions on Iraq: The Children Are Dying* (Farmington: Plough Publishing House, 1996).

193. Shafeeq N. Ghabra, "Iraq's Culture of Violence," *The Middle East Quarterly* VIII (3) (2001), http://www.meforum.org/article/101.

194. Edward L. Morse and James Richard, "The Battle for Energy Dominance," *Foreign Affairs* 81 (2) (2002), http://www.foreignaffairs.org/20020301faessay7969/edward-l-morse-james-richard/the-battle-for-energy-dominance.html.

195. Chrissie Hirst, "Executive Summary: The Arabian Connection: The UK Arms Trade to Saudi Arabia," written for CAAT (Campaign Against Arms Trade), http://www.caat.org.uk/information/publications/countries/saudi-arabia-intro.php; "BAE and Saudi Arabia: Bribery Alleged between the Kingdoms," *The Economist*, June 7, 2007, http://www.economist.com/world/britain/displaystory.cfm?story_id=9312370; "Arms Deals and Bribery—The Bigger Bang," *The Economist*, June 14, 2007, http://www.economist.com/opinion/displaystory.cfm?story_id=9339826.

196. "Palestine, Israel and the Arab-Israeli Conflict—A Primer," The Middle East Research and Information Project (MERIP), http://www.merip.org/palestine-israel_primer/toc-pal-isr-primer.html.

197. 1975 was also the year when the UN General Assembly declared that Zionism is a form of racism and racial discrimination, which was revoked 16 years later. "Elimination of all forms of racial discrimination," UN General Assembly Resolution 3379 of November, 1975, Rescinded in Resolution 46/86 of December 1991, http://domino.un.org/UNISPAL.NSF/0/761c1063530766a7052566a2005b74d1?OpenDocument and http://www.un.org/documents/ga/res/46/a46r086.htm.

198. Amnesty International, "Kuwait-Search www.amnesty.org," http://www.amnesty.org/.

199. Stephen Schwartz, "The Dysfunctional House of Saud—Compromised by Terror, the Saudi Regime Will Have to Change or Die," *The Weekly Standard* 8 (46), August 18, 2003, Volume 00, http://www.weeklystandard.com/Content/Public/Articles/000/000/002/978uclzj.asp.

200. "Background Note: Saudi Arabia," Bureau of Near Eastern Affairs, U.S. Department of State, September 2004, http://www.state.gov/r/pa/ei/bgn/3584.htm; "Saudi Arabia-EIA Country Analysis Briefs," Energy Information Administration, http://www.eia.doe.gov/emeu/cabs/saudi.html.

201. Find a complete account of the circumstances surrounding the September 11, 2001, terrorist attacks, including preparedness for and the immediate response to the attacks, created by an independent, bipartisan commission, at National Commission on Terrorist Attacks Upon the United States (also known as the 9/11 Commission), http://govinfo.library.unt.edu/911/report/index.htm.

202. "Why the Towers Fell," NOVA, Public Broadcasting Service (PBS), http://www.pbs.org/wgbh/nova/wtc/.
203. John Ford Shroder, "Afghanistan," Encarta Online Encyclopedia, http://encarta.msn.com/encyclopedia_761569370_5/Afghanistan.html.
204. Jimmy Carter, "State of the Union Address 1980," January 23, 1980, http://www.jimmycarterlibrary.org/documents/speeches/su80jec.phtml.
205. "The CIA's Intervention in Afghanistan," Interview with Zbigniew Brzezinski, President Jimmy Carter's National Security Adviser, *Le Nouvel Observateur*, January 1998, http://www.globalresearch.ca/articles/BRZ110A.html.
206. Joe Stephens and David B. Ottaway, "From U.S., the ABC's of Jihad, Violent Soviet-Era Textbooks Complicate Afghan Education Efforts," *Washington Post*, March 23, 2002, http://www.washingtonpost.com/ac2/wp-dyn/A5339-2002-Mar22?language=printer.
207. David Leigh and Richard Norton-Taylor, "House of Saud Looks Close to Collapse," *The Guardian*, November 21, 2001, http://www.guardian.co.uk/afghanistan/story/0,1284,602854,00.html.
208. "A Bitter Harvest—The Sufferings of Afghanistan Come to New York," *The Economist*, September 13, 2001, http://www.economist.com/displaystory.cfm?story_id=780398; Council On Foreign Relations, "Osama bin Laden," http://cfrterrorism.org/groups/binladen.html.
209. Council On Foreign Relations, "Al-Qaeda," http://cfrterrorism.org/groups/alqaeda.html; "Osama bin Laden's Network—The Spider in the Web," *The Economist*, September 20, 2001, http://www.economist.com/displaystory.cfm?story_id=788472.
210. Werner Daum, "Universalism and the West—An Agenda for Understanding," Harvard International Review-The Future of War, Volume 23, Issue 2, Summer 2001, http://hir.harvard.edu/articles/index.html?id=909&page=1.
211. "Today's New Cult Hero," *The Economist*, August 27, 1998, http://www.economist.com/displaystory.cfm?story_id=162711.
212. "Attack On The USS Cole," based on an official version of the USS *Cole* incident by the United States Navy, http://www.al-bab.com/yemen/cole1.htm.
213. "Bin Laden 9/11 Planning Video Aired," *CBC News*, September 7, 2006, http://www.cbc.ca/world/story/2006/09/07/al-qaeda-tape.html; "Reports of the Joint Inquiry into Intelligence Community Activities before and after the Terrorist Attacks of September 11, 2001," U.S. Senate Select Committee On Intelligence and U.S. House Permanent Select Committee On Intelligence, December 2002, http://www.gpoaccess.gov/serialset/creports/911.html.
214. Dana Priest, "CIA Holds Terror Suspects in Secret Prisons," *Washington Post*, November 2, 2005, http://www.washingtonpost.com/wp-dyn/content/article/2005/11/01/AR2005110101644.html; Jason Burke, "Secret World of US Jails," *The Observer*, June 13, 2004, http://observer.guardian.co.uk/international/story/0,6903,1237589,00.html.
215. Oliver Burkeman and Julian Borger, "The Ex-presidents' Club," *The Guardian*, October 31, 2001, http://www.guardian.co.uk/wtccrash/story/0,1300,583869,00.html; Simon English, "The Best-connected Investor in America—The Hugely Profitable Carlyle Group Has Become a Magnet for Conspiracy Rumours," *Telegraph*, May 27, 2003, http://money.telegraph.co.uk/money/main.jhtml?xml=/

money/2003/05/27/ccarly27.xml; Michael Moore, "What Fahrenheit 9/11 Says About The Relationship Between The Families Of George W. Bush And Osama Bin Laden," http://www.michaelmoore.com/warroom/index.php?id=5.

216. "Greed and Torture at the House of Saud," *The Observer*, November 24, 2002, http://observer.guardian.co.uk/worldview/story/0,11581,846600,00.html.

217. "Saudi Arabia—Adapt or Die," *The Economist*, March 4, 2004, http://www.economist.com/displaystory.cfm?story_id=2482168.

218. Robert Baer," The Fall of the House of Saud," *The Atlantic Monthly*, May 2003, http://www.theatlantic.com/doc/200305/baer.

219. "Plans For Iraq Attack Began On 9/11," *CBS News*, September 4, 2002, http://www.cbsnews.com/stories/2002/09/04/september11/main520830.shtml.

220. Ivo Daalder and James Lindsay, "Trust Clarke: He's Right about Bush," *The Globe and Mail*, March 26, 2004, http://www.theglobeandmail.com/servlet/ArticleNews/TPStory/LAC/20040326/COCLARKE26/TPComment/TopStories.

221. National Commission on Terrorist Attacks; Walter Pincus and Dana Milbank, "The Iraq Connection. Al Qaeda-Hussein Link Is Dismissed," *Washington Post*, June 17, 2004, http://www.washingtonpost.com/wp-dyn/articles/A47812-2004Jun16.html. Even later, in 2008, a much delayed report by the Senate Intelligence Committee accused President Bush and Vice President Cheney of taking the country to war in Iraq by exaggerating evidence of links between Saddam Hussein and Al Qaeda after the Sept. 11, 2001, terrorist attacks. The report cites instances in which public statements by senior administration officials were not supported by the intelligence available at the time, such as suggestions that Saddam Hussein's Iraq and Osama bin Laden's al Qaeda were operating in a kind of partnership, that the Baghdad regime had provided the terrorist network with weapons training, and that one of the Sept. 11 hijackers had met an Iraqi intelligence operative in Prague in 2001. Mark Mazzetti and Scott Shane, "Senate Panel Accuses Bush of Iraq Exaggerations," *New York Times*, June 5, 2008, http://www.nytimes.com/2008/06/05/washington/05cnd-intel.html?hp. The report can be downloaded at: http://intelligence.senate.gov/080605/phase2a.pdf and http://intelligence.senate.gov/080605/phase2b.pdf.

222. "Iraq's WMD-Secret Weapons—Did George Bush and Tony Blair wage war under false pretences?," *The Economist*, May 29, 2003, http://www.economist.com/displaystory.cfm?story_id=1812369; "Weapons of Mass Destruction: Iraq's Elusive Weapons," *The Economist*, May 29, 2003, http://www.economist.com/displaystory.cfm?story_id=1812042; Wolf Blitzer, "Did the Bush Administration Exaggerate the Threat from Iraq?," *CNN*, July 8, 2003, http://www.cnn.com/2003/ALLPOLITICS/07/08/wbr.iraq.claims/; "A Chart of Bush Lies about Iraq—Did George W. Bush Invade Iraq by Lying?," Northwestern University, http://faculty-web.at.northwestern.edu/music/lipscomb/BushLiesAboutIraq_BuzzFlash.htm; "Some Examples of Portrayals of Information about Iraq," Department of Political Science, Oregon State University, http://oregonstate.edu/Dept/pol_sci/fac/sahr/ps415/portray.pdf; National Commission on Terrorist Attacks Upon the United States.

223. "Iraq War Illegal, says Annan," *BBC News*, September 16, 2004, http://news.bbc.co.uk/1/hi/world/middle_east/3661134.stm.

224. "Iraq 'Will Accept' UN Resolution," *BBC News*, November 10, 2002, http://news.bbc.co.uk/1/low/world/middle_east/2433651.stm.

225. "The Accidental President," *The Economist*, December 14, 2000, http://www.economist.com/displaystory.cfm?story_id=S%26%28X8%2DP13%2A%0A.

226. Ramsey Clark, *The Impact of Sanctions on Iraq: The Children Are Dying* (Farmington: Plough Publishing House, 1996).

227. John King, "Bush Calls Saddam 'the guy who tried to kill my dad'," *CNN*, September 27, 2002, http://www.cnn.com/2002/ALLPOLITICS/09/27/bush.war.talk/.

228. Richard A. Clarke, *Against All Enemies: Inside America's War on Terror* (New York: Simon & Schuster, 2004).

229. "Mandela Condemns US Stance on Iraq," *BBC News*, January 30, 2003, http://news.bbc.co.uk/2/hi/africa/2710181.stm.

230. David Ignatius, "Bush's Fancy Financial Footwork," *Washington Post*, August 6, 2002, Page A15, http://www.washingtonpost.com/ac2/wp-dyn/A48301-2002-Aug6; Simon English," Bush on the Board Not Worth Much, Says Carlyle Founder," *Telegraph*, July 8, 2003, http://www.telegraph.co.uk/finance/2857132/Bush-on-the-board-not-worth-much-says-Carlyle-founder.html.

231. Greg Palast, "Secret US Plans for Iraq's Oil," BBC Newsnight, March 17, 2005, http://news.bbc.co.uk/2/hi/programmes/newsnight/4354269.stm.

232. Jeroen van der Veer, Chief Executive of Royal Dutch Shell, for instance, confirmed in 2008 that he had been in contact with Iraqi ministers about a range of onshore oil and gas projects there, but until the Iraq's new law governing the industry is passed, no final decision will be made. "We have to know the rules of the game," he said. The company's final decision to invest in a gas project in Iran meanwhile has been put off by "political considerations." Danny Fortson, "Oil fields of plenty?," *The Independent*, February 1, 2008, http://www.independent.co.uk/news/business/analysis-and-features/oil-fields-of-plenty-776776.html.

233. Michael T. Klare, "Mapping The Oil Motive," March 18, 2005, http://www.tompaine.com/articles/mapping_the_oil_motive.php.

234. Amnesty International, "Iraq: Secretary General Challenges Powell," January 26, 2003, http://www.amnesty.org/en/library/asset/MDE14/002/2003/en/dom-MDE140022003en.html.

235. Richard Horton, "The War in Iraq: Civilian Casualties, Political Responsibilities," *The Lancet* 364 (9448) (2004), http://www.thelancet.com/journal/vol364/iss9448/contents; Tim Parsons, "Updated Iraq Study Affirms Earlier Mortality Estimates," *The JHU Gazette*, October 16, 2006, http://www.jhu.edu/~gazette/2006/16oct06/16iraq.html.

236. Edward Epstein, "How Many Iraqis Died? We May Never Know: Some Observers Are Pressuring Pentagon to Put Forth an Informed Estimate," *San Francisco Chronicle*, May 3, 2003, http://www.sfgate.com/cgi-bin/article.cgi?f=/c/a/2003/05/03/MN98747.DTL.

237. "Iraq Body Count," iraqbodycount.org.

238. "The Iraqi War: Counting the Casualties," *The Economist*, November 4, 2004, http://www.economist.com/displaystory.cfm?story_id=3352814; "Iraq: The Perils of Imprecision—Whatever the Correct Figure, the Civilian Casualty Count is Far Too High—and Rising," *The Economist*, November 4, 2004, http://www.economist.com/displaystory.cfm?story_id=3353342; "War in Iraq, Forces: U.S. & Coalition/Casualties," *CNN*, http://www.cnn.com/SPECIALS/2003/iraq/forces/casualties/2005.02.html.

239. David Zucchino, "Army Stage-Managed Fall of Hussein Statue," *Los Angeles Times*, July 3, 2004, http://www.commondreams.org/headlines04/0703-02.htm.

240. "National Survey of Iraq, February 2004," Oxford Research International, http://news.bbc.co.uk/nol/shared/bsp/hi/pdfs/15_03_04_iraqsurvey.pdf.

241. "War and the Law in Iraq: Crime and Punishment," *The Economist*, May 6, 2004, http://www.economist.com/displaystory.cfm?story_id=2656537.

242. "Iraq: The Battle for Fallujah Now—and for Hearts and Minds Later," *The Economist*, November 11, 2004; America in Iraq: After Fallujah—America Still Faces a War on Many Fronts, *The Economist:* November 11, 2004.

243. "Iraq—Ever Bloodier: Despite the Insurgents' Efforts, the Election Looks Likely to Happen on Time," *The Economist*, January 6, 2005.

244. "Q&A: Saddam's capture—BBC News Online Looks at Some of the Issues Arising from the Capture of Former Iraqi Leader Saddam Hussein," *BBC News Online*, http://news.bbc.co.uk/1/hi/world/middle_east/3320289.stm.

245. Linda Bilmes and Joseph Stiglitz, "The Economic Costs Of The Iraq War: An Appraisal Three Years After the Beginning of The Conflict," Working Paper 12064, National Bureau of Economic Research, 2006, http://www2.gsb.columbia.edu/faculty/jstiglitz/download/2006_Cost_of_War_in_Iraq_NBER.pdf.

246. Thom Shanker and Eric Schmitt, "Pentagon Expects Long-Term Access to Four Key Bases," *New York Times*, April 20, 2003, http://www.nytimes.com/2003/04/20/international/worldspecial/20BASE.html.

247. With respect to North Korea's and Iran's nuclear ambitions, nearly all the world's nations agreed that they have to be curbed. "UN Security Council Demands that Iran Suspend Nuclear Activities," *UN News Centre*, July 31, 2006, http://www.un.org/apps/news/story.asp?NewsID=19353&Cr=iran&Cr1=, http://www.un.org/News/Press/docs//2006/sc8792.doc.htm.

248. Federation of American Scientists, "Iran and Libya Sanctions Act of 1996," House of Representatives, June 18, 1996, http://www.fas.org/irp/congress/1996_cr/h960618b.htm.

249. Energy Intelligence Group, "Pricing Differences Among Various Types of Crude Oil," in *The International Crude Oil Market Handbook (2004)*, E1, E287, E313, Updated July 2006, http://tonto.eia.doe.gov/ask/crude_types1.html.

250. "OPEC v Russia," *The Economist*, November 27, 2001, http://www.economist.com/research/backgrounders/displaystory.cfm?story_id=884790.

251. Find an analysis of historic oil prices at James L. Williams, "Oil Price History and Analysis," http://www.wtrg.com/prices.htm, http://news.bbc.co.uk/1/hi/world/3625207.stm.

252. The taxation of fuel varies widely among countries. Organization of the Petroleum Exporting Countries (OPEC), "Facts and Figures—Who Gets What from a Litre of Oil in the G7?," http://www.opec.org/home/PowerPoint/Taxation/taxation.htm.

BIBLIOGRAPHY TO PART IV

AEIOU, the Austrian cultural information system of the Federal Ministry for Education, Science and Culture (BMBWK). "Gräf & Stift." http://aeiou.iicm.tugraz.at/aeiou.encyclop.g/g638817.htm.

Agayev, Vagif, Fuad Akhundov, Fikrat T. Aliyev, and Mikhail Agarunov. "World War II and Azerbaijan." *Azerbaijan International* (3.2), Summer 1995. http://www.azer.com/aiweb/categories/magazine/32_folder/32_articles/32_ww22.html.

Amnesty International. "Iraq: Secretary General Challenges Powell." January 26, 2003. http://www.amnesty.org/en/library/asset/MDE14/002/2003/en/dom-MDE140022003en.html.

Amnesty International. "Kuwait-Search www.amnesty.org." http://www.amnesty.org/.

Austrian Press and Information Service (Washington, D.C.). "Siegfried Marcus-Mechanic and Inventor (1831–1898)." *Austrian Information*, Vol. 51, No. 8, (1998). http://www.austria.org/oldsite/aug98/marcus.html.

Baer, Robert. "The Fall of the House of Saud." *The Atlantic Monthly*, May 2003. http://www.theatlantic.com/doc/200305/baer.

Bairoch, Paul. *Economics and World History. Myths and Paradoxes.* Chicago: University Of Chicago Press, 1993.

Baker, Russ W. "Iraqgate-The Big One That (Almost) Got Away, Who Chased it—and Who Didn't." *Columbia Journalism Review* (March/April 1993). http://backissues.cjrarchives.org/year/93/2/iraqgate.asp.

Bamberg, J. H. *The History of the British Petroleum Company: Volume 2-The Anglo-Iranian Years, 1928–1954.* Cambridge: Cambridge University Press, 1994.

BBC News. "Iraq War Illegal, Says Annan." September 16, 2004. http://news.bbc.co.uk/1/hi/world/middle_east/3661134.stm.

BBC News. "Iraq 'will accept' UN Resolution." November 10, 2002. http://news.bbc.co.uk/1/low/world/middle_east/2433651.stm.

BBC News. "Laos: 'Most-heavily Bombed Place'." January 5, 2001. http://news.bbc.co.uk/2/hi/asia-pacific/1100842.stm.

BBC News. "Mandela Condemns US stance on Iraq." January 30, 2003. http://news.bbc.co.uk/2/hi/africa/2710181.stm.

BBC News Online. "Q&A: Saddam's Capture—BBC News Online Looks at Some of the Issues Arising from the Capture of Former Iraqi Leader Saddam Hussein." http://news.bbc.co.uk/1/hi/world/middle_east/3320289.stm.

BBC News. "Sellafield Towers are Demolished." September 29, 2007. http://news.bbc.co.uk/2/hi/uk_news/england/cumbria/7019414.stm.

BBC World Service. "The Story of Africa-Independence, Case Study: Congo." http://www.bbc.co.uk/worldservice/africa/features/storyofafrica/14chapter7.shtml.

Berril, Michael. *The Plundered Seas: Can the World's Fish Be Saved?* San Francisco: Sierra Club Books, 1997.

Bilmes, Linda, and Joseph Stiglitz. "The Economic Costs Of The Iraq War: An Appraisal Three Years After The Beginning Of The Conflict." Working Paper 12064, National Bureau of Economic Research, 2006. http://www2.gsb.columbia.edu/faculty/jstiglitz/download/2006_Cost_of_War_in_Iraq_NBER.pdf.

Blitzer, Wolf. "Did the Bush Administration Exaggerate the Threat from Iraq?" *CNN*, July 8, 2003. http://www.cnn.com/2003/ALLPOLITICS/07/08/wbr.iraq.claims/.

BP Statistical Review of World Energy. (Current edition.) http://www.bp.com/statisticalreview.

BP Statistical Review of World Energy June 2002. http://www.bp.com/downloads/1087/statistical_review.pdf. (No longer active.)

Bright, Chris. *Life Out of Bounds: Bioinvasion in a Borderless World.* New York: W.W. Norton, 1998.

Brimblecombe, Peter, and László Makra. "Selections from the History of Environmental Pollution, with Special Attention to Air Pollution. Part 2: From medieval times to the 19th century." *Int. J. Environment and Pollution* 23 (2005): 354. http://www.sci.u-szeged.hu/eghajlattan/makracikk/Brimblecombe%20Makra%20IJEP.pdf.

Brimblecombe, Peter. *The Big Smoke. A History of Air Pollution in London since Medieval Times.* London: Methuen, 1987.

Brown, Nathan J. "Iran-Iraq War." Microsoft Encarta Online Encyclopedia. http://encarta.msn.com/encyclopedia_761580640_2/Iran-Iraq_War.html.

Brzezinski, Zbigniew. *Out of Control: Global Turmoil on the Eve of the 21st Century.* New York: Touchstone, 1995.

Burden of Disease Unit, Center for Population and Development Studies at the Harvard School of Public Health. "Global Burden of Disease and Injury Series: Executive Summary." http://www.hsph.harvard.edu/organizations/bdu/summary.html. http://www.hsph.harvard.edu/organizations/bdu/gbdsum/gbdsum2.pdf.

Burke, Jason. "Secret World of US Jails." *The Observer*, June 13, 2004. http://observer.guardian.co.uk/international/story/0,6903,1237589,00.html.

Burkeman, Oliver, and Julian Borger. "The Ex-Presidents' Club." *The Guardian*, October 31, 2001. http://www.guardian.co.uk/wtccrash/story/0,1300,583869,00.html.

Carter, Jimmy. "State of the Union Address 1980." January 23, 1980. http://www.jimmycarterlibrary.org/documents/speeches/su80jec.phtml.

CBC News "Bin Laden 9/11 Planning Video Aired." September 7, 2006. http://www.cbc.ca/world/story/2006/09/07/al-qaeda-tape.html.

CBS News. "Plans For Iraq Attack Began On 9/11." September 4, 2002. http://www.cbsnews.com/stories/2002/09/04/september11/main520830.shtml.

CBS News. "We're Living Longer, Getting Fatter." September 12, 2002. http://www.cbsnews.com/stories/2003/03/10/health/main543315.shtml.

Center for Disease Control and Prevention, National Center for Health Statistics. http://www.cdc.gov/nchs.

Center for International Policy. "IMET: International Military Education and Training." http://www.ciponline.org/facts/imet.htm.

Chevron Learning Center. "A Petroleum Prospecting Primer." http://www.chevron.com/products/learning_center/primer.

CIA-The World Factbook. https://www.cia.gov/library/publications/the-world-factbook/.

CIA-The World Factbook. "United States." https://www.cia.gov/library/publications/the-world-factbook/geos/us.html.

Clark, Ramsey. *The Impact of Sanctions on Iraq: The Children Are Dying.* Farmington: Plough Publishing House, 1996.

Clarke, Richard A. *Against All Enemies: Inside America's War on Terror.* New York: Simon & Schuster, 2004.

CNN. "War in Iraq, Forces: U.S. & Coalition/Casualties." http://www.cnn.com/SPECIALS/2003/iraq/forces/casualties/2005.02.html.

Cohen, Daniel. *Fehldiagnose Globalisierung.* Frankfurt/New York: Campus Verlag, 1999.

Cohen, Joel E.. *How Many People Can the Earth Support?* New York: W.W. Norton & Company, 1995.

Cole, Leonard A. "Chemical and Biological Warfare." Microsoft Encarta Online Encyclopedia. http://encarta.msn.com/encyclopedia_761558349/Chemical_and_Biological_Warfare.html.

Coleman, James L. "The American Whale Oil Industry: A Look Back to the Future of the American Petroleum Industry?" *Natural Resources Research* Vol.4, 3 (1995). http://www.springerlink.com/content/7178248713276401/fulltext.pdf.

Cook, Earl. *Man, Energy, Society.* San Francisco: W.H. Freeman, 1976.

Cottrell, Fred. *Energy and Society: The Relation between Energy, Social Change, and Economic Development.* New York: McGraw-Hill Book Company, 1955.

Council On Foreign Relations. "Al-Qaeda." http://cfrterrorism.org/groups/alqaeda.html.

Council On Foreign Relations. "Osama bin Laden." http://cfrterrorism.org/groups/binladen.html.

Cumings, Bruce. "Korean War." Microsoft Encarta Online Encyclopedia. http://encarta.msn.com/encyclopedia_761559607_3/Korean_War.html#s14.

Daalder, Ivo, and James Lindsay, "Trust Clarke: He's Right about Bush." *The Globe and Mail*, March 26, 2004. http://www.theglobeandmail.com/servlet/ArticleNews/TPStory/LAC/20040326/COCLARKE26/TPComment/TopStories.

Dahl, Erik J. "Naval Innovation: From Coal to Oil." *Joint Force Quarterly* 27 (Winter 2000–01): 50–56. http://www.dtic.mil/doctrine/jel/jfq_pubs/1327.pdf. http://www.findarticles.com/p/articles/mi_m0KNN/is_2000_Winter/ai_80305799).

Daum, Werner. "Universalism and the West: An Agenda for Understanding." Harvard International Review-The Future of War, Volume 23, Issue 2, Summer 2001. http://hir.harvard.edu/articles/index.html?id=909&page=1.

Davies, Glyn. *A History of Money: From Ancient Times to the Present Day.* Cardiff: University of Wales Press, 1994.

Deffeyes, K.N. *Hubbert's Peak: The Impending World Oil Shortage.* Princeton: Princeton University Press, 2001.

DeLong, J. Bradford. "Slouching Towards Utopia?: The Economic History of the Twentieth Century, VII. The Pre-World War I Economy." University of California at Berkeley and NBER (National Bureau of Economic Research), January 1997. http://econ161.berkeley.edu/TCEH/Slouch_PreWWI7.html.

Denson, John V. *Reassessing the Presidency.* Auburn, AL: Ludwig von Mises Institute, 2001. http://books.google.com/books?id=hJGpAT7IWhwC&printsec=frontcover&dq=Reassessing+the+presidency.

Department of Defense. "Country Profiles-China, Ballistic Missile Defense Organization." April 1995. http://www.fas.org/nuke/guide/china/bmdo1995.pdf.

Department of Political Science, Oregon State University. "Some Examples of Portrayals of Information about Iraq." http://oregonstate.edu/Dept/pol_sci/fac/sahr/ps415/portray.pdf.

DeYoung, Karen. "Falling on His Sword: Colin Powell's Most Significant Moment Turned Out to Be His Lowest." *Washington Post*, October 1, 2006, p. W12. http://www.washingtonpost.com/wp-dyn/content/article/2006/09/27/AR2006092700106.html.

Dollar, David. "Globalization, Poverty, and Inequality since 1980." *World Bank Study* 3333 (2004). http://econ.worldbank.org/external/default/main?pagePK=64165259&piPK=64165421&menuPK=64166093&theSitePK=469372&entityID=000112742_20040928090739.

Doscher, Todd M. "Petroleum." Microsoft Encarta Online Encyclopedia. http://encarta.msn.com/encyclopedia_761576221_2____6/Petroleum.html#s6.

Drucker, Peter Ferdinand. *The Age of Discontinuity.* London: Heinemann, 1969.

Drucker, Peter Ferdinand. *Management: Tasks, Responsibilities, Practices.* New York: Harper & Row, 1973.

Drucker, Peter Ferdinand. *The Landmarks of Tomorrow.* New York: Harper, 1959.

Drucker, Peter Ferdinand. *The Post-Capitalist Society.* New York: HarperBusiness, 1993.

Duffy, Michael, ed. "Feature Articles: Military Casualties of World War One." First World War.com. http://www.firstworldwar.com/features/casualties.htm.

Duffy, Michael, ed., "Who's Who: Gavrilo Princip." First World War.com. http://www.firstworldwar.com/bio/princip.htm.

Economist, The. "The Accidental President." December 14, 2000. http://www.economist.com/displaystory.cfm?story_id=S%26%28X8%2DP13%2A%0A.

Economist, The. "Ageing: Parents, Telomeres and Lifespan." July 12, 2007.

Economist, The. "Air pollution: Driving Polluters from London." February 1, 2007. http://www.economist.com/world/britain/displaystory.cfm?story_id=E1_RGQGQSV.

Economist, The. "America in Iraq: After Fallujah—America Still Faces a War on Many Fronts." November 11, 2004.

Economist, The. "Arms Deals and Bribery: The Bigger Bang." June 14, 2007. http://www.economist.com/opinion/displaystory.cfm?story_id=9339826.

Economist, The. "BAE and Saudi Arabia: Bribery Alleged between the Kingdoms." June 7, 2007. http://www.economist.com/world/britain/displaystory.cfm?story_id=9312370.

Economist, The. "Big Oil's Biggest Monster." January 6, 2005. http://www.economist.com/displaystory.cfm?story_id=3545061.

Economist, The. "A Bitter Harvest: The Sufferings of Afghanistan Come to New York." September 13, 2001. http://www.economist.com/displaystory.cfm?story_id=780398.

Economist, The. "Chinese Aviation: On a Wing and a Prayer." February 23, 2006. http://www.economist.com/business/displaystory.cfm?story_id=E1_VVVSQRT.

Economist, The. "Energy Efficiency: The Elusive Megawatt." May 8, 2008. *http://*www.economist.com/displaystory.cfm?story_id=11326549.

Economist, The. "Energy Use: Less Intense." May 8, 2008. http://www.economist.com/daily/chartgallery/displaystory.cfm?story_id=11332762.

Economist, The. "Genetic Modification: Filling Tomorrow's Rice Bowl." December 9, 2006.

Economist, The. "The Global Car Industry: Extinction of the Predator." September 8, 2005. http://www.economist.com/business/displaystory.cfm?story_id=E1_QPGJSGN.

Economist, The. "Iraq: The Battle for Fallujah Now—and for Hearts and Minds Later." November 11, 2004.

Economist, The. "Iraq—Ever Bloodier: Despite the Insurgents' Efforts, the Election Looks Likely to Happen on Time." January 6, 2005.

Economist, The. "Iraq: The Perils of Imprecision—Whatever the Correct Figure, the Civilian Casualty Count Is Far Too High—and Rising." November 4, 2004. http://www.economist.com/displaystory.cfm?story_id=3353342.

Economist, The. "The Iraqi War: Counting the Casualties." November 4, 2004. http://www.economist.com/displaystory.cfm?story_id=3352814.

Economist, The. "Iraq's WMD—Secret Weapons: Did George Bush and Tony Blair Wage War under False Pretences?" May 29, 2003. http://www.economist.com/displaystory.cfm?story_id=1812369.

Economist, The. "Je ne texte rien." July 8, 2004. http://www.economist.com/displaystory.cfm?story_id=2908047.

Economist, The. "Oh, Sweet Reason: A Report Counts the Cost of Europe's Sugar Subsidies on Poor Countries." April 15, 2004. http://www.economist.com/displaystory.cfm?story_id=2599003.

Economist, The. "OPEC v Russia." November 27, 2001. http://www.economist.com/research/backgrounders/displaystory.cfm?story_id=884790.

Economist, The. "Osama bin Laden's Network: The Spider in the Web." September 20, 2001. http://www.economist.com/displaystory.cfm?story_id=788472.

Economist, The. "A Report from Congo: Africa's Great War." July 4, 2002. http://www.economist.com/displaystory.cfm?story_id=1213296.

Economist, The. "Saudi Arabia: Adapt or Die." March 4, 2004. http://www.economist.com/displaystory.cfm?story_id=2482168.

Economist, The. "Special Report: OPEC, Still Holding Customers over a Barrel." October 23, 2003. http://www.economist.com/displaystory.cfm?story_id=2155405.

Economist, The. "Today's New Cult Hero." August 27, 1998. http://www.economist.com/displaystory.cfm?story_id=162711.

Economist, The. "War, Coltan and Conservation: Digging a Grave for King Kong?" July 31, 2003. http://www.economist.com/displaystory.cfm?story_id=1956903.

Economist, The. "War and the Law in Iraq: Crime and Punishment." May 6, 2004. http://www.economist.com/displaystory.cfm?story_id=2656537.

Economist, The. "Weapons of Mass Destruction: Iraq's Elusive Weapons." May 29, 2003. http://www.economist.com/displaystory.cfm?story_id=1812042.

Ehrlich, Paul R., Anne H. Ehrlich. *The Population Explosion.* New York: Touchstone, 1991.

Encyclopædia Britannica, "Internet." http://www.britannica.com/EBchecked/topic/291494/Internet.

Encyclopaedia Britannica, 11th Edition (1911). "Petroleum." http://www.1911encyclopedia.org/Petroleum.

Energy Economics Research at the Bureau of Economic Geology, University of Texas at Austin. "Introduction to LNG." http://www.beg.utexas.edu/energyecon/lng/documents/Cee_Introduction_To_Lng_Final.pdf.

Energy Information Administration. (Statistical and analytical agency within the U.S. Department of Energy.) http://www.eia.doe.gov.

Energy Information Administration. "Crude Oil Production and Crude Oil Well Productivity, 1954–2007,." http://www.eia.doe.gov/emeu/aer/txt/ptb0502.html.

Energy Information Administration. Energy Basics 101. http://www.eia.doe.gov/basics/energybasics101.html.

Energy Information Administration. EIA Service Report: "Federal Energy Subsidies: Direct and Indirect Interventions in Energy Markets." DOE/EIA/SR/EMEU-92–02, November 20, 1992. Office of Energy Markets and End Use. http://tonto.eia.doe.gov/ftproot/service/emeu9202.pdf.

Energy Information Administration. "Energy in the United States: 1635–2000." http://www.eia.doe.gov/emeu/aer/eh/total.html.

Energy Information Administration. Energy Perspectives. (A graphical overview of energy history in the United States from 1949.) http://www.eia.doe.gov/emeu/aer/ep/ep_frame.html.

Energy Information Administration. "International Energy." Annual Energy Review. http://www.eia.doe.gov/aer/inter.html.

Energy Information Administration. "International Energy Outlook 2002, Highlights." http://www.eia.doe.gov/oiaf/archive/ieo02/index.html.

Energy Information Administration. "Saudi Arabia-EIA Country Analysis Briefs." http://www.eia.doe.gov/emeu/cabs/saudi.html.

Energy Information Administration. "US Crude Oil Field Production." http://tonto.eia.doe.gov/dnav/pet/hist/mcrfpus1A.htm.

Energy Information Administration. "U.S. Crude Oil Imports from All Countries." http://tonto.eia.doe.gov/dnav/pet/hist/mcrimus1A.htm.

Energy Information Administration. "U.S. Natural Gas Consumption by End Use." http://tonto.eia.doe.gov/dnav/ng/ng_cons_sum_dcu_nus_a.htm.

Energy Information Agency. "World Crude Oil Production, 1960–2007." http://www.eia.doe.gov/aer/txt/ptb1105.html.

Energy Information Administration. "World Net Electricity Generation by type." Office of Energy Markets and End Use, International Energy Statistics Team, Table posted on September 17, 2007. http://www.eia.doe.gov/emeu/international/RecentElectricityGenerationByType.xls.

Energy Intelligence Group. "Pricing Differences Among Various Types of Crude Oil." In *The International Crude Oil Market Handbook (2004)*, E1, E287, E313, Updated July 2006. http://tonto.eia.doe.gov/ask/crude_types1.html.

English, Simon. "The Best-connected Investor in America—The Hugely Profitable Carlyle Group Has Become a Magnet for Conspiracy Rumours." *Telegraph*, May 27, 2003. http://money.telegraph.co.uk/money/main.jhtml?xml=/money/2003/05/27/ccarly27.xml.

English, Simon. "Bush on the Board Not Worth Much, Says Carlyle Founder." *Telegraph*, July 8, 2003. http://www.telegraph.co.uk/finance/2857132/Bush-on-the-board-not-worth-much-says-Carlyle-founder.html.

Epstein, Edward. "How Many Iraqis Died? We May Never Know: Some Observers Are Pressuring Pentagon to Put Forth an Informed Estimate." *San Francisco Chronicle*, May 3, 2003. http://www.sfgate.com/cgi-bin/article.cgi?f=/c/a/2003/05/03/MN98747.DTL.

Federation of American Scientists (FAS). "India-Nuclear Weapons." http://www.fas.org/nuke/guide/india/nuke/index.html.

Federation of American Scientists (FAS). "International Military Education and Training (IMET)." http://www.fas.org/asmp/campaigns/training/IMET2.html.

Federation of American Scientists (FAS). "Iran and Libya Sanctions Act of 1996." House of Representatives, June 18, 1996. http://www.fas.org/irp/congress/1996_cr/h960618b.htm.

Federation of American Scientists (FAS). "Nuclear Propulsion." http://www.fas.org/man/dod-101/sys/ship/eng/reactor.html.

Federation of American Scientists (FAS). "Pakistan: Nuclear Weapons" http://www.fas.org/nuke/guide/pakistan/nuke/.

Federation of American Scientists (FAS). "Resolution 678 (1990)." http://www.fas.org/news/un/iraq/sres/sres0678.htm.

Federation of American Scientists (FAS). "Security Council Resolutions on Iraq." http://www.fas.org/news/un/iraq/sres/.

Flughistorische Forschungsgemeinschaft Gustav Weißkopf (FFGW). "Gustav A. Weißkopf." http://www.weisskopf.de/.

Food and Agriculture Organization (FAO) of the United Nations. *FAOSTAT* (FAO Statistical Databases). *Agricultural Data*. http://apps.fao.org/faostat/collections?version=ext&hasbulk=0&subset=agriculture.

Food and Agriculture Organization (FAO) of The United Nations. *FAOSTAT*. Time-series and cross sectional data relating to food and agriculture for some 200 countries. http://faostat.fao.org/default.aspx.

Food and Agriculture Organization (FAO) of the United Nations. "The State of Food Insecurity in the World 2006. Eradicating World Hunger-Taking Stock Ten Years After the World Food Summit." FAO Corporate Document Repository. http://www.fao.org/docrep/009/a0750e/a0750e00.htm.

Food and Agriculture Organization (FAO) of the United Nations. "Undernourishment Around The World. Counting the Hungry: Latest Estimates." The State of Food Insecurity in the World 2003. FAO Corporate Document Repository. http://www.fao.org/docrep/006/j0083e/j0083e03.htm.

Food and Agriculture Organization (FAO) of the United Nations. "Undernourishment Around the World: Depth of Hunger—How Hungry are the Hungry?" The State of Food Insecurity in the World 2000. http://www.fao.org/docrep/x8200e/x8200e03.htm.

Fortson, Danny. "Oil Fields of Plenty?" *The Independent*, February 1, 2008. http://www.independent.co.uk/news/business/analysis-and-features/oil-fields-of-plenty-776776.html.

Frank, Alison Fleig. "Galician California, Galician Hell: The Peril and Promise of Oil Production in Austria-Hungary." *Bridges* 10 (2006). http://www.ostina.org/content/view/1172/506/.

Frank, Alison Fleig. *Oil Empire: Visions of Prosperity in Austrian Galicia*. Cambridge: Harvard University Press, 2005.

Friends of the Earth. "Analysis of the Bush Energy Plan." http://www.foe.org/camps/leg/bushwatch/energyplan.pdf.

Gates to Jewish Heritage. "The Court Jews." http://www.jewishgates.com/file.asp?File_ID=53.

Gellatley, Juliet. "Murder, She Wrote." Viva! USA. http://www.vivausa.org/activistresources/guides/murdershewrote1.htm#fowlplay.

Ghabra, Shafeeq N. "Iraq's Culture of Violence." *The Middle East Quarterly* VIII (3) (2001). http://www.meforum.org/article/101.

Glancey, Jonathan. "Jaguar Halts Production of Venerable Daimler." *The Guardian*, March 2004. http://arts.guardian.co.uk/critic/feature/0,1169,1177372,00.html.

GlobalSecurity.org. "Tanker History." http://www.globalsecurity.org/military/systems/ship/tanker-history.htm.

Gonzalez, Henry B. "Kissinger Associates, Scowcroft, Eagleburger, Stoga, Iraq, and Bnl." Congressional Record, House of Representatives, April 28, 1992. http://www.fas.org/spp/starwars/congress/1992/h920428g.htm.

Goodman, Amy. "Interview with Eileen Welsome." May 5, 2004. http://www.democracynow.org/article.pl?sid=04/05/05/1357230.

Greenpeace. "The Subsidy Scandal—The European Clash Between Environmental Rhetoric and Public Spending." http://archive.greenpeace.org/~comms/97/climate/eusub.html.

"Gustave Whitehead's Flying Machines." http://gustavewhitehead.org/.

Hallinan, Conn. "Of Mice & Men." *San Francisco Examiner*, May 11, 2001. http://homepages.ihug.co.nz/~dcandmkw/ge/micemen.htm#top.

Haywood, John. *World Atlas of the Past, Vol. 1–4*. Oxford: Andromeda Oxford/ Oxford University Press, 1999. Danish edition: John Haywood, *Historisk Verdensatlas*. Köln: Könemann, 2000).

Herbert, Bob. "Ultimate Insiders." *New York Times*, April 14, 2003. http://www.commondreams.org/views03/0414-02.htm.

Herren, Ray V. *The Science of Animal Agriculture*. Albany: Delmar Publishers Inc., 1994.

Hersh, Seymour M. "A Case Not Closed." *The New Yorker*, November 1, 1993. http://www.newyorker.com/archive/content/?020930fr_archive02.

Herzog, Ze'ev. "Deconstructing the Walls of Jericho." October 29, 1999. http://mideastfacts.org/facts/index.php?option=com_content&task=view&id=32&Itemid=34.

Hirst, Chrissie. "Executive Summary: The Arabian Connection: The UK Arms Trade to Saudi Arabia." Written for CAAT (Campaign Against Arms Trade). http://www.caat.org.uk/information/publications/countries/saudi-arabia-intro.php.

Horton, Richard. "The War in Iraq: Civilian Casualties, Political Responsibilities." *The Lancet* 364 (9448) (2004). http://www.thelancet.com/journal/vol364/iss9448/contents.

House of Representatives. Joint Committee on Taxation, 105th Congress, 2nd Session, "Estimates of Federal Tax Expenditures for Fiscal Years 1999–2003 (JCS-7–98)." December 14, 1998. http://www.house.gov/jct/s-7-98.pdf. http://www.house.gov/jct/pubs_taxexpend.html.

Housen, Roger T. "Why Did The US Want To Kill Prime Minister Lumumba Of The Congo?" National Defense University. http://www.ndu.edu/library/n2/n025603K.pdf.

Hughes, Stephen. "The International Collieries Study." ICOMOS (International Council on Monuments and Sites) and TICCIH (The International Committee for the Conservation of the Industrial Heritage). International Council on Monuments and Sites, Paris, France. http://www.international.icomos.org/centre_documentation/collieries.pdf.

Hwang, Roland. "Money Down the Pipeline: The Hidden Subsidies to the Oil Industry." Union of Concerned Scientists. http://www.ucsusa.org/publication.cfm?publicationID=149.

Ignatius, David. "Bush's Fancy Financial Footwork." *Washington Post*, August 6, 2002, Page A15. http://www.washingtonpost.com/ac2/wp-dyn/A48301-2002Aug6.

International Court of Justice (The Hague). "Military and Paramilitary Activities in and against Nicaragua, Nicaragua v. United States, Decision of 27 June 1986." Collection of decisions, advisory opinions and rulings, page 14. Quoted in Monique Chemillier-Gendreau, "The International Court of Justice Between Politics and Law." *Le Monde diplomatique*, November 1996, Global Policy Forum. http://www.globalpolicy.org/wldcourt/icj.htm#fn1.

International Energy Agency (IEA). http://iea.org/. (The Paris-based IEA is an autonomous body established in November 1974 within the framework of the Organisation for Economic Co-operation and Development (OECD) to implement an international energy program.)

International Energy Agency. "Energy and Poverty." In *World Energy Outlook 2002*, p. 365. http://www.iea.org/textbase/nppdf/free/2000/weo2002.pdf.

International Energy Agency. "Key World Energy Statistics." http://www.iea.org/Textbase/publications/free_new_Desc.asp?PUBS_ID=1199.

International Energy Agency. "Key World Energy Statistics 2007." http://www.iea.org/textbase/nppdf/free/2007/key_stats_2007.pdf.

International Energy Agency. *World Energy Outlook 2006*. http://www.iea.org/textbase/nppdf/free/2006/weo2006.pdf.

Iraq Body Count. http://www.iraqbodycount.org/.

Japan Automobile Manufacturers Association (JAMA). "Japan's Auto Industry." http://www.jama.org/about/industry.htm.

Johnson, N.P., and J. Mueller. "Updating the Accounts: Global Mortality of the 1918–1920 "Spanish" Influenza Pandemic." *Bull Hist Med.* 76 (2002):105–15. http://www.ncbi.nlm.nih.gov/pubmed/11875246?dopt=Abstract.

Joint United Nations Programme on HIV/AIDS. http://www.unaids.org/.

Kapeller, H., C. Marschner, M. Weissenbacher, H. Griengl, "Synthesis of Cyclopentenyl Carbanucleosides via Palladium(0) Catalysed Reactions." *Tetrahedron* 54(8) (1998): 1439–1456. http://www.sciencedirect.com/science?_ob=ArticleURL&_udi=B6THR-3SH5D6B-1C&_user=10&_rdoc=1&_fmt=&_orig=search&_sort=d&view=c&_acct=C000050221&_version=1&_urlVersion=0&_userid=10&md5=a1e2f00d273c69e99ede1a89bbaaab92#bibl1.

Kaufman, Burton I. *The Oil Cartel Case: A Documentary Study of Antitrust Activity in the Cold War Era.* Westport, CT: Greenwood Press, 1978.

Kelly, Jack. "Estimates of Deaths in First War Still in Dispute." *Post-Gazette*, February 16, 2003. http://www.post-gazette.com/nation/20030216casualty0216p5.asp.

Keylor, William R. "World War I." Microsoft Encarta Online Encyclopedia. http://encarta.msn.com/encyclopedia_761569981/World_War_I.html#s1.

King, John. "Bush Calls Saddam 'the guy who tried to kill my dad'." *CNN*, September 27, 2002. http://www.cnn.com/2002/ALLPOLITICS/09/27/bush.war.talk/.

Klare, Michael T. "Mapping The Oil Motive." March 18, 2005. http://www.tompaine.com/articles/mapping_the_oil_motive.php.

Koplow, Doug, and John Dernbach. "Federal Fossil Fuel Subsidies and Greenhouse Gas Emissions: A Case Study of Increasing Transparency for Fiscal Policy." *Annual Review of Energy and the Environment* 26 (2001): 361–389. http://arjournals.annualreviews.org/doi/full/10.1146/annurev.energy.26.1.361.

Koplow, Doug, and John Dernbach. *Fueling Global Warming—Federal Subsidies to Oil in the United States.* http://archive.greenpeace.org/~climate/oil/fdsub.html.

Kurinsky, Samuel. "Siegfried Marcus, An Uncredited Inventive Genius." *Fact Paper* 32-I, Hebrew History Federation. http://www.hebrewhistory.info/factpapers/fp032-1_marcus.htm.

Lawrence University (Appleton, WI). "Human Rights Advocate Examines U.S. "School of the Assassins" in Lawrence Lecture." May 8, 2000, quoting a 1996 New York Times article on the School of the Americas. http://www.lawrence.edu/dept/public_affairs/media/release/9900/bourgeois.html.

Lefebvre, Jeffrey A. "The United States, Ethiopia and the 1963 Somali-Soviet Arms Deal: Containment and the Balance of Power Dilemma in the Horn of Africa." *The Journal of Modern African Studies* 36 (04) (1998): 611–643. http://journals.cambridge.org/bin/bladerunner?REQUNIQ=1078150879&REQSESS=48303&118000REQEVENT=&REQINT1=17499&REQAUTH=0.

Leigh, David, and Richard Norton-Taylor, "House of Saud Looks Close to Collapse." *The Guardian*, November 21, 2001. http://www.guardian.co.uk/afghanistan/story/0,1284,602854,00.html.

Library of Congress Country Studies. http://lcweb2.loc.gov/frd/cs/cshome.html. (Alternative site: http://countrystudies.us.) This site presents a comprehensive description and analysis of countries' historical setting, geography, society, economy, political system, and foreign policy. The country studies are on-line versions of books previously published in hard copy by the Federal Research Division of the Library of Congress as part of the Country Studies/Area Handbook Series sponsored by the U.S. Department of the Army between 1986 and 1998.

Library of Congress Country Studies. "Kuwait: Oil Industry." http://countrystudies.us/persian-gulf-states/21.htm.

Library of Congress Country Studies. "Mexico: Oil." http://countrystudies.us/mexico/78.htm.

Library of Congress Country Studies. *Persian Gulf States: A Country Study*, edited by Helen Chapin Metz. Washington: GPO for the Library of Congress, 1993. http://countrystudies.us/persian-gulf-states/.

Library of Congress Country Studies. "Persian Gulf States/ Kuwait: Treaties With The British." http://countrystudies.us/persian-gulf-states/13.htm or http://lcweb2.loc.gov/frd/cs/kwtoc.html#kw0015.

Library of Congress Country Studies. "Venezuela: -Energy—Petroleum." http://countrystudies.us/venezuela/30.htm.

Lilienthal, Otto. "Practical Experiments for the Development Of Human Flight." *The Aeronautical Annual* (Editor James Means), 7–20, 1896. http://invention.psychology.msstate.edu/i/Lilienthal/library/Lilienthal_Practical_Exp.html.

Linder, Doug. "An Introduction to the My Lai Courts-Martial." http://www.law.umkc.edu/faculty/projects/ftrials/mylai/Myl_intro.html.

Maugeri, Leonardo. "Not in Oil's Name," *Foreign Affairs* 82 (4) (2003). http://www.foreignaffairs.org/20030701faessay15412/leonardo-maugeri/not-in-oil-s-name.html.

Mazzetti, Mark, and Scott Shane, "Senate Panel Accuses Bush of Iraq Exaggerations." *New York Times*, June 5, 2008. http://www.nytimes.com/2008/06/05/washington/05cnd-intel.html?hp.

McNeill, John Robert. *Something New Under the Sun: An Environmental History of the Twentieth-century World*. New York: W. W. Norton & Company, 2001.

Melchior, Arne, Kjetil Telle and Henrik Wiig. "Globalisation and Inequality—World Income Distribution and Living Standards, 1960–1998." Norwegian Institute of International Affairs (NUPI), Royal Norwegian Ministry of Foreign Affairs, Studies on Foreign Policy Issues, Report 6B: 2000, October 2000. http://www.nupi.no/publikasjoner/boeker_rapporter/2004_2000/globalisation_and_inequality_world_income_distribution_and_living_standards_1960_1998.

Melosi, Martin V. "Automobile in American Life and Society: The Automobile Shapes The City." http://www.autolife.umd.umich.edu/Environment/E_Casestudy/E_casestudy.htm.

Microsoft Encarta Online Encyclopedia. "Ceramics." http://encarta.msn.com/encyclopedia_761565098/Ceramics.html.

Microsoft Encarta Online Encyclopedia. "Dresden." http://encarta.msn.com/encyclopedia_761552600/Dresden.html.

Middle East Research and Information Project (MERIP). "Palestine, Israel and the Arab-Israeli Conflict: A Primer." http://www.merip.org/palestine-israel_primer/toc-pal-isr-primer.html.

Milkov, Alexei V. "Global Estimates of Hydrate-bound Gas in Marine Sediments: How Much Is Really Out There?" *Earth-Science Reviews* Vol. 66, Issues 3–4 (2004): 183–197. http://www.sciencedirect.com/science?_ob=ArticleURL&_udi=B6V62-4BMTJ4G-1&_user=10&_rdoc=1&_fmt=&_orig=search&_sort=d&view=c&_version=1&_urlVersion=0&_userid=10&md5=c4a913455b5e27ed6935405cc476fced.

Moore, Michael. "What Fahrenheit 9/11 Says About The Relationship between the Families of George W. Bush and Osama Bin Laden." http://www.michaelmoore.com/warroom/index.php?id=5.

Morgan, Michael D., Joseph M. Moran and James H. Wiersma. *Environmental Science: Managing Biological & Physical Resources.* Dubuque: Wm. C. Brown Publishers, 1993.

Morse, Edward L., and James Richard. "The Battle for Energy Dominance." *Foreign Affairs* 81 (2) (2002). http://www.foreignaffairs.org/20020301faessay7969/edward-l-morse-james-richard/the-battle-for-energy-dominance.html.

Mount Holyoke College, MA. Vincent Ferraro, "Documents on the International Energy System." Includes "Achnacarry Agreement," "Red Line Agreement," "Draft Memorandum of Principles of 1934," "Memorandum for the Attorney General Relative to a Request for Grand Jury Authorization to Investigate the International Oil Cartel, June 24, 1952." http://www.mtholyoke.edu/acad/intrel/energy.htm.

Mountjoy, Alan B. "The Suez Canal at Mid-Century." *Economic Geography* 34 (2) (1958): 155–167. http://www.jstor.org/pss/142300.

Nader, Ralph. "Open Letter to Bill Gates." July 27, 1998. http://www.nader.org/index.php?/archives/1893-An-Open-Letter-to-Bill-Gates.html.

National Commission on Terrorist Attacks upon the United States (also known as the 9–11 Commission). http://govinfo.library.unt.edu/911/report/index.htm.

Natural Gas Supply Association. "Natural Gas: Uses In Industry." http://www.naturalgas.org/overview/uses_industry.asp.

National Security Archive of George Washington University. http://www.gwu.edu/~nsarchiv/NSAEBB/.

National Security Archive of George Washington University. "Brazil Marks 40th Anniversary of Military Coup: Declassified Documents Shed Light on U.S. Role." http://www.gwu.edu/~nsarchiv/NSAEBB/NSAEBB118/index.htm.

National Security Archive of George Washington University. "Chile and the United States: Declassified Documents relating to the Military Coup, 1970–1976." National Security Meeting on Chile Memorandum of Conversation, November 6, 1970. http://www.gwu.edu/~nsarchiv/NSAEBB/NSAEBB8/nsaebb8.htm.

National Security Archive of George Washington University. "CIA Acknowledges Ties to Pinochet's Repression." http://www.gwu.edu/~nsarchiv/news/20000919/.

National Security Archive of George Washington University. "CIA and Assassinations: The Guatemala 1954 Documents." National Security Archive Electronic Briefing Book No. 4. http://www.gwu.edu/~nsarchiv/NSAEBB/NSAEBB4/index.html.

National Security Archive of George Washington University. "Electronic Briefing Books: Latin America." http://www.gwu.edu/~nsarchiv/NSAEBB/index.html#Latin%20America.

National Security Archive of George Washington University. "The Kissinger Telcons." National Security Archive Electronic Briefing Book No. 123. http://www.gwu.edu/~nsarchiv/NSAEBB/NSAEBB123/index.htm#chile.

National Security Archive of George Washington University. "Kissinger To Argentines on Dirty War: 'The Quicker You Succeed the Better'." http://www.gwu.edu/~nsarchiv/NSAEBB/NSAEBB104/index.htm.

National Security Archive of George Washington University. "The Oliver North File." http://www.gwu.edu/~nsarchiv/NSAEBB/NSAEBB113.

National Security Archive of George Washington University. "The Oliver North File: His Diaries, E-Mail, and Memos on the Kerry Report, Contras and Drugs." http://www.gwu.edu/~nsarchiv/NSAEBB/NSAEBB113/index.htm#doc4.

National Security Archive of George Washington University. "Operation Condor: Cable Suggests US Role." http://www.gwu.edu/~nsarchiv/news/20010306/.

National Security Archive of George Washington University. "Saddam Hussein: More Secret History." http://www.gwu.edu/~nsarchiv/NSAEBB/NSAEBB107/.

National Security Archive of George Washington University. "Shaking Hands with Saddam Hussein: The U.S. Tilts toward Iraq, 1980–1984." The National Security Archive, *Electronic Briefing Book No. 82*, edited Joyce Battle, February 25, 2003. http://www.gwu.edu/~nsarchiv/NSAEBB/NSAEBB82/.

Nobel Foundation. http://nobelprize.org/.

Northwestern University. "A Chart of Bush Lies about Iraq—Did George W. Bush Invade Iraq by Lying?" http://faculty-web.at.northwestern.edu/music/lipscomb/BushLiesAboutIraq_BuzzFlash.htm.

Nottingham, Stephen. *Eat Your Genes—How Genetically Modified Food Is Entering Our Diet*. London: Zed Books Ltd., 1998.

Nouvel Observateur, Le. "The CIA's Intervention in Afghanistan." Interview with Zbigniew Brzezinski, President Jimmy Carter's National Security Adviser, January 1998. http://www.globalresearch.ca/articles/BRZ110A.html.

Nuclear Weapon Archive, The. "India's Nuclear Weapons Program—The Beginning: 1944–1960." http://nuclearweaponarchive.org/India/IndiaOrigin.html.

Nuclear Threat Initiative, The. "Iran—Chemical Chronology 1929–1983." http://www.nti.org/e_research/profiles/Iran/Chemical/2340.html.

Nye, David E. *Consuming Power: A Social History of American Energies*. Cambridge: MIT Press, 1999.

Observer, The. "Greed and Torture at the House of Saud." November 24, 2002. http://observer.guardian.co.uk/worldview/story/0,11581,846600,00.html.

O'Dwyer, William J. "The 'Who Flew First' Debate." *Flight Journal Magazine*, October 1998. http://www.flightjournal.com. http://findarticles.com/p/articles/mi_qa3897/is_199810/ai_n8815811.

Okeagu, Jonas E., Joseph C. Okeagu, Ademiluyi O. Adegoke, and Chinwe N. Onuoha, "The Environmental and Social Impact of Petroleum and Natural Gas Exploitation in Nigeria, *Journal of Third World Studies* (Spring 2006). http://findarticles.com/p/articles/mi_qa3821/is_200604/ai_n17179247/pg_1?tag=artBody;col1.

Organization of the Petroleum Exporting Countries (OPEC). "Facts and Figures: Who Gets What from a Litre of Oil in the G7?" http://www.opec.org/home/PowerPoint/Taxation/taxation.htm.

Osterfeld, David. "Voluntary and Coercive Cartels: The Case of Oil." *The Freeman: Ideas on Liberty* 37 (1987). http://www.fee.org/publications/the-freeman/article.asp?aid=2333.

Otto Lilienthal Museum. http://www.lilienthal-museum.de.

Oxford Research International. "National Survey of Iraq, February 2004." http://news.bbc.co.uk/nol/shared/bsp/hi/pdfs/15_03_04_iraqsurvey.pdf.

Palast, Greg. "Secret US Plans for Iraq's Oil," BBC Newsnight, March 17, 2005. http://news.bbc.co.uk/2/hi/programmes/newsnight/4354269.stm.

Paleontological Research Institution, Ithaca, NY. "Spindletop, Texas." History of Oil. http://www.priweb.org/ed/pgws/history/spindletop/spindletop.html.

Paleontological Research Institution, Ithaca, NY. "The Story of Oil in California." History of Oil. http://www.priweb.org/ed/pgws/history/signal_hill/signal_hill.html.

Parsons, Tim. "Updated Iraq Study Affirms Earlier Mortality Estimates." *The JHU Gazette*, October 16, 2006. http://www.jhu.edu/~gazette/2006/16oct06/16iraq.html.

Pennsylvania Historical and Museum Commission. "Edwin L. Drake and the Birth of the Modern Petroleum Industry." http://www.phmc.state.pa.us/ppet/edwin/page1.asp?secid=31.

PETA (People for the Ethical Treatment of Animals). http://www.peta.org/.

Petroleum History Society, The, 2001. "Canadian Petroleum Pioneers." http://www.petroleumhistory.ca/archivesnews/2001/01feb.html.

Pimentel, David, Laura Westra and Reed F. Noss, eds. *Ecological Integrity: Integrating Environment, Conservation and Health.* Washington: Island Press, 2000.

Pincus, Walter, and Dana Milbank. "The Iraq Connection. Al Qaeda-Hussein Link Is Dismissed." *Washington Post*, June 17, 2004. http://www.washingtonpost.com/wp-dyn/articles/A47812-2004Jun16.html.

Pomerantsev, A.P., N.A. Staritsin, Y.V. Mockov, L.I. Marinin. "Expression of Cereolysine ab Genes in Bacillus Anthracis Vaccine Strain Ensures Protection against Experimental Hemolytic Anthrax Infection." *Vaccine* 15(17–18) (1997):1846–1850.

Population Reference Bureau (PRB). "FAQ: World Population." http://www.prb.org/Journalists/FAQ/WorldPopulation.aspx.

Population Reference Bureau (PRB). "Population Growth." http://www.prb.org/Educators/TeachersGuPopulation Reference Bureau (PRB), ides/HumanPopulation/PopulationGrowth.aspx.

Powell, Colin L. "Remarks to the United Nations Security Council." New York City, February 5, 2003. http://www.state.gov/secretary/former/powell/remarks/2003/17300.htm.

Priest, Dana. "CIA Holds Terror Suspects in Secret Prisons." *Washington Post*, November 2, 2005. http://www.washingtonpost.com/wp-dyn/content/article/2005/11/01/AR2005110101644.html.

Princeton University. "Commemorating The 60th Anniversary of The Nanking Massacre, NANKING 1937." http://www.princeton.edu/~nanking/.

Public Broadcasting Service (PBS) Online. "The Rockefellers: People & Events: South Improvement Company / Cleveland Massacre, 1871–1872." and "The Rockefellers: The Ludlow Massacre." http://www.pbs.org/wgbh/amex/rockefellers/peopleevents/e_south.html. http://www.pbs.org/wgbh/amex/rockefellers/sfeature/sf_8.html.

Public Broadcasting Service (PBS). "Why the Towers Fell." NOVA. http://www.pbs.org/wgbh/nova/wtc/.

Reynolds, Paul. "US Ready to Seize Gulf Oil in 1973." *BBC News Online*, January 2, 2004. http://news.bbc.co.uk/1/hi/world/middle_east/3333995.stm.

Rhodes, Richard. *The Making of the Atomic Bomb.* New York: Simon & Schuster, 1986.

Richardson, Terry L. "Plastics." Microsoft Encarta Online Encyclopedia. http://encarta.msn.com/encyclopedia_761553604/Plastics.html.

Rosenberg, Rosalind. "Lecture Notes: American Civilization Since the Civil War." http://www.columbia.edu/~rr91/1052_2002/class_page/schedule_of_lectures_and_reading.htm.

Russell, Sabin. "Turning Point for HIV Infection Rate in Africa." *SF Chronicle*, November 21, 2006. http://sfgate.com/cgi-bin/article.cgi?f=/c/a/2006/11/21/BAGLTMHB0C12.DTL&type=health.

Sale, Richard. "Saddam Key in Early CIA Plot." *United Press International*, April 10, 2003. http://www.upi.com/print.cfm?StoryID=20030410-070214-6557r.

Sampson, Anthony. *The Seven Sisters.* New York: Viking Press, 1975.

San Joaquin Geological Society. "The History of the Oil Industry, with Emphasis on California and Kern County." http://www.sjgs.com/history.html.

Sanger, David E. "Pakistan Leader Confirms Nuclear Exports." *New York Times*, September 13, 2005, A10.

Sartor, Wolfgang. "International and Multinational and Multilateral Companies in the Russian Empire before 1914: The Integration of Russia into the World Economy," XIII Economic History Congress, Buenos Aires, July 2002, The International Economic History Association. http://eh.net/XIIICongress/cd/papers/43Sartor202.pdf.

Schatzker, Valerie. "Petroleum in Galicia." Drohobycz Administrative District. http://www.shtetlinks.jewishgen.org/drohobycz/history/petroleum.asp.

Schrödinger, Erwin. *What is Life?* Cambridge: Cambridge University Press, 1944.

Schwartz, Stephen. "The Dysfunctional House of Saud—Compromised by Terror, the Saudi Regime Will Have to Change or Die." *The Weekly Standard* 8 (46), August 18, 2003, Volume 00. http://www.weeklystandard.com/Content/Public/Articles/000/000/002/978uclzj.asp.

Science Museum, The. " New Materials: Plastics." *Making the Modern World.* http://www.makingthemodernworld.org.uk/stories/the_second_industrial_revolution/05.ST.01/?scene=6&tv=true.

Searight, Sarah. "Region of Eternal Fire—Petroleum Industry in Caspian Sea Region." *History Today* Vol.50, Issue 8 (2000): 45—51. http://www.historytoday.com.

Shanker, Thom, and Eric Schmitt. "Pentagon Expects Long-Term Access to Four Key Bases." *New York Times*, April 20, 2003. http://www.nytimes.com/2003/04/20/international/worldspecial/20BASE.html.

Shroder, John Ford. "Afghanistan." Encarta Online Encyclopedia. http://encarta.msn.com/encyclopedia_761569370_5/Afghanistan.html.

Sick, Gary. *October Surprise*. New York: Three Rivers Press, 1992.

Siebert, Detlef. "British Bombing Strategy in World War Two." August 1, 2001. http://www.bbc.co.uk/history/worldwars/wwtwo/area_bombing_05.shtml.

Simon Wiesenthal Center. "36 Questions About the Holocaust." http://motlc.wiesenthal.com/site/pp.asp?c=gvKVLcMVIuG&b=394663.

Simpson, Colin. *The Lusitania*. Boston: Little, Brown and Company, 1972.

Smil, Vaclav. *Essays in World History: Energy in World History*. Boulder, CO: Westview Press, 1994.

Smil, Vaclav. *Transforming the Twentieth Century: Technical Innovations and Their Consequences*. New York: Oxford University Press, 2006.

Stephens, Joe, and David B. Ottaway. "From U.S., the ABC's of Jihad, Violent Soviet-Era Textbooks Complicate Afghan Education Efforts." *Washington Post*, March 23, 2002. http://www.washingtonpost.com/ac2/wp-dyn/A5339-2002Mar22?language=printer.

Stern, N. *The Stern Review on the Economics of Climate Change*. Cambridge: Cambridge University Press, 2006. http://www.hm-treasury.gov.uk/stern_review_final_report.htm.

Sunshine Project, The. "An NGO Committed to Inform about the Dangers of New Weapons Stemming from Advances in Biotechnology. http://www.sunshine-project.org/.

Sunshine Project, The. "US Armed Forces Push for Offensive Biological Weapons Development." News Release, May 8, 2002. http://www.sunshine-project.org/.

Taylor, Robert E. *Scientific Farm Animal Production—An Introduction to Animal Science*. Upper Saddle River: Prentice Hall, 1995.

Telhami, Shibley, and Fiona Hill. "Does Saudi Arabia Still Matter? Differing Perspectives on the Kingdom and Its Oil." *Foreign Affairs* 81 (6) (November/December 2002). http://www.foreignaffairs.org/20021101faresponse10002/shibley-telhami-fiona-hill/does-saudi-arabia-still-matter-differing-perspectives-on-the-kingdom-and-its-oil.html.

Thompson, Eric V. "A Brief History Of Major Oil Companies In The Gulf Region." Petroleum Archives Project, Arabian Peninsula and Gulf Studies Program, University of Virginia. http://www.virginia.edu/igpr/apagoilhistory.html.

TNK. "History of the Oil Industry in the Soviet Union." http://www.tnk.com/company/history/industry.html.

"The Town of Petrolia Online." http://town.petrolia.on.ca/.

UN News Centre. "UN Security Council Demands that Iran Suspend Nuclear Activities." July 31, 2006. http://www.un.org/apps/news/story.asp?NewsID=19353&Cr=iran&Cr1=http://www.un.org/News/Press/docs//2006/sc8792.doc.htm.

United Nations Conference on Trade and Development (UNCTAD), "The Least Developed Countries Report, 2004." http://www.unctad.org/Templates/WebFlyer.asp?intItemID=3074&lang=1.

United Nations Department of Economic and Social Affairs. Population Division. Population Database. http://esa.un.org/unpp/.

United Nations Department of Economic and Social Affairs. Population Division. "World Population Prospects: The 2006 Revision." http://www.un.org/esa/pop

ulation/publications/wpp2006/wpp2006.htm. "Highlights." http://www.un.org/esa/population/publications/wpp2006/WPP2006_Highlights_rev.pdf.

United Nations Development Programme (UNDP). "Human Development Report 2004-Cultural Liberty in Today's Diverse World." 2004. http://hdr.undp.org/reports/global/2004/. http://hdr.undp.org/en/media/hdr04_complete.pdf.

United Nations General Assembly. "Elimination of All Forms of Racial Discrimination." UN General Assembly Resolution 3379 of November, 1975, Rescinded in Resolution 46/86 of December 1991. http://domino.un.org/UNISPAL.NSF/0/761c1063530766a7052566a2005b74d1?OpenDocument and http://www.un.org/documents/ga/res/46/a46r086.htm.

United Nations Organization. "The World at Six Billion." http://www.un.org/esa/population/publications/sixbillion/sixbilpart1.pdf.

United Nations Organization. "World Energy Requirements in 1975 and 2000." *Proceedings to the International Conference on the Peaceful Uses of Atomic Energy*, Geneva, 1955, Volume 1. New York: UNO, 1956.

United Nations Statistics Division (UNSD). Energy Statistics Database. http://unstats.un.org/unsd/energy/edbase.htm. Energy Statistics Yearbook. http://unstats.un.org/unsd/energy/yearbook/EYB_pdf.htm.

United States Centennial of Flight Commission. "Bombing During World War I." http://www.centennialofflight.gov/essay/Air_Power/WWI_Bombing/AP3.htm.

United States Centennial of Flight Commission. "Born of Dreams—Inspired by Freedom, Centennial of Flight Day: December 17, 2003." http://www.centennialofflight.gov/.

United States Centennial of Flight Commission. "Lilienthal—The "Flying Man."" http://www.centennialofflight.gov/essay/Prehistory/lilienthal/PH6.htm.

United States Centennial of Flight Commission. "The Zeppelin." http://www.centennialofflight.gov/essay/Lighter_than_air/zeppelin/LTA8.htm.

United States Department of Energy, "B Reactor-Hanford, Washington." http://www.energy.gov/about/breactor.htm.

United States Department of State. "Background Note: Saudi Arabia." Bureau of Near Eastern Affairs, September 2004. http://www.state.gov/r/pa/ei/bgn/3584.htm.

United States Geological Survey (USGS). Energy Resources Program. http://energy.usgs.gov/.

United States Navy. "Attack On The USS Cole." http://www.al-bab.com/yemen/cole1.htm.

United States Senate. Select Committee on Intelligence and US House Permanent Select Committee on Intelligence. "Reports of the Joint Inquiry into Intelligence Community Activities before and after the Terrorist Attacks of September 11, 2001." December 2002. http://www.gpoaccess.gov/serialset/creports/911.html.

United States Senate. Select Committee on Intelligence, 110th Congress, *Report on Whether Public Statements Regarding IRAQ by U.S. Government Officials Were Substantiated by Intelligence Information*, June 2008. http://intelligence.senate.gov/080605/phase2a.pdf. http://intelligence.senate.gov/080605/phase2b.pdf.

V2Rocket.com: The A-4/V-2 Resource Site. Photo of von Braun in SS uniform. http://www.v2rocket.com/start/chapters/vb-009.jpg.

Vienna Heeresgeschichtliches Museum (Museum of Military History). www.hgm.or.at.

Vogel, Hans Ulrich. "The Great Well of China.'" *Scientific American* 268 (1993): 116–121. http://www.sciamdigital.com/index.cfm?fa=Products.ViewIssuePreview&ARTICLEID_CHAR=169EA6F2-2CA5-4C50-8EBE-114-F3AF2991.

von Weizsäcker, Ernst Ulrich, Amory B. Lovins and L. Hunter Lovins. *Faktor vier. Doppelter Wohlstand—halbierter Naturverbrauch.* München: Droemersche Verlagsanstalt Th. Knaur, 1995, 1996. English language edition: Ernst Ulrich von Weizsäcker, Amory B. Lovins, and L. Hunter Lovins. *Factor Four. Doubling Wealth—Halving Resource Use.* London: Earthscan, 1997.

Weissenbacher, Manfred. "BSE Requires Stricter Feed Ban." *Toronto Star*, August 7, 2003. http://www.healthcoalition.ca/tstar-bse-aug7.pdf.

Weissenbacher, Manfred. *Rinderwahnsinn: Die Seuche Europas.* Vienna: Böhlau, 2001.

Wellspring Africa. "Hand Powered Percussion Drill." http://www.wellspringafrica.org/drildesc.htm.

Welsome, Eileen. *The Plutonium Files: America's Secret Medical Experiments in the Cold War.* New York: Dell Publishing, 2000.

White, Richard. *It's Your Misfortune and None of My Own: A History of the American West.* Norman: University of Oklahoma Press, 1991.

Williams, James L. "Oil Price History and Analysis." http://www.wtrg.com/prices.htm. http://news.bbc.co.uk/1/hi/world/3625207.stm.

Wooldridge, E.T. "Early Flying Wings (1870—1920)." http://www.century-of-flight.net/Aviation%20history/flying%20wings/Early%20Flying%20Wings.htm.

World Bank, The. http://www.worldbank.org/.

World Coal Institute. http://www.worldcoal.org/index.asp.

World Energy Council. (The World Energy Council is a London-based nongovernmental and noncommercial organization with official consultative status with the United Nations.) http://www.worldenergy.org/.

World Energy Council. "2007 Survey of Energy Resources (Executive Summary)." http://www.worldenergy.org/documents/ser2007_executive_summary.pdf. (Find the entire report at: http://www.worldenergy.org/publications/survey_of_energy_resources_2007/default.asp.)

World Health Organization. "AIDS epidemic update." UNAIDS/WHO, Joint United Nations Programme on HIV/AIDS (UNAIDS). Geneva: WHO, 2006. http://data.unaids.org/pub/EpiReport/2006/2006_EpiUpdate_en.pdf.

World Health Organization. *Cardiovascular Diseases.* Geneva: WHO, 2007. http://www.who.int/mediacentre/factsheets/fs317/en/index.html.

World Health Organization. "The Top 10 Causes of Death." Geneva: WHO, 2008. http://www.who.int/mediacentre/factsheets/fs310/en/.

World Health Organization. "The World Health Report 2002: Reducing Risks, Promoting Healthy Life." Geneva: WHO, 2002. http://www.who.int/whr/en/.

World Health Organization. "The World Health Report 2002: Reducing Risks, Promoting Healthy Life." "Annex Table 2 Deaths by cause, sex and mortality stratum in WHO Regions, estimates for 2001." Geneva: WHO, 2002. http://www.who.int/whr/2002/whr2002_annex2.pdf.

World Health Organization. "WHO Statistics, Mortality Database." "Table 1: Numbers of registered deaths, United States of America—1999." Geneva: WHO, 2000. http://www3.who.int/whosis/mort/table1.cfm?path=mort,mort_table1&language=english.

World Resources Institute. http://www.wri.org/.

World Resources Institute. "Fishery Production Totals (aquaculture and capture): Total for All Species." http://earthtrends.wri.org/searchable_db/index.php?theme=1&variable_ID=834&action=select_countries.

World Resources Institute. Searchable database "Agriculture and Food." http://earthtrends.wri.org/searchable_db/index.php?theme=8.

World Resources Institute. Searchable database "Coastal and Marine Ecosystems." http://earthtrends.wri.org/searchable_db/index.php?theme=1.

World Resources Institute. Searchable database "Energy and Resources." http://earthtrends.wri.org/searchable_db/index.php?theme=6.

World Resources Institute, United Nations Environment Programme, United Nations Development Programme, World Bank. "Diminishing Returns: World Fisheries Under Pressure." In *World Resources: A Guide to the Global Environment—Environmental Change and Human Health*, 195. New York/Oxford: Oxford University Press, 1998.

Würthle, Friedrich. *Dokumente zum Sarajevoprozeß. Ein Quellenbericht*, Mitteilungen des Österreichischen Staatsarchivs—Ergänzungsband 9. Horn: Verlag Berger, 1978. http://www.austria.gv.at/site/5212/default.aspx#a4.

Yale Law School. "The Versailles Treaty June 28, 1919: Part III." The Avalon Project—Documents in Law, History and Diplomacy, http://avalon.law.yale.edu/imt/partiii.asp.

Yergin, Daniel. *The Prize: The Epic Quest for Oil, Money & Power*. New York: Free Press, 1991.

Ziemke, Earl F. "World War II." Microsoft Encarta Online Encyclopedia. http://encarta.msn.com/encyclopedia_761563737/World_War_II.html.

Zucchino, David. "Army Stage-Managed Fall of Hussein Statue." *Los Angeles Times*, July 3, 2004. http://www.commondreams.org/headlines04/0703-02.htm.

PART V

Beyond the Oil Age

The future tends to be the harder to predict the longer we attempt to look ahead in time. Here is a notable exception. In about five billion years the sun will run out of fuel (hydrogen) and experience the fate typical for a star of its size: It will first expand and then explode and throw some of its matter into space, while a dense core will remain, dimly glowing for billions of years. For those worried that life on Earth will then lose its most essential source of energy, it is the wrong concern: our planet will at this stage cease to exist as it will be absorbed by the expanding sun.[1] However, in the more relevant, less distant future, the sun is fortunately expected to reliably keep providing energy to drive all life on Earth.[2] Solar radiation is also expected to play an increasingly important role in fueling the current human civilization in terms of technical energy. The transition will be slow, as we tend to prefer the concentrated energy provided by fossil fuels, which are still abundant. But concerns about both climate change caused by fossil fuel burning and potential supply shortages of oil-based liquid fuels have jump-started a trend towards energy diversification that in the long run may permit solar radiation to regain the central position it once held when it provided virtually all the energy that allowed for complex human civilizations to emerge and thrive during the Agricultural Age. The prime movers will be different, of course. Windmills and photovoltaic cells will provide electricity to run electric motors, and plant material will be converted into liquid or gaseous fuel for internal combustion engines or fuel cells. These technologies utilize current rather than ancient (fossil fuels) solar radiation, and there is plenty of it: Enough energy arrives from the Sun on the Earth's surface each minute to cover our total global energy needs for an entire year. We just need to learn how to harvest, concentrate, and store this energy. At the moment, we are not especially good at it. Renewables, except hydropower, comprise a minimal share of the total global energy mix, which is dominated by fossil fuels, and especially by oil, which accounts for nearly 40 percent of the total. But the Oil Age undoubtedly will come to an end, and the question remains what the next Energy Era will look like. What kind of energy source will define it? Which nations will become superpowers as they command most of this energy source? Will the transition into the new Energy Era be smooth, or will it involve severe hardships? Though there were signs that demand for oil has begun outstretching supply by the early 21st century, clear answers to these questions had yet to emerge. Much focus has been directed towards using existing energy more efficiently (with an efficiency revolution being the true new energy revolution); to turn to renewable forms of energy rather than to rely on those based on limited underground deposits; and to use energy sources that are available domestically to avoid dependence on imported fuel. The most pressing issue will be to come up with a substitute liquid fuel to serve the transportation sector. But energy for mobility, just like energy for stationary applications, now has to be evaluated in the context of climate change caused by fossil fuel burning.

NOTES

1. It has been speculated that our planet may escape this fate, because Earth will move away from the sun when it loses mass as an expanding red giant. However, the most recent models conclude that planet Earth will be engulfed anyway. And even if it was not, the planet would be fried, and lose its atmosphere, water, and life as we know it. K.-P. Schröder and Robert Connon Smith, "Distant Future of the Sun and Earth Revisited," *Monthly Notices of the Royal Astronomical Society* 386 (2008): 155–63, http://dx.doi.org/10.1111%2Fj.1365–2966.2008.13022.x.

2. Bacteria that thrive completely independent of oxygen and sunlight at hot deep-ocean vents, and their similar extremist cousins, are the one exception.

CHAPTER 36

THE PROBLEM OF CLIMATE CHANGE

When we burn fossil fuels, we release carbon dioxide that has slowly been removed from the atmosphere through plant growth millions of years ago. The carbon captured by this growth had been safely stored away underground until humans began mining and burning it in the Coal Age and the Oil Age. The problem is that even tiny changes around the current level of just 0.03 percent by volume (300 ppm) of carbon dioxide in the atmosphere may have a significant impact on the climate. Without carbon dioxide in the atmosphere, we would not be able to live on this planet. Wrapped around Earth like a blanket, carbon dioxide sensitively interacts with the heat that our planet is radiating into the cold space around it, and thus keeps the planet warm enough for life to flourish on its surface. However, atmospheric carbon dioxide is about much more than temperature: it is an essential part of life on Earth. All the carbon atoms in your body have previously been part of atmospheric carbon dioxide that was eventually captured by a growing plant. Removing carbon dioxide from the atmosphere would thus not only make the Earth a freezing place, it would also directly end life as we know it. But adding carbon dioxide to the atmosphere will have consequences, too.[3]

Towards the end of the Coal Age, in 1896, Swedish chemist Svante August Arrhenius (who in 1905 received a Nobel Prize) calculated how much the temperature on Earth would increase depending on the amount of carbon dioxide released into the atmosphere. According to his estimate, a doubling of the carbon dioxide concentration in the atmosphere would increase the global average temperature by about 5C to 6C, which is not too far from current estimates. To be sure, a difference of one degree Celsius in global average temperature is not to be confused with the temperature differences we experience between day and night or between seasons. In fact, if the global average temperature was 5C to 6C lower than it is now, it would be right at the level where it was during the last ice age 20,000 years ago, when large parts of Europe and North America were covered by a massive sheet of ice, several kilometers thick, while the sea level was 120 meters lower because so much water was bound as ice. The connection between carbon dioxide and the climate suggests that humanity, by burning more and more coal, oil, and natural gas, conducted an uncontrolled experiment on its own habitat. This was pointed out in 1957 by Austrian Hans Suess and American Roger Revelle, working at the Scripps Institute of Oceanography in California. The same year routine measurements of atmospheric carbon dioxide were started at the observatory on Mauna Loa, Hawaii. (This site is isolated enough to avoid interference of local carbon dioxide emissions with the measurements.) After just a few years, these measurements clearly demonstrated a gradual increase of carbon dioxide present in the atmosphere.[4] Scientists were alarmed, but there were no obvious or measurable signs that this increase in atmospheric carbon dioxide was indeed causing a change in the planet's climate.

ANALYZING ICE CORES

From 1987 the issue gained momentum, because scientists restored a full record of the planet's environmental temperature and atmospheric composition for the past 200,000 years by drilling ice cores out of the polar ice caps. This works because the polar ice caps are permanently replenished by snow falling onto them, while the lowest ice layers melt under the pressure of the weight above them. Ice cores drilled from a depth of about 2.5 kilometers are some 200,000 years old, while those drilled (in 1998) from a depth of 3,623 meters are over 425,000 years old. Each layer of the ice caps reveals the atmospheric conditions at the time of precipitation, because small bubbles of air are trapped within the layers of ice. Meanwhile the layers' water molecules themselves reveal the temperature at the time of precipitation according to their content of deuterium, the heavier version of hydrogen, and O-18, the heavier version of oxygen: This content varies sensitively with the oceans' temperature during condensation (when water entered the atmosphere before snowing down over the polar caps).

Ice core analysis showed that during the past 400,000 years atmospheric temperature and carbon dioxide concentration moved up and down in concert. The global average temperature varied considerably (by 5 C to 6 C), while the atmospheric carbon dioxide concentration moved from 180 ppm (at the coldest periods) to 280 ppm (at the warmest). This was initially considered quite sensational, but for two reasons these results were insufficient to jump-start a serious global climate protection movement in the 20th century. One was that a closer look at the ice core studies showed that changes in the atmospheric concentration of carbon dioxide lagged perhaps a thousand years behind the changes in temperature: The record thus revealed that historic carbon dioxide changes were the result of natural climate changes, not the cause of it.[5] The other issue was that atmospheric temperature and carbon dioxide concentration actually did *not* move in concert during the most recent 1,000 years. It was warmer in previous centuries than in the mid-Oil Age, even though carbon dioxide concentrations were much lower in pre-industrial times. Temperatures reached a peak in the Middle Ages, but then the climate began to cool despite increasing carbon dioxide emissions. By 1960, when the Coal Age had released a lot of carbon dioxide, and the Oil Age was in full swing, the atmospheric carbon dioxide concentration had reached 317 ppm (compared to 280 ppm around the year 1000), and by 2007 it had reached around 380 ppm (according to direct measurements on Hawaii), which is a lot higher than at any time during the past 400,000 years. Yet, temperatures in the early 21st century had hardly risen above medieval levels.[6]

WARM MIDDLE AGES—COLD FOSSIL FUEL AGE

The temperature record of the past two millennia restored from ice cores shows that average temperatures were initially quite stable, but slowly increased to reach a peak between about 800 C.E. and 1300 C.E.: this was the time when Viking farmers spread to Greenland and Iceland, and vineyards were thriving in England. The warming period was then interrupted by a kind of global cooling, with temperatures decreasing for several centuries, from about 1400 C.E. to 1850 C.E. The former period has traditionally been referred to as the Medieval Warm Period (though it was in fact part of a much longer warming era), while the latter period is often called the Little Ice Age, because rivers such as London's Thames would frequently freeze over in the winter-time. There is as yet no good explanation for these climate variations, but anthropogenic (human-caused) carbon dioxide emissions can certainly not be held accountable.

After 1850 the global average temperature remained low and relatively constant, but reached a new minimum around 1910, even though atmospheric carbon dioxide concentrations had now soared to a historic high of around 300ppm.[7] Thereafter both the global average temperature and atmospheric carbon dioxide rose fast. The carbon dioxide concentration increased steadily throughout the Oil Age, while the average global temperature rose towards medieval standards to reach a hitherto Oil-Age maximum in 1998, followed by a quite constant 10-year period 1998–2007, and 2008 being the coldest year since 2000.[8]

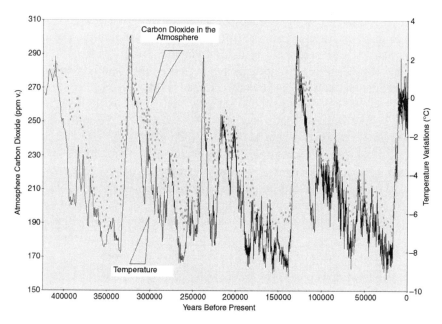

The Past Four Ice Age Cycles: Atmospheric Temperature and Carbon Dioxide Content as Retrieved from Vostok Ice Cores Not that sensational. The close correlation between temperature levels (shown is the temperature difference with respect to the mean recent time value) and atmospheric carbon dioxide concentration as derived from Antarctic ice cores drilled at Vostok was initially dubbed the Vostok sensation. However, as the graph reveals, atmospheric carbon dioxide concentrations were the *result* of climate change, *not* the cause of it. Various studies have shown that carbon dioxide in the atmosphere adjusts with a delay of about 800 years or more to natural climate variations. Very likely the oceans are behind this. They are the planet's largest carbon reservoir, holding about twenty times as much carbon as the continents, and forty times as much as the atmosphere. The oceans slowly release carbon dioxide into the atmosphere when they are warming, and absorb carbon dioxide when they are cooling. (Based on data in J.R. Petit et al., "Climate and Atmospheric History of the Past 420,000 years from the Vostok Ice Core, Antarctica," *Nature* 399 (1999), 429–436, IGBP PAGES/World Data Center for Paleoclimatology Data Contribution Series #2001–076, NOAA/NGDC Paleoclimatology Program, Boulder, CO, ftp://ftp.ncdc.noaa.gov/pub/data/paleo/icecore/antarctica/vostok/deutnat.txt, and J.M. Barnola et al., "Historical CO_2 Record from the Vostok Ice Core," January 2003, Carbon Dioxide Information Analysis Center (CDIAC), http://cdiac.ornl.gov/ftp/trends/co2/vostok.icecore.co2.

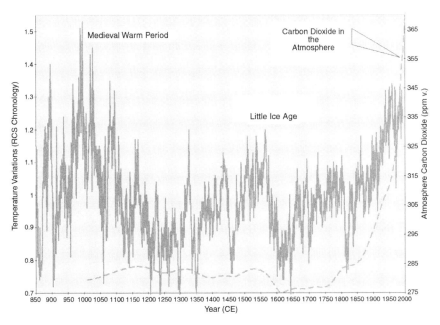

A Northern Hemisphere Temperature Reconstruction from the Year 800 C.E. If atmospheric carbon dioxide from fossil fuel burning is causing warm temperatures in the Oil Age, what caused them during the Agricultural Age? Temperatures in the Middle Ages were around the same level as they were at the turn of the 21st century. In medieval times vineyards flourished in England and Scotland, and Viking farmers moved to Greenland. This was followed by a cooler period, referred to as the Little Ice Age, that lasted to about 1850. Was part of the temperature increase in the 20th century a natural upward adjustment following this long cool period? (Graph based on northern hemisphere extratropical tree ring temperature reconstruction. Presented is the Regional Curve Standardization (RCS) Chronology data from J. Esper, E.R. Cook, and F.H. Schweingruber, "Low-Frequency Signals in Long Tree-Ring Chronologies for Reconstructing Past Temperature Variability," *Science* 295, 5563 (2002): 2250–2253, http://www.sciencemag.org/cgi/content/short/295/5563/2250, IGBP PAGES/World Data Center for Paleoclimatology Data Contribution Series #2003–036, NOAA/NGDC Paleoclimatology Program, Boulder, CO, ftp://ftp.ncdc.noaa.gov/pub/data/paleo/treering/reconstructions/n_hem_temp/esper2002_nhem_temp.txt. Carbon dioxide data for 1010 C.E. to 1964 C.E. from D.M. Etheridge, et al., "Historical CO_2 record derived from a spline fit (75 year cutoff) of the Law Dome DSS, DE08, and DE08-2 ice cores," June 1998, Carbon Dioxide Information Analysis Center (CDIAC), http://cdiac.ornl.gov/ftp/trends/co2/lawdome.smoothed.yr75. Carbon dioxide data for 1965 C.E. to 2007 C.E. from R. F. Keeling, et al., "Atmospheric CO_2 values (ppmv) derived from in situ air samples collected at Mauna Loa, Hawaii, USA," May 2008, Carbon Dioxide Information Analysis Center (CDIAC), http://cdiac.ornl.gov/ftp/trends/co2/maunaloa.co2.

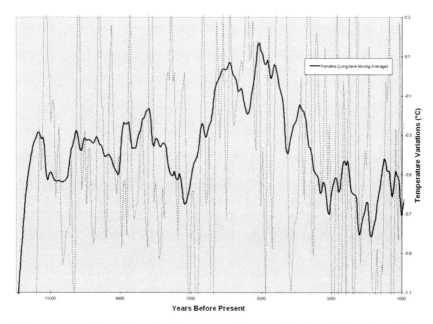

A Temperature Record for the Past 12,000 Years as Retrieved from Vostok Ice Cores During the early Agricultural Age it was even warmer than in the Middle Ages. The temperature maximum that lasted for a few millennia around 3000 B.C.E. probably promoted the emergence of the early, complex, grain-based civilizations of high population densities. (Based on data from J.R. Petit et al., "Climate and Atmospheric History of the Past 420,000 years from the Vostok Ice Core, Antarctica," *Nature* 399 (1999): 429–436, IGBP PAGES/World Data Center for Paleoclimatology Data Contribution Series #2001-076, NOAA/NGDC Paleoclimatology Program, Boulder, CO, ftp://ftp.ncdc.noaa.gov/pub/data/paleo/icecore/antarctica/vostok/deutnat.txt.)

The temperature increase during the Oil Age looks especially large, because it set out at the historic global-temperature minimum of several millennia, which had been reached in the late Coal Age. The United Nations' Intergovernmental Panel on Climate Change (IPCC) stated that the global average surface temperature increase over the 100-year period from 1906–2005 was 0.74 C, and that eleven of the twelve years 1995 to 2006 ranked among the 12 warmest years in the instrumental record of global surface temperature. (The instrumental temperature record refers to the time period since people started to use more or less accurate thermometers to routinely track environmental temperatures. In the northern hemisphere, this began around 1850. Earlier climate records exist as observations in written accounts from the earliest times of literacy.) However, even though global temperatures have been approaching medieval levels towards the end of the 20th

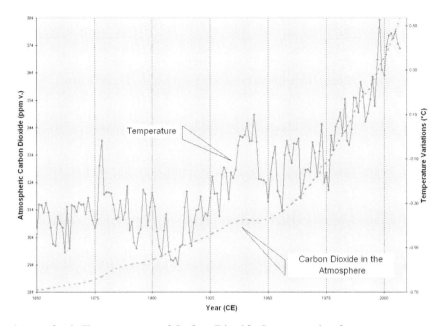

Atmospheric Temperature and Carbon Dioxide Concentration from 1850 to 2007 as Retrieved from Vostok Ice Cores (carbon dioxide concentrations from 1965 from direct atmosphere measurements on Hawaii) After the Little Ice Age temperatures started to rise again in the mid-Coal Age. The 20th century witnessed four decades of warming, followed by three decades of cooling, followed by three decades of warming. Why did the temperature rise so much in the early 20th century, when carbon dioxide emissions were much smaller than in the later 20th century? (There is now much agreement that the sun, not greenhouse gas, was in fact behind this warming. See the main text under solar activity and climate models that combine the effects of greenhouse gas, solar activity, and particles in the atmosphere.) And why did temperatures fall from 1940 until around the time of the First Oil Price Shock, despite the strong rise in carbon dioxide emissions during the post-World War II economic boom? (Again, factors other than greenhouse gases will have to serve for an explanation.) Finally, during the past three decades of the 20th century, atmospheric carbon dioxide, rather than solar output, is supposed to have been the major driver of climate changes. At least temporarily, the warming seems to have stopped in 1998: no year has been warmer than 1998 in terms of average global surface temperature in the 11-year period 1998–2008. (Based on data from P. D. Jones, et al., "Global Monthly and Annual Temperature Anomalies (degrees C), 1850–2007 (Relative to the 1961–1990 Mean)," September 2008, Carbon Dioxide Information Analysis Center (CDIAC), http://cdiac.ornl.gov/ftp/trends/temp/jonescru/global.dat; carbon dioxide data for 1850 C.E. to 1964 C.E. from D.M. Etheridge, et al., "Historical CO_2 record derived from a spline fit (20 year cutoff) of the Law Dome DE08 and DE08-2 ice cores," June 1998, Carbon Dioxide Information Analysis Center (CDIAC), http://cdiac.ornl.gov/ftp/trends/co2/lawdome.smoothed.yr20; carbon dioxide data for 1965 C.E. to 2007 C.E. from R. F. Keeling, et al., "Atmospheric CO_2 values (ppmv) derived from in situ air samples collected at Mauna Loa, Hawaii, USA," May 2008, Carbon Dioxide Information Analysis Center (CDIAC), http://cdiac.ornl.gov/ftp/trends/co2/maunaloa.co2.

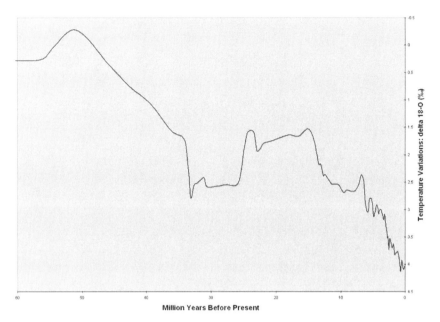

The Very Big Picture: The Global Climate of the Past 60 million Years, Reconstructed from Sediment Cores   It was a lot warmer than it is now before humans were around. For instance, 55 million years ago, when dinosaurs had become extinct and mammals were on the rise, the Earth's North Pole had no ice at all, and the ocean there was as warm as 24 degrees C. Gradual trends of warming and cooling have been driven by tectonic processes, orbital processes, and rare rapid aberrant shifts and extreme climate transients. (James Zachos et al., "Trends, Rhythms, and Aberrations in Global Climate 65 Ma to Present," *Science* 292 (5517) (2001): 686–93, http://www.sciencemag.org/cgi/content/abstract/292/5517/686, http://pangea.stanford.edu/research/Oceans/GES206/readings/Zachos2001.pdf, http://geosci.uchicago.edu/~archer/classes/GeoSci238/zachos.2001.since_65M.pdf. Graph based on data from Jan Veizer, "Isotope Data," http://www.science.uottawa.ca/~veizer/isotope_data/. It is estimated that each 1 part per thousand change in $\delta 18O$ represents roughly a 1.5–2 °C change in tropical sea surface temperatures. J. Veizer, Y. Godderis, and L.M. Francois, "Evidence for decoupling of atmospheric CO_2 and global climate during the Phanerozoic eon," *Nature* 408 (2000): 698–701. See also: J. Veizer, et al. "87Sr/86Sr, $\delta 13C$ and $\delta 18O$ evolution of Phanerozoic seawater," *Chemical Geology* 161 (1999): 59–88.)

century, the trend was not the same in all regions. In 2007 NASA corrected its U.S. temperature records: The hottest year on record was no longer 1998, as previously stated, but 1934, the year the infamous Dust Bowl devastated the Midwest.[9] Half of the 10 warmest years on record in the United States (where temperatures had been quite well recorded since 1880) were between 1920 and 1939, only one of the ten was in the 21st century, and the 15 hottest years were spread across seven decades, apparently indicating a natural variation rather than a warming trend. (The year 1934 was the hottest one on record, 1998 the second hottest, 1921 in third hottest, 2006 the fourth hottest, followed by 1931, 1999, 1953, 1990, 1938 and 1939.[10]) In short, by the early 21st century it was not possible to prove that the increased carbon dioxide concentration had indeed changed the climate. No substantial change in weather patterns had been observed that was significant in comparison to weather fluctuations experienced in the past. But this did not mean that the radically increased atmospheric carbon dioxide concentrations would not change the climate in the future. Or would they?

MODELING THE CLIMATE

To clarify this question, scientists put together elaborate computer programs to model the climate system. This could draw upon various elements used in weather forecasting, and was also of great interest for military strategists, who would like to know all about the (weather-dependent) geographic distribution of radioactive material entering the upper atmosphere after nuclear bombings.

The climate system is, however, extremely complex and not entirely understood. This is true for the system as a whole as well as for the various interacting carbon reservoirs and the feedbacks that follow changes of atmospheric carbon dioxide or the atmospheric temperature. There are quick feedbacks, slow feedbacks, some are enhancing the initial warming effect of added CO_2, others are slowing or reducing it. Many, but not all, plant species, for instance, respond to higher atmospheric CO_2 concentrations by accelerated growth, which removes some additional CO_2 from the atmosphere. (This effect is utilized in commercial greenhouses that are purposely filled with carbon dioxide.) On the other hand, rising temperatures may reduce the growth of plankton in the oceans, because less water and nutrients from deeper layers would reach the warmed top layers of the oceans. Similarly, more CO_2 is expected to be absorbed by the oceans if the concentration of carbon dioxide in the atmosphere increases (according to Henry's Law[11]), but warmer water would release more CO_2, just like carbonated lemonades become 'flat' when left in an open glass outside the refrigerator. (That's also why the Vostok sensation is not all that sensational: in a warming climate the oceans would release more carbon dioxide, while a cooling of the oceans would allow for more CO_2 absorption. Atmospheric carbon dioxide concentrations would thus be expected to follow temperatures changes in the same direction.) A related question is how much water will evaporate from the (warmed) oceans under new conditions. This is highly critical, because water vapor in the atmosphere works as a greenhouse gas similar to carbon dioxide: the atmosphere may contain up to two percent of water, with water vapor constituting up to 95 percent of all greenhouse gas in the atmosphere and thus exhibiting a much greater effect than carbon dioxide. (Greenhouse gas refers to the property of a gas to let incoming solar radiation pass through, while capturing outgoing heat radiation—just like the glass walls of a greenhouse.) An initial warming would in this respect result in a further warming, but, on the other hand, additional water vapor in the atmosphere may also increase cloud formation. And this is where climate modeling gets really tricky.[12] Clouds may either enhance or reduce the initial warming: On the one hand, they shelter the planet from solar radiation (sending it straight back into space); on the other hand, they also keep heat in the atmosphere. (Clear nights tend to be colder than cloudy nights.) Which effect predominates is critical, because this is a major feedback with strong influence on the overall climate result. The question is subject to ongoing discussions, with the type and altitude of clouds determining which effect outweighs the other. (In general, low-altitude clouds are assumed to result in a cooling, clouds at higher altitudes in a warming effect.) And it is not known how much of the water vapor present in the atmosphere will end up being bound in clouds, because the mechanisms behind cloud formation are not well understood. This uncertainty then turns into a major modeling problem, because a change of but a few percent in cloud cover is comparable to the expected changes due to a doubling of the atmospheric CO_2 concentration (excluding the consequent feedbacks).

The system becomes even more complicated when computer models are expected to account for small particles (from 0.001 to 0.01 mm in diameter) present in the atmosphere. These are cooling the climate, because they scatter solar radiation straight back into space like clouds, but they also act as seeds to promote cloud formation. Naturally occurring particles include dust, sea salt crystals, spores, and so on, plus the large quantities of particles that are occasionally injected into higher atmosphere by volcanic eruption to cool the global climate significantly for several years. Anthropogenic sources of similar particles include fossil fuel burning, which generates soot and sulfate particles (which are actually droplets) resulting in a regional cooling effect concentrated around industrial regions.

Various Greenhouse Gases

However, fossil fuel burning is mainly associated with the release of climate-warming carbon dioxide, and so is the burning of forests to create fields or residential areas, though deforestation (and overgrazing) also increases the planet's general reflectivity, which is a cooling effect.[13] There are also several other greenhouse gases that are released due to anthropogenic activities. Methane (natural gas) acts as a far more potent greenhouse gas than carbon dioxide (in fact, it has 23 times the warming potential), but is present in the atmosphere in much lower concentration, even though it has more than doubled since pre-industrial times. Methane is mainly released by natural wetlands, while anthropogenic sources include rice paddy fields, leakage from natural gas pipelines and from oil wells, decay of rubbish in landfills, wood and peat burning, and enteric fermentation (belching) from cattle and other livestock. In fact, the livestock sector by itself is responsible for 37 percent of anthropogenic methane emissions. All in all, animal farming produces 18 percent of the world's combined greenhouse gas emissions, a lot more than global transport fuels, which account for 13.5 percent.[14] The livestock sector accounts for 65 percent of anthropogenic nitrous oxide (N_2O), for instance, which is 300 times as potent a greenhouse gas as carbon dioxide. (This potency makes nitrous oxide a very relevant greenhouse gas, even though it is present in the atmosphere at levels no more than a fifth above pre-industrial concentrations.) Sources of nitrous oxide include agriculture (fertilizers), fossil fuel burning, deforestation, and the chemical industry (nylon production), while animal farming is the worst polluter (mainly from manure). Chlorofluorocarbons, produced for use in refrigerators and in insulation manufacturing, are even more potent greenhouse gases than nitrous oxide: They have up to 10,000 times the effect of CO_2 on a per molecule basis. Chlorofluorocarbons have largely been phased out, as it was realized that they destroy the stratosphere's thin ozone layer (they created the so-called ozone hole), but they remain in the atmosphere between 100 and 200 years. For comparison, a nitrous oxide molecule survives some 120 years, the average lifetime of methane in the atmosphere is about 12 years, and an average CO_2 molecule is believed to stay in the atmosphere for no more than four years before being absorbed by some carbon reservoir such as the continental or oceanic biosphere, or the top level of oceans.

The Vast Size of the Oceans

There is actually an extreme difference in size between the carbon reservoirs in the oceans (ca. 40,000 Gt [gigatons] of carbon), on land (ca. 2,200 Gt carbon), and in the atmosphere (less than 1,000 Gt). This, too, is a challenging issue for climate modelers. If the atmospheric CO_2 concentration were to double, the oceans would have to absorb no more than an extra two percent of the amount of carbon dioxide they already store in order to get the atmosphere back to normal. On the other hand, if conditions were to change in a way that would cause the oceans to release just a tiny part of their CO_2 content, this would alter the atmospheric CO_2 concentration by a lot. But again, things are complicated by a timeline question: If carbon dioxide is added to the atmosphere, the oceans' upper layers may become saturated with dissolved CO_2 quite fast, while it takes a long time to pass this CO_2 on to lower, cooler ocean layers. (If it takes too long, the top layer may warm up and release the captured CO_2 into the atmosphere again.) In fact, it takes centuries for the lower ocean layers to reach the CO_2 equilibrium expected according to the gas concentration present in the atmosphere.

Compared to the atmosphere, the oceans are also very large in terms of their heat capacity. The fact that a large quantity of heat is needed to raise the temperature of the oceans by just a bit adds considerable inertia to the climate system as a whole. Accordingly, the thermal expansion of the oceans is slow, but it is substantial. In a warming climate, this expansion would eventually contribute more to sea level rise than all the melting ice from the polar caps and all glaciers combined. (To be sure, the shrinkage of glaciers and the polar ice caps would result in a further warming, because it reduces the planet's general reflectivity.) Only ice melting from landmasses such as Antarctica and Greenland will actually raise the sea level, while a melting of the floating Arctic ice cap around the north pole does not. Nevertheless, melting glaciers may contribute more to sea level rise in the short run, as the ocean temperature changes much more slowly than the atmosphere, on the order of several decades. The extent of sea level rise due to the oceans' thermal expansion is in turn not all that easy to calculate, because the extent to which warming water expands depends on its previous temperature. At zero degrees the thermal expansion of water is negligible, while thermal expansion at 25 C is three times larger than at 5 C.

Fitting the Models

As computer programs modeling the climate have to take so many interacting factors, feedbacks, and timelines into account, they are large and complex, and consume weeks of calculation time for single runs (scenarios) even on the world's fastest supercomputers. What is more, a lot of factors remain unknown, and various assumptions have to be made. Many of the factors involved are fitted to give realistic outcomes. Numerical values that describe the exchanges of water, heat, and momentum at the interfaces between the oceans, the continents, and the atmosphere, for instance, are adjusted until programs fit what is observed in nature. This way models can simulate the climatic effects of such past events as the volcanic eruption of Mount Pinatubo in the Philippines in 1991 (which cut solar radiation by 2 percent due to particle emission and resulted in a significant decrease in global average temperature during the following two years), or anomalies associated with El Nino episodes (that is, the abnormal warming of the surface waters of the eastern equatorial Pacific Ocean that occurs at irregular intervals of 2–7 years). This has established a certain credibility, but the fitting approach does not necessarily lead to perfectly correct models, because the same result can be achieved in different ways in a complex system. Besides, some controversy remains around various critical factors that introduce a lot of uncertainty, because they sensitively change modeling results by a lot even if they are themselves changed by only a tiny bit. In this context, a lot of attention has been directed towards cloud formation, where assumptions have to be made on the level of humidity required for clouds to form, the speed at which raindrops fatten, the ease with which moist air in the tropics travels into the upper atmosphere, and the like. All of these contribute to the type, location, and durability of clouds assumed by computer models to be present in the atmosphere, with a strong impact towards climate cooling or climate warming.

Once computer models are deemed fit to describe past or present climate conditions, things certainly do not get easier when they are supposed to model the future. Now the human factor comes into play, and human behavior is notoriously difficult to predict. Assumptions have to be made on population growth, economic development, willingness to use less or more expensive energy, and perhaps even willingness to change various life-style features, including nutrition. In the end, this is not called forecasting or predicting, but rather scenario modeling, with results being presented as a wide range of different projections.

Results of Computer Modeling and Consequences of Climate Change

In its Third Assessment Report, "Climate Change 2001," for instance, the Intergovernmental Panel on Climate Change, IPCC, established by the United Nations to assess scientific, technical, and socio-economic information relevant for the understanding of climate change, stated that the average surface temperature is projected to increase by 1.4 to 5.8 Celsius degrees over the period 1990 to 2100, depending on different emissions scenarios. (Remember, a difference in average surface temperature by 5 to 6 C is equivalent to the difference between a full blown Ice Age and current temperatures.) If that is correct, even in the minimum scenario the warming in the 21st century would be twice as fast as the observed warming during the 20th century.

The range of projections is then delivered with possible consequences, which are also uncertain. For instance, what will happen to fresh water resources, agriculture and food supply, natural ecosystems, and human health, if the global average temperature rises by 2 C? Nobody knows, but various general trends are expected in response to global warming. Due to changes in the global water cycle, rainfall patterns would be expected to be altered, with some regions experiencing less rainfall to deliver water for agriculture, industry, and residential use. Fortunately, there is no strong evidence that the effect of climate change on global food supply would be large. It should not be all that difficult to match crops to changing climatic conditions, simply because we have so much knowledge on crops currently grown under widely varying climatic conditions all over the world. In some regions, such as parts of Siberia and northern Canada, the growing season would be expected to become longer. Around the world, the growth of so-called C_3 plants, including wheat, rice, and soy bean, will be promoted by higher atmospheric carbon dioxide concentrations, as this enables such plants to fix carbon at a higher rate. (The effect is less relevant to C_4 plants including maize, sorghum, sugarcane, millet, and many pasture and forage grasses.) The analysis of satellite data (images) actually showed that the later Oil Age experienced a Global Greening. The United Nations Environment Programme summarized in 2003 that net primary production (NPP), which is the amount of solar energy (and thus atmospheric carbon dioxide) captured by plants through photosynthesis minus what they use in respiration, rose globally by about six per cent during the last two decades of the 20th century.[15] Increasing atmospheric carbon dioxide levels, temperatures, precipitation, and nitrogen deposition, plus changes in cloud cover and land use, have all been implicated in this global greening, but we do not know exactly how much each factor may have contributed. Ecosystems in tropical zones and in the high latitudes of the Northern Hemisphere accounted for 80 percent of the increase in global net primary production, with the tropical rain forests in the Amazon alone contributing nearly 40 percent of the global increase. The latter is attributed to a decline in cloud cover that increased the amount of solar radiation reaching the surface. Meanwhile changes in monsoon dynamics in the 1990s resulted in more rainfall, causing an increase in vegetation over the Indian sub-continent and the African Sahel zone. Generally, the impact of climate change on natural ecosystems would be expected to be larger than the impact on agriculture, simply because the former are not under direct human management. In some of the proposed scenarios the warming would be a lot faster than species have been prepared for by nature in terms of evolutionary experience. Besides, even relatively minor climate-related changes, such as those well within the range known from the natural climate variabilities of the past, may alter the competitive position of species in relation to each other, creating winners and losers under new conditions.

Direct impacts on human health would probably not be all that drastic, simply because most health aspects are well managed. If the occurrence of weather extremes truly becomes more frequent, with a larger number of very warm days, and an intensified hydrological cycle causing more droughts, floods, and storms (which is altogether uncertain on global average), wealthier regions will be better able to deal with it than less wealthy areas. (In Europe, for instance, a climate warming would actually be associated with an instant overall reduction of deaths directly related to temperature: more people die during cold winters than during summer heat waves.) Wealthier regions would also be better able to fight certain diseases speculated to spread from tropical regions to mid-latitudes. (Many insect carriers of disease thrive better in warmer and wetter conditions.)

Sea level rise is considered a serious and measurable consequence in a global warming scenario. It is a problem because a lot of people live in low coastal areas in countries as diverse as Bangladesh, the Netherlands, and various Pacific Island nations. (In Bangladesh, well over six million people live on land less than one meter above sea level, and in parts of China the situation is similar.) What is more, many low coastal areas are important for agriculture and sensitive to groundwater contamination by salt water. (In Egypt, a sea level rise of one meter will affect about 12 percent of the arable land.) On the other hand, sea level rise is a slow feedback, and there is time to react to it. The IPCC's 2001 report projected the sea level to rise by between 10 and 90 centimeters over the period 1990 to 2100, between 0.09 cm and 0.8 cm per year. Breakdown of the West Antarctic and/or Greenland ice sheets would contribute significantly, but in the case of Greenland, for instance, the temperature threshold for breakdown is believed to be about 1.1 to 3.8c above today's global average temperature, which is not expected to happen before 2100 according to a IPCC likely scenario (in the report of 2007).[16] On the other hand, there have been indications that some ice might melt faster than expected. In September 2007 the European Space Agency reported that the shrinking of Arctic ice had (for a short time during the summer) opened the fabled Northwest Passage, the long-sought, but until recently almost impassable, northern sea route between Europe and Asia: satellite photos showed that the ice along Greenland, northern Canada, and Alaska, had retreated to its lowest level since such images were first taken in 1978. For comparison, the IPCC summarized respective climate-related changes observed during the entire 20th century as follows: global mean sea level increased at an average annual rate of 1 to 2 mm during the 20th century, 10 to 20 cm in total. The duration of ice cover of rivers and lakes decreased by about two weeks over the 20th century in mid-and high latitudes of the Northern hemisphere. The Arctic sea-ice cover thinned by 40 percent in recent decades during late summer to early autumn and decreased by 10 to 15 percent since the 1950s in spring and summer.

Extensive studies conducted in Greenland, where some glaciers have been retreating very rapidly in recent years, have cast doubts on the theory that a global warming (alone) is behind this. A warming of the atmosphere would be expected to affect all glaciers, while only those that reach deep into the sea water are retreating at the fastest rate. It has in turn been measured that ocean layers in medium depths have been warming along Greenland's western coast in recent years. This explains the rapid post-1997 thinning of Jakobshavn Isbræ, a glacier that thickened substantially from 1991 until 1997. Previous periods of rapid thinning had been observed in 1902–13, and in 1930–59. During the 1920s and 1930s, there was a marked warming of the northern North Atlantic Ocean that included an enhanced Atlantic inflow in northern regions. It is not fully understood why this is happening, and what is behind the changes of the North Atlantic ocean circuit.[17] Another effect observed on Greenland's glaciers is the massive precipitation of soot, which darkens the snow, reducing its albedo and increasing its solar energy absorption. Soot is also altering the structure of the ice, making it more prone to melting. Soot is being transported over long distances through the atmosphere. China's economic boom, much of which was coal-fueled, is being held responsible for a large share of the problem in recent years, though soot dating back to 1788 has been detected in ice cores from Greenland.[18] Currently about 20 percent of the particle pollution on Greenland's inland ice comes from East Asia, and some researchers believe that soot might be responsible for over 90 percent of temperature changes in the Arctic.[19] In short, not all melting ice is a sign of global climate change.

Doomsday Scenario: Slowing the Gulf Stream

Doomsday or surprise scenarios can be modeled as well, usually under the assumption of a substantial temperature increase in the 22nd century following the emission of large amounts of carbon dioxide that indeed remain in the atmosphere. One such scenario would be for ocean currents to be greatly altered. The more northern parts of Europe owe their inhabitability to the heat the Gulf Stream transports northward across the Atlantic Ocean from the Gulf of Mexico. Like a conveyor belt, the stream sinks to lower ocean levels in the far north of the Atlantic and flows back towards America at great depths to surface again in the Gulf of Mexico where it is warmed up. If increased precipitation in the North Atlantic rendered surface water there less salty (and hence less dense) to a degree that would decrease the sinking of such water (deep water formation) by 30 percent, the Gulf Stream circulation would slow down, and the climate in Europe would cool markedly. A report commissioned by a Pentagon think tank created a storm of controversy by drawing such an extreme climate change scenario for the short run. The report suggested a number of dire consequences in a scenario in which the current period of global warming ends in 2010, followed by a period of abrupt cooling. (Such a climate event is believed to have happened 8,200 years ago, lasting for a century.) By 2020, after a decade of cooling, Europe's climate would have become "more like Siberia's," and "mega-droughts" would have hit southern China and northern Europe for 10 years. All this would result in a global catastrophe, costing millions of lives due to wars and natural disasters within 20 years. However, this doomsday scenario included a severe change of the Gulf Stream circulation patterns, which is not expected to happen, at least not during the 21st century.[20]

Where Has All the Carbon Gone?

Obviously, one of the central questions in scenario modeling of the future is how much carbon dioxide will pile up in the atmosphere. The current flow of anthropogenic CO_2 into the atmosphere is about three percent compared to the natural carbon dioxide flows in and out of the atmosphere. How much carbon dioxide will we release in the future, and how much of it will stay in the atmosphere?

Currently, the system absorbs a lot of the anthropogenic CO_2 emissions. This is relatively easy to estimate, as we know quite accurately the amount of fossil fuels we are burning (and the approximate magnitude of our other CO_2-releasing activities), and we can directly measure how the concentration of carbon dioxide is increasing in the atmosphere. The difference, which turns out to be about half of what is released by human activity, is absorbed by carbon reservoirs. Ultimately, the more important question will be how much additional carbon dioxide stays in the atmosphere in comparison to the amount already there. Based on current observations this turns out to be under half a percent per year. It would thus take more than two centuries to double the amount of carbon dioxide in the atmosphere at current emission and absorption rates.

Flows of carbon dioxide in and out of the atmosphere are usually expressed in terms of carbon equivalent, stating the weight of carbon contained in carbon dioxide. (One ton of carbon, C, is equivalent to about 3.67 tons of actual carbon dioxide, CO_2.) The idea behind carbon equivalents is to create a common measure to compare quantities of CO_2 to quantities of carbon stored in different organic molecules in various terrestrial or oceanic carbon reservoirs. Due to the magnitudes involved, figures are usually given in gigatons, Gt, of carbon equivalent, with one gigaton equaling one billion (metric) tons.

Towards the end of the 20th century, roughly 5.5 Gt worth of carbon were released yearly into the atmosphere in form of CO_2 due to fossil fuel burning and cement production. An additional 1.6 Gt of carbon equivalent were released by other human activities, such as changes in tropical land use (that is, burning and decay of forests). The yearly total was thus about 7.1 Gt of carbon. For comparison, carbon dioxide worth 800 Gt of carbon equivalent is already present in the atmosphere, and some 210 Gt are being annually cycled in and out of the atmosphere by natural chemical and physical processes. Terrestrial plants capture about 120 Gt per year through photosynthesis during their growth, but they also release some 60 Gt through respiration, which is the reverse reaction to photosynthesis.[21] This indicates that about half the carbon dioxide fixed during plant growth is released again, but the extent of respiration depends strongly on temperature and light variations during the seasons in different latitudes.[22] The other half, another 60 Gt of carbon equivalent, is released annually through decomposition of terrestrial organic material (plants, animals, detritus). This has to be so: otherwise, the natural terrestrial plant growth cycle would be out of balance and remove more and more carbon dioxide from the atmosphere. Meanwhile about 90 Gt of carbon in form of CO_2 are absorbed and released by the oceans every year, through both biological and physical processes.

In light of the system cycling carbon dioxide worth 210 Gt of carbon out of the atmosphere each year, we would perhaps expect that it would be easy for it to remove an extra 7.1 Gt of carbon equivalent added from anthropogenic sources. After all, this is no more than an additional three percent or so. But apparently it is not that simple. Direct measurements of atmospheric CO_2 showed that compared to 7.1 Gt of carbon equivalent released into the atmosphere per year by human activity, about 3.3 Gt of carbon equivalent remained there. This implies that only 3.8 Gt of carbon equivalent (51 percent) of the carbon dioxide released by human activity were indeed being absorbed by the system. It is not that easy to tell where the extra carbon dioxide actually goes. Presumably, about 2.0 Gt of carbon equivalent of the extra portion are annually absorbed by the oceans, and 1.3 Gt by terrestrial sinks. The latter includes 0.5 Gt absorption by northern hemisphere forest regrowth. (One way to get hints about carbon absorption is to study the satellite photos showing global greening[23], and to check the respiration levels in the regions experiencing accelerated plant growth.)

Even though anthropogenic CO_2 emissions have increased, the system has proved remarkably robust in removing about half of the anthropogenic CO_2 emissions from the atmosphere year after year after year. As of 2005, the cumulative total of all CO_2 thus far released through fossil fuel burning and land use change was estimated at 480 Gt of carbon equivalent, while 210 Gt, or 44 percent remained in the atmosphere. The same average was found for the 1959–2005 period,[24] while the percentage of total anthropogenic CO_2 emissions remaining in the atmosphere averaged 48 percent in the 2000 to 2005 period.[25]

To be sure, the release of additional carbon dioxide (and the absorption of CO_2) varies quite a bit from year to year, not just because of changes in fossil fuel burning (which varies with economic growth).[26] Towards the end of the 20th century the years 1997–1998 saw an unusual high amount of carbon dioxide being added to the atmosphere, which is attributed to the Indonesian wildfires of those years. (Incidentally, the year 1998 was then recorded as the warmest year since the start of the Fossil Fuel Age in terms of global average surface temperature, and the second warmest year, after 1933, in the United States.) Later on, the years 2002 and 2003 saw high increases of atmospheric CO_2 associated with droughts: The severe U.S. drought of 2002 left an extra 0.36 Gt of carbon in the atmosphere (an amount equivalent to the annual emissions of 200 million cars!), and the drastic European drought of 2003 left more than 0.50 Gt of carbon equivalent in the air.[27] Such incidents of foregone removal of carbon dioxide may be interpreted as a disturbing and accelerating feedback to initial climate change, but there is no statistical evidence that such droughts are now occurring in the United States or Europe more often than expected due to natural weather variance. Nevertheless, the 1997/98, 2002, and 2003 carbon dioxide incidents have undoubtedly helped to make the first decade of the 21st century a warm period. On the other hand, the year 2008 was expected to become the coolest year in a decade due the development of a strong La Niña in the tropical Pacific Ocean.[28]

El Niño and La Niña

La Niña, the cooling of the equatorial Eastern Central Pacific Ocean, is the counterpart to El Niño. Both are natural occurring events that irregularly once every few years cause a marked weather variability in many areas around the globe. El Niño events were initially observed by fishermen off the west coast of South America: the flow of unusually warm surface waters towards (and along) the Pacific coast prevents the upwelling of nutrient-rich cold deep water and thus negatively affects fishing in the region. (Peruvian fishermen dubbed the phenomenon, or rather the warm northerly current, El Niño, because it was most notable around Christmas time.[29]) It is not known for sure, which processes in the climate system trigger these events that first lead to the diminishing of equatorial westerly winds, and consequently to El Niño occurrences. Notably, the most recent marked El Niño was in 1997/98. It was the strongest El Niño since accurate data gathering had started in the mid-20th century. Together with the Indonesian wildfires of that time, this El Niño helped making 1998 what it was: a record year in terms of global average surface temperature. It was followed by a lengthy La Niña, lasting until 2001, which contributed to a serious drought in much of the West.[30] La Niña events occur after some, but not all, El Niño events. The warm year of 2007 witnessed a weak El Niño (that was credited with partially shutting down summer Atlantic hurricane activity), and was set to be followed by a strong La Niña.

Solar Activity Rather Than Carbon Dioxide?

A major obstacle in attempts to model possible climates of the future is that future solar activity levels are unknown. Some scientists maintain the view that variations in solar activity, rather than atmospheric greenhouse gas concentrations, have been the principal factor behind climate variability on Earth. Solar activity can be monitored by observing sunspots, which appear as dark spots on the sun's surface when viewed with the unaided eye (through a filter) or with a telescope. The number of sunspots rises and falls as part of the solar cycle (solar magnetic activity cycle), over periods of approximately 11 years, but there are also long-term variations. (During the 11-year cycle the energy released by the Sun changes by only about a tenth of a percent from maximum to minimum, but a larger percentage of high-energy radiation is put out during the maximum, which is believed to result in more ozone formation in the Earth's upper atmosphere, which in turn becomes warmer.)

During the past four centuries people have observed and recorded sunspot numbers, and a record for the past 11,000 years has been established through indirect methods.[31] As it turned out, observed sunspot activity correlates well with the observed temperature variations of the past.[32] The coldest part of the Little Ice Age, for instance, coincides with the rare occurrence of sunspots during what is called the Maunder Minimum (roughly from 1645 to 1715).[33] Notably, the 20th century has witnessed an unusually high solar activity from the mid-1930s. This was apparently so exceptional that we would have to go 8,000 years back in time to find a previous period of equally high solar activity.[34] Exactly how variations in solar activity are influencing the climate on Earth is still debated. Supposedly, solar wind sent towards Earth interacts with cosmic radiation that constantly pours into the atmosphere and hits water vapor to promote cloud formation. (The source of the cosmic rays is assumed to be super novae-stellar explosions.) A reduction of such cloud formation due to high solar activity would then have a warming effect, given that such clouds generally have a cooling effect.[35]

Shaky Conclusions? Combining Solar Activity with Greenhouse Gas and Aerosols

In response to the research on solar activity, climate models have been expanded to include solar activity in addition to greenhouse gas emissions. These advanced models also include the effect of particles emitted into the atmosphere from various natural (volcanoes) as well as anthropogenic (fossil fuel burning) sources. (Particles have a direct cooling effect as they scatter solar radiation, but they also promote cloud formation.) According to such models, the substantial warming during the first half of the 20th century, when atmospheric carbon dioxide concentrations were still relatively low, was mainly due to strong solar activity.[36] This is quite remarkable, as the warming in the first half of the 20th century was comparable in magnitude to the warming during the second part of the 20th century. The global climate cooling period from about 1940 to the early 1970s is less well explained. This period coincided with substantial anthropogenic greenhouse gas emissions during the oil-driven post-World War II economic boom, while neither solar activity was especially low, nor volcanic activity was especially high. (One notable eruption was the one of Mount Agung, Indonesia, in 1963–64).[37]

In an attempt to explain this cooling period in the midst of the Oil Age, it has been theorized that increased concentrations of aerosols (particularly sulfates), released into the atmosphere by industrial processes (most importantly the combustion of coal), may be behind it, because sulfurous emissions around the world increased sharply between about 1945 and 1989, and allegedly declined markedly thereafter.[38] However, while sulfurous emissions did indeed peak in North America and Europe during the 1970s as pollution controls were introduced (to counter acid rain), it is not certain if the claim can be sustained that these emissions did decline globally: The United Nation's IPCC stated in 2006 that emissions of sulfur dioxide from 25 countries in Europe and from the United States have been reduced over the 1982–2001 period, but "over the same period SO_2 emissions have been increasing significantly from Asia . . . and from developing countries. The net result of these combined regional reductions and increases leads to uncertainty in whether the global SO_2 has increased or decreased since the 1980s."[39] Besides, much uncertainty remains in regards to how sulfurous emissions spread in the atmosphere, how they influence cloud formation, and to what extent they contribute to a cooling or warming, locally and globally.[40] The United Nations' IPCC calls aerosols the largest uncertainty in climate modeling. (The term aerosol refers to a suspension of small solid or liquid particles in gas.) Fossil fuel burning releases sulfates, black carbon, and organic aerosols which in turn all have complex effects on clouds: their formation, longevity, and reflectivity. Cloud brightness is increased, because aerosols lead to a larger number and smaller size of cloud droplets (a cooling effect). The cloud cover is increased, because smaller droplets inhibit rainfall and increase cloud lifetime (a cooling effect). However, some aerosols are heating the atmosphere directly and decrease cloud brightness if present within cloud drops. (Both are warming effects.). Black carbon, for instance, reduces aerosol albedo, reduces cloud cover, and reduces cloud particle albedo. (All of these effects cause a warming, which can be reduced when black carbon emissions from diesel fuel and coal are reduced.) But altogether, (anthropogenic) aerosols are assumed to have a substantial cooling effect, possibly in the same magnitude as (anthropogenic) carbon dioxide has a warming effect. On global average, the effect of fossil fuel burning on the climate is therefore somewhat neutralized due to the partial offsetting of aerosol and greenhouse gas forcings.[41] (According to IPCC data published in 2007, anthropogenic global radiative forcing from carbon dioxide accounts for plus 1.66 W/m2, while radiative forcing from anthropogenic aerosol accounts for minus 1.20 W/m2.[42])

The 20th century's second climate warming period, between the early 1970s and 1998, has apparently been quite well described by combined models: There was a strong solar activity level, but it was neutralized (countered) by negative volcanic forcing (El Chichón erupted in Mexico in 1982, Pinatubo in the Philippines in 1991). Thus, the increased greenhouse gas effect became determining: The net warming effect in the later 20th century is usually ascribed to anthropogenic greenhouse emissions.[43] However, the authors of a widely quoted 2004 study of this kind warned against overinterpreting the good correspondence between the model simulations and the observed temperature.[44] Another study concluded in 2007 that the slow, but steady, increase in solar output during the 20th century was reversed about 1985.[45] If so, we do indeed have no other good explanation for the warming between

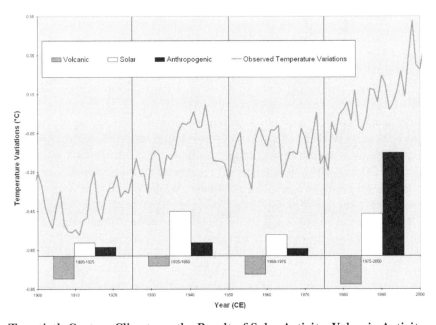

Twentieth-Century Climate as the Result of Solar Activity, Volcanic Activity, and Anthropogenic Greenhouse Gas Forcing and Aerosols Climate models that take into account variations in solar and volcanic activity conclude that the warming during the first part of the 20th century was caused by increased solar activity. The cooling period of about three decades during the post–World War II economic boom is not well explained. Relative low solar and volcanic activity offset one another, while the atmospheric carbon dioxide concentration was increasing. It is thus attempted to explain this cooling period through anthropogenic sulfate aerosol. In the later 20th century relatively strong volcanic activity offset nearly the entire effect of the relatively strong solar activity, and greenhouse gas radiative forcing became determining. The upper part of the graph shows the temperature development observed during the 20th century, while the lower part shows the modeled relative contributions by volcanic, solar, and anthropogenic factors (greenhouse gas and aerosols) to climate change (radiation forcing), averaged over 25-year periods. To be sure, the global average temperature during the first decade of the 21st century did not increase over the record year of 1998, which was marked by a powerful (natural) El Niño event. (Based on data from Caspar M. Ammann et al., "A Monthly and Latitudinally Varying Volcanic Forcing Dataset in Simulations of 20th Century Climate," *Geophysical Research Letters* 30, No. 12 (2003):1657, http://www.ncdc.noaa.gov/paleo/pubs/ammann2003/ammann2003.html, and P. D. Jones, et al., "Global Monthly and Annual Temperature Anomalies (degrees C), 1850–2007 (Relative to the 1961–1990 Mean)," September 2008, Carbon Dioxide Information Analysis Center (CDIAC), http://cdiac.ornl.gov/ftp/trends/temp/jonescru/global.dat.)

1985 and 1998 than increased radiative forcing by greenhouse gases in the atmosphere. But there is no total consensus. Some scientists still insist that the sun appears to be the main forcing agent in global climate change.[46]

Future Carbon Dioxide Concentration in the Atmosphere

Since we cannot predict future levels of solar and volcanic activity, it is not possible to predict what the climate will be like in the future. What is more, we do not know how much carbon dioxide will be released by fossil fuel burning in the future, and if the carbon reservoirs will keep removing added carbon dioxide from the atmosphere at the current pace. (To be sure, this could go both ways: some carbon reservoirs may become saturated, others might become more accessible when the partial pressure of CO_2 in the atmosphere increases.) If the added (and prevailing) amount of atmospheric carbon dioxide amount remained constant at the current level of about 4 Gt of carbon equivalent per year, it would take one century to increase the atmosphere's 800 Gt of carbon equivalent worth of carbon dioxide by half (to 1200 Gt). However, if the growth trends observed in the early 21st century continued until 2100, we would have to expect carbon dioxide levels in the atmosphere to become a lot higher. At the start of the Coal Age, the concentration of CO_2 in the atmosphere was about 280 ppm. By 1960 it had reached 317 ppm, by 2003 376 ppm, and by 2007 some 380 ppm. For comparison, it is expected that the concentration of carbon dioxide in the atmosphere may increase to somewhere between 650 ppm and 970 ppm by the year 2100 if strong policies against fossil fuel burning are absent. (The latter figure would be two and a half times the 2007 level.) Meanwhile, many scientists warn we should not go above 550 ppm.

Much carbon dioxide will be released in countries that have not yet fully entered the Oil Age, especially those populous ones (China, India) that may reach industrialization and motorization standards similar to those now prevalent in the West. According to the reference (base) case of the 2007 International Energy Outlook (published by the U.S. Department of Energy's Energy Information Administration), world carbon dioxide emissions are projected to rise from 7.3 Gt carbon equivalent in 2004 to 9.2 Gt in 2015 and 11.7 Gt in 2030, with much of the increase coming from developing countries.[47] In 2004, energy-related carbon dioxide emissions from non-OECD[48] countries (low per capita income) exceeded those from OECD countries (high per capita income) for the first time, and by 2030 carbon dioxide emissions from non-OECD countries are projected to exceed those from the OECD countries by 57 percent. In 2006 China has passed the United States as the world's biggest CO_2 emitter from fossil fuels and cement production, with Chinese CO_2 emissions amounting to 1.7 Gt of carbon equivalent, and U.S.

CO$_2$ emissions to 1.6 Gt of carbon equivalent.[49] (Cement production consumes large amounts of energy, accounting for about 4 percent of the global CO$_2$ emissions released by fuel burning and industrial sources. As of 2007, China's fast growing cement industry produced 44 percent of world supply, thus accounting for almost 9 percent of China's CO$_2$ emissions.) Nevertheless, China's per capita CO$_2$ emissions remained relatively low, about a quarter of that released on average per U.S. citizen. Thus, there remained a lot of potential for China to catch up, and China's energy-related emissions of carbon dioxide have been projected to exceed U.S. emissions by 41 percent in 2030.[50] China's oil-related carbon dioxide emissions were expected to grow at 3.5 percent per year until 2030, faster than in any other country, to meet soaring demand for road transport energy. Nevertheless, the United States was expected to remain the largest source of oil-related emissions: 66 percent above those of China in 2030.

POLITICS, ECONOMICS, OR SCIENCE?

The issue that people in full-blown Oil Age countries produce a lot more climate-altering gases than people in less industrialized regions has led to a serious moral conflict. As these emissions have global, not regional, consequences of uncertain magnitude, ordinary cost-benefit analyses did not work, and the "polluter pays" principle was distorted. To be sure, utilitarian ethics is deeply embedded in Western thought. English Enlightenment thinker John Locke, for instance, stated in 1690 in *The Second Treatise of Civil Government* that "no one ought to harm another in his life, health, liberty, or possessions."[51] Similarly, the work of German philosopher Wilhelm von Humboldt, upon whose *On the Limits of State Action* (1810) John Stuart Mill's essay *On Liberty* (1859) drew, contained the harm principle: each individual has the right to act as she or he wants, so long as these actions do not harm others. But in case of global climate change, utilitarian ethics was largely brushed aside, partly because major vested interests were at stake, partly because the uncertainties involved made it easy to pretend that nothing was wrong.[52] How would we otherwise explain that Americans would gladly accept the homeland of Pacific Islanders sinking in the ocean in exchange for being allowed to hold on to their beloved gasoline-guzzling sports-utility vehicles? And how would we explain this to Pacific Islanders who may have never owned or driven an automobile? How would we compensate them for their loss? American policy makers closed their eyes and allowed oil corporations and automobile manufacturers to lobby successfully in the interest of their shareholders: Despite tremendous technological advances with respect to building efficient engines and cars, the average vehicle on American streets in 2001 used more gasoline per mile driven than it did during the 20 previous years. (This was made possible by carefully creating legislation that exempted large and heavy SUVs from the Corporate Average Fuel Economy (CAFE). By 2008 the tide had finally changed. Various manufacturers, including Toyota, General Motors, Ford, Chrysler, and Porsche were hybridizing SUVs and pickups in response to a historic 40 percent jump in mileage standards, the largest in decades. This time SUVs were not exempt: manufacturers' car and truck fleets must average 35 miles per gallon by 2020, without breaks benefiting heavier sport utilities, pickups, and minivans.[53])

To understand why policy makers have been able to defend nearly any position on the climate issue, from dismissive to entirely concerned, it is worth remembering what a summary by an adviser might have looked like in the early 21st century:

- We have a fundamental understanding of how CO_2 interacts in the atmosphere with heat radiation
- We can directly measure that CO_2 in the atmosphere is increasing, and we know that about half of the current anthropogenic carbon dioxide emissions are being absorbed by the planet's natural carbon reservoirs
- We do not know how much CO_2 will be removed from the atmosphere by natural feedbacks under new conditions. Possibly the global carbon/climate system would absorb larger amounts of carbon dioxide when its concentration in the atmosphere is increasing, but we do not understand the processes behind it well enough to judge how this would work in comparison to the unprecedented speed of carbon dioxide release by human activities.
- With continuing population growth and industrialization in developing countries, where most of the world's people are living, anthropogenic carbon dioxide emissions in the 21st century are expected to rise well above the levels of the 20th century.
- We know that historically atmospheric CO_2 concentrations were the result of temperature change, not the cause of it. We are not entirely sure what caused (often very pronounced) climate changes in the past.
- We know that historically temperatures have been at a similar level to what they are now, notably in pre-industrial times just a few centuries ago, even though atmospheric CO_2 concentrations were much lower back then. This implies that factors others than greenhouse gas concentrations are at work to influence the climate.
- We have to contemplate that the global average surface temperature rise of the later 20th[h] century might continue towards new record levels, well above those of the Middle Ages, in the 21st century.
- The actual consequences of such record temperature levels are uncertain. Some regions are expected to be better off under new conditions, others will be worse off, but the overall effect is expected to be negative.
- The global mean temperature remained constant (did not increase) between 1998 and 2008. (For 10 years no year has been warmer on global average than 1998 despite rising CO_2 concentrations in the atmosphere.)

- The glaciers of the northern hemisphere tend to shrink a lot faster than has been observed in recent history, while Antarctica, which comprises about 90 percent of total global ice, is gaining ice at the moment.
- Thus far, there is no evidence that any person on the planet has been killed, injured, or harmed due to anthropogenic climate change. No weather extremes have been observed that can be seriously claimed to be outside the natural range of variability.

Initially, the uncertainties involved made things difficult for politicians willing to tackle the problem and enforce CO_2 emission cuts. After all, voters in Western countries had now become used to a very energy-intense lifestyle. People enjoyed living far away from their work-place, and commuted by car every day. Many would not accept that their holiday flights or drives would become expensive. Trucks had a central role in economies to transport raw materials, manufactured products, and food. The cement and other industries produced goods that were now indispensable and enjoyed by every single citizen. And most importantly, high per capita energy consumption was positively correlated with high per capita income, and thus overall prosperity. On the other hand, it was quite obvious even for the last politician that it was not a good idea to conduct an uncontrolled experiment on our planet. Thus, the national governments of the world in the early 1990s decided to take action, at least formally.

The Kyoto Protocol

In 1992, 154 nations, including the United States, signed the United Nations Framework Convention on Climate Change (UNFCCC), an international environmental treaty to reduce emissions of greenhouse gas. However, this treaty set no mandatory limits on greenhouse gas emissions for individual nations and contained no enforcement provisions. Mandatory emission limits would instead be set by updates, called protocols. The principal update, the Kyoto Protocol, was installed in 1997.[54] It committed industrialized countries to reduce their collective emission of (six different) greenhouse gases by at least 5 percent below 1990 levels in the commitment period 2008 to 2012. (This represented a nearly 30 percent cut in terms of emissions expected for 2010 without the Kyoto Protocol. The six greenhouse gases concerned were: Carbon dioxide (CO_2), methane (CH_4), nitrous oxide (N_2O), hydrofluorocarbons (HFCs), perfluorocarbons (PFCs), and sulfur hexafluoride (SF_6).) Individual national targets varied quite widely, from 8 percent reductions for the European Union, 7 percent for the United States, and 6 percent for Japan, to zero percent for Russia. Australia was even permitted an increase of 8 percent, for instance. If a country was to fail to meet its targets, it was to be allowed to buy emission credits from countries staying below their target. (The Netherlands, for instance, bought emissions rights from Poland,

Romania, and the Czech Republic in 2001.) Alternatively, countries were allowed to increase their emission rights by creating so-called carbon sinks, by planting forests, for instance. Developing countries were not included in the system at all: they should have no limitations on their emissions in order not to slow their economic growth.

But there was a principal problem involved. For the Kyoto Protocol to enter into force as a legally binding international treaty, it had to be ratified by nations responsible for at least 55 percent of total anthropogenic carbon dioxide emissions for the year 1990. The prospect for this to happen was set back when the United States, the world's largest emitter of carbon dioxide, made clear that it would not ratify the Protocol, even though it had signed it. President Clinton hailed the global warming pact, but his vice president, Al Gore, stated that the administration would not seek a ratification vote in the Senate before participation by key developing nations has been achieved.[55] President George W. Bush argued along similar lines, complaining in 2001 that China, the world's second-largest emitter of greenhouse gases, was entirely exempt from the requirements of the Kyoto Protocol.[56] Of course, this argument omits that China is a large emitter only because of its large population, not because of large energy per person consumption. In fact, American per person CO_2 emissions were then nearly nine times as large as those of China, and nearly two and a half times those of the European Union.

The European Union ratified the Protocol in May 2002, but lagged far behind in the implementation of emission reductions. As of 2003, only two member states, Sweden and the UK, were in line with the agreed Kyoto goals, while the other member states lagged from 1.3 percent (Germany) to 37.8 percent (Denmark) behind the target. Nevertheless, the EU had managed to reduce its greenhouse gas emission to a level 3.5 percent below the 1990 level, and, when compared to the United States, produced a quarter less of CO_2 per unit of GDP, 60 percent less in CO_2 emissions per person, while using half the amount of primary energy per person. In the period 1990 to 2004, Germany, for instance, reduced its greenhouse gas emissions by 17.2 percent, while the United States increased its emissions by 15.8 percent.[57] But even without support from the United States, the Kyoto Protocol eventually entered into force, after Russia had followed the example of well over 100 other countries (including Japan, China, Canada, and India) and ratified the treaty in February 2005. (To be sure, China and India were not obliged to meet any targets, but to report their emissions.)

Corrupted by Interests?

Nevertheless, the Kyoto Protocol kept on being criticized. It may be considered the most far-reaching agreement on the environment and sustainable development ever adopted, but environmentalists complained that the measures did not go far enough, while economists complained that the measures are too expensive in comparison to the damages they avoid, and that the Protocol may slow economic growth. What is more, the oil and automobile industries, represented by several of the world's largest corporations, would not stay put and watch some new legislation being enforced to cut their revenues. Thus, the whole discussion about climate change remained politically charged.

Much criticism focused on American Kyoto neglect and the George W. Bush administration, which was known for its close ties to the oil industry. After assuming office in 2001, Bush stated: "We do not know how much our climate could, or will change in the future," and, "We do not know how fast change will occur, or even how some of our actions could impact it." Bush made Phil Cooney, a former oil industry lobbyist (working for the American Petroleum Institute), chief-of-staff at the White House Council on Environmental Quality. Cooney, educated as a lawyer, not a scientist, in turn edited the reports by the government's scientific advisors on climate change in his own hand: "Earth is undergoing rapid change" turned into "may be undergoing change;" "Uncertainty" turned into "significant remaining uncertainty;" and a statement concluding that energy production contributes to warming was simply crossed out, for instance.[58] Eventually Cooney went on to work for ExxonMobil, the corporation that had given the Republican party $1.4 million in the 2000 election cycle. (ExxonMobil is a true giant. It's 2007 revenue was over $400 billion, topping the GDP even of such countries as Sweden or Austria, and its net income was over $40 billion, a record in U.S. corporate history.) However, this did not end the government's efforts to distort the opinion aired by America's climate scientists, according to James Hansen, head of the NASA Goddard Institute for Space Studies in New York City. Hansen, a vehement proponent of climate protection measures,[59] complained that NASA no longer allowed him to communicate freely with the press and the public, after he had said in a talk at the University of Iowa that he found in the Bush administration "a willingness to listen only to those portions of scientific results that fit predetermined inflexible positions."[60] Such inflexible positions, maintained to protect the oil and automobile industries, are not all that surprising. After all, these industries had long been among the principal employers in America, and they represented in many ways the oil-energy-based U.S. rise to superpowerdom in the 20th century. Thus, the U.S. government has had a long tradition of cooperating closely with them.

Environmentalists claimed that oil interest groups gave billions of dollars to scientists who could show that carbon dioxide and other greenhouse emissions were not behind climate change. The issues worth investigating from that standpoint were the higher-than-now temperature levels in pre-industrial times; the cooling of the climate in the decades after World War II (despite a rapid increase in CO_2 emissions); and the fact that greenhouse warming theory expected that the troposphere, the layer of the atmosphere about 10 to 15 km above the surface, should heat up faster than the surface of the planet: Data collected from satellites and weather balloons did not seem to support this (though there is now some evidence that trends in satellite-observed tropospheric temperatures, when properly analyzed, agree with trends in surface temperature observations).[61] Perhaps most importantly, various scientists maintained the view that variations in solar activity, rather than atmospheric greenhouse concentrations, were the principal factor behind climate variability on Earth.

The Iron Lady and Environmentalism

Industries that would be disadvantaged by climate protection measures countered environmentalist criticism by pointing out that billions of dollars have been spent to provide evidence of a link between global warming and human activities. The scientific community is well aware that reporting threats and potential problems opens up funding channels. Real money starts pouring in as soon as something truly alarming is reported. (Think about AIDS research, to counter a health threat, or space research, to win the Cold War against the communist threat.) On the other hand, which government would spend millions of dollars per year on climate research, computer modeling, and such, if it was clear from the start that the warming trend was caused by a natural cyclicality of solar output? Much funding of research into alternative energy solutions would be cut as well if the climate problem was to be dismissed, though such research would still be important for countries to achieve more energy self-sufficiency. In fact, it has been shown that the quest for energy independence was behind the initial boost for funding of research expected to discourage fossil fuel burning. As the story goes, it all began with conservative Margaret Thatcher, the Iron Lady, who assumed office as prime minister in Britain in 1979, the year of the Second Oil Price Shock. She set out to privatize national industries and utilities, and confronted the labor unions. The National Union of Mineworkers, however, maintained substantial political power, not least because the oil price remained high at the time and seemed to guarantee the importance of coal energy. But the Iron Lady had different plans: She attempted to wean Britain from both coal miner power and Middle Eastern oil, and to strive for British energy self-sufficiency by means of nuclear power. In 1984 the conservative government attempted to close 20 collieries, but some 165,000 British coal miners took to the streets in what was modern Britain's most bitter industrial dispute. And right then, in the mid-1980s, research results were published that established a link between historic temperature and carbon dioxide levels from ice cores.[62] As it was obvious that this would advance her nuclear power agenda, Thatcher quickly reacted by funneling funds towards research that would establish the danger imposed by carbon dioxide emissions. The UK Met Office set up a small modeling unit which in turn provided the basis for the founding of the IPCC, the United Nations' Intergovernmental Panel on Climate Change, in 1988.[63] Simultaneously, the Cold War came to a close, and peace activists turned into environmental activists, with environmentalism arguably becoming a new outlet for anti-capitalism. What is more, according to Greenpeace co-founder Patrick Moore, environmentalism became more extreme, because environmental goals such as clean air and clean water had become so mainstream by the mid-1980s that activists had to adopt more extreme positions to remain anti-establishment. Later on various environmental groups, founded for all the right reasons, turned into corporations themselves, with their existence depending on the continued existence of environmental problems. Without these, their foundation would be undermined, their members would leave, and their organizational structures erode.[64]

With climate alarmism going mainstream, politicians soon realized that outspoken advocacy of climate protection can win votes. Before long, members of the establishment made declamatory remarks on the issue, and scientists were under pressure to conform with the prevailing paradigm of climate alarmism if they wished to receive funding for their research.[65] Those who dissent from the alarmism have seen their work derided, and themselves libeled as industry stooges, scientific hacks, or worse.[66] Compared to the European Union, such trend has been delayed in the United States. Former Vice President Al Gore deserves credit for directing American attention to the issue of climate change, but he sometimes seems to have used somewhat questionable methods to achieve this. In 1992 then-Senator Gore, who has no educational background in science, ran two congressional hearings during which he tried to bully dissenting scientists into changing their views and supporting his climate alarmism.[67] And his 2006 movie *An Inconvenient Truth* misrepresented or dramatized various climate issues. He equated global warming with 9/11 terrorism, for instance, and presented the threat visually by showing Manhattan and other parts of the world on a satellite map being rapidly flooded due to melting Antarctic and Greenland ice. In reality, sea level rise would be a very slow feedback, and there would be a lot of time to prepare for such period of gradual change.[68]

Eventually, even the largest oil companies had to realize that climate worries had become so mainstream that it would hurt sales if they would not play along. They all adopted an increasingly greener corporate image, began calling themselves energy (rather than oil and gas) companies, and started to invest heavily into alternative energy to divert the idea that their core business is centered around what was increasingly viewed as the largest environmental threat ever. (To be sure, investing "heavily" is a relative notion: the amounts spent by oil firms on alternative energy, though large in absolute figures, remained meager in comparison to what these corporate giants keep reinvesting into the oil and gas sector, their core business.)

Few Critics Left

So it happened that by the start of the 21st century there were relatively few outspoken critics left to question mainstream global warming ideology. While newspapers were filled with articles on melting polar ice, increased tropical storms and hurricanes, and even mass extinctions of species (frogs and toads), few pointed out, for instance, that more ice than ever was surrounding Antarctica (in 2008), and that all of the IPCC's models of Antarctica in the 21st century forecast a gain in ice, because a warmer surrounding ocean evaporates more water, which subsequently falls in the form of snow when it reaches the Antarctic continent. (The IPCC noted that there is "a lack of warming reflected in atmospheric temperatures averaged across the region," and other studies showed that Antarctica, which comprises 89.5 percent of total global ice, has experienced an actual cooling in recent decades. Despite a warming Southern Ocean, the amount of ice surrounding Antarctica was in early 2008 at the highest level ever measured for that time of the year since satellites first began to monitor it almost 30 years ago.)[69] Similarly, reports of rapid disintegration of Greenland ice, which comprises just under ten percent of total global ice, ignore the fact that the region was warmer than it is now for several decades in the early 20th century, when atmospheric carbon dioxide concentrations were still relatively low. (Rapid disintegration here means loss of eight-thousandths of a percent of Greenland ice per year, translating into a sea level rise of two-hundredths of an inch per year.) Besides, the rate of temperature increase on Greenland was about 50 percent higher during the 1920–1930 warming period than during the 1995–2005 period.[70] What is more, global warming affects hurricanes in both positive and negative fashions, and there is no relationship between the severity of storms and ocean-surface temperature, once a commonly exceeded threshold temperature is reached. Reports of massive species extinction were flawed as well.[71]

The Costs Involved in Global Warming

A firm stand against suggested climate protection measures has been taken by various economists, who would typically calculate the costs that an economy or society would face if its use of fossil fuel energy was restricted, and compare this to the cost of damages expected due to unrestricted carbon dioxide emissions. William Nordhaus of Yale University, now considered something of a father to climate-change economics, calculated in the early 1990s that it would cost the United States a GDP reduction of some 200 billion dollars a year if an attempt was made to stabilize CO_2 emissions (according to the goals outlined in Toronto in 1987).[72] Environmentalists harshly criticized the work of Nordhaus, and the huge figure at which he was arriving, by pointing out that the assumptions underlying the calculation were flawed.[73] However, given Nordhaus's reputation as a leading economist, the debate was pretty much settled in America before it had started, and attempts to introduce an eco-tax ("Climate Change Levy") on fossil fuels were stalled.

If the cost of restricting carbon dioxide emissions was difficult to estimate, the task became no easier when we turn towards the damages to be expected. For starters, it remained unclear what exactly the consequences of global warming would be. One consequence for which it would be possible to estimate a cost in damages, would be a higher frequency of tropical storms (as the damage such storms cause is known from the present). However, it is disputed whether global warming would actually lead to more storms. There is much, though not unanimous, consensus that the *intensity* of tropical storms will increase with higher oceanic temperature, but not their *frequency*. In fact, model runs support the theory that the frequency of storms in general will decrease, because global warming reduces the temperature differences between the poles and the equator, causing less excitation of extratropical storms.[74] However, even if consensus could be assumed on all claimed expected climatic consequences, the cost calculation would still be tricky, because it is extremely difficult to put a price tag on the potential damages to people, property, and the environment. After all, global warming may affect agriculture, forestry, fisheries, energy, water supply, infrastructure, hurricane damage, drought damage, coast protection, land loss caused by a rise in sea level, loss of wetlands, forest loss, loss of species, loss of human life, pollution, and migration.[75] In the long run, these effects may easily add up to overall costs of several trillion US$. (Of course, any figure, astronomical or not, is somewhat bizarre in this context, as homeland lost to sea level rise, for instance, may well be priceless for environmental refugees.)

Taking into account the risk of catastrophes, William Nordhaus' model anticipates that 3 percent would be wiped off of global GDP, or GWP (gross world product), if the global average surface temperature increases by four degrees Celsius.[76] Three percent of the GWP, which amounted to some $66 trillion (at PPP) in 2007, is a lot of money. Even just one percent of GWP amounts to $660 billion. Does that mean it is worthwhile to invest heavily into climate protection measures (and to accept a potential reduction in economic output)? One school of thought argues that all resources potentially spent on climate protection have to be evaluated in terms of the number of lives that can be saved if the same resources were to be used for other purposes. According to a United Nations estimate, $75 billion used per year could solve all the world's major basic problems: it would provide everyone in the world with clean drinking water, sanitation, basic health care, and education right now. Just dealing with malaria alone could provide economic boosts to the order of 1 percent extra GWP growth per capita per year.[77] Removing all the mentioned basic problems would then make the world a lot richer in the future, and thus better prepared to deal with the effects of climate change (which, arguably, is not that deadly anyway). We would thus have to conclude that it would be unwise to spend money on CO_2 emission cuts rather than on health, nutrition, education, and so on, for the poor. (When an expert panel of economists in 2004 investigated great global challenges, as identified by the United Nations, in order to prioritize where investments should be made so money spent would achieve the greatest benefit, their conclusion rated Disease Control of HIV/AIDS first, and put climate protection measures into the category Bad, ranking last.[78])

Is this line of argumentation reasonable? Yes, but no. There are two major flaws involved. First, the world is not centrally governed: If it was in the interest of national governments (and the people they represent) to work towards maximizing the well-being and prosperity of all people on Earth, they would spend a lot more on foreign aid in the first place. The second problem is that people, at least in the West, tend to be guided by utilitarian ethics and acceptance of the polluter pays principle. They might accept high taxes on fossil fuels as a measure to make them drive and fly less, with the proceeds ideally being used to pay for the environmental damage they cause. They would be less inclined to accept that energy taxes would in turn be used to finance clean water in Africa, for instance. (Or the other way around: It might serve as incentive to drive more, if a clear link was to be established between high CO_2 emissions and high foreign aid spending.) To be sure, taxpayers are hardly being informed at the moment what their energy taxes are being used for (finance ministers blend them into their overall fiscal budget.), and tax levels on energy vary widely between countries. Western Europeans pay a lot more tax on gasoline than American drivers, which contributes to the latter's much larger gasoline consumption.

Fossil Fuels—The Best Investment Ever Made

Another way to impose a tax on energy would be to remove the tax cuts and other open and hidden subsidies that the energy sector is enjoying (in the United States and the European Union, as well as many other regions).[79] In the United States, subsidies to the oil industry include allowance of special accounting procedures, for instance. But here we are back at the energy-equals-power issue. Benefits have been handed to the energy sector, because it constitutes the most fundamental base of economic, political, and military strength. In many countries the energy sector remains entirely state-owned or state-controlled. Fossil fuel burning, which is at the center of current industrial societies, has been promoted by governments, as there is no other fuel that compares in price, performance, and availability. The carbon dioxide levels now observed in the atmosphere bear witness to the great energy-based achievements of the Coal Age and the Oil Age. The 20th century saw a doubling of life expectancy in the industrialized world, and a 10-fold increase in real personal wealth. At no other point in time has there been so much human life: the population explosion, and the agricultural revolution supporting it, have been as much fossil fuel-based as the increased specialization and knowledge accumulation that in result allowed us to detect, measure, and investigate global climate change. In short, accepting the accumulation of anthropogenic carbon dioxide in the atmosphere during the Fossil Fuel Age was a worthwhile investment.

The Hidden Costs of Energy Security

With critics arguing that climate protection measures have little effect despite their huge costs, and that the rate of planetary warming now seems to fall in line with the low end of 21st century projections made by the UN's IPCC (and thus should not be expected to be extreme in comparison to the rate of warming during the 20th century),[80] the greatest windfall to support climate protection (economics) will perhaps come from the increasing cost to secure access to remaining overseas oil reservoirs. If we factor security risks associated with foreign oil dependence into the equation, the price of alternative energy investments (and climate protection measures) may suddenly seem a lot more reasonable. What is a gallon of gasoline worth in the United States, when most oil consumed in America is imported, and access to Middle Eastern oil is achieved by the loss of life of American soldiers in Iraq? Such high hidden cost that is not directly reflected in the price of gasoline might convince more Americans that they should drive smaller cars, or drive less, than any talk about the intensity of tropical storms or a sea level rise by 2100. The money spent on American troops stationed in various Middle Eastern countries would certainly carry research into renewable energy a very long way. (On the other hand, it might be argued that the difficulties to secure access to foreign oil also have their economic benefits: Preparing for war has historically proven to provide a substantial boost for the American economy through expansion of the military-industrial complex. The same has been observed in other countries.)

What is more, development of alternative energy technology and efficiency improvements in fossil fuel technology may create economic growth that can compensate for a bit of the cost incurred by climate protection measures. Japan and Germany have been at the forefront of alternative energy technology development. Notably, both these countries never had domestic oil resources. For them, a general change in the energy regime might not only be a way to stimulate the economy (while protecting the environment), but also a ticket towards more international power. Oil nations and oil corporations, on the other hand, have reason to militate against such change. For oil industry employees the situation now is not all that different than it was for the weavers rioting in the Coal Age, when mechanized weaving was introduced: changes in the energy regime will always create winners and losers, and thus be politically charged. However, as energy command equals power, it is clear who is in charge[81] in an Oil Age setting: Those who release the most carbon dioxide are also on the top of the pyramid of political, economic, and military power. They are the richest, and thus best capable of accepting the cost of climate change. And they have the power to decide for themselves if they would rather ensure access to the remaining overseas oilfields, protect the climate, or promote the transition into a new Energy Era.

SHEEP IN A WOLF'S SKIN? DELAYING THE NEXT ICE AGE

One curious aspect of climate change has been put forward in the early stages of the climate debate. In the early 1970s, the mainstream idea was that the principal climate threat was the arrival of a new ice age.[82] The temperature record restored from ice cores has in turn provided support for this theory, suggesting that we are currently indeed at the brink of a new ice age. The record shows that shorter-term, less pronounced, irregular climate variability has been overlapped by a marked long-term climate cycle: the planet's mean temperature has been rising and falling quite regularly for the past hundreds of thousands of years, with glaciers waxing and waning. Each cycle was roughly 100,000 years long, comprising a long period of slow cooling being followed by a short period of rapid warming. The total duration of these cycles seems to have become longer: a cycle of about 80,000 years was followed by cycles of 86,000 years, 95,000 years and 108,000 years. (The latter consisted of a 88,000-year cooling followed by a 20,000-year warming.) The cycle that we are currently in is even longer. The last cooling period began about 120,000 years ago and lasted for about 100,000 years. 20,000 years ago the climate was at a temperature minimum, started to warm, and has been warming ever since, according to this record.[83] (Other methods to establish the planet's recent climate past indicate that temperatures started to rise about 13,000 years ago.) Accept for the Foraging Age, all Energy Eras of the human civilization have taken place in the current inter-glacial period. But as it seems, this warming period has already lasted unusually long, and temperatures should soon again begin their slow decline that will last tens of thousands of years.

The difference between the temperature maximum and minimum within these cycles is about 5 to 6 degrees Celsius. Some 15,000 years ago, nearly a third of the present land area (including large parts of North America and

Europe) lay under several kilometers of ice, and the sea level was some 120 meters lower than it is today (because so much water was stored away as ice on the continents). It certainly is not a very attractive thought that we are headed that way, but due to the time scale involved—millennia rather than centuries—there is no reason to panic.

Supposedly, the intensity of these cyclical climate variations is the result of numerous sensitive feedback mechanisms in the climate system. The initial source of the cyclical variation, however, is theorized to be periodical changes in the distribution of the incoming solar radiation due to variations in the Earth's orbit. The ice age cycles have been associated with three such variations. First, the eccentricity of the ellipse along which Earth is orbiting varies regularly over a time period of about 100,000 years. Second, the tilt of the Earth's axis, that is, the extent to which Earth's axis is leaning sideward, is changing cyclically over a period of about 41,000 years. Third, the time of the year when the Earth is closest to the Sun changes cyclically over a period of about 22,000 years. All these variations change the local distribution of incoming solar radiation rather than the total amount of energy received.[84]

When the climate was cooling between about 1940 and the early 1970s (despite measurements at the observatory on Mauna Loa, Hawaii, showing from 1957 that atmospheric carbon dioxide was increasing fast), it was a common concern that we might have reached the end of the current interglacial warming period. However, a young Swedish meteorologist, Bert Bolin, then tentatively suggested that the industrial carbon dioxide emissions might turn the trend around.[85] In this respect, carbon dioxide emissions would be positive, delaying the onset of the next Ice Age. (When temperatures started to rise, Bolin became the founding chairman [1988–97] of the UN's International Panel on Climate Change.[86]) Along the same lines, we may even speculate that such interruption has already happened in the late Coal Age, when temperatures began to rise at the end of what consequently became known as the Little Ice Age.

Today, we still do not know how current (anthropogenic) warming will interact or intervene with the overall climate cycle scheme. Perhaps an extended warming period will be followed by a more rapid cooling episode? Some research indicates that with or without human perturbations, the current warm climate may last another 50,000 years, because the eccentricity of Earth's orbit around the Sun is at a minimum.[87] However, the temperature record of the past has not been regular enough to predict precisely when a general cooling will set in again, and the exact mechanisms behind the climate system have not yet been fully understood. Critics are warning that even if it turned out that the release of CO_2 by fossil fuel burning would have a beneficial effect in terms of slowing or reducing the natural path towards a new ice age, concerns must remain that a rapid human-induced climate change (on the scale of decades and centuries rather than millennia) would be an experiment with effects that are difficult to predict and cannot be expected to be beneficial. The carbon now released into the atmosphere by the burning of coal, oil, and natural gas had been stored away from the climate system for more than 300 million years. Hence, it was captured by photosynthetic plant growth long before humans had appeared on the scene. In fact, no mammals at all were around back then, and even birds and dinosaurs had yet to emerge. The natural locking away of carbon back then may have helped to bring about the climate conditions that allowed for mammals and humans to emerge in the first place. Perhaps we owe it to ourselves as a species, as much as to the planet and nature in general, to be careful about how we release that carbon back into the atmosphere.

WHAT SHOULD WE DO? MEASURES AGAINST GLOBAL WARMING

As a growing number of people started to agree that anthropogenic climate change might be a serious problem, various solutions have been proposed to fix the problem. In principle, these fall into two categories: one is to stabilize greenhouse gas concentrations in the atmosphere, the other to find (other) means to cool the climate. Stabilizing atmospheric greenhouse gas concentrations does not necessarily mean to tackle carbon dioxide emissions alone, though their share is largest among anthropogenic greenhouse gases in terms of both amount released and radiative forcing, that is, contribution to the enhanced greenhouse effect. In terms of such anthropogenic greenhouse effect, carbon dioxide contributes about 72 percent, methane 21 percent and nitrous oxide 7 percent, if contributions by chlorofluorocarbons and tropospheric ozone are left aside.[88] Between pre-industrial times (ca. 1750) and 2005, atmospheric carbon dioxide increased by 35 percent (from around 280 parts per million to 379 parts per million), methane by 148 percent (from around 715 parts per billion to 1774 parts per billion,[89]) and nitrous oxide by 18 percent (from around 270 parts per billion to 319 parts per billion). To be sure, methane has 23 times the greenhouse gas (radiative forcing) properties of carbon dioxide, and nitrous oxide is 300 times as powerful a greenhouse gas as carbon dioxide.

Reduce Your Meat Consumption

Methane (CH_4) emissions can be reduced by collecting (and utilizing) methane emitted from landfills, and by minimizing leakage from natural gas pipelines and petrochemical operations. Both methane (CH_4) and nitrous oxide (N_2O) emissions may be substantially reduced by changes in agriculture. While a reduction of rice cultivation, which is an important anthropogenic source of both atmospheric nitrous oxide and methane, is virtually impossible to target

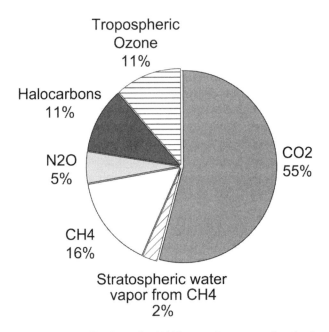

Radiative Forcing: Contributions by Different Gases to the Anthropogenic Greenhouse Effect, 2005 Carbon dioxide is not the only greenhouse gas released by human activity. Methane and nitrous oxide combined account for about a quarter of the total anthropogenic greenhouse effect. This includes anthropogenic methane that undergoes chemical destruction in the stratosphere, producing a small amount of water vapor, which is a potent greenhouse gas. The percentages are based on global average radiative forcing estimates. (Based on data from S. Solomon et al., "Summary for Policymakers-Climate Change 2007: The Physical Science Basis," ed. H.L. Miller, Contribution of Working Group I to the Fourth Assessment Report of the Intergovernmental Panel on Climate Change, IPCC, 2007, http://www.ipcc.ch/pdf/assessment-report/ar4/wg1/ar4-wg1-spm.pdf.)

in light of a growing world population, a decrease in cattle farming holds a great promise. A general reduction of meat consumption would substantially contribute to reduce greenhouse gas emissions for several reasons: modern agriculture is energy-intensive; eating meat is principally inefficient from the energy standpoint; livestock directly produces methane (flatulence); and land-use associated with animal raising increases carbon dioxide emissions through deforestation. (Livestock represent the largest of all anthropogenic land uses.) By the end of it all, the livestock sector is responsible for 18 percent of greenhouse gas emissions (measured in CO_2 equivalent), which is a higher share than transport (14 percent). Animal farming accounts for 9 percent of all anthropogenic CO_2 emissions (mainly due to land-use changes, predominantly

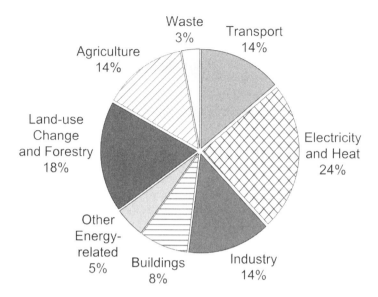

Radiative Forcing: Contributions by Different Sectors to Anthropogenic Greenhouse Effect, 2000 When emissions are presented according to the sector from which they are directly released, electricity production contributes the most to the anthropogenic greenhouse effect, followed by land use changes (deforestation), while transport, industry, and agriculture all contribute at the same level. However, if land use changes (forest cleared for pasture) and fossil fuel used in the production process are factored in, animal farming is in itself responsible for more greenhouse gas emissions (18 percent) than all the world's transportation fuels (13.5 percent). (H. Steinfeld et al., "Livestock's Long Shadow—Environmental Issues and Options," *LEAD-Livestock Environment and Development Initiative*, coordinated by the UN FAO's Animal Production and Health Division, (Rome: FAO, 2006), http://www.virtualcentre.org/en/library/key_pub/longshad/A0701E00.htm, http://www.virtualcentre.org/en/library/key_pub/longshad/A0701E00.pdf. Chart based on data from N. Stern, *The Stern Review on the economics of Climate Change* (Cambridge: Cambridge University Press, 2006), http://www.hm-treasury.gov.uk/stern_review_final_report.htm.)

deforestation); for 37 percent of anthropogenic methane emissions (mostly from enteric fermentation by ruminants, cattle and sheep); and for 65 percent of anthropogenic nitrous oxide (the great majority from manure). Thus, eating a beef hamburger patty contributes about 13 times the CO_2 equivalent compared to eating a veggie burger, and eating one kilogram of beef creates as much climate alteration as driving an average car for 250 kilometers.[90]

(Besides, livestock are responsible for almost two-thirds of anthropogenic ammonia emissions, which contribute significantly to acid rain and acidification of ecosystems.[91]) Politicians, if uninfluenced by the agrarian lobby, should favor dietary change towards vegetarianism as a contribution to climate protection, because it is an inexpensive measure (and it will reduce dependency on imported oil). What is more, people who consume less meat tend to live longer and healthier lives, which implies savings for the health system and the economy in general. (The saturated fatty acids contained in meat products are associated with the formation of plaque that narrows or blocks arteries. Cardiovascular diseases are now by far the largest killer in the world.)

Halting Deforestation

Another way to help stabilizing the atmospheric CO_2 concentration is to reduce deforestation and increase afforestation. According to the Oxford-based Global Canopy Programme,[92] an alliance of leading rain forest scientists using United Nations data, deforestation now accounts for somewhere between 18 percent and 25 percent of global carbon emissions, second only to electricity production (24 percent). (Transport and industry account for 14 percent each, while aviation makes up 3 percent of the total.[93]) In fact, deforestation releases within 24 hours as much CO_2 into the atmosphere as 8 million people flying from New York to London. Arguably, stopping the logging would thus be the fastest and most cost effective solution to reduce carbon emissions,[94] while the international debate on how to mitigate climate change has been focusing on reducing emissions from the energy and transport sectors.[95] Indonesia has become the third biggest emitter of greenhouse gases (behind China and the United States) on the back of deforestation[96], which contributes 85 percent of its emissions. Similarly, Brazil follows as a close fourth, with nearly 75 percent of its carbon emissions deriving from deforestation. (In the past 40 years, close to 20 percent of the Amazon has been cut down. The Brazilian government says it has cut the rate of deforestation by 59 percent in just three years, but there are now signs of problems ahead. Land cleared for cattle is the leading cause of deforestation, but the growth in soy bean production has become increasingly significant, and illegal logging remains a factor.)[97] As countries at mid-latitudes have removed much of their forest cover a long time ago to make room for agriculture, the largest remaining forests now lie in developing countries, where population pressures have increased the rate of deforestation during past decades. Every square kilometer of growing forest fixes between about 200 and 500 tons of carbon per year, while deforestation by burning or other destruction releases approximately two third of the carbon stored in forest's biomass into the atmosphere. (Typical tropical forests contain about 10,000 to 25,000 tons of carbon per square kilometer.)

Using Less Energy

In terms of reducing carbon emissions from energy use there are various options; many of them are quite obvious, some easy to achieve, and others require substantial investments or life-style changes. One such measure would be to shift the fossil fuel mix from coal to natural gas, because electricity production from natural gas releases just about half the amount of CO_2 per unit of energy generated when compared to coal. (But unfortunately natural gas is a lot more scarce than coal.) Even better, the energy mix should be switched from fossil fuel to renewable energy (and, arguably, nuclear energy). The other principal approach is to improve efficiencies throughout the entire energy chain, from energy generation to energy transmission to application of energy in factories, homes, and offices, plus the agricultural, and the transport sector. Driving less, or in a lighter, more fuel-efficient car with properly inflated tires will help, and so will a proper public transport system. Flying less will reduce carbon emissions substantially: A return transatlantic flight releases roughly 1.3 tons of CO_2 per passenger, as much as an average British home will use in roughly one hundred days.[98]

In your home, you can use energy-saving light bulbs; switch off electrical devices (unplugging those that remain on stand-by); keep walls, doors, and windows properly insulated; heat or cool a little less; shorten your hot showers; eat more raw and less cooked food; eat less meat; and eat more locally produced food. (The latter is not necessarily right, as land transport is more energy-intense than water-borne transport: One study showed, for instance, that in terms of minimizing carbon emission it would be better for New Yorkers to drink wine from Bordeaux shipped by sea than wine from California sent by truck. Similarly, it might be better to eat apples in northern Europe or New York that are imported from New Zealand, where the climate is especially apple-friendly and uses electricity produced by renewable sources, than apples raised 50 miles away.[99])

Capturing Carbon Dioxide

A climate protection measure that involves the use of more rather than less energy is to capture carbon dioxide created by fossil fuel burning. This is already being tested at various coal-fired power plants, which is the right place to start, as electricity generation produces more greenhouse gas emissions than any other industry, and coal is the dirtiest fuel in this respect. Coal contributes a bit over forty percent to the electricity produced, but generates nearly seventy percent of the carbon dioxide emitted by global electricity production. In fact, coal-fired plants are responsible for carbon dioxide worth perhaps 2.2 out of the 7.6 Gt tons of carbon equivalent released every year.[100] What is more, coal-based electricity production is on the rise:

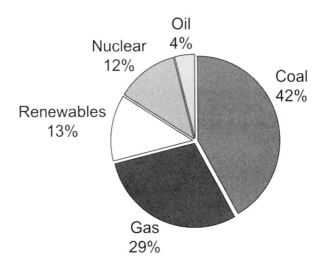

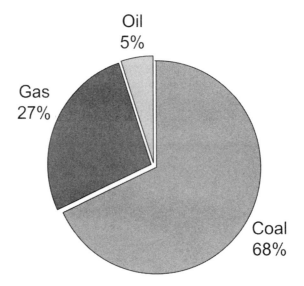

Global Contributions to Electricity Production (upper pie chart) and to Carbon Dioxide Emissions from Electricity Production (lower pie chart) by Fuel Coal is the most important fuel for electricity generation, but it is also the dirtiest: it generates more carbon dioxide per unit of electricity produced than other fossil fuels. Nuclear energy and renewable energy do not contribute to carbon dioxide emissions. Hydropower accounts for nearly the entire renewables segment. (Based on data from N. Stern, *The Stern Review on the Economics of Climate Change* (Cambridge: Cambridge University Press, 2006), http://www.hm-treasury.gov.uk/stern_review_final_report.htm.)

in America, where over half of all electricity derives from coal, some 150 coal power plants were on drawing boards as of 2006.

Capturing carbon dioxide from the exhaust stream, and storing it, can easily consume quite a bit of energy, and is therefore expensive. As gases fill a lot of volume, carbon dioxide is usually liquefied before being pumped underground into former oilfields, gasfields or coal beds.[101] Attempts have been made to pump carbon dioxide straight into the oceans but CO_2 has to be forced into a depth of over 1,000 meters in order for it to sink further by itself.[102] While underground injection provides a good starting point for carbon dioxide disposal, the availability of such sites will be limited in the longer run. One suggestion to replace geological storage, once the necessary capacity fails, is to react carbon dioxide with magnesium rich silicates (which can easily be mined as they are readily accessible and abundant) to yield magnesium carbonate and silica, that is, end products that are naturally occurring. (This approach is based on the observation that carbon dioxide in the environment reacts with a variety of minerals to form carbonates.)[103] Another principal thought is to direct carbon dioxide through cultures of algae which utilize sunlight and CO_2 for their growth. Some research has focused on the creation of genetically engineered microorganisms that consume a lot of carbon dioxide in their metabolism. The biomass may in turn be removed from the algae tanks, dried and used as a fuel. As some algae species build up oil substances as their energy reserve, such algae could also be used to make biodiesel. (A similar idea on a larger scale uses algae to remove carbon dioxide that has already entered the atmosphere. The concept involves adding iron to the oceans to promote algae growth, because iron is a limiting growth factor in the natural oceanic system. However, this idea has met much skepticism.)

If carbon dioxide is to be liquefied (by applying pressure or by cooling), it needs to be separated from the nitrogen in the flue gas, because nitrogen turns liquid at a much lower temperature than carbon dioxide, which would make the liquefaction even more energy-expensive. (Oxygen, the actual reagent, comprises merely 21 percent of air used to combust coal. At 78 percent, nitrogen is the main constituent of air and thus also present in the exhaust gas in high percentages.) There are three principal technologies that are currently favored for nitrogen separation. The simplest, which can easily be added to existing power plants, passes flue gas through a solution of chemicals (called amines), which absorbs carbon dioxide but no nitrogen. Later heating of the solution will release the carbon dioxide for subsequent liquefaction and storage. However, the amines are expensive, and heating them to release the captured carbon dioxide consumes energy, which reduces a state-of-the-art coal power plant's overall efficiency by some 10 percent. Another approach is to combust coal in pure oxygen rather than air to avoid getting nitrogen into the flue gas in the first place. However, the energy used to separate oxygen from air before burning is almost as large as that needed to remove nitrogen afterwards, leading to a similar loss of efficiency. Nevertheless, this approach might be cheaper than using amines, and proponents of oxy-fuel plants maintain that existing modern plants can easily adopt this technology. The third technology involves coal gasification as known principally from traditional coal gas (town gas) production. ($3C + O_2 + H_2O \rightarrow H_2 + 3\ CO; C + 2\ H_2O \rightarrow CO_2 + 2\ H_2; CO + H_2O \rightarrow CO_2 + H_2$.) In the integrated gasification combined cycle (IGCC) the obtained hydrogen is burnt in a modified furnace, while carbon dioxide is absorbed in amine solution. In this case, the amine scrubbing is cheaper than usual, because carbon dioxide is generated in a more concentrated form. Besides, it might be possible to separate hydrogen from carbon dioxide by passing the gas mix through membranes. A handful of pilot plants have demonstrated IGCC technology in Europe and America, but capital costs are high compared to traditional coal power plants. (Part of the attraction is that it produces hydrogen, which can either be combusted or used for other applications, including fuel cells. President Bush in 2003 unveiled a subsidized scheme to build a zero-emissions IGCC plant called FutureGen by 2013.)[104]

To be sure, capturing and sequestering carbon dioxide is likely a solution for stationary fossil fuel burning only. In terms of mobile applications, ships could perhaps release their exhaust gas underneath the water surface through a towed pipe, or throw blocks of dry ice, that is, solid carbon dioxide, overboard. (Obviously, the production and handling of dry ice would be expensive.) A solution to capture carbon dioxide emitted from cars and planes would be even more difficult and expensive.[105]

Artificially Cooling the Climate

Another approach to deal with climate change would be to focus less on restricting fossil fuel burning and other activities that release greenhouse gases, and to attempt to actively cool the climate to compensate for any observed global warming. The principal strategy would be to imitate the exhausts of major volcanic eruptions, which are known to cool the climate for several years. More specifically, this would involve injecting sulfur into the stratosphere, perhaps by burning S_2 or H_2S carried into the stratosphere on balloons or by artillery guns to produce SO_2. In the stratosphere, chemical and micro-physical processes would then convert SO_2 into sub-micrometer sulfate particles. (Reflective particles remain in the stratosphere for perhaps two years, compared to a week in the troposphere.) To enhance the residence time of the material in the stratosphere and minimize the required mass, the reactants might be released near the tropical upward branch of the stratospheric circulation system.[106]

In principle, this idea has been around since the 1970s,[107] but it has not caught on, because climate change did not yet cause any severe harm, and because there is general agreement that lowering greenhouse gas emissions would be the better solution from the technical standpoint, mainly because it is less intrusive. Besides, it would be difficult to achieve international consensus on such geo-engineering attempts affecting global climate, not least because any sort of climate change, anthropogenic or not, will likely create winners and losers, with nations being better off under new, warmer climate conditions certainly doing all they could to avert such a sulfur experiment. Nevertheless, in light of the failed attempts to reduce global greenhouse emissions, and the prospect that fossil fuels will defend their central position to enable the world's less developed regions to advance towards higher energy consumption (with all its social, economic, and other benefits), it is somewhat reassuring that there remains a sort of technological trump card, if that is what it is, up the sleeve in case a rapid rate of global climate warming needs to be countered. The very attraction of the sulfur idea is that it can be employed quickly, and that the effects remain for only a few years—long enough to make such attempt worthwhile, and short enough to avoid some sort of unexpected adverse effect getting a chance to stick around long-term. Nevertheless, a lot more research will be necessary to clarify if such approach is truly feasible and responsible.

Permissions Trading

An economic tool to reduce carbon emissions that has gained wide international acceptance is the cap-and-trade mechanism. This is similar to a fishing quota, for instance, where a government or a central organization decides the maximum amount of fish to be caught, and distributes (weight-limited) fishing permits between fishermen (or nations). If these permits can be traded, the more efficient fishermen will eventually pay the less efficient fishermen for staying off the water, and everybody is better off in economic terms while overfishing is avoided. Similarly, defining a total carbon emissions cap and distributing tradable carbon emissions permits leads to an increase in energy efficiency. Essentially, a price is put on industrial CO_2 emissions, and rising CO_2 prices accelerate energy innovations: the buyer of allowances is paying a charge for emitting, while the seller is being rewarded for having reduced emissions. The European Union Emissions Trading Scheme (EU ETS) was launched on January 1, 2005 as the world's first large experiment with an emissions trading system for carbon dioxide.[108] Three years later, by January 2008, this enormous multi-national emissions trading scheme and cornerstone of the EU's strategy for fighting climate change covered the 27 EU Member States (plus Norway, Iceland, and Liechtenstein) with more than 10,000 installations in the energy and industrial sectors that are collectively responsible for close to half of the EU's emissions of CO_2. (For the five-year trading period 2008–2012 the national emissions from EU ETS sectors have been capped at an average of around 6.5 percent to help ensuring that the EU as a whole, and member states individually, deliver on their Kyoto commitments.)[109]

Nevertheless the trading of greenhouse gas emissions was in principle an American innovation, pioneered when corporations began voluntarily trading greenhouse gas emission allowances on the Chicago Climate Exchange (CCX) in 2003. As of 2008, this was North America's only legally binding multi-sectoral, integrated greenhouse gas emission registry, reduction and trading system, and the only available mechanism through which U.S. based entities may engage in the integrated carbon market with a linked reduction and trading system. The CCX covers emissions of all six major greenhouse gases (GHGs), with offset-projects worldwide. Eligible emission offset-projects include carbon sequestration, reforestation, landfill and agricultural methane combustion, and switching to lower-emitting fuels such as biomass-based fuels. CCX members range from large industrial concerns to utilities to universities to non-governmental organizations to cities to farmers to single states. (The State of New Mexico was the first U.S. state to join CCX.)[110] Buying and selling emissions rights was expected to top $70 billion by the end of 2008. Meanwhile the Climate Security Act of 2007, a "bill to direct the Administrator of the Environmental Protection Agency to establish a program to decrease emissions of greenhouse gases, and for other purposes," was under consideration.[111] (The purpose of this bill was to create a national cap-and-trade scheme for greenhouse gas emissions.)

An international carbon permissions trading scheme had not been installed as of 2008. According to the United Nations, this would be essential to reduce global carbon dioxide emissions, but it involves the difficult issue of having to distribute carbon emission permits among countries of widely varying industrialization and carbon emission levels. The majority of all people on Earth currently live in developing countries, where more (albeit also more efficient) energy consumption has to be promoted, rather than discouraged or restricted.

Renewable Energy

Ultimately, a large part of energy consumption should, and will, be shifted to carbon-neutral renewable energy sources, which replenish themselves.[112] Fossil fuels are a non-renewable source of energy, because they cannot be replenished in any time-frame meaningful to human civilizations. (It takes hundreds of millions of years for fossil fuels to be generated underground.) Uranium, used in current nuclear power plants, and even abundant deuterium, the potential raw material for future nuclear technology, are principally non-renewables as well, since they occur on Earth in limited quantities. Most renewable energy sources, in contrast, rely on solar radiation that constantly arrives on Earth from the sun. They will thus be there as long as the sun is there, and there is no need to worry: the material that serves as fuel for the nuclear reactions that make the sun shine will last for another 4.5 billion years or so. The solar energy that arrives on the planet is actually enormous. About 11,000 times as much energy as we consume annually worldwide in commercial energy (assumed as 84 billion barrels of oil equivalent) hits the atmosphere each year. (A popular statement at the end of the 20th century was that every day more solar energy arrives at the Earth than the total amount of energy the planet's 6 billion inhabitants would consume in 30 years.) But this energy is difficult to harness. Twenty-one percent of the incoming radiation is reflected right back into space, nineteen percent is absorbed by the atmosphere, 25 percent evaporates water, 27 percent heats the surface, three percent is reflected by the surface, and 0.06 percent goes into photosynthesis.[113] (Thus less than a tenth of a percent of solar radiation that actually reaches the surface drives the growth of plants and algae.) Hydroelectric power schemes recover some of the solar energy that evaporates the planet's water; windmills and wave power stations recover some of the solar energy that heats up the atmosphere (unevenly); biomass directly burned or processed into gaseous or liquid fuels captures some of the solar energy that drives photosynthesis; and solar cells that produce electricity or collectors that heat water directly utilize sunshine.[114]

Tidal power stations rely on non-solar renewable energy as they are driven by the gravitational interaction between Earth and moon. Geothermal energy systems are considered quasi-renewable as they are driven by a very large energy reservoir, the Earth's hot interior. (The heat inside Earth stems from collisions that turned kinetic energy [speed energy] into heat when Earth initially consolidated from smaller pieces that gravitationally attracted one other. It is also sustained by nuclear reactions occurring in the Earth's core: the planet's interior will stay hot throughout the planet's lifetime.) Renewable energy currently accounts for a very small share of commercial energy consumption. Even if large hydroelectric schemes are included, this share is currently no more than seven (according to some estimates up to nine) percent, with the percentage expected to rise to eight percent (or, according to some estimates, even to slightly decrease) by the year 2030.[115] (To be sure, non-commercial biofuels from plant and animal sources are an important source of energy in many developing countries. Some 2.5 billion people depend on traditional biomass as their main fuel for cooking, for instance.)[116]

So far, hydropower is the only renewable energy source of global industrial importance. All other renewable energy sources combined account for about two percent to the world's commercial energy mix. Three quarters of the non-hydro renewable energy is comprised by modern biomass, while one quarter is being shared among solar energy, wind energy, geothermal, and small hydro-electric schemes. (The term modern biomass differentiates biomass for commercial energy production from traditional biomass.) Growth of the renewable energy share depends on the cost of fossil fuels, and support by government policies. Much of the overall growth in renewable energy consumption until 2030 is expected to be driven by new, medium-or large-scale hydroelectric projects, particularly in developing Asia and in South America. In fully industrialized countries, growth in renewables is concentrated in the non-hydroelectric sector, including wind, solar, geothermal, municipal solid waste, and biomass. However, this growth is relatively low in terms of production capacity installed despite increases in fossil fuel prices and government incentives. Governments support renewable energy, because it would otherwise be too expensive to compete with fossil fuels. (Hydropower is the principal exception.) Reasons for this support tend to include both, the more publicized reason of climate protection and the less advertised goal to reduce dependency on imported energy (in form of Middle Eastern, Russian, and other oil and gas).

Unfortunately, renewables tend to share a serious problem: they deliver energy intermittently. Photovoltaic cells will not work well on a cloudy day and not at all during the night; windmills stand still when the breeze winds down; wave power fades away when the ocean is calm; tidal stations produce energy less than half a day; and hydroelectric power plants deliver the least energy during cold winter days (when much water is bound as ice while energy requirements for heating are high) or during warm summer periods (when much water evaporates and the water line is low while electricity needs for cooling are high). Thus, we either need to combine different renewable energy sources in a way to offset their intermittent nature (either stormy weather for windmills or sunny weather for photovoltaic cells), or much of the future of renewables will depend on ways to store the energy they produce and to deliver it efficiently whenever people require it. Energy storage may take many different forms. The energy obtained from renewable sources, usually in the form of electricity, may be stored as electrochemical energy by charging a battery; it may be used to produce an energy carrier such as hydrogen; it may be used to lift water from a lower to a higher reservoir; or it may be used to turn a flywheel that will at some later time power a generator. The problem is, however, that energy conversions are never fully efficient, and every time energy is converted, some of it is lost for the end use. Wind, for instance, may be converted from mechanical energy (moving rotor blades) to electricity (via a generator) to electrochemical storage energy (by charging a battery), and back to electricity (by discharging the battery) and to mechanical energy (via an electric motor driving the wheels of an electric vehicle), with every step incurring an energy loss. In addition to conversion losses, current energy storage systems often involve a considerable loss of stored energy. However, rechargeable batteries are constantly being improved and advances in material science have allowed for the construction of very efficient flywheels, which store electricity as kinetic energy.[117]

NOTES

3. Much of the basic information in the climate change chapter, though not conclusions and interpretations, derives from John Houghton, *Global Warming: The Complete Briefing* (Cambridge: Cambridge University Press, 1997).

4. C. D. Keeling et al., "Atmospheric Carbon Dioxide Variations at Mauna Loa Observatory, Hawaii," *Tellus 28* (1976 and updates): 538–51.

5. The time-period after which temperature changes are followed by changes in atmospheric carbon dioxide is usually given as 800 years. "The Great Global Warming Swindle—The Arguments," *Channel 4*, http://www.channel4.com/science/micro sites/G/great_global_warming_swindle/arguments_2.html.

6. A quite silly debate has emerged around the question if it was slightly colder or slightly warmer in the Middle Ages than it was in the early 21st century. Why should it matter? The point is that we are not exactly sure what has been driving the climate in the past, and if these drivers are still at work. The National Academy of Sciences' Board on Atmospheric Sciences and Climate in 2006 commented on the issue as follows: "It can be said with a high level of confidence that global mean surface temperature was higher during the last few decades of the 20th century than during any comparable period during the preceding four centuries." It continues: "Less confidence can be placed in large-scale surface temperature reconstructions for the period from A.D. 900 to 1600." "The main reason that our confidence in large-scale surface temperature reconstructions is lower before A.D. 1600 . . . is the relative scarcity of precisely dated proxy evidence." "Presently available proxy evidence indicates that temperatures at many, but not all, individual locations were higher during the past 25 years than during any period of comparable length since A.D. 900."

The Board also stated that the conclusion that the late 20th century warmth in the Northern Hemisphere was unprecedented during the last 1,000 years has been supported by an array of evidence. (Note, that this statement concerns the Northern Hemisphere, not the planet as a whole.) "Not all individual proxy records indicate that the recent warmth is unprecedented, although a larger fraction of geographically diverse sites experienced exceptional warmth during the late 20th century than during any other extended period from A.D. 900 onward." National Academy of Sciences-Board on Atmospheric Sciences and Climate (BASC), *Surface Temperature Reconstructions for the Last 2,000 Years* (Washington: The National Academies Press, 2006), http://books.nap.edu/openbook.php?record_id=11676&page=3.

The United Nation's Intergovernmental Panel on Climate Change comments similarly in the Fourth Assessment Report of 2007: "Average Northern Hemisphere temperatures during the second half of the 20th century were very likely higher than during any other 50-year period in the last 500 years and likely the highest in at least the past 1300 years." UN Intergovernmental Panel on Climate Change, *Fourth Assessment Report: Climate Change 2007: "The Physical Science Basis, Summary for Policymakers*, http://www.ipcc.ch/SPM2feb07.pdf. "20th Century Climate Not So Hot," Press Release by the Harvard-Smithsonian Center for Astrophysics, March 31, 2003, http://cfa-www.harvard.edu/press/archive/pr0310.html.

7. The cold period of the late Coal Age is often incorrectly presented as a sort of climate base-line. Accordingly, the notion "increase above pre-industrial temperature," as used by the United Nation's Intergovernmental Panel on Climate Change, can be misleading. It refers to cold Coal Age climates, while the warm climate of the Middle Ages was also "pre-industrial." See, for instance, "Equilibrium Global Mean Temperature Increase above Pre-industrial (°C)" in *Contribution of Working Group III to the Fourth Assessment Report of the Intergovernmental Panel on Climate Change*, Summary for Policymakers, May 2007, page 17, http://www.ipcc.ch/pdf/assessment-report/ar4/wg3/ar4-wg3-spm.pdf.

8. The particular climatic conditions at the end of 2007 actually allowed this to be predicted ahead of time. See Mathew Carr, "Global Climate in 2008 May Be Coolest Since 2000," *Bloomberg*, January 3, 2008, http://www.bloomberg.com/apps/news?pid=20601081&sid=aBBw.Bw5TOpg&refer=australia.

9. Richard Luscombe, "Blogger Gets Hot and Bothered over Nasa's Climate Data error," *The Guardian*, August 16, 2007, http://www.guardian.co.uk/environment/2007/aug/16/1; Paul Driessen, "Global warming insanity," *The Washington Times*, September 12, 2007, http://washingtontimes.com/article/20070912/COMMENTARY/109120009/1012.

10. "GISS Surface Temperature Analysis-August 2007 Update and Effects," NASA Goddard Institute for Space Studies, http://data.giss.nasa.gov/gistemp/updates/200708.html.

11. According to Henry's Law, the amount of a given gas dissolved in a given type and volume of liquid is directly proportional to the partial pressure of that gas in equilibrium with that liquid. Exactly how much gas is dissolved, depends on the value of the Henry constant, which is temperature-dependent.

12. Among the factors that influence cloud formation and characteristics are the ease at which moist air in the tropics travels into the upper atmosphere, the speed with which raindrops fatten and the level of humidity required for clouds to form. Each aspect has a big impact on the degree of warming predicted by models. "Climatology: Grey-sky thinking," *The Economist*, July 5, 2007, http://www.economist.com/science/displaystory.cfm?story_id=9433721

13. The albedo, or reflective power, of the planet is enhanced because bare light surfaces act more like a mirror to solar radiation than dark forest areas do.

14. This is in terms of radiative forcing (contribution to global warming). H. Steinfeld et al., "Livestock's Long Shadow: Environmental Issues and Options," *LEAD-Livestock Environment and Development Initiative*, coordinated by the UN FAO's Animal Production and Health Division, (Rome: FAO, 2006), http://www.virtualcentre.org/en/library/key_pub/longshad/A0701E00.htm, http://www.virtualcentre.org/en/library/key_pub/longshad/A0701E00.pdf.

15. United Nations Environment Programme, *Global Environment Outlook, GEO Year Book 2003, International Environmental Agenda*, Box 4: Greening of the biosphere (Nairobi: UNEP, 2003), http://new.unep.org/geo/yearbook/yb2003/box7a.htm.

16. Intergovernmental Panel on Climate Change, http://www.ipcc.ch/. "IPCC Fourth Assessment Report," Working Group I Report "The Physical Science Basis,"

http://www.ipcc.ch/ipccreports/ar4-wg1.htm; "Fourth Assessment Report: Climate Change 2007: Synthesis Report," Summary for Policymakers, http://www.ipcc.ch/pdf/assessment-report/ar4/syr/ar4_syr_spm.pdf; "Climate Change 2007: The Physical Science Basis," Summary for Policymakers, http://www.ipcc.ch/SPM2feb07.pdf.

17. David M. Holland et al., "Acceleration of Jakobshavn Isbræ Triggered by Warm Subsurface Ocean Waters," *Nature Geoscience* 1 (2008): 659–664, http://www.nature.com/ngeo/journal/v1/n10/full/ngeo316.html.

18. James E. Hansen, "As Pure as Snow," *Science Briefs*, December 2003, http://www.giss.nasa.gov/research/briefs/hansen_10/.

19. Journal of Atmospheric Chemistry and Physics, quoted in Nicolai Østergaard, "Kinesiske kraftværker soder indlandsisen," *Ingeniøren*, September 19, 2008, http://ing.dk/artikel/91415?bund.

20. Peter Schwartz and Doug Randall, "An Abrupt Climate Change Scenario and Its Implications for United States National Security—A Report Commissioned by the U.S. Defense Department," October 2003, http://www.ems.org/climate/pentagon_climatechange.pdf; "Storm over Pentagon climate scenario," *MSNBC News*, February 26, 2004, http://msnbc.msn.com/id/4379905/.

21. The difference between the two is net primary production (NPP). Respiration provides a plant with chemical energy that had been stored in organic molecules. It uses oxygen from the air and converts carbohydrates, or rather sugars, back into CO_2 and water. Respiration is crucial in the absence of light. Thus, it is important during the night and dark seasons. Trees that shed their leaves, for instance, survive during the winter through respiration.

22. Find more information on respiration in Hans Lambers and Miquel Ribas-Carbo, eds., "Plant Respiration: From Cell to Ecosystem," in *Advances in Photosynthesis and Respiration Series* 18 (Dordrecht: Springer, 2005).

23. United Nations Environment Programme, *Global Environment Outlook*, *GEO Year Book 2003*, *International Environmental Agenda*, Box 4: Greening of the biosphere (Nairobi: UNEP, 2003), http://new.unep.org/geo/yearbook/yb2003/box7a.htm.

24. Michael R. Raupach et al., "Global and Regional Drivers of Accelerating CO_2 Emissions," *PNAS* (*Proceedings of the National Academy of Sciences of the United States of America*) 104 (24) (2007): 10288–10293, http://www.pnas.org/cgi/reprint/0700609104v1.pdf, http://www.pnas.org/content/104/24/10288/suppl/DC1.

25. Ibid.

26. In 2005, for instance, the anthropogenic CO_2 emissions amounted to some 7.6 Gt worth of carbon equivalent.

27. Robert Lee Holz, "Tracking Carbon Dioxide," *The Wall Street Journal*, December 28–30, 2007, Science Journal.

28. "El Niño/Southern Oscillation (Enso) Diagnostic Discussion," Climate Prediction Center/Ncep, National Weather Service, National Oceanic and Atmospheric Administration, United States Department of Commerce, January 10, 2008, http://www.cpc.noaa.gov/products/analysis_monitoring/enso_advisory/ensodisc.html; Mathew Carr, "Global Climate in 2008 May Be Coolest Since 2000," *Bloomberg*, January 3, 2008, http://www.bloomberg.com/apps/news?pid=20601081&sid=aBBw.Bw5TOpg&refer=australia.

29. El Niño, the boy child (that is, Jesus), arrives at Christmas according to Christian tradition.

30. Seth Borenstein, "There Goes El Nino, Here Comes La Nina: Forecasters Expect Nino's 'Evil Twin Sister' To Bring More Hurricanes, More Drought," *The Associated Press,* February 28, 2007, Washington, http://www.cbsnews.com/stories/2007/02/28/tech/main2523483.shtml.

31. Find a reconstruction of the sunspot number covering the past 11,400 years based on dendrochronologically (growth rings in trees and aged wood) dated radiocarbon concentrations in S.K. Solanki et al., "Unusual Activity of the Sun during Recent Decades Compared to the Previous 11,000 Years," *Nature 431* (2004): 1084–1087, http://www.ncdc.noaa.gov/paleo/pubs/solanki2004/solanki2004.html.

32. Jan Veizer, "Celestial Climate Driver: A Perspective from Four Billion Years of the Carbon Cycle," *Geoscience Canada* 32, 1 (2005), http://www.gac.ca/publications/geoscience/GACV32No1Veizer.pdf. Henrik Svensmark, "A brief summary of cosmoclimatology," *Astronomy & Geophysic* 48, 1 (2007), Danish National Space Center (DTU), http://www.spacecenter.dk/research/sun-climate/cosmoclimatology/a-brief-summary-on-cosmoclimatology.

33. John E. Beckman and Terence J. Mahoney, "The Maunder Minimum and Climate Change: Have Historical Records Aided Current Research?," *Library and Information Services in Astronomy III, ASP Conference Series* 153 (1998), http://www.stsci.edu/stsci/meetings/lisa3/beckmanj.html.

34. S.K. Solanki et al, "Unusual Activity"; Henrik Svensmark and Eigil Friis-Christensen, "Variation of Cosmic Ray Flux and Global Cloud Coverage—A Missing Link in Solar-climate Relationships," *Journal of Atmospheric and Solar-Terrestrial Physics* 59, 11 (1997):1225–1232, http://www.sciencedirect.com/science?_ob=ArticleURL&_udi=B6VHB-3SW03B6-H&_user=10&_coverDate=07%2F31%2F1997&_rdoc=1&_fmt=summary&_orig=browse&_cdi=6062&_sort=d&_docanchor=&view=c&_ct=1&_acct=C000050221&_version=1&_urlVersion=0&_userid=10&md5=b2e3b6773a039b6b02396b1b02df3492.

Torben Stockflet Jørgensen and Aksel Walløe Hansen, "Comments on 'Variation of Cosmic Ray Flux and Global Cloud Coverage'" *Journal of Atmospheric and Solar-Terrestrial Physics* 62, 1 (2000): 73–77, http://www.sciencedirect.com/science?_ob=ArticleURL&_udi=B6VHB-3YN9385-8&_user=10&_rdoc=1&_fmt=&_orig=search&_sort=d&view=c&_acct=C000050221&_version=1&_urlVersion=0&_userid=10&md5=be6aef97b4a3f3cc8e9bb63edc59d5e9; E. Friis-Christensen and K. Lassen, "Length of the Solar Cycle: An Indicator of Solar Activity Closely Associated with Climate," *Science* 254 (1991): 698–700; K. Lassen and E. Friis-Christensen, "Variability of the Solar Cycle Length during the Past Five Centuries and the Apparent Association with Terrestrial Climate," *Journal of Atmospheric and Terrestrial Physics* 57, 8 (1995): 835–845.

35. Possibly, such clouds would have a warming effect in the winter, when they keep heat in the atmosphere, and a cooling effect in the summer, when the effect predominates that clouds reflect solar radiation directly back into space. Find a summary of this and other arguments to contradict the greenhouse gas theory of global warming at: "The Great Global Warming Swindle—The Arguments," *Channel 4,* http://www.channel4.com/science/microsites/G/great_global_warming_swindle/arguments.html; "The Great Global Warming Swindle," ABC Television, http://www.abc.net.au/tv/swindle/; John Houghton, "The Great Global Warming Swindle-Programme directed by Martin Durkin, on Channel 4 on Thursday 8 March 2007-Critique," http://www.jri.org.uk/news/Critique_Channel4_Global_Warming_Swindle.pdf.

36. To be sure, the role of the sun as the principal driver of global warming in the first part of the 20th century is appreciated by the United Nation's IPCC. See, for instance, the remarks by John Houghton, former Chairman of the Scientific Assessment Working Group of the Intergovernmental Panel on Climate Change, at John Houghton, "The Great Global Warming Swindle-Programme directed by Martin Durkin, on Channel 4 on Thursday 8 March 2007-Critique," http://www.jri.org.uk/news/Critique_Channel4_Global_Warming_Swindle.pdf.

37. A list of "Major Volcanic Eruptions Since 1900," though with focus on casualties rather than particles emitted into the atmosphere, shows six major volcano eruptions in the first four decades of the 20th century: Santa María, Guatemala, 1902, Pelée, Martinique, 1902, Taal, Philippines, 1911, Kelut, Java, Indonesia, 1919, Merapi, Indonesia, 1930, Rabaul Caldera, Papua New Guinea, 1937. This compares well to the seven major outbreaks recorded after 1970: St. Helens, United States, 1980, El Chichón, Mexico, 1982, Nevado del Ruiz, Colombia, 1985, Lake Nyos, Cameroon, 1986, Pinatubo, Luzon, Philippines, 1991–1996, Unzen, Japan, 1991, Mayon, Philippines, 1993. However, there have only been three major volcano outbreaks in the period 1940–1975: Lamington, Papua New Guinea, 1951, Hibok Hibok, Philippines, 1951, Agung, Indonesia, 1963. "Major Volcanic Eruptions Since 1900," United States Geological Survey in Microsoft Encarta Online Encyclopedia, http://encarta.msn.com/media_701500557_761570122_-1_1/Major_Volcanic_Eruptions_Since_1900.html. Find a more complete list of volcanic eruptions in: "Large Holocene Eruptions," Global Volcanism Program, Department of Mineral Sciences, National Museum of Natural History, Smithsonian Institution, http://www.volcano.si.edu/world/large eruptions.cfm.

38. Sulfur emissions data can be found in D. I. Stern, "Global Sulfur Emissions from 1850 to 2000," *Chemosphere* 58 (2005): 163–175. Stern used sulfur emissions data from 1850 to 1990 for most countries in the world from the A.S.L. & Associates database (http://www.asl-associates.com/sulfur1.htm) and developed a time series of sulfur emissions by country for 1990. The abstract is available at these addresses: http://www.ncbi.nlm.nih.gov/pubmed/15571748?ordinalpos=2&itool=EntrezSystem2.PEntrez.Pubmed.Pubmed_ResultsPanel.Pubmed_RVDocSum; http://ideas.repec.org/p/rpi/rpiwpe/0311.html.

The following paper describes a climate model that used estimates of anthropogenic SO_2 emissions taken from Orn et al. (1996) for 1860–1970, the Global Emissions Inventory Activity (GEIA) 1B dataset for 1985, and the preliminary SRES datasets for 1990 and 2000 (Nakic´enovic´ and Swart, 2000) and linearly interpolated between these times. Peter A. Stott, Gareth S. Jones, and John F. B. Mitchell, "Do Models Underestimate the Solar Contribution to Recent Climate Change?" *Journal of Climate* 16 (2003):4079–4093, http://climate.envsci.rutgers.edu/pdf/StottEtAl.pdf; G. Orn, U. Hansson, and H. Rodhe, "Historical Worldwide Emissions of Anthropogenic Sulfur: 1860–1985," *Tech. Rep.* CM-91 (1996), Department of Meteorology/International Meteorological Institute, Stockholm University. N. Nakićenović and R. Swart, ed., *Special Report on Emission Scenarios* (Cambridge: Cambridge University Press, 2000).

Sulfur dioxide emissions decreased markedly in such Western countries as the United States (by ca. 17 percent between 1970 and 1980). China emitted 25,376 thousand tons of sulfur dioxide in 1990, and 34,205 tons in 2000, according to emissions data by the Netherlands Environmental Assessment Agency, with the latter figure probably being about the same level as in the United States in the 1960s. Apparently China increased its sulfur dioxide emissions by another 27 percent between 2000 and 2005. "Emissions Data," Netherlands Environmental Assessment Agency, http://www.mnp.nl/edgar/model/; "Sulfur Dioxide," U.S. Environmental Protection Agency, http://www.epa.gov/air/airtrends/sulfur.html; "China Has its Worst Spell of Acid Rain," *United Press International*, September 22, 2006, http://www.upi.com/NewsTrack/Science/2006/09/22/china_has_its_worst_spell_of_acid_rain/7049/.

The following paper states that "Significant changes in sulfur emissions have occurred over the last decades with decrease in the Unites States and Europe and increase in Southeast Asia. U.S., European, and Chinese SO2 emissions have changed by -17.6%, -47.5%, and +93%, respectively." Tore F. Berglen et al., "A Global Model of the Coupled Sulfur/oxidant Chemistry in the Troposphere: The Sulfur Cycle," *Journal of Geophysical Research* 109, D19310, October 9, 2004, http://www.agu.org/pubs/crossref/2004/2003JD003948.shtml.

39. "IPCC Fourth Assessment Report, Second Draft chapter 2," February 28, 2006, page 30. Both, the IPCC report and the Stern paper are quoted by Steve McIntyre, "RMS and Sulphate Emissions," Climate Audit, May 6, 2007, http://www.climateaudit.org/?p=1536.

40. Sulfur dioxide and dust emissions from anthropogenic sources at industrial centers are generally expected to result in local cooling, while volcano outbreaks may have a more regional or even global effect, as their emissions reach higher in the atmosphere. D. D. Lucas and H. Akimoto, "Contributions of Anthropogenic and Natural Sources of Sulfur to SO_2, $H_2SO_4(g)$ and Nanoparticle Formation," *Atmos. Chem. Phys. Discuss* 7 (2007): 7679–7721, www.atmos-chem-phys-discuss.net/7/7679/2007/, http://www.atmos-chem-phys-discuss.net/7/7679/2007/acpd-7-7679-2007-print.pdf; P. J. Crutzen, "Albedo Enhancement by Stratospheric Sulfur Injections: A Contribution to Resolve a Policy Dilemma?," *Climatic Change* 77 (3–4) (2006): 211, http://www.springerlink.com/content/t1vn75m458373h63/. "Trends in Aerosols," Climate Change 2001: Working Group I: The Scientific Basis, IPCC, http://www.grida.no/climate/ipcc_tar/wg1/180.htm.

41. James Hansen et al., "Global Warming in the Twenty-First Century: An Alternative Scenario," *PNAS (Proceedings of the National Academy of Sciences of the United States of America)* 97, 18 (2000): 9875–9880, http://www.pnas.org/cgi/content/full/97/18/9875.

42. S. Solomon et al., "Summary for Policymakers-Climate Change 2007: The Physical Science Basis," ed. H.L. Miller, Contribution of Working Group I to the Fourth Assessment Report of the Intergovernmental Panel on Climate Change, IPCC, 2007, http://www.ipcc.ch/pdf/assessment-report/ar4/wg1/ar4-wg1-spm.pdf.

43. Peter A. Stott, Gareth S. Jones, and John F. B. Mitchell, "Do Models Underestimate the Solar Contribution to Recent Climate Change?" *Journal of Climate* 16 (2003):4079–4093, http://climate.envsci.rutgers.edu/pdf/StottEtAl.pdf.

44. "The good correspondence between the model simulations and the observations should not be overinterpreted. If the assumed forcings were correct, then this agreement would indicate that the model's climate sensitivity was realistic. Forcing uncertainties, however, admit a quite wide range of sensitivity possibilities." Gerald A. Meehl et al., "Combinations of Natural and Anthropogenic Forcings in Twentieth-

Century Climate," *Journal of Climate* 17 (2004) 3721–3727, American Meteorological Society, http://www.cgd.ucar.edu/ccr/publications/meehl_additivity.pdf; Gerald A. Meehl et al., "Solar and Greenhouse Gas Forcing and Climate Response in the Twentieth Century," *Journal of Climate* 16 (2003) 426, http://www.cgd.ucar.edu/ccr/publications/meehl_solar.pdf.

45. Mike Lockwood and Claus Fröhlich, "Recent Oppositely Directed Trends in Solar Climate Forcings and the Global Mean Surface Air Temperature," *Proc. R. Soc. A* (2007), http://publishing.royalsociety.org/media/proceedings_a/rspa20071880.pdf; Richard Black, "No Sun Link to Climate Change," *BBC News*, July 10, 2007, http://news.bbc.co.uk/2/hi/science/nature/6290228.stm.

46. H. Svensmark and E. Friis-Christensen, "Reply to Lockwood and Fröhlich—The Persistent Role of the Sun in Climate Forcing," *Scientific Report* 3, Danish National Space Center, Danish Technical University, 2007, http://www.spacecenter.dk/publications/scientific-report-series/Scient_No._3.pdf.

47. Energy Information Administration, *International Energy Outlook 2007*, Office of Integrated Analysis and Forecasting, http://www.eia.doe.gov/oiaf/archive/ieo07/index.html. Chapter 7: Energy-Related Carbon Dioxide Emissions, http://www.eia.doe.gov/oiaf/archive/ieo07/pdf/emissions.pdf.

48. OECD (Organization for Economic Cooperation and Development) member countries as of the end of 2007: Australia, Austria, Belgium, Canada, Czech Republic, Denmark, Finland, France, Germany, Greece, Hungary, Iceland, Ireland, Italy, Japan, Korea, Luxembourg, Mexico, Netherlands, New Zealand, Norway, Poland, Portugal, Slovak Republic, Spain, Sweden, Switzerland, Turkey, United Kingdom, United States, http://www.oecd.org.

49. Netherlands Environmental Assessment Agency, http://www.mnp.nl/en/index.html; John Vidal and David Adam, "China Overtakes US as World's Biggest CO_2 Emitter," *The Guardian*, June 19, 2007, http://www.guardian.co.uk/environment/2007/jun/19/china.usnews. Note that some of China's carbon dioxide emissions represent consumption in other nations: China has become a leading exporter of various products.

50. Energy Information Administration. *International Energy Outlook 2007*. Office of Integrated Analysis and Forecasting. http://www.eia.doe.gov/oiaf/archive/ieo07/index.html. Chapter 7: Energy-Related Carbon Dioxide Emissions, http://www.eia.doe.gov/oiaf/archive/ieo07/pdf/emissions.pdf.

51. John Locke, *The Second Treatise of Civil Government*, first published in 1690. Text of the sixth edition of 1764 available at: http://ebooks.adelaide.edu.au/l/locke/john/l81s/.

52. A somewhat similar situation evolved around the issue of passive smoking. Powerful industrial interests worked against the introduction of smoking bans in public places, until it was shown that involuntary (second hand) smokers are exposed to the same carcinogens as smokers. World Health Organization: International Agency for Research on Cancer, "Tobacco Smoke and Involuntary Smoking: Summary of Data Reported and Evaluation, Tobacco Smoking and Tobacco Smoke, Involuntary Smoking," *IARC Monographs on the Evaluation of Carcinogenic Risks to Humans* 83 (Geneva: WHO, 2002), http://monographs.iarc.fr/ENG/Monographs/vol83/volume83.pdf.

53. Lawrence Ulrich, "Facing New Mileage Rules, Porsche Preps a Hybrid S.U.V.," *New York Times*, January 6, 2008, http://www.nytimes.com/2008/01/06/automobiles/06HYBRID.html?_r=2&oref=slogin&ref=automobiles&pagewanted=print&oref=slogin.

54. "Kyoto Protocol to the United Nations Framework Convention on Climate Change," http://unfccc.int/kyoto_protocol/items/2830.php; http://unfccc.int/essential_background/kyoto_protocol/items/1678.php.
55. "Clinton Hails Global Warming Pact," *CNN*, December 11, 1997, http://www.cnn.com/ALLPOLITICS/1997/12/11/kyoto/.
56. "President Bush Discusses Global Climate Change," The White House, Press Release June 11, 2001, http://www.whitehouse.gov/news/releases/2001/06/20010611-2.html.
57. "Changes in GHG Emissions from 1990 to 2004 for Annex I Parties," United Nations Framework Convention on Climate Change, http://unfccc.int/files/essential_background/background_publications_htmlpdf/application/pdf/ghg_table_06.pdf.
58. Catherine Herrick and Bill Owens, "Rewriting The Science-Scientist Says Politicians Edit Global Warming Research," *60 Minutes*, CBS, March 19, 2006, http://www.cbsnews.com/stories/2006/03/17/60minutes/main1415985.shtml.
59. J. Hansen, "Can we defuse the global warming time bomb?" *NaturalScience* (2003), http://pubs.giss.nasa.gov/docs/2003/2003_Hansen.pdf.
60. Catherine Herrick and Bill Owens, "Rewriting The Science.
61. "Tropical Tropospheric Trends," RealClimate-Climate science from climate scientists, December 12, 2007, http://www.realclimate.org/index.php/archives/2007/12/tropical-troposphere-trends/.
62. H. Friedli et al, "Carbon-13/carbon-12 Ratios in Carbon Dioxide Extracted from Antarctic Ice," *Geophysical Research Letters* 11 (1984), http://chemport.cas.org/cgi-bin/sdcgi?APP=ftslink&action=reflink&origin=npg&version=1.0&coi=1:CAS:528:DyaL2MXivV2ltw%3D%3D&pissn=0028-0836&pyear=1985&md5=09da79795cc62c731c8e31d535ae29d8; C. Lorius et al., "A 150,000-year Climatic Record from Antarctic Ice," *Nature 316* (1985): 591–596, http://www.nature.com/nature/journal/v316/n6029/abs/316591a0.html.
63. To be sure, climate modeling had already been relatively advanced by this stage. In the United States, coupled atmosphere-ocean models had been set up in the 1965–1975 period.
64. Steven Milloy, "Must-See Global Warming TV," *Fox News*, March 19, 2007, http://www.foxnews.com/story/0,2933,258993,00.html.
65. Bob Carter, "There IS a Problem with Global Warming . . . It Stopped in 1998," *Telegraph*, April 9, 2006, http://www.telegraph.co.uk/opinion/main.jhtml?xml=/opinion/2006/04/09/do0907.xml&sSheet=/news/2006/04/09/ixworld.html.
66. Richard Lindzen, "Climate of Fear: Global-warming Alarmists Intimidate Dissenting Scientists into Silence," *The Wall Street Journal*, April 12, 2006, The Opinionjournal Archives, http://www.opinionjournal.com/extra/?id=110008220. Richard Lindzen is Alfred P. Sloan Professor of Atmospheric Science at MIT.
67. Ibid.
68. Gore has also been criticized for failing to practice what he preaches. At the end of *An Inconvenient Truth*, viewers are asked, "Are you ready to change the way you live?," and the movie's website has various suggestions on how you can "reduce your impact at home," including using less heating and air conditioning, using less hot water, etc. But the Tennessee Center for Policy Research reported that Gore's Nashville mansion consumed more than 20 times the electricity of the national average. In August 2006, the Gore mansion burned more than twice the electricity in a

single month as the average American family uses in an entire year. Gore's heated pool house alone uses more than $500 in electricity every month. Steven Milloy, "Al Gore's Inconvenient Electric Bill," *Fox News*, March 12, 2007, http://www.foxnews.com/story/0,2933,257958,00.html.

69. Patrick J. Michaels, "More Ice Than Ever," *American Spectator*, February 5, 2008, http://www.cato.org/pub_display.php?pub_id=9136. Patrick J. Michaels is a research professor of environmental sciences at the University of Virginia and a past president of the American Association of State Climatologists. He was program chair for the Committee on Applied Climatology of the American Meteorological Society, and is a contributing author and reviewer of the United Nations Intergovernmental Panel on Climate Change; P. T. Doran et al., "Antarctic Climate Cooling and Terrestrial Ecosystem Response," *Nature* 415 (2002): 517–20.

70. P. Chylek et al., "Greenland Warming of 1920–30 and 1995–2005," *Geophysical Research Letters* 33 (2006).

71. Patrick J. Michaels, "Is the Sky Really Falling? A Review of Recent Global Warming Scare Stories," *Trade Policy Analysis* 576 (2006), Cato Institute, http://www.cato.org/pubs/pas/pa576.pdf.

72. William Nordhaus, *Managing the Global Commons: The Economics of Climate Change* (Cambridge: MIT Press, 1994); William Nordhaus, "Count Before You leap: Economics of Climate Change," *The Economist*, July, 1990, http://www.uwmc.uwc.edu/geography/globcat/globwarm/nordhaus.htm.

73. This included the assumption that only taxes on carbon fuels would make consumers buy energy from environmentally-friendly sources, and the neglect of a scenario in which eco-taxes would be redistributed as focus investments that raise GDP. Ernst Ulrich von Weizsäcker, Amory B. Lovins, and L. Hunter Lovins, *Faktor vier. Doppelter Wohlstand—halbierter Naturverbrauch* (Munchen: Droemersche Verlagsanstalt Th. Knaur, 1995, 1996). English language edition: Ernst Ulrich von Weizsäcker, Amory B. Lovins, and L. Hunter Lovins, *Factor Four. Doubling Wealth—Halving Resource Use* (London: Earthscan, 1997). See also the Internet site of the Wuppertal Institute for Climate, Environment and Energy at http://www.wupperinst.org/FactorFour/).

74. Richard Lindzen, "Climate of Fear. Richard Lindzen, Alfred P. Sloan Professor of Atmospheric Science at MIT, argued in 2006 that the claim of an increased tropical storminess, as presented by the Intergovernmental Panel on Climate Change, is flawed. The claim rests on the assumption that a warmer world would have more evaporation, with latent heat providing more energy for disturbances. However, the ability of evaporation to drive tropical storms relies not only on temperature but also on humidity, with less humid air promoting the effect, while claims for starkly higher temperatures are based upon there being more humidity, not less. Richard S. Lindzen, "Global Warming: The Origin and Nature of the Alleged Scientific Consensus," *Regulation-The Cato Review of Business & Government* 15, No. 2 (1992), http://www.cato.org/pubs/regulation/regv15n2/reg15n2g.html.

75. According to one IPCC business-as-usual scenario, about 150 million people would become displaced by the impacts of global warming by 2050, that is, about 3 million per year on average. About 100 million would have to relocate due to sea-level rise and coastal flooding, and 50 million due to dislocation of agricultural production (areas of drought). If a resettling cost of $3,000 and $5,000 per person is assumed, the

costs will add up to $9 billion per year, excluding human cost associated with displacement, and the cost of social and political instabilities.

76. Bjørn Lomborg, "Stern Review: The Dodgy Numbers behind the Latest Warming Scare," *The Wall Street Journal*, November 2, 2006, The Opinionjournal Archives, http://www.opinionjournal.com/extra/?id=110009182.

77. Ibid.

78. Here is the entire rating list, divided into categories. Very good: 1. Disease-Diseases Control of HIV/AIDS, 2. Malnutrition-Providing micro nutrients, 3. Subsidies and Trade-Trade liberalization. Good: 4. Diseases-Control of malaria, 5. Malnutrition-Development of new agricultural technologies, 6. Sanitation & Water-Small-scale water technology for livelihoods, 7. Sanitation & Water-Community-managed water supply and sanitation, 8. Sanitation & Water-Research on water productivity in food production, Fair: 8. Government-Lowering the cost of starting a new business, 10. Migration-Lowering barriers to migration for skilled workers, 11. Malnutrition-Improving infant and child nutrition, 12. Malnutrition-Reducing the prevalence of low birth weight, Bad: 13. Diseases-Scaled-up basic health services, 14. Migration-Guest worker programs for the unskilled, 15. Climate-Optimal carbon tax, 16. Climate-The Kyoto Protocol, 17. Climate-Value-at-risk carbon tax. Copenhagen Consensus 2004, http://www.copenhagenconsensus.com/Default.aspx?ID=158.

Critics have argued that the conclusion that nothing should be done about climate change on the grounds that the costs of doing anything would exceed the benefits is flawed. The problem is that we are comparing costs of prevention to be paid now with possible damages or benefits in the far future. This involves choosing a discount rate, a rate of return demanded on invested capital. The Copenhagen Consensus conference chose to demand 5 percent, which critics said was much too high for such a long time-frame (one century). With a smaller chosen rate, investments into climate protection would make sense. Besides, revenues from eco-taxes could be used to solve other, more pressing problems in developing countries.

However, here is another, somewhat similar, study that concludes that "Halting climate change would reduce cumulative mortality from various climate-sensitive threats, namely, hunger, malaria, and coastal flooding, by 4–10 percent in 2085, while increasing populations at risk from water stress and possibly worsening matters for biodiversity. But according to cost information from the UN Millennium Program and the IPCC, measures focused specifically on reducing vulnerability to these threats would reduce cumulative mortality from these risks by 50–75 percent at a fraction of the cost of reducing greenhouse gases (GHGs). Simultaneously, such measures would reduce major hurdles to the developing world's sustainable economic development, the lack of which is why it is most vulnerable to climate change. The world can best combat climate change and advance well-being, particularly of the world's most vulnerable populations, by reducing present-day vulnerabilities to climate-sensitive problems that could be exacerbated by climate change rather than through overly aggressive GHG reductions." Indur Goklany, "What to Do about Climate Change," *Policy Analysis* 609, February 5, (2008), Cato Institute, http://www.cato.org/pub_display.php?pub_id=9125.

79. EIA Service Report: "Federal Energy Subsidies: Direct and Indirect Interventions in Energy Markets," DOE/EIA/SR/EMEU-92–02, November 20, 1992, En-

ergy Information Administration, Office of Energy Markets and End Use, http://tonto.eia.doe.gov/ftproot/service/emeu9202.pdf; Roland Hwang, "Money Down the Pipeline: The Hidden Subsidies to the Oil Industry," Union of Concerned Scientists, http://www.ucsusa.org/publication.cfm?publicationID=149; Doug Koplow and John Dernbach, "Federal Fossil Fuel Subsidies and Greenhouse Gas Emissions: A Case Study of Increasing Transparency for Fiscal Policy," *Annual Review of Energy and the Environment* 26 (2001): 361–389, http://arjournals.annualreviews.org/doi/full/10.1146/annurev.energy.26.1.361; Doug Koplow and John Dernbach, *Fueling Global Warming—Federal Subsidies to Oil in the United States*, http://archive.greenpeace.org/~climate/oil/fdsub.html; Friends of the Earth, "Analysis of the Bush Energy Plan," http://www.foe.org/camps/leg/bushwatch/energyplan.pdf; Greenpeace, "The Subsidy Scandal—The European Clash between Environmental Rhetoric and Public Spending," http://archive.greenpeace.org/~comms/97/climate/eusub.html.

80. As of early 2008, the rate of planetary warming was falling in line with the low end of 21st century projections made by the UN's Intergovernmental Panel on Climate Change. Patrick J. Michaels, "Carbon Copies," *American Spectator*, February 27, (2008), http://www.cato.org/pub_display.php?pub_id=9242.

81. The word "charge" ultimately derives from the Latin word carrus, a wheeled vehicle.

82. In the following paper NASA scientists argued in 1971 that "An increase by only a factor of 4 in global aerosol background concentration may be sufficient to reduce the surface temperature by as much as 3.5° K. If sustained over a period of several years, such a temperature decrease over the whole globe is believed to be sufficient to trigger an ice age." S. I. Rasool and S. H. Schneider, "Atmospheric Carbon Dioxide and Aerosols: Effects of Large Increases on Global Climate," *Science* 173 (3992) (1971):138–141, http://www.sciencemag.org/cgi/content/abstract/173/3992/138.

83. The described record is based on data from Vostok ice cores. Deep-sea cores can reveal the temperature record even further into the past than ice cores can. The methods are similar as cores drilled from the sea bed contain shells of microscopic marine organisms consisting of calcium carbonate: variations in the ratio of two oxygen isotopes in the carbonate of these shells serve as a sensitive indicator of sea temperature at the time the organisms were living. A sea-core temperature record has been established for the past two million years. The earlier of the 20 or so recorded cycles were different from the more recent ones, as both the warming and cooling periods were roughly 40,000 years long.

84. The cycles are known as Milankovitch Cycles, named for the initial developer of this theory. "Milutin Milankovitch," NASA Earth Observatory, http://earthobservatory.nasa.gov/Library/Giants/Milankovitch/.

85. To be sure, there were quite a few scientists who were concerned about anthropogenic climate change, including a warming caused by anthropogenic carbon dioxide emissions, in the early 1970s, when the measurements at Mauna Loa had shown the consistent increase in atmospheric CO_2 year after year from the late 1950s. The following 1970 study, prepared as input to the UN Conference on the Human Environment (Stockholm, June 5–16, 1972), discusses both carbon dioxide and particles in the atmosphere: Study of Critical Environmental Problems, *Man's Impact on the Global Environment* (Cambridge, MA: MIT Press, 1970). Published a year later, *Study*

of Man's Impact on Climate: Inadvertent Climate Modification (Cambridge, MA: MIT Press, 1971) was prepared as input to the same UN conference.

86. It was Bolin, who was supposed to accept in Sweden the 2007 Nobel Prize for Peace that the Intergovernmental Panel of Climate Change was sharing with Al Gore for the "efforts to build up and disseminate greater knowledge about man-made climate change, and to lay the foundations for the measures that are needed to counteract such change," but he was already ill and passed away in Stockholm on December 30, 2007. Bolin was a leading figure at the now legendary meeting in Villach, Austria, when scientists discussed the future threat from climate change in 1985. The meeting proved a catalyst, resulting in 1988 in the formation of the IPCC by the UN. Fred Pearce, "Bert Bolin: Meteorologist and First Chair of the IPCC Who Cajoled the World into Action on Climate Change," *The Independent*, January 5, 2008, http://www.independent.co.uk/news/obituaries/bert-bolin-meteorologist-and-first-chair-of-the-ipcc-who-cajoled-the-world-into-action-on-climate-change-768355.html; The Nobel Peace Prize 2007, The Nobel Foundation, http://nobelprize.org/nobel_prizes/peace/laureates/2007/index.html.

87. A. Berger and M. F. Loutre, "Climate: An Exceptionally Long Interglacial Ahead?," *Science* 297 (5585) (2002):1287–1288, http://www.sciencemag.org/cgi/content/summary/297/5585/1287.

88. P. Forster et al., "Changes in Atmospheric Constituents and in Radiative Forcing," in *Climate Change 2007: The Physical Science Basis. Contribution of Working Group I to the Fourth Assessment Report of the Intergovernmental Panel on Climate Change*, ed. S. Solomon et al. (Cambridge/New York: Cambridge University Press, 2007). "Table 2.1. Present-day concentrations and RF for the measured LLGHGs," page 141, http://www.ipcc.ch/pdf/assessment-report/ar4/wg1/ar4-wg1-chapter2.pdf.

89. Find more data on historic atmospheric CH_4 in D.M. Etheridge et al., "Historic CH_4 Records from Antarctic and Greenland Ice Cores, Antarctic Firn Data, and Archived Air Samples from Cape Grim, Tasmania-Period of Record: 1008–1995," in *Trends: A Compendium of Data on Global Change*, CDIAC (Carbon Dioxide Information Analysis Center), Oak Ridge National Laboratory, U.S. Department of Energy, 2002, http://cdiac.ornl.gov/trends/atm_meth/lawdome_meth.html.

90. Akifumi Ogino, Hideki Orito, Kazuhiro Shimada, and Hiroyuki Hirooka, "Evaluating Environmental Impacts of the Japanese Beef Cow-calf System by the Life Cycle Assessment Method," *Animal Science Journal* 78 (2007): 424, http://www.blackwell-synergy.com/doi/abs/10.1111/j.1740-0929.2007.00457.x; Vegetarian Society, "Information Sheets," http://www.vegsoc.org/info; People for the Ethical Treatment of Animals (PETA), "Factsheets," http://peta.com/mc/facts.asp.

91. H. Steinfeld et al., "Livestock's Long Shadow"; Vegetarian Society, "Why It's Green to Go Vegetarian," http://www.vegsoc.org/environment/why%20its%20green%20final%20small.pdf.

92. The Global Canopy Programme is an alliance of 29 scientific institutions in 19 countries, which lead the world in forest canopy research, education and conservation: http://www.globalcanopy.org/.

93. N. Stern, *The Stern Review on the Economics of Climate Change* (Cambridge: Cambridge University Press, 2006), http://www.hm-treasury.gov.uk/stern_review_final_report.htm, quoted in Andrew W. Mitchell, Katherine Secoy, and Niki Mardas, "Forests First in the Fight Against Climate Change-The Vivo Carbon Initiative,"

Global Canopy Programme, June (2007), http://www.globalcanopy.org/themedia/file/PDFs/Forests%20First%20June%202007.pdf.

94. Daniel Howden, "Deforestation: The Hidden Cause of Global Warming," *The Independent*, May 14, 2007, http://www.independent.co.uk/environment/climate-change/deforestation-the-hidden-cause-of-global-warming-448734.html.

95. Andrew W. Mitchell, Katherine Secoy, and Niki Mardas, "Forests First in the Fight Against Climate Change—The Vivo Carbon Initiative," *Global Canopy Programme*, June (2007), http://www.globalcanopy.org/themedia/file/PDFs/Forests%20First%20June%202007.pdf.

96. The problem in Indonesia is that huge areas of wet peatland forests are drained and logged. Drainage starts a rapid process of decomposition, made worse by annual peat fires that last for months. The global demand for hardwood, paper pulp, and palm oil, and local economic development, are driving the destruction of dense lowland rainforest that used to cover much of the marshy areas of South-east Asia. In the soaking wet soil, plant material decomposed very slowly, and the layer of peat was too moist to burn. After drainage and clearance, the peat dries, starts decomposing, and emits carbon dioxide. In the tropics this process takes place very rapidly and is often accelerated by fires. Wetlands International, "Shocking Climate Impact of Wetland Destruction in Indonesia," November 2006, http://www.wetlands.org/news.aspx?ID=2817de3d-7f6a-4eec-8fc4-7f9eb9d58828; Wetlands International, "Peat CO_2," December 21, 2006, http://www.wetlands.org/publication.aspx?id=51a80e5f-4479-4200-9be0-66f1aa9f9ca9.

97. Gary Duffy, "Pressures Build on Amazon Jungle," *BBC News*, 14 January 2008, http://news.bbc.co.uk/2/hi/americas/7186776.stm.

98. "Top Ten Eco Things You Can Do," *Channel 4*, September 2006, http://www.channel4.com/science/microsites/S/science/society/eco_topten.html.

99. Michael Specter, "Big Foot: In Measuring Carbon Emissions, It's Easy to Confuse Morality and Science," *The New Yorker*, February 25, 2008, http://www.newyorker.com/reporting/2008/02/25/080225fa_fact_specter.

100. This was the level of CO_2 emissions around 2005. "Can Coal Be Clean?" *The Economist*, November 30, 2006, http://www.economist.com/printedition/displaystory.cfm?story_id=E1_RPTNPTN.

101. Geological storage of CO_2 by injection into saline aquifers appears most practical, in terms of capacity and widespread geographical availability, but this method means that the stored CO_2 will be underground in supercritical phase for centuries, with potential of a catastrophic release of CO_2. SRI Consulting, "Carbon Capture From Natural Gas Power Generation," *Process Economics Program Report 180A*, December 2004, http://www.sriconsulting.com/PEP/Public/Reports/Phase_2004/RP180A/#xml=http://www.sriconsulting.com/cgi-bin/texis/webinator/search/pdfhi.txt?query=180A&pr=super&prox=page&rorder=500&rprox=500&rdfreq=500&rwfreq=500&rlead=500&sufs=0&order=r&id=47d2f15f7; http://www.sriconsulting.com/PEP/Private/Reports/Phase_2004/RP180A/RP180A.pdf

102. Read about various suggested methods to inject CO_2 into the deep ocean, with the goal to minimize costs, leakage, and environmental impacts, in Howard J. Herzog, Ken Caldeira, and Eric Adams, "Carbon Sequestration via Direct Injection," *The Carbon Capture and Sequestration Technologies Program at MIT*, http://sequestration.mit.edu/pdf/direct_injection.pdf.

103. Klaus Lackner, "Future of Coal Depends on CO_2 Disposal," *Statement at the U.S. Senate Energy and Natural Resources Committee Coal Conference*, March 10, 2005, http://energy.senate.gov/public_new/_files/Question2SelectedParticipantsSubmittals.doc; U.S. Senate Energy and Natural Resources Committee, "Coal Conference 2005," http://energy.senate.gov/public_new/index.cfm?FuseAction=Conferences.Detail&Event_id=befd9d70-fcc9-4a81-8517-e3cd45c70ec5&Month=3&Year=2005; U.S. Senate Energy and Natural Resources Committee, "Past Conferences," http://energy.senate.gov/public_new/index.cfm?FuseAction=Conferences.Home; H.-J. Ziock and D. P. Harrison, "Zero Emission Coal Power, a New Concept," http://www.netl.doe.gov/publications/proceedings/01/carbon_seq/2b2.pdf; "ZEC-Zero Emission Coal Technology, CO_2 Sequestration, Chemistry of the Environment," *CHM333 web page projects*, Princeton University, Fall 2002, http://www.princeton.edu/~chm333/2002/fall/co_two/minerals/zec.htm#_ftnref5.

104. "Can Coal Be Clean?" Read details on carbon dioxide capture and storage in Bert Metz et al., *IPCC Special Report on Carbon Dioxide Capture and Storage*, prepared by Working Group III of the Intergovernmental Panel on Climate Change, (Cambridge: Cambridge University Press, 2005), http://www.ipcc.ch/pdf/special-reports/srccs/srccs_wholereport.pdf.

105. Find a few more thoughts on this topic in Klaus Lackner, "Future of Coal Depends on CO_2 Disposal."

106. The eruption of Mount Pinatubo in June 1991 injected some 10 Tg S, initially as SO_2, into the tropical stratosphere. Consequently, the enhanced reflection of solar radiation to space by the particles cooled the earth's surface on average by 0.5 C in the year following the eruption. P. J. Crutzen, "Albedo Enhancement."

107. M. I. Budyko, *Climatic Changes* (Washington: American Geophysical Society, 1977), 261, quoted in P. J. Crutzen, "Albedo enhancement."

108. A. Denny Ellerman and Barbara K. Buchner, "The European Union Emissions Trading Scheme: Origins, Allocation, and Early Results," *Rev Environmental Economics and Policy* 1(1) (2007): 66–87, http://reep.oxfordjournals.org/cgi/content/abstract/1/1/66, 2007.

109. The European Commission, "Questions and Answers on the Commission's Proposal to Revise the EU Emissions Trading System," *MEMO/08/35*, Brussels, January 23, 2008, http://europa.eu/rapid/pressReleasesAction.do?reference=MEMO/08/35&format=HTML&aged=0&language=EN&guiLanguage=en.

110. "About CCX," Climate Exchange Plc, http://climateexchangeplc.com/home/about-ccx; "Overview," Chicago Climate Exchange (CCX), http://www.chicagoclimatex.com/content.jsf?id=821.

111. U.S. Sen Joseph I. Lieberman et al., "America's Climate Security Act of 2007," *U.S. Library of Congress* (2007): 2191, http://thomas.loc.gov/cgi-bin/bdquery/z?d110:S.2191:.

112. Steven Chu, José Goldemberg et al., "Lighting the way: Toward a Sustainable Energy Future," InterAcademy Council, 2007, http://www.interacademycouncil.net/?id=12161.

113. "Exergy Flow Charts," Stanford University: Global Climate and Energy Project (GCEP), http://gcep.stanford.edu/research/exergycharts.html; Conversion from TW to barrels of oil equivalent: 1 ton of oil equivalent=4.2 E10 J, 1 ton of oil=7.33 barrels.

114. To be sure, the energy that arrives from the sun is not evenly distributed. The tropics receive over 250 per m2, while polar regions receive some 100 watt per m2. An average of 180 watt per m2 is assumed for the world as a whole. James R. Craig, David J. Vaughan, Brian J. Skinner, *Resources of the Earth: Origin, Use, and Environmental Impact* (Upper Saddle River: Prentice Hall, 1996) cited in Bjørn Lomborg, *Verdens Sande Tilstand* (Viby: Centrum, 2002).

115. Energy Information Administration. *International Energy Outlook 2007*. Office of Integrated Analysis and Forecasting. http://www.eia.doe.gov/oiaf/archive/ieo07/index.html.

116. Noncommercial fuels and dispersed renewables (renewable energy consumed on the site of production, such as energy from solar panels used to heat water) are not included in the Department of Energy's projections. The Department of Energy argues that comprehensive data on their use are not available. Energy Information Administration. *International Energy Outlook 2007*. Office of Integrated Analysis and Forecasting. http://www.eia.doe.gov/oiaf/archive/ieo07/index.html.

117. Made of strong, lightweight composite materials, flywheels can spin in a vacuum at up to 200,000 revolutions per minute, with the potential to store and release energy at an efficiency of more than 90 percent. They receive or deliver energy through an integrated electric motor and generator system and have virtually frictionless electromagnetic bearings which allow them to store energy for weeks and to last years before wearing out. Flywheels are suited for stationary applications, but also for electric vehicles. Find a discussion of different types of batteries in: "In Search of the Perfect Battery," *The Economist*, March 6, 2008, http://www.economist.com/displaystory.cfm?story_id=10789409.

CHAPTER 37

TRANSPORT FUEL FOR THE FUTURE

The main challenge in terms of future energy will be to provide fuel for transportation purposes. Demand for oil is driven by demand for transportation, and both the U.S. Energy Information Administration (EIA) and the Paris-based International Energy Agency (IEA)[118] estimate that about 118 million barrels of oil a day will be demanded globally in 2030, compared with roughly 85 million barrels a day currently. It is highly doubtful that enough oil can be supplied to meet this demand at a reasonable price. But what's the alternative? Transport fuel mainly equals liquid fuel, simply because the fuel that powers cars and aircraft needs to flow smoothly and continuously from storage tank to combustion chamber (or, more generally, to the on-board power plant). Gases are an option, too, but they fill more volume and hence have a much lower energy density than liquids even when they are compressed. Thus, the current choices for transport fuel include (the remaining) conventional oil, unconventional oil, liquid biofuels, coal liquids, natural gas (methane), and perhaps hydrogen (or methanol) to power fuel cell vehicles. Generally the question is how much energy has to be invested to produce and distribute these fuels, and how much carbon dioxide (or other greenhouse gas) is released during production, distribution, and the actual burning of the fuel.

HOW MUCH CONVENTIONAL OIL IS LEFT?

The immediate question is how much conventional oil is left, and how much conventional oil is left outside the politically charged Middle East. The question can be tackled by looking at the known reserves, and how the discovery rate of new reserves changed over time, but this is complicated by various issues. For once, technology is improving to recover more oil from existing oil fields. Oil occurs in porous underground rock reservoirs, typically under pressure that forces some of the oil to the surface as soon as it is reached by the bore hole. Still in 1979 (the time of the Second Oil Price Shock) as little as 22 percent of the total oil contained in the field was recovered, but the recovery rate increased to 35 percent within the following 20 years. One way to recover more oil is to inject hot water or steam, which increases the pressure and decreases the viscosity of the oil. Another approach is horizontal (sideways) drilling from (deep) vertical boreholes. Deepwater drilling technology has improved as well. Generally, such advanced methods may be more expensive, but they are justified as soon as the oil price is sufficiently high. Thus, proved reserves have been pragmatically defined as those discovered quantities of oil that can economically be recovered under existing price and technological conditions.

Another issue is that oil corporations and oil countries are for various reasons inclined to overstate (or, occasionally, to understate) their proven oil reserves. Oil firms have repeatedly overstated the capacity of oil fields under their management to increase their book (and shareholders') value. (Royal Dutch / Shell in January 2004 suddenly reduced its previously overstated proven reserves of oil by a whopping 20 percent, amounting to nearly four billion barrels of oil. These reserves did not entirely disappear, but Shell had to reclassify reserves as probable or possible, and the management was hoping that 85 percent of the downgraded reserves would be booked as proven again, although only within the long period of 10 years.[119]) However, oil companies may at times also have understated their reserves to justify restraining output to keep prices high. OPEC countries may overstate their oil reserves to receive higher production quotas. The OPEC cartel agrees to a certain production ceiling, but in turn needs to figure out how much of the total each member country is allowed to produce. Up until the earlier 1980s, this was based on production capacity, but thereafter the system was changed to take each member-state's oil reserves into consideration, which led to a curious leap in oil reserves reported by some OPEC member states in the late 1980s. Generally, proved oil reserves are an asset on which banks, including the World Bank, will give loans, and just about everyone with vested interests in the oil industry is interested in overstating reserves to keep people from looking for energy alternatives.

Interestingly, information on global oil reserves, as used by policy makers and widely reported by the media, is not collected by an international institution, but by an industry journal (*Oil and Gas Journal*) that yearly sends a questionnaire to all governments, asking them for a guess on their national oil reserves. This adds uncertainty to reserves data that are already troubled by the principal difficulty to estimate the size of oil fields, which are not great underground lakes, but consist of oil trapped in porous subsurface rocks.

Another discrepancy has emerged around the world's ultimately recoverable oil resources, which is the sum of all oil that has already been recovered and all oil resources that can ultimately be produced through a well bore in the future. This had long been estimated at around 2,000 billion barrels, but the US Geological Survey (USGS) kept on increasing its estimates of ultimately recoverable oil, making a jump to over 3,000 billion barrels in the year 2000.[120] (Some geologists have called this "implausibly large estimates of world oil."[121]) This is quite relevant, as some 900 billion barrels of oil had been cumulatively produced by the beginning of the 21st century, while reported proven reserves amounted to about a thousand billion barrels of oil: If the global oil production curve was a perfectly bell-shaped curve, it would peak when half of ultimately recoverable oil has been produced, and production rates would then decrease in the same way they had increased before the peak. This simple model of oil production over time is quite popular, because it was applied by geophysicist M. King Hubbert in 1956 to predict with remarkable accuracy that U.S. oil production would peak in 1970.[122] However, the United States (48 states) had been thoroughly surveyed by 1956 and fit the conditions required by the model better than the world as a whole does. The global oil production curve has not been rising smoothly. It is influenced by such fluctuating factors as economic growth, oil price, politically or economically motivated supply variations, exploration efforts, and production methods, which all influence one another. Several studies in the late 1990s predicted the global Hubbert peak to occur between 2003 and 2006, and some were later revised to shift the peak to around 2010. According to a Hubbert forecast published by the U.S. Energy Information Administration, based on the USGS's 3,003 billion barrels estimate for the total amount of recoverable oil, and a production growth (before the peak) as well as a decline of 2 percent per year (after the peak), the peak would occur in 2016. The peak can, however, be shifted to 2037, if continued production growth is assumed in exchange for a rapid production decline after the peak.[123] Similarly, Shell said the peak would occur in 2025 at a production rate of about 46 billion barrels per year, almost double the 2001 level, while critics were wondering how such high production capacity could be realized.

To be sure, the peaking of the global oil production curve, even if it were strongly distorted towards "the right" (to peak later than a symmetrical curve), will not mean that almost all oil is gone by this time. Rather, this is the point when demand, though it should by this stage have weakened due to increased oil prices, will no longer be met by supply, because no more oil production capacity can be added. Production will decline, and the gap between supply and demand widen. Besides, nearly all oil still left is expected to be located in the Middle East (or, more generally, in OPEC countries). Thus, if the

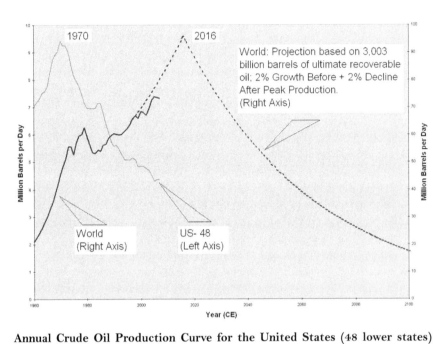

Annual Crude Oil Production Curve for the United States (48 lower states) and the World, plus a Hubbert-type Projection for Future Global Oil Production Though Hubbert curve modeling was successful in predicting the oil production peak for the United States (lower 48 States), it is less applicable for the world as a whole. The shape of the future global oil production curve strongly depends on assumptions on production growth and decline before and after the peak, as well as on the amount of ultimately recoverable oil. The latter is assumed to be 3,003 billion barrels in the depicted scenario of 2 percent production growth and decline before and after the peak. (This is based on a projection by the Energy Information Administration, EIA.) Under the assumption that growth would continue for a longer time, the peak would occur later and at a higher peak production volume, but would be followed by a more rapid production decline, as the area underneath the curve remains the same. (Based on data from Energy Information Administration, "World Petroleum Consumption, 1960–2006," http://www.eia.doe.gov/emeu/aer/txt/ptb1110.html, "Crude Oil Production and Crude Oil Well Productivity, 1954–2007," http://www.eia.doe.gov/emeu/aer/txt/ptb0502.html; projection based on data from Robert L. Hirsch, Roger Bezdek and Robert Wendling, "Peaking of World Oil Production: Impacts, Mitigation, & Risk Management," February 2005, page 70, http://www.netl.doe.gov/publications/others/pdf/Oil_Peaking_NETL.pdf, and Energy Information Administration, "Long Term World Oil Supply (A Resource Base/Production Path Analysis)," 2000, Slide 14, http://www.eia.doe.gov/pub/oil_gas/petroleum/presentations/2000/long_term_supply/sld014.htm.)

world wishes to avoid an oil-shock-like scenario when the peak is reached, it needs to prepare itself in good time to make sure enough moderately-priced fuel alternatives are in place to serve transportation and to maintain the standard of living that people have become accustomed to.

Another part of the Hubbert-type analysis is to look at the discovery rates, as those will necessarily peak before production does.[124] Indeed, the worldwide discovery rate of new major fields peaked as early as 1962. Some 100 giant oil fields, each holding ultimately recoverable reserves exceeding 500 million barrels of oil, had been discovered by the 1960s, another 30 giant fields were found in the 1980s, and another 29 were discovered during the 1990s. From about 1980 on new discoveries could no longer balance world consumption of oil. Much of the confidence that there was enough fuel left to power the Oil Age well into the 21st century had thus to come from reserves growth in oil fields already discovered. Most important were the OPEC hopes: reported proven reserves between 1985 and 1986 jumped from 33 billion barrels to 97 billion barrels in the United Arab Emirates, and from 59 billion barrels to 93 billion barrels in Iran; from 1986 to 1987 from 72 billion barrels to 100 billion barrels in Iraq, and from 1987 to 1988 from 170 to 255 billion barrels in Saudi Arabia.[125] Thus, reported global proved reserves jumped from 679 billion barrels in 1981 to 1,001 billion barrels in 1991, a level were they still remained in 2001 (1,050 billion barrels).[126] Meanwhile the increase in global oil production was comparatively low, from 21.092 billion barrels per year in 1981 (right after the Second Oil Price Shock) to 26.278 billion barrels per year in 2001. However, only 62 percent of the oil produced between 1990 and 1994, and only 52 percent of the oil produced between 1995 and 1999 was replaced by newly reported volumes in proven oil. And then another curious leap in reported proven reserves was taken between 2001 (1,050 billion barrels) and 2006 (1,208 billion barrels), which kept the reserves per annual production ratio at 40.5 years. ("At current production rates reserves of proven oil will last for 40.5 years.") This 15 percent reserves increase, which can in part be explained by high oil prices that allow for less accessible oil to be produced economically, occurred in the absence of any major new oil field discoveries. It mainly owed to the rise in reported reserves in Iran (from 89.7 billion barrels to 137.5 billion barrels), Russia (from 48.6 billion barrels to 79.5 billion barrels), Nigeria (from 24.0 billion barrels to 36.2 billion barrels), Libya (from 29.5 billion barrels to 41.5 billion barrels), Canada (6.6 billion barrels to 17.1 billion barrels), and, notably, Kazakhstan (from 8.0 to 39.8 billion barrels).[127]

Kazakhstan, an enormously large country (four times the size of Texas), is located at the northeastern Caspian Sea, the only new major oil province discovered after 1980. It is safe to say that nearly all of the world has now been thoroughly searched for oil basins. Perhaps unexplored Arctic basins may hold oil, and in parts of the South China Sea oil prospecting has been delayed due to legal disputes over drilling rights among neighboring countries. Most of the 29 giant oil fields discovered during the 1990s were located in regions already known to be rich in oil, but seven were discovered in water depths greater than one thousand meters. Deep offshore technology is relatively recent (an offshore production record was set at 1,800 meters in Brazil in 1998), and major oil fields might still be discovered in deep waters. However, production of oil in the deep waters off the coast of Brazil and Angola costs about seven times as much as producing oil in the Saudi desert. (And production in ultra deep offshore, in depths beyond 2,000 meters, will be substantially more expensive, because deep offshore technology cannot be directly adapted to such great depths.)

In a speech in January 2008, Shell's Chief Executive Jeroen van der Veer predicted that "easily accessible supplies of oil and gas probably will no longer keep up with demand" after 2015.[128] Shell was perhaps in a better position than any other of the oil giants to make such prediction of the "Peak of Easy Oil," as its reserves debacle in 2004 forced the company to shift into high gear to rebuild its reserves and resource base. (Shell received approvals from governments around the world to explore for new fields in 2007, but production fell by 1 percent in 2007 and was expected to fall just as much in 2008 and 2009, before some large projects, including Sakhalin 2 in Russia and the massive Pearl project in Qatar, were expected to begin boosting production at Shell within five to seven years. Note, that both mentioned undertakings are gas-to-liquids projects, not oil projects.) The oil corporations enjoyed record revenues in the early 21st century due to high oil prices, but it became clear that the struggle to keep up with rising demand was getting harder, and that costs of extraction were also soaring (in part due to inflation). In 2000 Shell devoted about $10 billion to capital expenditure, or money spent on developing new projects, but that figure rose to $23.8 billion in 2007, and was expected to rise to $25 billion in 2008. The situation was similar for Shell's competitors, who were also struggling to replace their reserves. In 2002 oil companies spent a combined $250 billion to produce 30 million barrels of oil per day (according to research from Deutsche Bank). By 2006, the industry spent $550 billion to get just 20 million barrels per day, indicating that most of the oil that is relatively straightforward to extract has been found.[129]

The other problem for private-sector oil corporations was a rising tide of resource nationalism. As of 2007 the five largest private oil corporations combined, though they are true giants, owned no more than some four percent of global oil reserves. Most of the remainder was controlled by governments or state-owned companies. In December 2006, Russia, through its state gas monopoly Gazprom, took the majority stake in the Sakhalin-II gas field. (The Sakhalin-II project has estimated reserves of 150 million tons of oil and 500 billion cubic meters of natural gas.) In January 2008, Kazakhstan forced the international oil corporations on the massive Kashagan project to sell some of their ownership stakes to the government. And in Nigeria the government holds 55 per cent of the joint venture with Shell, for instance. Here, unrest and violence continued, with state authorities failing to keep up on their obligation to pay their share of the costs.[130]

Most of the known remaining oil reserves are actually located in places where oil is cheapest to produce. By the end of 2006 the Middle East accounted for over 60 percent of reported proven reserves. Add another five percent for north African Muslim countries Libya and Algeria, 10 percent for Russia and Kazakhstan, and another seven percent for Venezuela. (OPEC had three quarters of the reported proven reserves, OPEC and the Former Soviet Union together over 85 percent.) Most of these countries are somewhat difficult politically, and many did not have traditionally warm relations to the industrialized West. Hence, the fact that the remaining oil is increasingly more regionally concentrated (in the Middle East), has been more of an incentive to diversify into other liquid fuels than the imminence of an upcoming oil shortage or climate change. And the prospect of the end of "easy oil" as accessible to private-sector oil corporations has motivated such firms to invest into more expensive and technologically challenging unconventional oil.

UNCONVENTIONAL OIL

One alternative to conventional oil is oil recovered from unconventional sources. The border line between the two is somewhat blurry as far as heavy oil is concerned, as this can be recovered (through a well bore) by Enhanced Oil Recovery methods such as steam injection. (Heavy oil has a higher density than conventional oil and therefore will not flow to the surface under typical reservoir conditions). On the other hand, the distinction between conventional and unconventional oil becomes entirely obvious as soon as mined oil is considered, which cannot be produced through a wellhead. Such mined oil derives either from tar sands or from oil shales. Tar sands (also called oil sands or bituminous sands) are even more viscous than heavy oil, but located closer to the surface. They are deposits of bitumen, a heavy black viscous oil, created when porous rock became impregnated with hydrocarbon residue, while the more volatile parts of the oil escaped. Oil sands recovery is energy-expensive. The process includes extraction and separation systems to remove the bitumen from sand and water. Once the oil has been retrieved, it still has to be cleaned and upgraded before it will behave more like conventional refinery feedstocks. Bitumen deposits found close to the surface can be recovered by open-pit mining techniques, but some two tons of oil sands must be dug up, moved and processed to produce one barrel of oil. (Bitumen deposits buried more than 75 meters, but more typically beyond 400 meters, under the ground are recovered in situ, without digging the sand up. This involves injection of steam and light organic solvents.) Producing a barrel of oil from tar sand currently costs between six and twelve times as much as producing a barrel of conventional crude in the Saudi desert. Nevertheless oil production from tar sands has already begun. Total proven deposits are quite large (a few hundred billion barrels of oil, compared to Saudi Arabia's proven 240 billion barrels of conventional oil reserves), with much of it located in Canada (164 billion barrels in 2006[131]). Venezuela has similarly large reserves in heavy oil deposits. (Perhaps 270 billion barrels are considered technically recoverable.) Both Canada and Venezuela have begun to produce these resources at great cost to develop the necessary technology, but production volumes thus far remain minimal. Nevertheless, Canadian oil sand reserves have at times been added to total global proven oil reserves, which is somewhat questionable. The idea is that rising oil prices would justify more expensive production methods, but the energy inputs required for processing, would make the overall climate-burden substantially higher than if conventional oil is used for transport.

Producing oil from oil shale is even less economical. First of all, there is not really oil in oil shale. Oil shale is a fine-grained sedimentary rock that contains an bituminous organic liquid called kerogen. After mining, the shale has to be heated to about 480 C (900 F), and hydrogen has to be added, before kerogen decomposes to yield oil. Obviously, this consumes quite a bit of energy. What is more, the shale pops like popcorn during the process, and the resulting volume of waste is larger than what has been mined in the first place. Oil shale deposits occur worldwide, but oil shale is not really expected to ever substitute for conventional oil. (The largest shale oil deposits in the world are in the Colorado Plateau. Unfortunately, this is a water poor region, while it takes several barrels of water to produce one barrel of oil from shale.)

LIQUID BIOFUELS

An entirely different approach to produce energy for the transport sector is to make liquid biofuels. Biofuels have the advantage that they can be produced domestically; that they are renewable; and that they, at least in theory, do not contribute to global warming, because they capture as much carbon dioxide during plant growth as is being released when they are burned. Liquid biofuels include first-generation bioethanol, produced from sugar cane or grain; second-generation or cellulosic bioethanol, produced from stalks, wood trimmings, and the like; biodiesel, produced from oil-rich seeds and perhaps from algae; and biobutanol, which can be made from the same feedstocks as bioethanol.[132]

Ethanol has been quite well received as an alternative fuel, because it does not require much alteration of established gasoline engines. Most gasoline-powered cars produced towards the end of the 20th century can use E20, that is, gasoline mixed with 20 percent ethanol,[133] and much of the automobile industry agreed in the beginning of the 21st century to make those small design changes that would allow newly produced cars to burn E85. Saab, Volvo, and Ford were among the early car producers offering so-called Flexi Fuel Vehicles (FFV), which can use E85. (To be sure, it will become obvious with such large ethanol percentage that ethanol contains some 34 percent less energy per unit volume than gasoline, and hence produces lower mileage per stop at the gas station. Another disadvantage of ethanol compared to gasoline is that ethanol absorbs moisture from the atmosphere.)

Ethanol, C_2H_5OH, is regular alcohol of the kind found in alcoholic beverages. People have fermented starchy grains (and sweet fruits) into intoxicating liquids since the early Agricultural Age. This involves several steps. First, starch has to be transferred into free sugars (so-called C_6-sugars, such as glucose and fructose). In beer brewing, for instance, this is called mashing, which involves adding malt to grain in water. Malt, or malted grain, is grain induced to germinate and then quickly dried, which triggers the production of enzymes capable of breaking the grain's starch into sugars. (To be sure, wine-making does not need this step, as grapes already contain C_6 sugars to start out with.) The next step is for (naturally-occurring) yeasts to ferment sugar into ethanol, C_2H_5OH, which can be separated from water by distillation. First-generation bioethanol is made the same way, typically either from maize (in temperate zones) or from cane sugar (tropical zones). Brazil is the country with the lowest production cost for first-generation bioethanol. Located in the tropics, Brazil has devoted large plantations of sugarcane to the production of clean-burning ethanol ("E100") for the transportation sector since the 1970s. (Currently Brazil replaces 40 percent of its gasoline consumption with bioethanol.) In the United States, the federal government has subsidized the (maize-based) bioethanol industry with billions of dollars in the early 21st century. In 2004, 75 distilling plants were operating, while it was 127 in 2007, with another 81 under construction. (Consequently, 27 percent of America's 2007 record harvest of 12.46 billion bushel of corn has been used by the biofuel industry.[134]) The United States and Brazil, who produce similar amounts, combined account for nearly ninety percent of global ethanol production, but this share is sinking as output in the rest of the world grows rapidly. The United States and Brazil, who produce similar amounts, combined account for nearly ninety percent of global ethanol production, but this share is sinking as output in the rest of the world grows rapidly. (Global growth was over ninety percent in 2006, for instance, with a third of that growth coming from China and another third coming from Europe.) Worldwide, fuel ethanol production reached 20.2 million tons of oil equivalent in 2006, which is almost three times the amount produced in 2001 and five times the amount produced in 1996.[135]

In the United States, demand for ethanol in recent years was actually not driven by use of ethanol as fuel, but as oxygenating gasoline additive. The Clean Air Act mandated the use of cleaner burning fuels in the dirtiest U.S. cities in the early 1990s, which was to be achieved by adding oxygen to gasoline through additives. MTBE (methyl tert-butyl ether; $C_5H_{12}O$) was the additive of choice, but in January, 2004, California, eventually followed by other states, banned the use of MTBE due to groundwater contamination. This opened the way for ethanol (and ethanol-derived ETBE, ethyl tert-butyl ether) to be used as gasoline additive. And in turn the national renewable fuel standard (RFS) program began helping ethanol to be integrated into the American gasoline pool.[136]

Production of ethanol from cane sugar is a lot cheaper than production from corn (maize). In fact, the question has been raised if ethanol production is at all advisable in temperate zones as far as climate protection is concerned. A 2007 review (at the University of California at Berkeley) of six studies on the issue concluded that, gallon for gallon, corn ethanol is probably 10–15 percent better than petrol in terms of greenhouse gas emissions.[137] For comparison, CO_2 emissions reductions in fueling a vehicle with ethanol from sugar cane are estimated to be 50–90 percent on a life-cycle basis compared with petrol use. The meager CO balance, and the related question if bioethanol production from corn actually yields net energy, derives from the fact that modern agriculture relies on large inputs of technical energy. Natural gas is the feedstock for fertilizer production via the Haber-Bosch process; gasoline fuels tractors, and electricity powers irrigation pumps, for instance. In 2002 the U.S. Department of Agriculture (USDA) stated that corn-ethanol yields 34 percent more energy than it takes to produce it, including growing the corn, harvesting it, transporting it, and distilling it into ethanol. However, some researchers claimed that it is the other way around: they found that making bioethanol from corn actually takes 29 percent more energy than it yields. The bottom line is that the energy balance depends sensitively on the regional growing conditions. It may well be energy-efficient to make ethanol in eastern Minnesota, where rainfall is abundant and some of America's cheapest corn is grown, while in dryer Nebraska around 80 percent of the corn has to be irrigated, normally by natural-gas powered pumps. The International Energy Agency (IEA) found that recent large-scale production practices do pretty well, providing a net energy gain of 30–40 percent. (This net energy gain includes credit for co-products such as corn oil and protein feed on top of energy value of ethanol.)[138]

To be sure, bioethanol production would be a good idea one way or another if improving energy security is the goal. Inputs of coal-derived electricity or gas-derived fertilizer may well be accepted to produce a liquid fuel for transport that reduces oil imports. Otherwise it would actually make more sense from both the energy and the climate protection standpoint to directly burn (solid) biomass. This would also save the large amounts of water used in bioethanol production, but it would burn all the protein content (which may otherwise used as animal feed), and there is a limit to how much biomass can be added to the feedstock in coal-fired power plants, for instance.

However, the most severe reputation damage for bioethanol came with increasing food prices. U.S. President Bush's January 2007 call for a massive increase in the use of ethanol over the next decade is said to have led to an increase of corn prices, then meat prices, and to Tortilla Riots in Mexico. Nevertheless, it cannot be claimed that the global increase in food prices was due to ethanol production. It was rather a combination of population growth, bad harvest seasons, and increased demand for milk and meat products from populous Asian regions in which people become more wealthy and their diet less vegetarian. Indeed, there is a large energy saving and climate protection potential if only we would eat less meat. In countries such as the United States, most of the maize harvested on the fields has been fed to animals for decades. This habit reduces the overall nutritional energy available to humans and promotes excessive meat consumption associated with cardiovascular diseases, now the number one killer worldwide. On the other hand, even shifting corn from animal feed to bioethanol production, or using all maize for nothing else but bioethanol, would do little to reduce oil imports: Even if all maize currently consumed in the United States went into ethanol production, no more than 15 percent of all gasoline could be replaced.

Because of the above reasons, much focus has changed towards so-called second-generation ethanol, which is also referred to as cellulosic ethanol as differentiated from (first-generation) starch ethanol. Cellulose-based feedstocks include everything from corn stover, wheat or rice straw, and sawdust, to dedicated energy crops such as fast-growing switchgrass that tolerates poor soil and climatic conditions. The production of cellulosic ethanol obviously has the advantage that it does not directly compete for edible biomass, and it may dramatically reduce the land base as well as the fertilizer and water requirements necessary to produce meaningful amounts of biofuel. The initial technical difficulty to be overcome is to develop enzymes that can break down tough cellulose into sugars, which is much more difficult than breaking down starch. Danish company Novozymes has been first to deliver enzymes for the hydrolysis step in the production of second-generation bioethanol, albeit initially in small quantities only.[139] (In 2007, Novozymes expected second-generation bioethanol technology to reach full production status and commercial availability in the course of the following four to five years.) Another issue involved is that cellulose, once it is broken down, delivers both C_6 sugars and C_5 sugars, while traditional yeast can only ferment C_6 sugars into ethanol. Efforts have therefore been directed towards the development of a genetically modified yeast equipped with both C_6 and C_5-fermenting capabilities.[140]

Another bioalcohol, which is (due to its lesser polarity) actually more similar to gasoline than ethanol is biobutanol, C_4H_9OH. Biobutanol can be produced with the Acetone-Butanol-Ethanol process which ferments starchy biomass with help of the bacterium Clostridium acetobutylicum. (This is produce the process used by Chaim Weizmann during World War I to make acetone that England needed to make explosives.) The feedstocks for biobutanol would be thus be similar to those for first-generation ethanol.

In the European Union, the liquid biofuel that perhaps received most attention (and funding) is biodiesel produced from vegetable oil.[141] Pure biodiesel is available at many gas stations in Germany, but typically it is mixed in relatively low percentages to conventional diesel, because biodiesel costs about twice as much to produce. Biodiesel may generate about 60 percent less net lifecycle carbon dioxide emissions than petroleum-based diesel, but the issues in regards to the CO_2 and energy balance are the same as in bioethanol production. In Europe biodiesel is mainly made from rapeseed, though it may also be produced from the oils derived from soybeans, sunflowers, or oil palms, for instance. (It may also be made from waste stream sources such as used cooking oils or animal fats.) In 2007 the European Union announced that 10 percent of its transport fuel should be comprised by biofuels as of 2020, which would require some 15 percent of all arable land in Europe. As this requirement could also be met by imports, a controversial boom in palm oil production set in. Palm oil is rivaling soybean oil as the most widely produced vegetable oil in the world, with Indonesia and neighboring Malaysia accounting for 85 percent of the global production. Demand for palm oil was now soaring to meet demand for biodiesel production as well as to replace soybean oil (mainly for processed foods), because U.S. farmers grew more corn than soybeans due to the bioethanol boom. This had negative consequences in Indonesia, as people were deprived of their traditional land, and natural forest areas were cleared (which contributes to climate change), to create new oil palm plantations. To be sure, the whole issue of fuel crops competing with food crops for arable land and other resources such as water and fertilizer will change for the better as soon as cellulosic materials serve as fuel feedstock. However, it would be quite a challenge to substitute biomass for all the current energy consumption: It is estimated that about a quarter of the planet's yearly capture in photosynthetic energy would be necessary to replace the annual global consumption in fossil fuels. For comparison, only about one percent of all biomass is currently used as food by humans and animals. Besides, the cost of energy production from biomass for the moment remains about twice that of energy generated from fossil fuel sources.

COAL LIQUIDS

Coal can be turned into liquid fuel for transportation as well. Coal liquefaction is rooted in the coal gasification process known since the early Coal Age: coke production yields a flammable gas known as coal gas or water gas. The actual composition of the obtained gas depends on how much air (oxygen) and water steam is directed over red-hot coke or coal, but the main constituents are always hydrogen, carbon monoxide, and methane. ($3C + O_2 + H_2O \rightarrow H_2 + 3\ CO$ and $C + 2\ H_2 \rightarrow CH_4$.) A gas rich in carbon monoxide and hydrogen is generally called synthesis gas or syngas, and oil-lacking Germany during World War II developed a process to turn coal-derived syngas (in presence of an iron or cobalt catalyst) into diesel and jet fuel. The process is named for Franz Fischer and Hans Tropsch, who initially described it in 1923. (Another German, Friedrich Bergius, described the hydrogenation of coal in 1910, and had shown the large-scale practicality of coal liquefaction by 1927.) After World War II the Fischer-Tropsch process was commercialized in coal-rich South Africa, whose government due to its apartheid policy faced an international trade embargo that included oil and gas. The U.S. government heavily funded coal liquefaction technology after the oil price shocks of the 1970s, but these efforts waned away with the oil price collapse of the mid-1980s. In the early 1990s a large-scale coal liquefaction plant in Texas produced 6.1 barrels of oil equivalent, of which 5.0 barrels were gasoline, for each metric ton of coal, but the process was uneconomical in light of low oil prices.[142] However, since the remaining coal stocks are far larger than those of oil and gas, coal liquefaction is undoubtedly a technology of the future.[143]

NATURAL GAS

Natural gas can be turned into liquids via the Fischer-Tropsch process as well, because it can be reacted into synthesis gas. ($CH_4 + H_2O \rightarrow CO + 3\ H_2$: the reaction takes place at high temperatures (around 850°C) and moderate pressures (10–20 atm) in presence of water steam and a nickel catalyst.) This is routinely done around the world, because natural gas (CH_4, methane) is the chief source of hydrogen gas, which is used to produce various important chemicals, including fertilizers via the Haber-Bosch process. Syngas derived from natural gas is therefore more readily available in large quantities than syngas derived from coal. Several companies have worked on a low-temperature Fischer-Tropsch process to produce Gas-to-liquids (GTL) technology, but it is still only marginally economical. Shell has operated a medium scale GTL plant at Bintulu, Malaysia, since 1993, and invested in a large plant in Qatar, the nation that announced plans to establish facilities capable of producing 200 million barrels of GTL per year to become the GTL capital of the world. Sasol, South Africa's former state energy company that is now privately owned, has substantial know-how in Fischer-Tropsch technology and initiated a project to shift syngas feedstock from domestic coal to natural gas imported from neighboring Mozambique to reduce costs. In 2003 Sasol started to build the world's largest GTL plant in Qatar in a 49–51 joint-venture with Qatar Petroleum. Overall costs of establishing the plant that came online in 2007 was around $950 million. It was designed to deliver about 12.5 million barrels of liquid per year, comprising 8.8 million barrels of diesel, 3.3 million barrels of naphtha and 0.4 million barrels of liquefied petroleum gas (LPG). Naphtha is a mix of hydrocarbons chiefly used as solvents, while LPG is a mix of propane (C_3H_8) and butane (C_4H_{10}), the gases recovered in oil and gas fields (and during oil refining) and can be easily compressed into LPG, used for household applications (heating, cooking) and as clean-burning autogas. (Toyota built various motor vehicles that ran on LPG during the 1970s, and now several car manufacturers are offering LPG models.)

To be sure, it may be questioned if it really makes sense to convert natural gas into liquid fuel for transportation. Natural gas can easily be compressed to power vehicles directly, while syngas contains less energy than natural gas, and the subsequent liquefaction consumes energy as well. However, investments in GTL are extensive, because they serve current vehicle fleets and vested interests, adding gradually to the existing flows of gasoline and diesel. Though natural gas was the fuel used to develop the earliest internal combustion engines, cars that run on natural gas are rare for the moment (though most auto makers have now brought dedicated natural gas vehicles to market or developed them to pilot status). Unlike propane and butane, natural gas (methane) would need extreme cooling to be liquefied, and is thus stored in heavy pressure tanks onboard vehicles. The energy density of the gas is low, which reduces mileage, but on the upside methane burns very cleanly and can be supplied through the pipelines already in place to serve the other applications of natural gas. What is more, conventional gasoline-powered cars can quite easily be converted into natural gas vehicles. (Argentina, Brazil, and Pakistan combined had well over four million natural gas vehicles on the road as of mid-2007.)[144] But there is yet another option to utilize natural gas for transportation: It may be turned into syngas first, but in turn not be liquefied: Instead, its hydrogen content may be utilized either by directly burning it in an internal combustion engine, or, more attractively, by feeding it to a fuel cell.

HYDROGEN AND METHANOL FUEL CELLS

Fuels cells can power a vehicle via an electric motor. Employing a catalyst, they typically react hydrogen with oxygen (from the air), silently and cleanly, yielding nothing but electricity, heat, and water.[145] Theoretically, fuel cells may actually be much more efficient in terms of energy conversion when compared to internal combustion engines, though this potential has not yet been fully realized. (Only stationary fuel cells employing waste heat currently achieve efficiencies of around 90 percent.) If hydrogen gas is to be used as fuel onboard vehicles, it either has to be highly pressurized or liquefied by cooling to minus 253C, which consumes as much as one third of the energy content of hydrogen. Compressed hydrogen is likely the better solution, especially because the traditional, heavy pressure tanks of steel have now been replaced with lightweight composite tanks. However, even in liquid form hydrogen has only about a quarter of the energy density per volume when compared to gasoline, which translates to relative low mileage.[146]

Yet much of the technological challenge lies directly within fuel cells. The electrodes are made of graphite, but need to contain a precious metal catalyst such as platinum, which is expensive. (Platinum tends to cost nearly twice as much as gold.) A lot of development work therefore goes into reducing fuel cell cost by reducing the platinum load. This could in theory be done by raising the operating temperature, but PEM (proton exchange membrane) cells, the preferred fuel cell type for vehicle propulsion, have material limitations in terms of temperature. (They are low-temperature cells, typically operating at about 80C/175F.) Efforts have therefore been directed towards the development of high-temperature PEM cells, which would also tolerate less clean input gas.[147] The requirement of very clean hydrogen and oxygen gas is actually the main obstacle to mass-market introduction of another fuel cell system, the highly efficient alkaline fuel cell as used in space crafts. Less expensive yet robust phosphoric acid fuel cells, on the other hand, are heavy, less efficient, and use a corrosive electrolyte, which makes them more applicable for stationary electricity generation than mobile use. Yet another type of fuel cell category are real high-temperature systems. Stationary molten carbonate fuel cells and solid oxide fuel cells work at temperatures up to 1,000C to provide high power for industrial applications or decentralized electricity generating stations.

Most practical fuel cells run on hydrogen, but they are often quite fuel-flexible, as they may be delivered with a reformer, that is, an integrated unit capable of converting one or more different hydrocarbon or alcohol fuels such as natural gas, ethanol, methanol, propane, gasoline, or diesel into hydrogen gas that is in turn fed into the fuel cell. A notable exception running on a non-hydrogen fuel without the need of a reformer is the Direct-Methanol Fuel Cell (DMFC), which is of the PEM type. (The active agent is still hydrogen, but it is generated catalytically inside the fuel cell from methanol.) DMFCs require more (expensive) catalysts, and for the moment perform worse than true hydrogen cells, but they seem to be the better option for vehicle propulsion, because methanol is liquid under normal environmental temperatures and contains more energy per volume than hydrogen. Liquid methanol can be easily stored and distributed with existing gasoline infrastructure, while a proposed hydrogen economy would require substantial investments that could only in part draw on existing natural gas infrastructure. (Like ethanol, methanol may even be mixed with gasoline to power internal combustion engines, but the energy density of methanol is even lower than that of ethanol.) Methanol is as readily available as hydrogen, because large-scale production of both is based on syngas produced from natural gas. (Syngas can quite easily be reacted into methanol: $CO + 2\,H_2 \rightarrow CH_3OH$.) Using syngas produced from coal would be an option, too, but notably methanol, also known as wood alcohol, may be generated sustainably from biomass (wood), municipal solid wastes and sewage, that is, from feedstocks known as renewable sources of hydrogen, but in which methanol is really the intermediary product. (Similarly, more recently developed Direct-Ethanol Fuel Cells could be fueled by bioethanol. However, these cells do not yet perform as well as DMFCs.)

Direct-methanol, and other fuel cells running on a carbon fuel, do release climate-altering carbon oxides during operation, but hydrogen is not necessarily climate-neutral in terms of production either. Almost all of the vast amounts of hydrogen now produced for the chemical industry is made from natural gas, which generates carbon oxides. Thus, hydrogen as a transport fuel (or rather an energy carrier) would in the shorter term serve the purpose of weaning the world off Middle Eastern oil, not so much for healing the climate (though it may be argued that carbon monoxide or dioxide obtained during the production of hydrogen from fossil fuels can be centrally sequestered so it will not enter the atmosphere). Production of hydrogen by the electrolysis of water ($2\,H_2O \rightarrow 2\,H_2 + O_2$) may be more environmentally friendly in this respect, but this is currently unimportant and only feasible if very cheap electricity is available (in large hydro schemes, for instance). Besides, electricity could be directly used to charge battery-driven electric vehicles. As with fuel cells, almost all larger car manufacturers have demonstrated vehicles powered by batteries, but the high cost and limited range of such cars has deterred consumers. There are quite a few different types of rechargeable batteries on the market, and they all have their advantages and disadvantages in terms of energy density, power output, cost, environmental friendliness, cycle life, shelf life, and so on. None has emerged as the obvious and viable choice for vehicle propulsion. Toyota picked nickel-metal-hydrate (NiMH) batteries for its first- and second-generation hybrid vehicles, in which the internal combustion engine intermittently takes over before the battery-pack is too deeply discharged to reduce performance in terms of cycle-life. (The Toyota Prius of 2004 consumed roughly half as much petrol, and so releases half as much climate-changing carbon dioxide, compared with a new American gasoline car of the same size.) However, more expensive lithium ion batteries, which perform better, but are arguably less safe, would probably be necessary for plug-in hybrids (whose batteries can be charged directly from the grid as well as by the internal combustion engine) or high-performance electric-only vehicles. But just like electrolytically produced hydrogen, such vehicles would depend on environmentally friendly electricity from renewable, domestic sources, if they are expected to decrease both the impact on the climate and the dependency on imported oil.

Early Fuel Cell Vehicles
A long time coming. Experimental fuel cell vehicles have been on the road for several decades. The top picture shows Karl Kordesch in New York City in 1967 riding the world's first fuel cell motorcycle, silently powered by a hydrazine-air fuel cell system. Kordesch, who supervised the author's doctoral thesis at the Technical University of Graz, Austria, built this vehicle at Union Carbide Corporation in Ohio. Based on a Puch motorcycle, it ran up to 25 miles an hour and traveled 200 miles on a gallon of hydrazine. The bottom picture shows Kordesch's 1970 Alkaline fuel cell vehicle, based on an Austin A40. Kordesch drove it for his own personal transportation needs for over three years. He was thus the first person to produce and drive a practical Fuel Cell/Battery Electric Car. The fuel cell was installed in the trunk of the car and hydrogen tanks on the roof, leaving room for 4 passengers in the 4-door car. It had a driving range of 180 miles (300 km). (The vehicle is now located at the Technical Museum in Vienna, Austria.) The power unit was initially developed by Union Carbide Corporation for General Motors' Electrovan of 1966, which is considered the world's first fuel cell vehicle. (At Union Carbide, Kordesch also developed hydrazine and alkaline fuel cells in cooperation with the U.S. Army, U.S. Navy and NASA. Kordesch is otherwise known for inventing the Alkaline Primary Battery Cell, the "alkalines" that largely replaced zinc-carbon flashlight batteries. "Karl Kordesch," Famous Scientists, Institute of Chemistry, The Hebrew University of Jerusalem, http://chem.ch.huji.ac.il/history/kordesch.html; "Easy Rider," Technisches Museum Wien, http://www.tmw.ac.at/default.asp?id=758&al=Deutsch; "Die Brennstoffzelle-Die 'kalte Verbrennung'," Technisches Museum Wien, http://www.tmw.ac.at/default.asp?id=662&al =Deutsch. Photographs courtesy of Karl Kordesch.)

The Lohner-Porsche Electric Vehicle of 1900 Electric vehicles have been around even longer than fuel cell cars. In 1900 the Austrian Lohner company sent its then 24-year-old employee Ferdinand Porsche to the Exposition Universelle (World's Fair) in Paris to demonstrate an electric vehicle with electric motors placed in the front wheels. The year after Porsche added an internal combustion engine to create the world's first hybrid car.

NOTES

118. The International Energy Agency (IEA) is an autonomous body which was established in November 1974 within the framework of the Organisation for Economic Co-operation and Development (OECD) to implement an international energy program. http://iea.org/.

119. Leonardo Maugeri, "The Shell Game," *Newsweek*, February 16, 2004, http://www.msnbc.msn.com/id/4402707/site/newsweek/hl/da/.

120. U.S. Geological Survey, "World Petroleum Assessment and Analysis" (Washington, DC, various years), quoted in U.S. Department of Energy-Energy Information Administration, "Are We Running Out of Oil?," http://www.eia.doe.gov/oiaf/archive/ieo99/boxtext.html.

121. K.N. Deffeyes, *Hubbert's Peak: The Impending World Oil Shortage* (Princeton: Princeton University Press, 2001), 134.

122. Earl Cook, *Man, Energy, Society* (San Francisco: W.H. Freeman, 1976); M. King Hubbert, *Energy Resources: A Report to the Committee on Natural Resources* (Washington D.C.: National Academy of Sciences, 1962); K.N. Deffeyes, *Hubbert's Peak: The Impending World Oil Shortage* (Princeton: Princeton University Press, 2001); L.F. Ivanhoe, "Get Ready For Another Oil Shock!," http://dieoff.org/page90.htm.

123. Robert L. Hirsch, Roger Bezdek and Robert Wendling, "Peaking of World Oil Production: Impacts, Mitigation, & Risk Management," February 2005, page 70, http://www.netl.doe.gov/publications/others/pdf/Oil_Peaking_NETL.pdf.
124. L.F. Ivanhoe, "Get Ready For Another Oil Shock!,"
125. "Oil-Proved Reserves History," BP Statistical Review—June 2008, http://www.bp.com/liveassets/bp_internet/globalbp/globalbp_uk_english/reports_and_publications/statistical_energy_review_2008/STAGING/local_assets/downloads/spreadsheets/statistical_review_full_report_workbook_2008.xls. A study of 186 giant oil fields revealed that from 1981 to 1986 estimates of the oil in those fields jumped on average by more than 25 percent. Leonardo Maugeri, "Oil: Never Cry Wolf—Why the Petroleum Age Is Far from Over," Science 304 (2004): 1115; Eugene Gholz and Daryl G. Press, "Energy Alarmism: The Myths That Make Americans Worry about Oil," April 5, 2007, http://www.cato.org/pubs/pas/pa589.pdf.
126. The BP Statistical Review of World Energy, June 2002, http://www.bp.com/downloads/1087/statistical_review.pdf. Oil production here includes crude oil as well as oil from 'nonconventional resources', such as shale oil, oil sands, and natural gas liquids (the liquid content of natural gas where this is recovered separately). It excludes liquid fuels from other sources such as coal derivatives. BP's current Statistical Review of World Energy can be downloaded at www.bp.com/statisticalreview.
127. "BP Statistical Review of World Energy June 2007," http://www.bp.com/liveassets/bp_internet/globalbp/globalbp_uk_english/reports_and_publications/statistical_energy_review_2007/STAGING/local_assets/downloads/pdf/statistical_review_of_world_energy_full_report_2007.pdf.
128. Jeroen van der Veer, "Two Energy Futures: On Shell's New Energy Scenarios to 2050, Scramble and Blueprints," January 25, 2008, Speeches 2008, http://www.shell.com/home/content/media-en/news_and_library/speeches/2008/jvdv_two_energy_futures_25012008.html. Jeroen van der Veer is Chief Executive of Royal Dutch Shell plc and Energy Community leader of the World Economic Forum energy industry partnership in 2007–2008. The Royal Dutch Shell group's 2007 profit was $27.6 billion.
129. Danny Fortson, "Oil Fields of Plenty?" *The Independent*, February 1, 2008, http://www.independent.co.uk/news/business/analysis-and-features/oil-fields-of-plenty-776776.html.
130. Ibid.
131. The issue of resource nationalism in the oil arena extends to unconventional oil. Canada recently pushed through a significant tax increase on oil sand revenues. Danny Fortson, "Oil fields of plenty?"
132. Find information on most biofuels, though not on biobutanol, in World Energy Council: "Survey of Energy Resources," http://www.worldenergy.org/wec-geis/publications/reports/ser/overview.asp.
133. To be sure, Germany is not all that enthusiastic about ethanol additions to gasoline. Apparently, some 3.3 million cars in Germany could not even handle E10 in early 2008, due to ethanol corrosion that attacks aluminum parts.
134. Worldwide, 4.5 percent of the grain consumed in 2007 was used to make liquid biofuel.
135. BP, "Ethanol: The Global Ethanol Industry Continued to Grow Rapidly in 2006, with Output Increasing by 22%," http://www.bp.com/sectiongenericarticle.do?categoryId=9017929&contentId=7033479.

136. C. Berg, "World Fuel Ethanol Analysis and Outlook," 2004, http://www.distill.com/World-Fuel-Ethanol-A&O-2004.html.

137. This is a low estimate. Even though benefits of CO_2 emissions reduction are partly offset by the energy-intensive production process needed to produce ethanol, fueling a vehicle with corn-based ethanol will still reduce CO_2 emissions by 20–30 percent on a life-cycle basis, compared with petrol use, according to the United Nations Environment Program: "Alternative Fuels," http://www.unep.org/tnt-unep/toolkit/Actions/Tool14/index.html.

138. "Ethanol: Dirty as Well as Dear?," *The Economist*, January 15, 2004, http://www.economist.com/displaystory.cfm?story_id=2352788; Kurt A. Rosentrater, "Economics And Impacts of Ethanol Manufacture, *BioCycle*, 47 (12) (2006): 44, http://www.jgpress.com/archives/_free/001205.html; "American Coalition for Ethanol," http://www.ethanol.org/.

139. Another Danish company, Genencor, is Novozymes' closest competitor. Novozymes: http://www.novozymes.com/; Genencor—A Danisco Division: http://www.genencor.com/.

140. Danish company Terranol (http://www.terranol.com/) and Canadian company Iogen (http://www.iogen.ca/) are leading the way in this field. This approach targets a one-step fermentation process. In a less economical process, the C_5 and C_6 streams can be separated and processed in two steps.

141. U.S. Department of Energy, "Energy Efficiency and Renewable Energy: Biodiesel Production," http://www.afdc.energy.gov/afdc/fuels/biodiesel_production.html; Union of Concerned Scientists, "Biodiesel: FAQ," http://www.ucsusa.org/clean_vehicles/big_rig_cleanup/biodiesel.html.

142. Michael D. Morgan, Joseph M. Moran, and James H. Wiersma, *Environmental Science: Managing Biological & Physical Resources* (Dubuque: Wm. C. Brown Publishers, 1993)

143. James T. Bartis, "Policy Issues for Coal-to-Liquid Development," Testimony presented before the Senate Energy and Natural Resources Committee, May 2007, http://www.rand.org/pubs/testimonies/2007/RAND_CT281.pdf.

144. International Association for Natural Gas Vehicles (IANGV), http://www.iangv.org/statistics.html.

145. "Hydrogen-Future Fuel? An Infinite Charge: Fuel Cell Chemistry Is a Simple Dance. The Trick Is to Make It Cheaper," Penn State University, http://www.rps.psu.edu/hydrogen/infinite.html; U.S. Department of Defense ERDC-CERL Fuel Cell Project, "Fuel Cell Information Guide," http://www.dodfuelcell.com/fcdescriptions.html; MIT Fuel Cell Laboratory, "About Fuel Cells," Department of Mechanical Engineering, Massachusetts Institute of Technology, http://web.mit.edu/afs/athena.mit.edu/org/m/mecheng/fcp/about%20f%20cells.html.

146. Frederick E. Pinkerton, and Brian G. Wicke, "Bottling the Hydrogen Genie," *The Industrial Physicist*, American Institute of Physics, February/March 2004, http://www.aip.org/tip/INPHFA/vol-10/iss-1/p20.html.

147. A leading supplier of (low temperature) PEM cells is Canadian company Ballard, http://www.ballard.com/. Danish firm Serenergy, http://www.serenergy.dk/, develops high-temperature PEM cells. German company BASF Fuel Cells, www.basf.com/fuelcell, is a leading supplier of various fuel cell components.

CHAPTER 38

ELECTRICITY FOR THE FUTURE

There was a distinct reason why the later Oil Age witnessed the emergence of a strong hydrogen lobby, even though various issues surrounding the distribution and storage of hydrogen remained unresolved. It was obvious to everyone that hydrogen was a versatile energy carrier, capable of fueling internal combustion engines, fuel cells, gas burners, and gas turbines alike, but industrialist and environmentalists liked the hydrogen idea for different reasons. The former sympathized with the concept of a hydrogen-based economy, because hydrogen can be easily produced from extremely abundant coal and moderately abundant natural gas, while the latter romanced with the idea that clean, emissions-free, renewable energy based on water, wind, wave, tidal, or solar power systems would deliver electricity to generate hydrogen from water through electrolysis. However, the demand for electricity has been rising sharply during the Oil Age, and electricity production tends to be either inexpensive or environmentally friendly, but hardly ever both. Thus, the more important question is, where clean and abundant electricity for the future should come from, not to generate hydrogen, but to electrify those parts of the world, where a quarter of the global population is still waiting for their energetic liberation, as they have to do without electric lighting and cooking, without refrigeration (for food and medicine storage), and without advanced communication systems (telephones). Global demand for electricity is expected to grow at about 2.6 percent a year on average during the next three decades, which translates into a doubling of demand within well under 30 years. But what kind of source should deliver energy for such large electricity demand, and how environmentally friendly can energy generation on this scale possibly be?

COAL

Coal is the most obvious contestant to provide for electricity in the future, because coal-fired power plant technology is well established, and coal is abundant, widely distributed over the continents, and (hence) cheap. Unfortunately, coal does not burn cleanly, releasing soot and various noxious chemicals in addition to climate-altering carbon dioxide. Coal-fired plants actually produce roughly twice as much carbon dioxide per unit of electricity generated compared to those that run on natural gas. Further efficiency improvements (such as higher operating temperature and pre-drying of wetter coals) could take coal-fired plants into the same range as gas-fired ones, but installations that truly deal with the emissions generated at coal-fired power plants are expensive, both financially and in terms of energy: most technology to capture and store carbon-dioxide involves liquefying it and pumping it underground into former oilfields, gasfields or coal beds.[148]

One way or another, since energy security is prioritized over climate protection, coal has a bright future in electricity generation. The International Energy Agency (IEA) estimated that consumption of coal will increase by 71 percent between 2004 and 2030. Coal provides some three quarters of electricity in both China and India, and over half in America, where some 150 coal-fired power plants were on drawing boards in 2006. The situation was similar in Germany, where a lot of older power plants were about to retire, while the government had vowed to phase out nuclear energy.

WATER

Hydroelectric power plants will play an important part in future electricity generation as well. Hydropower is the oldest and best established form of commercial renewable electricity, and it is economically competitive with electricity generated by other means. Best of all, hydropower schemes do not release climate-altering gases. As about 20 percent of the sun energy that arrives on Earth is fueling the evaporation of water that eventually precipitates back onto the planet's surface, there should be plenty of hydropower potential. However, most practical proposals had already been realized in the industrialized world by the early 21st century, while several huge dam projects were under construction in such countries as China, India, and Malaysia. Worldwide, perhaps less than a quarter of the total practical hydropower potential as been tapped, with the largest potential being left in Africa and Asia. Currently just five countries, Brazil, Canada, China, Russia, and the United States, account for more than half of global hydroelectricity production. (In 2001 hydropower accounted for about 15 percent of the world's electricity production, which was equivalent to some 4.4 billion barrels of oil consumption per year. For comparison, some 25.7 billion barrels of oil, coal worth some 16.5 billion barrels of oil equivalent, and natural gas worth 15.1 billion barrels oil equivalent were consumed globally in 2001 for all applications, not just electricity production.)

A principal problem of hydroelectric schemes is that rivers flow day and night, while electricity consumption peaks during the day. Thus, some facilities use part of the off-peak energy to pump water from a lower to a higher reservoir. (Alternatively, energy harnessed during off-peak hours may be used to produce hydrogen or to charge batteries.) In cooler areas it is also a problem that much water is bound as snow and ice during the colder season, right when energy demand is the highest. Apart from these general issues, very serious local environmental and social problems are often associated with the larger hydropower schemes. State authorities have relocated a substantial number of people for China's Three Gorges project, for instance, while Malaysia's highly controversial Bakun project involved the flooding of vast areas of rain forest and the displacement of some 10,000 tribespeople. Besides loss of land, environmental impacts include disruption of migration routes of fish (such as salmon, for instance) as well as more complex disturbances in the river ecosystem, upstream and downstream. Sediment transport is interrupted as well, which may result in sediment shortage causing erosion in the estuaries of the river. Following the construction of the Assuan Dam, fertility of the soils along the Nile decreased as the river sediments were no longer distributed to the fields. Such effect has in turn to be offset with more use of energy-intense fertilizers. (To be sure, there are sometimes direct positive environmental effects associated with dams as well, because artificial lakes provide new habitats for birds, for instance.)

Another form of hydropower is tidal energy. The tides are not driven by sun energy, but are the result of the gravitational interaction of the sun, moon and Earth, and the centrifugal force of the Earth-moon system. A lot of energy is stored in the tides, but it is difficult to harness it, because it is not very concentrated. The best location to put a tidal power plant is at the entrance of a large bay, estuary, or delta with a large tidal head difference. The entrance of such bay can be relatively conveniently blocked off by a barrage with turbines spinning when the tide goes in as well as when it goes out. Truly well-suited locations are quite rare, but tidal power stations are currently operating in Canada, France, Russia, and China, and potential sites have been identified in various coastal countries. Unfortunately, barrage tidal power stations need to be large even at the good spots. They are therefore expensive to build, and may be competitive in long term cost only. Even the largest tidal power plant in the world, at La Rance in France, generates only quite a small amount of electricity, but it has been working reliably for decades. One problem is that a barrage will only provide power for about 10 hours per day, part of which is during off-peak hours (in the night), with seasonal variations. Besides, tidal stations also affect the habitat of seabirds, fish, and other animals. Tidal stream generators, which are less expensive than whole barrage tidal stations, have less impact on ecosystems around them. In contrast to barrage tidal stations, which drive generators based on the different water level on each side, tidal stream systems directly utilize the kinetic energy of the water current and may be installed under water as single units like windmills.

Wave power plants are another option. Some designs are similar to tidal plants, but for the moment wave power plants are even more uncommon. The enormous amount of energy contained in waves derives from sun energy, because waves are the result of wind and weather. Various schemes have been advanced since the 1973 oil price shock, but wave power schemes suffer some of the same flaws as tidal power plants as they will not produce electricity at a steady rate and not necessarily at times of peak demand. Major design breakthroughs will be necessary for wave and tidal power to contribute significantly to the world's future electricity mix.[149]

WIND

Wind energy is now the world's fastest growing power source. The contribution of wind energy to the world's commercial energy mix is currently still minimal, but it may rise to around 10 percent of all electricity as soon as 2020, because governments around the world subsidize this form of electricity generation. Germany, though not especially windy, emerged as the number one producer of wind energy, but major investments have now been made in such large countries as the United States, China, and India. Economies of scale have brought down the cost of wind electricity slowly to approach the cost of electricity generated from fossil fuels. The potential of wind energy is enormous. Between 1 and 2 percent of the solar energy arriving on Earth is converted through atmospheric circulation into wind energy, and there are hardly any principal restrictions as to where wind energy can be harnessed, which is in sharp contrast to water or tidal energy, for instance. Thus, the world's relatively windy areas combined could potentially deliver several times the current global commercial energy demand without impacting the composition of the atmosphere. On the downside, this is also an intermittent energy source that depends on the weather conditions, and many people feel that the current tall slim windmills with large propeller-rotors do not esthetically fit into the landscape, especially when many windmills are set up close to each other to form a dense windmill park. If the optimistic goals set by various governments in terms of wind energy expansion are to be met, the issue that windmills are land intensive will become more and more of a problem (unless they are put far off shore). Many of the current modern windmills are 2 megawatt generators (with rotors of about 85 meters in diameter). A lot of such mills would have to be installed to match a large hydropower plant of several thousand megawatt capacity. Besides, a windmill needs quite a bit of wind to deliver its nominal output. For a typical windmill to deliver its maximum of 2 megawatt, it will need at least a wind speed of 15 meters per second, while it needs two thirds of that wind speed (10 meters per second) to deliver half the maximum output (1 megawatt). Due to this relation even windmills set up in rather windy areas will on yearly average reach capacity factors that are only around a third of the nominal capacity. Currently, the trend goes towards even larger windmills (6 megawatt with a propeller diameter of 126 meters at an overall height of 198 meters was the record in 2007), but in some non-electrified areas such as Mongolia people use small mobile or stationary wind turbines, usually combined with an energy storage system.

BIOMASS

Biomass has in recent years become a significant source of commercial renewable energy as well. While contributions to the liquid fuel market (in form of bioethanol and biodiesel) are relatively new, biofuel has been used for electricity production for a while. There are three main sources of biofuels (in addition to agricultural output). The first is forestry, that is, the production of fuel wood from fast-growing tree species. The second is dry organic waste such straw, wood, peat, and so on, which can be sourced from household refuse as well as industrial and agricultural wastes. This organic waste can either be directly combusted, partially combusted (which due to restricted oxygen supply yields syngas), or put into landfills, where it is subjected to bacterial activity to decompose into roughly equal amounts of carbon dioxide and methane over the course of about 10 to 100 years. However, the landfill option is more common for biofuels of the third principal kind: wet wastes such as sewage sludge and farm slurries and manures. Such wet wastes may generate a biogas that contains up to 60 percent methane. Animal feedlots may install microbial gasifiers, for instance, to convert waste to gaseous fuel used to heat farm buildings or to generate electricity via gas turbines. In principle, methane (natural gas) recovered from landfills may also be resold into gas markets.

The good part about biomass as an energy source is that it is carbon neutral, releasing no more carbon dioxide when it is burned than is captured during plant growth. But this sounds better than it is. In reality energy inputs during biomass production, conversion and distribution may degrade the carbon and energy balance. (Besides, methane escaping from landfills has a greater impact on the atmosphere than when the same amount of carbon dioxide is released.) Nevertheless, substituting biomass for some of the coal burned in power plants would reduce the impact of electricity production on the climate. Perhaps to up to 20 percent biomass could be used alongside coal, otherwise the ash from biomass burning would gum up the works of most furnaces. Some two-thirds of the world's population still cooks by burning biomass (usually wood), while technical, commercial biomass is currently far less important. A principal problem of biomass as an energy source is that relatively small amounts of the solar radiation arriving on the surface go into photosynthesis. In fact, it is less than a 10th of one percent, which is far less than the principal water or wind power potential, for instance. In the United States, energy consumption has become so extreme that all annual nationwide photosynthesis captures the equivalent of less than 60 percent of the nation's annual energy use. And even globally all photosynthesis on land and in the oceans captures no more than some six times our worldwide energy consumption.

SOLAR

Direct utilization of solar energy for electricity production through photovoltaic solar cells may be the most attractive option for future electricity production. After all, even the part of solar radiation that makes it yearly through the atmosphere to the planet's surface is over 6,000 times as large as all the energy we use globally per year. But direct conversion of solar radiation into electricity is not that easy. Photovoltaic cells are known from small devices such as calculators or watches, as well as from spacecrafts which are covered with solar panels. More than 80,000 Japanese households installed solar roof panels after the government started to offer generous subsidies in 1994, and a German law of 2000 regulated the price for solar energy to launch the installation of 100,000 solar panels on homes and businesses. Such initiatives helped to introduce economies of scale to reduce the cost of energy from solar cells, but electricity from such units can thus far still not compete in price with conventional electricity. Silicon photovoltaic cells consist of a thin slice of silicon into which appropriate impurities have been introduced to create a system that generates electricity when exposed to sunlight. The most efficient solar cells use crystalline silicon as the base material and convert about 20 percent of solar radiation into electricity. Cheaper amorphous silicon is used for cells that have an efficiency of only about 10 percent. (Other semiconductor material, such as cadmium-telluride and copper-indium, with similar photovoltaic properties may become important in the future as well.) This relative inefficiency means that quite large areas have to be covered by solar panels to achieve appropriate outputs. However, there is easily enough space on the roof of a house to place solar panels capable of powering light systems, radios and televisions, water pumps, refrigerators, and so forth This has significance in rural areas of developing countries that lack a national power grid, and has been demonstrated in industrialized countries that appreciate the contribution of solar energy to reduce the generation of greenhouse gases. Unfortunately, solar cells do not work all that well in cloudy weather, and the electricity collected on a wide area of solar panels would have to be concentrated to achieve truly high power outputs. When the German photovoltaic power plant at Göttelborn, then one of the largest in the world, opened in 2004, it covered an area of about 80.000 m^2, equivalent to ten soccer fields, to provide a peak nominal power of 4.0 Megawatt. (The module area was ca. 30,000 m^2, delivering some 3.8 million kWh a year at an efficiency of 15 percent), sufficient to meet the demand of about 1,600 average German households.) Later on, Germany planned a 40 MW photovoltaic power station, while Portugal planned a 62 MW station, and Australia one with a capacity of 154 MW.

The truth is, however, that solar radiation can be more efficiently utilized (at nearly 100 percent) when it is converted into heat, rather than electricity. This is well known from the solar heaters used in various sunny regions to warm water for domestic and industrial use. (Such equipment usually consists of a black panel that contains pipes in which water (or air) is heated by sunshine and circulated either by thermal convection or by a pump.) Conversion of solar radiation into heat is efficient enough to produce water steam to drive a conventional turbine for electricity generation. The world's first solar-thermal power station to be connected to a national grid opened in 1991 at Adrano, in Sicily. Scores of giant movable mirrors follow the sun throughout the day, focusing the rays into a boiler. A similar prototype, called Solar 1, had been built in the Mohave desert near Daggett, California. It consisted of 1,818 computer-controlled mirrors arranged in circles around a central boiler tower which is 91 meters (300 feet) high. However, higher efficiency is achieved with long paraboloidal troughs (rather than dishes), which focus sun radiation onto a tube (running the length of the mirror), which is filled with a heat transfer fluid (usually oil). The heat transfer fluid is then used to heat steam in a standard turbine generator. The Kramer Junction Company built three such parabolic trough solar-thermal power stations filling an enormous area within 40 miles of one another in the Mojave Desert. Combined, they generate about 354 megawatts at peak output, easily the largest solar energy facility in the world. Conveniently, its output is the largest when the region's electricity demand is the highest due to air conditioning. But in addition to being land intensive, and weather-dependent, such solar power plants suffer from high capital cost to be invested upfront.[150]

GEOTHERMAL

A rather attractive source of energy for electricity production is the planet's hot interior.[151] This is a virtually inexhaustible energy reservoir that stems from impact energy from the time the nascent planet was still consolidating by gravitationally attracting matter from its vicinity. In addition, heat is continuously added by the decay of radioactive isotopes of underground rocks.

Geothermal energy can be directly harnessed in areas where hot springs reach the surface, with hot water being used for heating purposes or to drive steam turbines for electricity production. More interestingly, emissions-free geothermal energy may also be harnessed elsewhere through Hot Dry Rock technology (also known as Enhanced Geothermal Systems). This involves drilling at least two boreholes, one to pump water underground, the other to collect hot water after it has passed through porous, hot underground rock formations. However, successful heat mining depends on local geologic conditions and may result in low frequency stress waves similar to those caused by small earthquakes, because the injected water slightly moves the rock formation. In Basel, Switzerland, the first commercial geothermal power plant of this kind was to be constructed to deliver some 30 MW in thermal and some 3 MW in electrical energy. However, when the Swiss engineers injected water into their 5,000 meter deep borehole on 9 December 2006, a tremor measuring 3.4 on the Richter scale caused widespread fear. Nevertheless, similar projects are underway in California, France, Germany, Japan, and Australia, where temperatures reach around 235C at a depth of 3.5 kilometers. To be sure, geothermal energy may not be truly renewable at any given site, as the heat contained in rock formations is ultimately limited. However, such rock formations would eventually recover some heat from the vast heat reserves of the planet's mantle.

NUCLEAR

Nuclear energy is another option to generate electricity for the future.[152] Nuclear (fission) energy is not renewable energy, as it relies on ore, but it does not generate greenhouse gas emissions. Nuclear energy is well established and can be expanded, as there is plenty of uranium left. Uranium resources are reported by confidence levels, from reasonably assured to speculative, and by production cost, from under $40 per kg to under $130 per kg. In 2005, when demand for uranium amounted to 68 thousand tons, global reasonably assured uranium resources to be produced at a cost under $40 per kg were reported at a level of 1.9 million tons, enough for 28 years at current consumption levels. (To be sure, demand may increase to 100 thousand tons a year by 2030.) And if all confidence and production levels are included, reported uranium resources amounted to 14.8 million tons. (Besides, spent fuel can be reprocessed to extract plutonium and uranium for use as new reactor fuel.) These deposits are reported by 43 different countries, though the top 20 countries account for 96 percent of the total. Uranium resources are fairly widely distributed among the continents. In terms of reasonably assured and inferred uranium resources Australia has 1.1 million tons, Kazakhstan 0.8 million, Canada 0.4 million, South Africa, the United States, Namibia, and Brazil 0.3 million each, Niger and Russia 0.2 million each, and Uzbekistan

0.1 million. Canada is the world's biggest producer of uranium (29 percent of the total), and Australia a close second (22 percent of the total).[153]

The bad news is that nuclear power plants are expensive to build (A single plant requires an initial investment between two and three and a half billion dollars.); that they might be attractive targets for terrorist attacks; that they might promote the proliferation of nuclear weapons; and that the long-term storage of nuclear waste remains unresolved. Nevertheless, the nuclear industry has recently experienced a considerable upswing in the name of climate protection, despite the fact that many countries moved away from nuclear energy. Recent projections hold that the share of nuclear power in the world's total electricity supply will decrease from the current 16 percent to 10 percent in 2030 (World Energy Outlook 2006, Reference Scenario). While the German government decided in 2000 to rather quickly phase out the country's 19 operating nuclear power plants, most of the U.S. nuclear plants indicated an intention to seek license renewals, and over half of the world's 34 nuclear power plants under construction were located in Asia.

A potentially much more attractive alternative to the established nuclear fission technology (which splits large atomic nuclei) is nuclear fusion. The fusion of small, colliding atomic nuclei into larger atoms releases a lot of energy: This is what powers the sun and the other stars in the universe. However, these nuclear reactions are difficult to achieve, because the particles that have to collide are very small and repel one another, as they all are positively charged. The fusion reaction usually targeted is between deuterium and tritium, the two heavier versions of hydrogen. Deuterium, with a core consisting one proton and one neutron, occurs in nature, replacing a low percentage of hydrogen in water molecules. (It is quite easy, though, to separate deuterium-containing heavy water from normal water.) Tritium, on the other hand, does not occur in nature. It is radioactive, decays relatively fast, and needs to be produced from lithium, a metal that is plentiful in the Earth's crust. In theory, deuterium and tritium can merge into helium (a harmless gas that occurs naturally in the atmosphere) in a reaction that releases a lot of energy, plus a free neutron. However, reactors with high enough temperatures and particle speeds to collide and merge these atoms, and the capability of harnessing the generated energy, have yet to be constructed. Significant technical barriers must be overcome, but the amounts of energy to be harvested would be enormous: Known lithium reserves would be sufficient for the deuterium/tritium fusion reaction to supply the entire current global electricity production for more than 1,000 years. This prospect has attracted 10s of billions of dollars since the 1950s. From 1986, researchers in different countries have collaborated planning ITER, the International Thermonuclear Experimental Reactor. In November 2006 the European Union, the United States, China, Japan, India, and South Korea finally agreed to build ITER in Cadarache, southern France, over the course of a decade (starting in 2008) at a cost of $12.8 billion (€10 billion).[154] The reactor will be of the tokamak type, an abbreviation of the Russian expression for "toroidal magnetic chamber," which indicates that the reactor confines a very high-temperature plasma within a doughnut-shaped vessel by magnetic fields. The interior of the plasma has to reach a temperature of around 100 million C in order for particles to move fast enough to collide. ITER will have a plasma volume of 800 m3, 10 times the volume of the largest tokamak now in existence. This research facility is not designed to actually deliver electricity, but will supposedly be used to study plasma fusion for about 15 years. Constructing a commercial fusion reactor would probably take at least another 15 years, but skeptics argue that the engineering difficulties involved are too extreme to be overcome. This includes potentially prohibitive costs of building the reaction vessel, and the difficulties of repairing and maintaining it: it needs to consist of lithium and rare metals (to absorb the neutrons emitted by the plasma), but will degrade and become radioactive over time, requiring regular dismantling and replacement.[155] In short, nobody can as yet be sure that nuclear fusion will ever turn into a productive energy source.[156]

NOTES

148. James Katzer et al., "The Future of Coal," Massachusetts Institute of Technology, 2007, http://web.mit.edu/coal/.

149. One company developing technology to exploit tidal currents for large-scale power generation is Marine Current Turbines Ltd. in Bristol, http://www.marineturbines.com/. This firm quotes the Carbon Trust, an independent company set up by the government, in terms of a 2006 estimate that tidal stream energy could meet 5 percent of the U.K.'s present electrical energy needs.

150. To be sure, there are also some rather daring solar proposals, including giant solar reflectors put into space to harness solar energy that would be beamed to Earth in the form of microwaves.

151. Jefferson Tester et al., "The Future of Geothermal Energy," Massachusetts Institute of Technology, 2006, http://geothermal.inel.gov/publications/future_of_geothermal_energy.pdf.

152. John Deutch et al., "The Future of Nuclear Energy," Massachusetts Institute of Technology, 2003, http://web.mit.edu/nuclearpower/.

153. International Energy Agency, *World Energy Outlook 2006*, http://www.iea.org/textbase/nppdf/free/2006/weo2006.pdf.

154. International Thermonuclear Experimental Reactor (ITER), http://www.iter.org/; Gia Tuong Hoang and Jean Jacquinot, "Controlled Fusion: The Next Step," *Physics World*, January 2004, http://physicsweb.org/article/world/17/1/6; "Green Light for Nuclear Fusion Project," *New Scientist*, November 21, 2006, http://technology.newscientist.com/article/dn10633-green-light-for-nuclear-fusion-project.html.

155. David L Chandler, "No Future for Fusion Power, Says Top Scientist," *New Scientist*, March 9, 2006, http://www.newscientist.com/channel/fundamentals/dn8827-no-future-for-fusion-power-says-top-scientist.html; William E. Parkins, "Energy:

Fusion Power: Will It Ever Come?," *Science* 311 (5766) (2006), 1380, http://www.sciencemag.org/cgi/content/summary/311/5766/1380.

156. Other nuclear fusion technologies than the one described here have been discussed as well. One involves the use of helium-3, a non-radioactive isotope of helium that is rare on Earth, as raw material. Some scientists believe that it would be possible to economically procure helium-3 from regolith, the moon's soil. Another technology approach is to trigger nuclear fusion reactions with the help of lasers, which is potentially a lot cheaper than ITER technology, though it would face the same material challenges. "Nuclear Fusion: Firing New Shots," *The Economist*, April 19, 2007, http://www.economist.com/research/backgrounders/displaystory.cfm?story_id=9033026.

CHAPTER 39

EXTENDED OIL AGE?—ENERGY
MIX UNTIL 2030

An educated guess of how various energy alternatives may contribute to the future energy mix can be found in the projections published by different governmental and private energy agencies. Such reports cannot be truly accurate, as they are necessarily founded on a lot of assumptions, but the more comprehensive ones consider everything from future population growth, economic growth, climate and other policies, and energy prices, to advances in energy technology, to arrive at a most likely scenario. The results are fairly similar. The "International Energy Outlook 2007" (published by the U.S. Department of Energy's Energy Information Administration) projected that the total world consumption of marketed energy will increase by 57 percent between 2004 to 2030, for instance, while the "World Energy Outlook 2006" (published by the OECD's International Energy Agency) projected that global primary energy consumption will increase by 53 percent during the same time period. About half of the increase goes to electricity generation and about a fifth to transport needs. Most (nearly three-quarters) of this enormous projected growth in energy consumption within just two-and-a-half decades is growth in developing countries. China and India together account for as much as 45 percent of the increase. But how will the global energy mix change in the process?

THE FOSSIL FUEL WORLD REMAINS

According to the International Energy Agency fossil fuels accounted for 80 percent of the total primary energy mix in 2004. Oil accounted for

35.2 percent, coal for 24.8 percent, natural gas for 20.5 percent, biomass and waste (including non-commercial) for 10.5 percent, nuclear energy for 6.4 percent, hydropower for 2.1 percent, and other renewables for 0.5 percent. A generation later, in 2030, the share of fossil fuels as a whole is projected to remain unchanged (81 percent). The share of oil is supposed to slightly decline (to 32.6 percent), but remains the single largest fuel, while the shares of coal (to 26.0 percent) and natural gas (to 22.6 percent) are slightly increasing. As the role of fossil fuel energy remains fully intact during the next quarter of a century according to the International Energy Agency reference case, alternative energy will—despite much talk about climate protection—remain relatively unimportant in the near-term future. (The share of nuclear energy, which also provides energy without carbon dioxide release, is similarly projected to slightly decline [to 5.0 percent].) As fossil fuels are supposed to maintain their market share despite an overall expansion in energy consumption, all three, oil, coal, and natural gas, must be produced at rates more than 50 percent above current rates 25 years from now, if the International Energy Agency's reference scenario is to become reality.

COAL

Such production increase can well be achieved for coal. Coal has been the fastest growing major fuel in recent years, and the International Energy Agency has revised its projections for coal upward to increase to 28 percent of world energy consumption by 2030. Over 80 percent of this increase arises in China and India, and most (80 percent) of the increase is for electricity production. (Already now, China and India combined account for 45 percent of global coal consumption.) Coal consumption in Western industrialized countries, on the other hand, grows only very slowly until 2030. This growth is driven by the United States, where production increases are expected to account for 8 percent of the global total. Meanwhile hard coal output in the European Union, where production costs are high, decreases in line with remaining subsidies being phased out. (Brown coal production is expected to remain constant in the European Union.)

Coal is the most abundant fossil fuel, with proven reserves of over 900 billion tons (909 billion tons at the end of 2005), equivalent to 164 years of production at current rates. Well over half of these reserves are located in just three countries, the US (27 percent), Russia (17 percent), China (13 percent), and over 80 percent of all reserves are accounted for if another three countries, India (10 percent), Australia (9 percent), and South Africa (5 percent), are added to the list. Nevertheless, there are well over 20 countries that each hold substantial reserves of at least one billion tons.

Its large coal reserves will help China to meet its rapidly growing energy demand, which is expected to more than double between 2005 and 2030.

China has four times the population of the United States, and will soon, probably shortly after 2010, pass the United States to become the world's largest energy consumer. China actually needs to add more electricity-generating capacity to its current base than now exists in all of the United States, if the projected electricity demand is to be met. (Thus, projected cumulative investment in China's energy-supply infrastructure would be $3.7 trillion in year-2006 dollars over the period 2006 to 2030, with three-quarters going to the power sector.) Interestingly, China will, despite its vast domestic coal resources, in part use imported coal to meet its growing electricity demand. (China actually became a net coal importer in the first half of 2007.) This is due to the ancient, but unchanged, reality that water-borne transport is a lot cheaper than road or even rail transport: More than 90 percent of China's coal resources are located in inland provinces, while energy is needed at the booming coastal regions where imported coal can easily be received.[157]

The situation is similar in India, where many power plants are located close to ports, but India also imports coal for quality reasons for the steel sector. Overall, India's primary energy demand is projected to more than double by 2030, but coal consumption is supposed to nearly triple between 2005 and 2030. The number of Indians relying on biomass for cooking and heating is expected to decline from 668 million in 2005 to some 470 million in 2030, with the share of people having access to electricity rising from 62 percent to 96 percent, mainly based on coal energy. (India's power-generation capacity needs to triple between 2005 and 2030 to meet demand, with newly installed capacity being equal to the current combined electricity-production capacity of Japan, Korea, and Australia.)[158]

NATURAL GAS

Global natural gas production is also expected to be able to keep track with the rise in demand over the next decades. Here, too, much of the increased demand, over half, derives from increased demand for electricity. Global natural gas production is projected to increase by two-thirds between 2004 and 2030, with the Middle East and Africa contributing most. Proven natural gas reserves amounted to 180 trillion cubic meters at the end of 2005, enough to guarantee supply for 64 years at current rates. Over half of these reserves are found in just three countries: Russia, Iran, and Qatar, while gas reserves in Western industrialized countries represent less than a tenth of the global total. North Sea gas production is expected to peak around 2012, and Russia's large reserves will be technically difficult to extract and transport to market. Worldwide proven gas reserves have increased by more than 80 percent since 1985, with large additions in Russia, Central Asia, and the Middle East. Much of this has been discovered while exploring for oil, but, as with oil, the gas fields found since 2000 are smaller on average than those found previously.

Europe will import even more gas by 2030, and Africa is expected to replace Russia as Europe's largest supplier in natural gas. North America is now largely self-sufficient in terms of gas, but will emerge as a major importer, covering some 16 percent of consumption by LNG (liquefied natural gas) imports. Chinese gas imports are expected to grow as much as 56-fold between 2004 and 2030, and yet gas will meet no more than 5 percent of Chinese energy needs, up from 3 percent today. (China's first LNG terminal was commissioned in 2006.)

OIL

Oil is expected to remain the single largest fuel until 2030. The Oil Age thus will not be over anytime soon. Nevertheless, it is perhaps surprising that the International Energy Agency expects the world's remaining oil resources to be sufficient to meet rising demand until 2030, which is supposed to reach 116 million barrels per day, 38 percent above the 2006 level (84 million barrels per day).[159] Some 42 percent of the increase in demand comes from China and India alone. Their combined oil consumption is expected to more than double from 9.3 million barrels per day in 2005 to 23.1 million barrels per day in 2030. Demand for oil will continue to be driven by the transport sector: while some 900 million cars and trucks were on the world's roads in 2005, the number is expected to be above 2.1 billion by 2030. In China, the total number of light-duty vehicles on the roads is expected to increase from about 22 million in 2005 to more than 200 million in 2030, and in India from 11 million to 115 million in India. China's total vehicle fleet is expected to expand sevenfold to reach nearly 270 million in 2030, and new vehicle sales in China are expected to exceed those of the United States around 2015. However, China has became a net oil importer already in 1993, passed Japan to become the world's second-largest oil consumer (after the United States) in 2004, and will see its oil production peak (at 3.9 million barrels per day) around 2012. Thus, China's share of imported oil in total consumption will increase from 50 percent to 80 percent, and from 3.5 million barrels per day to 13.1 million barrels per day (2030) in absolute terms. Meanwhile, India, with very small indigenous oil resources, will increase its imports to 6 million barrels per day by 2030, and will, before 2025, pass Japan to become the world's third-largest importer of oil, after the United States and China. Thus, China's and India's combined oil imports are expected to surge from 5.4 million barrels a day in 2006 to 19.1 million barrels a day in 2030, which is more than the current imports of Japan and the United States combined.

Here is how the import shares will change for the main importers: in 2004, the United States imported 64 percent of the oil it consumed, the European Union 79 percent, India 69 percent, and China 46 percent. By 2030 the United States is expected to cover 74 percent of its oil consumption by imports, the European Union 92 percent, India 87 percent, and China 77 percent. As non-OPEC (conventional) oil is expected to peak around 2015 (at a level of about 47 million barrels per day), the world will increasingly have to rely on oil exports from Middle Eastern OPEC nations plus Russia. (By 2030, the share of OPEC oil in total world oil supply will have increased from the current 42 percent to 52 percent. OPEC output including natural gas liquids and gas-to-liquids is projected to rise from 36 million barrels a day in 2006 to 61 million barrels per day in 2030.) However, the International Energy Agency warns that substantial investments will be necessary to keep the oil from these countries flowing: "A supply-side crunch in the period to 2015, involving an abrupt escalation in oil prices, cannot be ruled out."[160]

NOTES

157. "China's Quest For Resources: A Large Black Cloud," *The Economist*, March 13, 2008, http://www.economist.com/specialreports/displaystory.cfm?story_id=10795813.
158. International Energy Agency, *World Energy Outlook 2007*, http://www.iea.org/textbase/npsum/WEO2007SUM.pdf.
159. International Energy Agency, "World Energy Outlook 2007: Fact Sheet-Oil," http://www.iea.org//textbase/papers/2007/fs_oil.pdf.
160. All information from the International Energy Agency's *World Energy Outlook*, http://www.worldenergyoutlook.org, http://www.worldenergyoutlook.org/2008.asp; International Energy Agency, *World Energy Outlook 2006*, http://www.iea.org/textbase/nppdf/free/2006/weo2006.pdf. International Energy Agency, "World Energy Outlook 2007: Fact Sheet-Oil," http://www.iea.org//textbase/papers/2007/fs_oil.pdf; International Energy Agency, "World Energy Outlook 2007: Fact Sheet Global Energy Demand," http://www.worldenergyoutlook.org/2007.asp, http://www.iea.org/Textbase/npsum/WEO2007SUM.pdf.

CHAPTER 40

WHO GOT THE POWER?

Perhaps the most obvious sign for the Oil Age to come to an end is that all the world's major existing and emerging power blocs do not have oil or are rapidly running out of it. The 20th century had been different. The United States and the Soviet Union were the world's top oil producers and firmly maintained their position as the planet's two and only superpowers for several decades after World War II. Now a set of four main centers of power seems to be emerging, consisting of the United States, the European Union, China, and India. All of these consist of large populations and internal markets, and all of them possess nuclear weapons, but all of them also depend on oil imports, for which they will eventually have to compete more fiercely.

United States

The United States is slightly larger than China, twice as large as the European Union, and nearly three times as large as India, but at 301 million (2007) its population is actually the smallest of the four. In terms of economic output ($13.86 trillion, 2007 at PPP), the United States is second only to the European Union, but in terms of economic growth (2.2 percent, 2007 in real terms) the it currently ranks last among these four. What is more, the United States is being plagued by sizable trade and budget deficits. (The merchandise trade deficit reached a record $847 billion in 2007.) But the United States has repeatedly pulled a simple trick to fix its balance of trade: Due to its large internal market, it can afford to allow its currency, the dollar, to devaluate

Table 40.1 Comparison of Key Economic and Oil Supply Indicators for the United States, the European Union, China, and India

	Population / million	Economic output / trillion US$	Economic growth / percent	Per capita GDP / US$	Oil consumption / million barrels per day	Oil production / million barrels per day	Oil reserves (proved) / billion barrels
United States	301	13.86	2.2	46,000	20.59	6.87	29.9
European Union	490	14.45	3.0	32,900	14.87	2.31	6.7
China	1,322	7.04	11.4	5,300	7.45	3.68	16.3
India	1,130	2.97	8.5	2,700	2.58	0.81	5.7

Population: estimates for mid-2007; Economic Output: 2007 Gross domestic product (GDP) at purchasing power parity (PPP). (A nation's GDP at PPP exchange rates is the total value of all goods and services produced in the country valued at prices prevailing in the United States.); Economic growth: 2007 GDP growth adjusted for inflation; Per capita GDP: 2007 GDP on a purchasing power parity basis divided by population as of July 1 for the same year; Oil production: 2006 estimates, including crude oil, shale oil, oil sands, and NGLs (the liquid content of natural gas where this is recovered separately), but excluding liquid fuels from other sources such as biomass and coal derivatives; Oil consumption: 2006 estimates; Oil reserves, proved: at end 2006. Proved reserves of oil are those quantities that with reasonable certainty can be recovered in the future from known reservoirs under existing economic and operating conditions; European Union: oil data for 25 member states.

Economic data: "CIA—The World Factbook" (https://www.cia.gov/library/publications/the-world-factbook/); Oil data: "BP Statistical Review of World Energy 2007" (http://www.bp.com/liveassets/bp_internet/globalbp/globalbp_uk_english/reports_and_publications/statistical_energy_review_2007/STAGING/local_assets/downloads/pdf/statistical_review_of_world_energy_full_report_2007.pdf; BP's current Statistical Review of World Energy can be downloaded at www.bp.com/statisticalreview).

against other currencies, which promotes U.S. exports and cuts imports. Most significantly, the U.S. per capita GDP remains very large ($46,000, 2007 at PPP), and the United States may by this stage certainly still be called the world's one and only economic, political, and military superpower: It has the best weapons systems in the world, maintains a large standing army, and has over and over again shown its readiness to go to war.

In terms of oil, the United States is still somewhat better positioned than the other three power blocs. It has by far the largest oil deposits left (29.9 billion barrels), it produces the most oil (6.9 million barrels a day), and though it covers two-thirds of its oil consumption by imports, this share is yet larger in the European Union and India. (In absolute terms, the U.S. imports amounts similar to the European Union, some 13 million barrels a day.) On the other hand, no other major nation has been consuming oil as rampantly as the United States, and it may be most difficult for America to rid itself of high oil consumption habits and to fully realize that the very domestic resource base that carried it to superpowerdom is rapidly declining. (Germans, for instance, consume less than half the amount of oil per person compared to Americans.) Between 1996 and 2006, U.S. oil imports rose by nearly half (45 percent), and the United States, with four and a half percent of the global population, now still consumes about a quarter (24 percent) of all the world's oil. To hedge against supply risks, the United States has long been attempting to spread its imports. In 2001, it received 24 percent of its oil imports from the Middle East, 22 percent from South and Central America (Venezuela), 15 percent from Canada, 12 percent from Mexico, and 12 percent from West Africa (Nigeria). Nevertheless, it will be a new situation for the United States to lose its position as a significant oil producer and to rely on ever larger shares of imported oil. Thus, the question remains, if the United States, with a declining productive oil-based energy base but intact destructive energy resources, will cling to power by all cost, and keep going to war.

To be sure, many decades of superpowerdom have left the United States with a firm grip on the globe and a worldwide network of allies on all continents. The United States de facto controls various oil producers in the Middle East and elsewhere, and America has traditionally maintained good relations with the European Union (and especially Britain) as well as India, while relations with communist China have remained strained politically, though they have lately been intensifying economically. What is more, the opinion persists in many circles that U.S. superpowerdom is preferable to other powers imposing their views on the world, because America at least formally upholds values such as personal freedom and democracy, which, at least under Western bias, are universally desirable. On the other hand, the United States is marked by many decades of superpower ballast and power abuse. The relatively minor misdoings are on the economic side, where free trade and market economics are propagated, while tariffs have been imposed (on Japanese cars and Brazilian steel, for instance), and subsidies been handed out domestically (U.S. cotton farmers), which has disadvantaged people outside America. More drastically, it is well documented that America has, whenever it has suited its interests, overthrown, or helped to overthrow, democratically elected leaders (Allende in Chile or Mossadegh in Iran); supported murderous dictators (Marcos in the Philippines, Suharto in Indonesia, various leaders in Latin America); and assassinated leaders of other nations (President Eisenhower's order to assassinate Patrice Lumumba, prime Minister of the Republic of the Congo, but also the missile attacks against the Iraqi headquarters in a targeted killing attempt against Saddam Hussein). Recently, America's credibility has suffered when President George W. Bush sent his secretary of state, Colin Powell, to mislead the United Nations' Security Council about evidence on Iraq's possession of nuclear and biological weapons.[161] What is more, human rights abuses by U.S. soldiers, in Afghanistan and in Iraq, for instance, have made headlines around the world.[162] Many have therefore concluded that America has long since lost any claim to be a moral leader of the world. And arguably, only credible leadership in terms of an accepted set of values can foster a superpower position in the longer run.

The European Union

Whether the European Union will be able to assume a role as some kind of superpower in the near future remains unclear. The European Union started out in 1951 as a regional economic agreement between six countries, Belgium, France, West Germany, Italy, Luxembourg, and the Netherlands. (This was known as the European Coal and Steel Community.) In 1957 this cooperation turned into the European Economic Community, attempting to eliminate trade barriers among its members, and in 1967 the European Community, with a Commission, a Council of Ministers, and the European Parliament, was founded. The Community began growing in 1973, when Denmark, Ireland, and the United Kingdom joined, followed by Greece (1981), and Spain and Portugal (1986). The European Union as such came into being on the basis of the 1992 Treaty of Maastricht, laying the basis for cooperation in foreign and defense policy, in internal affairs, and in the creation of a monetary union. Austria, Finland, and Sweden joined the EU in 1995, and 10 new countries followed in 2004: Cyprus, the Czech Republic, Estonia, Hungary, Latvia, Lithuania, Malta, Poland, Slovakia, and Slovenia. And when Bulgaria and Romania joined in 2007, membership rose to 27 nations.

To be sure, the large and wealthy European Union is not truly a federation, and internal cohesion has not yet materialized to the extent that some had hoped for. When the new common currency, the euro, was introduced in 1999, Britain, Sweden, and Denmark refused to participate, and of the 12 most recent member states only Slovenia (2007), Cyprus, and Malta

(2008) adopted the euro within short time. Efforts to establish an EU constitution failed in 2004 and 2007, and great differences remain in terms of wealth among different member states: per capita incomes vary from $7,000 to $81,000, that is, by more than a factor of 10. What is more, the European Union often does not seem to be a unity when dealing with the rest of the world. Germany, for instance, refused to go to war against Iraq as long as the work of the United Nations' weapons inspectors had not been finalized, while Britain was eager to support the U.S.-led invasion of Iraq. Nevertheless, it is expected that the European Union will proceed towards more common internal and external policies over time.

In terms of oil resources, the European Union is not exactly well-endowed. (The European Union does not come close to the United States, China, and India in terms of coal deposits either.) Britain is the richest country within the EU as far as oil goes, but British oil production began to decline in 1999. (Nevertheless, UK production was still 1.64 million barrels a day at the end of 2006, well above domestic UK consumption.) Norway is also a major oil nation, producing as much as 2.78 million barrels a day (2006) while having no more than four million inhabitants. (With just 10 percent of the Norwegian work force being employed in the oil industry that generates about half of Norway's gross domestic income, the country has become very wealthy.) However, Norway refused to join the European Union. Besides, Norwegian production peaked in 2001, and both Norway and Britain were set to produce about half as much oil in 2010 than they did in 2000. Meanwhile Germany, France, Italy, and Spain are among the world's top 10 oil importers, and imports into the EU as a whole rose by nearly a third (29 percent) within 10 years between 1996 and 2006. At proven reserves of 6.7 billion barrels of oil, a domestic production of 2.31 million barrels per day, and a consumption of nearly 15 million barrels per day, the balance does not look good. The share of imported oil is nearly 80 percent already now, and it is set to soar above 90 percent within one to two decades.

The European Union, too, has attempted to hedge against supply risks by diversifying its oil imports. In 2002 the EU received 28 percent of its oil imports from eastern Europe (mainly Russia), 25 percent from the Middle East, 21 percent from Africa and 20 percent from Norway. (If imports are calculated for western Europe including Norway, then import shares from the Former Soviet Union and the Middle East amounted to about 31 percent each.) Perhaps its geographic position close to Russia, Africa, the Middle East, and the countries around the Caspian Sea, will help the EU to secure access to oil from these regions. The struggle to make Turkey a member of the European Union is certainly associated with this issue. It is Turkey's geographic location that makes the country strategically important. Turkey borders oil-rich Iraq,[163] and its Mediterranean port of Ceyhan is the primary terminal through which Iraq's northern oil exports pass. Turkey also transits oil arriving from Russia, the Caucasus, and the Caspian region. This oil, if arriving in the Black Sea to be shipped to the Mediterranean, has to be shipped through Turkey's Bosporus Straits. For alternative transport of such oil, a pipeline is being constructed from Samsun, at Turkey's northern Black Sea coast, to Ceyhan, at Turkey's southern Mediterranean coast. (At a length of 550 km, and a maximum daily capacity of 1.5 million barrels of crude oil, this pipeline will potentially transport some 70 million tons of oil, chiefly from Russia and Kazakhstan, per year cost-effectively and safely to relieve oil transits through the highly-populated area along the Bosporus Straits.[164]) What is more, Turkey's Mediterranean Ceyhan port is the end-point of the Baku-Tbilisi-Ceyhan (BTC) Pipeline, the first transnational pipeline that transports Caspian oil without crossing Russian soil.[165] (This pipeline crosses from Azerbaijan into Georgia, which unfortunately is politically unstable. The West is now eager to make Georgia a NATO member, which Russia is trying to prevent.)

There has been somewhat of a struggle between the European Union and the United States over influence in Turkey. NATO-member Turkey became the third largest recipient of U.S. military aid in the early 1990s (behind Israel and Egypt), while the Turks attempted to achieve European Union membership, which has so far been denied. European policy makers are well aware of Turkey's strategic location and geopolitical importance, and the European Union in 2005 began the formal accession procedure for Turkey to become a full member of the EU. However, many EU citizens feel that this Muslim nation, which is technically almost entirely situated on the Asian continent, does not truly belong into the community of European states. The EU has asked Turkey to work towards further economic reform and liberalization, and has at times complained about the human rights situation in Turkey.

China

China has experienced tremendous economic growth in recent years, and on a country-by-country comparison, China's economic output is second only to the United States. Likely, China's amazing economic growth (11.4 percent in 2007, compared to the United States's 2.2 percent) will ease a little during the next few years, but it is expected to remain well above eight percent[166]. Living standards have improved dramatically, but the economic boom has been concentrated in the highly-populated coastal regions, while many of the internal areas did not join this untethered rush into the Oil Age and remained in their traditional rural settings. (Hundreds of millions of peasants live in the interior, and some 200 million rural laborers have relocated to urban areas to find work.) What is more, the rapid economic transformations are associated with environmental damage and social strife, and (as a result of the one child policy) China is now one of the most rapidly aging nations. Meanwhile, the communist government still exerts tight political control and maintains a human rights situation that is grossly inadequate, though room for personal choice has expanded compared to previous decades.

To the outside, China is now visible as a colossal economic force, absorbing raw materials and energy resources from all over the globe, and selling its products to consumers worldwide. (In 2007 China had a trade surplus of $261 billion, resulting from enormous merchandise exports worth $1.2 trillion, and imports worth $956 billion.) With about a fifth of the global population, China now consumes half of the world's cement, a third of all steel, and over a quarter of the world's aluminum. China's copper imports have risen more than 20-fold within eight years, and shipments of iron ore by an average of 27 percent for several years. As China is securing much of these resources in Africa and Latin America, these, and other regions that supply China, have enjoyed faster economic growth than they have for decades. Thus, China's economic growth is likely to lift more people out of poverty than all the West's aid schemes put together. By the end of 2007, over 5,000 domestic Chinese enterprises had established direct investments in 172 countries and regions around the world. Chinese entrepreneurs have set up numerous copper processing plants in the Congo, for instance, and in late 2007 the Congolese government announced that Chinese state-owned firms would build or refurbish various railways, roads, and mines around the country at a cost of $12 billion, in exchange for the right to mine copper ore of an equivalent value. Meanwhile oil-exporter Angola, after receiving plenty of aid and investment from China, in 2006 decided it had no need of the International Monetary Fund's conditional billions. In Europe and America this development has been followed with skepticism. For one, China's enormous appetite for commodities has driven up prices for raw materials, which squeezes the margins of Western producers. And then, some are concerned that the West is losing its political influence in the developing world. Westerners have therefore claimed and complained that China is coddling dictators, despoiling poor countries, and undermining Western efforts to spread democracy.[167]

The focus on heavy industries has pushed China's energy consumption to new levels. Steelmaking consumes 16 percent of China's electricity, compared with 10 percent for all household needs. More power capacity has come on line in 2006 with the completion of the enormous Three Gorges Dam across the Yangtze River, and in 2007 China decided to purchase five modern nuclear reactors from Western companies.[168] However, most electricity is still produced from coal energy. Much coal is being imported (from Australia, for instance), and domestic production is on the rise. (The increase in demand for coal has in part been blamed for the terrifying safety situation in

Chinese coal mines. In 2004 China produced 35 percent of the world's coal, but reported 80 percent of the global mining accident deaths. Officially, the mining deaths amounted to 5,000, but independent reports put them more likely towards 20,000.)

In terms of oil, the Chinese government handles affairs mainly through three state-owned companies. The China National Petroleum Corporation (CNPC) and the China Petroleum and Chemical Corporation (Sinopec) were reorganized into vertically integrated firms starting in 1998 and operate virtually all of China's oil refineries and the domestic pipeline network, while the China National Offshore Oil Corporation (CNOOC) handles offshore exploration and production, and accounts for roughly 15 percent of China's domestic crude oil production.[169] All three companies have been listed on the stock market, but only minority stakes (around 15 percent) have been sold: the government remains the majority shareholder and thus in control of the oil infrastructure. PetroChina, created as a subsidiary in early 2000 by separating out most of CNPC's high quality assets, is now China's biggest oil and gas company, and actually became the world's first company with a $1 trillion market capitalization in November 2007. (This was a market value far beyond that of the global second, ExxonMobil Corp., with a market cap of $488 billion. In terms of reserves, PetroChina stood at 20.5 billion barrels at the end of 2006, compared to ExxonMobil's 22.7 billion barrels.)

In the future China will have to rely increasingly on imported oil. China became a net oil importer in 1993, and in 2003 passed Japan as the second largest consumer of oil after the United States. Between 2000 and 2006 China alone accounted for one-third of the increase in world oil consumption, and in 2006 China imported petroleum and related products worth $84 billion.[170] The International Energy Agency now expects China's imports of oil to triple by 2030, and the Chinese government does whatever it can to secure future oil imports. By 2005 CNPC had acquired exploration and production interests in over 20 countries on four continents and announced its intentions to invest a further $18 billion in foreign oil and gas assets until 2020. CNPC invested more than $8 billion in Sudan alone, which included investments in a 900-mile pipeline to the Red Sea. Sinopec in 2006 acquired a 97 percent stake in Udmurtneft, which holds 1 billion barrels of proven reserves in Russia, while CNOOC has become the largest operator in the offshore Indonesian oil sector. In Kazakhstan, which is on its way to become one of the world's principal oil exporters, CNPC purchased PetroKazakhstan, whose assets include 11 oil fields, and a 1,000 km (620 mile) pipeline that opened in late 2006 to become the first pipeline ever to go to mainland China.[171] Half the oil delivered through this channel will originate from Kazakhstan, the other half from Russia. Russian oil will also reach China through the 2,500 mile pipeline that is being constructed by Russian state-owned oil giant Transneft from Eastern Siberia towards the Pacific Coast. Elsewhere, China's investments into the oil sector have been less welcome. America's congressmen, citing nebulous national security concerns, in 2005 scuppered the proposed takeover of Unocal, an American oil firm, by the CNOOC.[172] (At a time when General Motors had become the largest carmaker in China, and most of Unocal's reserves were in Asia, the purported national security concerns were largely bogus. The Chinese foreign ministry demanded that Congress "correct its mistaken ways of politicizing economic and trade issues."[173])

In early 2006 Angola passed Saudi Arabia as China's largest source of crude oil imports, with Iran and Russia ranking third and fourth. In total, China receives nearly half its crude oil imports from the Middle East, and around a third from Africa.[174] (This reflects quite an achievement in terms of diversifying oil imports. Previously, 60 percent of Chinese oil imports had come from the Persian Gulf.)

India

India has been a democracy ever since the nation gained independence in 1947. This second most populous country in the world, which is set to pass China to become first,[175] is situated half in the subtropics and half in the tropics (while China is mostly situated in temperate climates). The population in the tropical part is concentrated in coastal areas, while much of the population in the northern, subtropical part is concentrated hundreds of miles from the nearest sea port, living in the great plain along the Ganges. India's economy has been booming in the early 21st century, growing at nearly nine percent in 2007, for instance, and over 7 percent on average between 1997 and 2006. By 2007 India's economic output (GDP at PPP) was fourth behind the United States, China, and Japan, but ahead of Germany's. As in China, this growth has improved living standards in the fast-growing urban areas, while extreme poverty remains in the countryside. India also has to address such problems as environmental degradation and ethnic and religious strife. As much as three-fifths of India's work force is still in agriculture, but the production sector is growing, with several industries being government-owned. India's economy also includes a very modern service sector, which employs a third of the labor force to generate over half of India's output. (India has been able to capitalize on its large population of well-educated, English-speaking people to become a major exporter of software and other services.)[176]

Unlike China, India actually runs a trade deficit ($62 billion in 2006). (The United States remains India's largest trading partner, while China is India's leading supplier and its third-largest export market.) Energy imports are partly to blame, though India's economic output is less energy-based than

China's (due to less focus on heavy industries and a large contribution by the service sector). India has the fourth-largest coal reserves in the world, but is a coal importer and will supposedly consume three times as much coal in 2030 than in 2005. In terms of oil, India imports some 80 percent of what it consumes, and the share will rapidly increase. (Nevertheless, India's 5.7 billion barrels in proven oil reserves are the second-largest in the Asia-Pacific region, second only to China.) The Indian oil sector is dominated by two state-owned firms. The Oil and Natural Gas Corporation (ONGC) is India's largest oil company, and the country's largest company overall in terms of market capitalization. This firm controls India's upstream sector, accounting for roughly three-fourths of the country's oil output during 2006. The Indian Oil Corporation (IOC), on the other hand, is the largest state-owned company in the downstream sector, operating 10 of India's 17 refineries and controlling about three-quarters of the domestic oil transportation network. Some private companies have expanded their activities in the oil and gas sector in recent years as well. Notably, the Indian firm Reliance Industries opened the country's first privately-owned refinery in 1999 and secured licenses to seven deepwater blocks in the Krishna-Godavari and Mahanadi basins, which are considered among India's most promising offshore hydrocarbon basins. In 2000 India also fully opened the oil sector to foreign investments by allowing foreign companies to hold 100 percent equity ownership in oil and natural gas projects for the first time. Meanwhile, the state-owned companies have invested around the world. ONGC Videsh, Ltd., the overseas investment arm of ONGC, held interests in 25 oil and natural gas projects in 15 countries in Africa, Asia, Latin America, and the Middle East as of January 2007. ONGC Videsh holds a 20 percent stake in the ExxonMobil-led consortium that operates the Sakhalin-I project in Russia, for instance, and a 25 percent stake in the Greater Nile Petroleum Operating Company (GNPOC) that operates in Sudan. (Forty percent of GNPOC is held by the China National Petroleum Company, thirty percent by Malaysian Petronas, and 5 percent by the Sudan National Oil Company.)

THE OIL EXPORTERS

While the world's populous current or emerging centers of economic and technological power, the United States, the European Union, China, and India, are running out of oil, the question remains whether any of the world's oil-rich, oil-exporting nations will be able to assume a role of some kind of superpower. A list of these nations, ranked by export volume, reads as follows: Saudi Arabia, Russia, Norway, Iran, Venezuela, the United Arab Emirates, Nigeria, Iraq, Kuwait, Mexico, Libya, and Algeria. These nations tend to be relatively small in terms of population size and territory, often with a small domestic industrial base, though there are major exceptions in one way or another, including Russia, Nigeria, Iran, and Kazakhstan.

Russia

Russia, the largest country in the world, but with a relatively small population of 141 million, is perhaps the most obvious contestant to achieve an exceptionally powerful position based on oil (and gas) resources. Once at the core of the former Soviet Union, Russia is a nuclear power with a superpower past. Its economic output ($2.1 trillion at PPP in 2007) now ranks seventh in the world (after the United States, China, Japan, India, Germany, and the UK). The Russian economy is growing fast (at over 8 percent in 2007, and averaging 7 percent per year 1998–2007), but much of this growth has been based on energy exports and current high energy prices. Russia holds the world's largest natural gas reserves (between a quarter and a third of the total), the second largest coal reserves, and the eighth largest oil reserves. The country is also the world's largest exporter of natural gas, the second largest oil exporter, and sometimes produces even more oil than Saudi Arabia. Besides, Russia is the world's third largest energy consumer, though Russia has not performed well in terms of turning its high energy consumption into technological advance. Today, the Russian Federation does not seem to be a technological leader in anything, except perhaps in mining, while oil, natural gas, metals, and timber account for more than 80 percent of exports and 30 percent of government revenues. Obviously, this leaves the country vulnerable to swings in world commodity prices, but in recent years these prices went up substantially, and consumer demand has contributed significantly to Russia's growth. Poverty has declined, the middle class expanded, and reforms in the areas of tax, banking, and labor have raised business and investor confidence. (Foreign direct investment amounted to $45 billion in 2007.) On the other hand, market-oriented policies in Russia remain paired with state intervention and state control over strategic sectors. Corruption is widespread, and the Kremlin has in recent years overseen a recentralization of power that has undermined democratic institutions.[177]

Russia's relations to the West are perhaps closer than ever, but they remain strained. Arguably Russia is using its energy might to restore the influence the country lost after the collapse of the Soviet Union. Russia has put pressure on former Soviet republics Ukraine and Georgia not to join NATO and the EU, for instance, and has half pulled out of a treaty governing the movement of troops and tanks in Europe. On the other hand, Russia and the West have increasingly been cooperating in counter-terrorism and the fight against drug smuggling (in Afghanistan and beyond), and in curbing the nuclear ambitions of Iran and North Korea. Russia and the United States have also jointly upgraded security at some of Russia's most sensitive nuclear sites, and are cooperating in returning fresh and spent Russian-originated nuclear fuel from reactors overseas. Besides, Russia's and America's civilian space programs are closely intertwined.[178]

As a major non-OPEC supplier of oil and natural gas to both Europe and South and East Asia (and to some extent even the United States), Russia's power is bound to increase in future years. (The EU is expected to import 75 percent of its gas requirements from Russia by 2020.) On the other hand, the same argument is true for Russia as for many of the Middle Eastern oil exporters: As long as the Russian economy is not more diversified, and energy exports play such as major role, the country's economic well-being depends on energy sales, and Russia will be interested in maintaining good relations with its customers. Besides, at 80 billion barrels (at the end of 2006), set to last for 22 years at current production rates, Russia's proved oil reserves are not truly enormous. (Much of Russia's future energy might will thus stem from natural gas rather than oil.) A theoretic possibility, though it is unthinkable at the moment, would be an integration of Russia into the European Union: this would create an energy-rich Eurasian superpower of unprecedented dimensions. However, at the moment it seems more likely that Russia will intensify its cooperation with China to challenge the position of the West.

The Middle East

None of the other major oil producers are nuclear powers, have a substantial industrial base, or have developed any high technology in modern times. The Middle East now produces nearly a third of all oil and has enormous proven oil reserves, 743 billion barrels as of the end of 2006, which would last for 80 years at current production rates. The largest reserves are in Saudi Arabia (with 264 billion barrels lasting for 67 years), followed by Iran (138 billion barrels), Iraq (115 billion barrels), Kuwait (102 billion barrels), and the United Arab Emirates (98 billion barrels). Forming a kind of Oil Crescent around the Persian Gulf, these five neighboring nations combined account for 60 percent of the world's proven oil reserves.

Saudi Arabia has been a top oil producer alongside the United States and Russia (the Soviet Union) for decades and is now, despite its relatively small population of 27.6 million (2007, including 5.6 million foreign workers), the largest economy in the Middle East. The petroleum sector accounts for some 90 percent of export earnings, 45 percent of GDP, and 75 percent of budget revenues. Saudi Arabia's burgeoning population is very young (with 40 percent being under 15 years old), and unemployment is high, especially among the young population that generally lacks the education and technical skills the private sector needs.[179] Saudi Arabia remains an unfree Muslim country, but the United States keeps arming the nation with the most modern weapons, and contributes to the oppression of democratic development by keeping the pro-American royal family and the hereditary monarchy in place. In exchange, Saudi Arabia, with more than a fifth of the world's proven oil reserves, remains a reliable partner and oil supplier. The situation is similar in Kuwait, and to some extent in the United Arab Emirates, while it has been a lot more difficult in Iran, from the American point of view. The Islamic Republic of Iran, with a population of 65 million (2007), has been ruled by anti-Western leaders since 1979. Iran has been designated a state sponsor of terrorism for its involvement in Lebanon and elsewhere and thus remains subject to U.S. and UN economic sanctions and export controls. Nevertheless, Iran endures as the world's fourth largest oil exporter with nations such as Japan, China, India, South Korea, Italy, Turkey, and France accepting large shipments of Iranian oil. The oil sector provides 85 percent of government revenues, but despite high oil prices in recent years the economy has seen moderate growth only, and the population is burdened by double-digit unemployment and inflation. With Iran being suspected of having pursued the construction of a nuclear bomb, the international community passed resolutions 1737 and 1747 in December 2006 and March 2007 respectively, after Iran failed to comply with UN demands to halt the enrichment of uranium or to agree to full IAEA oversight of its nuclear program. This halted foreign investments even from countries that had previously still been willing to do business in Iran. And foreign investments in the oil and gas sector are urgently needed to maintain production volumes at current levels. Meanwhile, things did not go that well in neighboring Iraq either. If Americans thought the invasion of Iraq would bring the country's oil resources under U.S. control, they will probably be disappointed. Instead, even a partition of Iraq may be lurking (into a Shia, Sunni, and Kurdish segment), with possibly none of the three parts ending up with a pro-American government.

Due to their internal structure and technological standards no Middle Eastern country, no matter how rich in oil, will become a superpower within the foreseeable future. Nevertheless, this region will represent considerable economic power due to oil and gas revenues, and strategic power, because the world's remaining oil resources will be increasingly concentrated in the region. This may sound problematic to those viewing the Middle East as an unstable, undemocratic, underdeveloped, overpopulated, food-importing region with a substantial share of Islamic fundamentalism and anti-Westernism, but the region has little else than energy to sell, and will thus uphold oil and gas exports in order to maintain or improve its standard of living. Eventually, when overall global oil supply truly fails to meet demand, it will become a political issue that Middle Eastern countries can choose whether to sell to the West or to China, for instance.

Kazakhstan, Venezuela, and Others

Compared to the Middle East, oil reserves are meager everywhere else. A relative newcomer on the scene is the former Soviet republic of Kazakhstan, with 40 billion barrels in proven reserves (at the end of 2006), lasting for 77 years at current production rates. These are by far the largest proven reserves at the Caspian shelf, and the Republic of Kazakhstan, earning over half of its export revenues from oil and gas, has within short time become a major player in the oil industry. (Kazakhstan is a very large country, nearly four times the size of Texas, but with a population of just 15.3 million, about half of them Muslims and almost all of the other half Russian Orthodox Christians.) Oil has been produced in non-OPEC member Kazakhstan by Western companies from the early 1990s, but the Kashagan field, discovered in 2000 off the northern shore of the Caspian Sea at a depth of 4,500 meters underneath an ancient coral atoll, was the largest oil field found in three decades (since the discovery of the Alaskan Prudhoe Bay field in 1970). The Caspian Consortium pipeline has been transporting Kazakh oil from the Tengiz oilfield to the Black Sea from 2001, and the Atasu-Alashankou portion of the pipeline to China was completed in 2006. Kazakh production, with the help of substantial foreign investments, had reached 1.43 million barrels a day, of which only 15 percent were for domestic consumption. (Production in the two smaller former Soviet republics that border the Caspian Sea, Azerbaijan (0.65 million barrels a day) and Turkmenistan (0.16 million barrels a day), is comparatively meager.) The United States has built strong relations with Kazakhstan. Direct U.S. assistance since Kazakhstan's independence is now in the hundreds of millions of dollars rather than tens of millions of dollars. U.S. assistance in 2003 alone amounted to $92 million.[180] This includes arms sales and military education and training. Following September 11, 2001, Kazakhstan allowed the United States to use its air space and air bases freely, but the United States did not establish a true American base in Kazakhstan as it did in neighboring Uzbekistan.

Compared to Kazakhstan's 40 billion barrels in proved reserves, those of nearly all other non-Middle-Eastern oil exporters look small: Canada has 17 billion barrels, Mexico 13 billion barrels, Algeria and Brazil 12 billion barrels each, and Angola and Norway 9 billion barrels each (all end of 2006 figures). Thus, just three other countries remain in a higher reserves category: Nigeria, with 36 billion barrels, Libya, with 42 billion barrels, and Venezuela, with 80 billion barrels in proved (conventional) oil reserves. None of them has superpower potential. West African nation Nigeria is twice the size of California and Africa's most populous country (135 million inhabitants). Nigeria has long been plagued by political instability, corruption, ethnic and religious tensions, inadequate infrastructure, and overdependence on the oil sector, which provides 95 percent of foreign exchange earnings, and some 80 percent of budgetary revenues. Seventy percent of Nigerians work in agriculture, and on a list of 230 countries, Nigeria ranks as number 174 in terms of per capita GDP ($2,200 at PPP, 2007), behind such nations as Mongolia, Yemen, Cameroon, and Moldova.[181]

Libya, a large North African desert country of just 6 million people, has been ruled by military leader Gaddafi (Qadhafi) since a 1969 coup. Gaddafi sponsored terrorism, and the Islamic nation was isolated through UN sanctions from 1992 following the downing of Pan AM Flight 103 over Lockerbie, Scotland. However, after Gaddafi accepted responsibility for the Lockerbie bombing, renounced terrorism, and announced that Libya will reveal and end its programs to develop weapons of mass destruction, UN sanctions (December 2003), and U.S. unilateral sanctions (April 2004), were lifted, and international oil and gas companies rushed into Libya. These investments are expected to lift Libyan production from 1.8 million barrels a day (2006) to 3 million barrels a day by 2013. To be sure, Libya is deeply dependent on its oil sector, which provides some 95 percent of export earnings. Large parts of the population do not seem to have benefited from this income, and Libya imports about 75 percent of its food.[182]

Venezuela, a South American nation of 26 million (2007), has been ruled by democratically elected governments since 1959. However, Hugo Chavez, president from 1999, sought to implement "21st Century Socialism;" condemned globalization, imperialism, and United States foreign policy; and repeatedly claimed that the United States had planned and attempted to assassinate him. Nevertheless, Venezuela remained a top supplier of oil to the United States (next to Canada, Mexico, and Saudi Arabia). As the largest reserves holder outside the Middle East, and with a daily production of 2.8 million barrels (a rate that could be sustained for 78 years according to currently known reserves levels), the strategic position of Venezuela can only increase, especially because of its geographic proximity to North America. Following nationwide strikes in 2002 and 2003, Chavez (after his re-election in December 2006) nationalized firms in the petroleum, communications, and electricity sectors, and economic output began to grow again based on government spending and high oil prices. GDP growth was an astounding 9 percent in 2006, and 8 percent in 2007, but per capita GDP (at PPP) is still well under a third that of the United States. Among various problems faced by Venezuela, the nation's overdependence on the petroleum industry is considered most problematic: oil revenues account for some 90 percent of export earnings and more than 50 percent of federal budget revenues.[183]

NOTES

161. "We've really got to make the case" against Hussein, Bush told Colin Powell, "and I want you to make it." Bush believed that only Powell had the "credibility to do this": "Maybe they'll believe you." Karen DeYoung, "Falling on His Sword: Colin

Powell's Most Significant Moment Turned Out to be His Lowest," *Washington Post*, October 1, 2006, Page W12, http://www.washingtonpost.com/wp-dyn/content/article/2006/09/27/AR2006092700106.html.

162. R. Jeffrey Smith, "Army Files Cite Abuse of Afghans: Special Forces Unit Prompted Senior Officers' Complaints," Washington Post, February 18, 2005, Page A16, http://www.washingtonpost.com/wp-dyn/articles/A33178-2005Feb17.html.

163. Two great oilfields, around Mosul and Kirkuk, are located within 150 km of Turkey's border.

164. "Eni and Çalik Enerji—the Samsun-Ceyhan Oil Pipeline Project," *EIC Online Newsletters*, June 2006, http://www.the-eic.com/News/Archive/2006/Jun/Article2232.htm; "Samsun—Ceyhan Crude Oil Pipeline Project," Calik Enerji-Oil & Gas Projects, Natural Gas & Oil Transportation & Distribution, http://www.calikenerji.com.tr/oilgas.php?ID=5.

165. "Turkey," *Country Analysis Briefs*, Energy Information Administration, http://www.eia.doe.gov/emeu/cabs/Turkey/Background.html; Lutz C. Kleveman, "Der Kampf ums Kaspische Öl," *Der Spiegel*, August 2002, http://www.spiegel.de/archiv/dossiers/0,1518,245770,00.html.

166. "China—Economic Data," *Country Data*, The Economist Intelligence Unit, March 7, 2008, http://www.economist.com/countries/China/profile.cfm?folder=Profile%2DEconomic%20Data.

167. "China—The New Colonialists," *The Economist*, March 13, 2008, http://www.economist.com/displaystory.cfm?story_id=10853534; "China's Quest for Resources: A Ravenous Dragon," *The Economist*, March 13, 2008, http://www.economist.com/displaystory.cfm?story_id=10795714

168. "China," CIA-The World Factbook, https://www.cia.gov/library/publications/the-world-factbook/geos/ch.html.

169. "China—Oil," *Country Analysis Briefs*, Energy Information Administration, http://www.eia.doe.gov/emeu/cabs/China/Oil.html.

170. "China—Economic structure," The Economist Intelligence Unit, September 7, 2007, http://www.economist.com/countries/China/profile.cfm?folder=Profile%2DEconomic%20Structure.

171. Natalia Antelava, "China's Increasing Hold over Kazakh Oil," *BBC News*, 20 August 2007, http://news.bbc.co.uk/2/hi/asia-pacific/6935292.stm.

172. "China's quest for resources."

173. "Tug of War Goes Beyond Chinese Oil Bid," *The Washington Post*, July 10, 2005, Page F02, http://www.washingtonpost.com/wp-dyn/content/article/2005/07/09/AR2005070900109.html.

174. "China—Oil."

175. By 2025, the population of India is projected to surpass that of China, and the two countries will then account for about 36 per cent of the world population. "World Population Prospects: The 2006 Revision," United Nations, Department of Economic and Social Affairs-Population Division, http://www.un.org/esa/population/publications/wpp2006/wpp2006.htm; "Highlights," http://www.un.org/esa/population/publications/wpp2006/WPP2006_Highlights_rev.pdf.

176. "India," CIA-The World Factbook, https://www.cia.gov/library/publications/the-world-factbook/geos/in.html.

177. "Russia," CIA-The World Factbook, https://www.cia.gov/library/publications/the-world-factbook/geos/rs.html.
178. "The West and Russia: Lights, Camera and a Different Ending," *The Economist*, February 14, 2008, http://www.economist.com/displayStory.cfm?story_id=10696195.
179. "Saudi Arabia," CIA-The World Factbook, https://www.cia.gov/library/publications/the-world-factbook/geos/sa.html.
180. US Department of State, "U.S. Assistance to Kazakhstan—Fiscal Year 2003," http://www.state.gov/p/eur/rls/fs/29487.htm
181. "Nigeria," CIA-The World Factbook, https://www.cia.gov/library/publications/the-world-factbook/geos/ni.html.
182. "Libya," CIA-The World Factbook, https://www.cia.gov/library/publications/the-world-factbook/geos/ly.html.
183. "Venezuela," CIA-The World Factbook, https://www.cia.gov/library/publications/the-world-factbook/geos/ve.html.

CHAPTER 41

NUCLEAR AGE OR SECOND COAL AGE?

Current projections clearly indicate that oil will remain the world's most critical fuel for at least two decades to come. As it seems, we are still deep inside the Oil Age, and it has not yet become clear what the next Energy Era will look like. If the lessons learned from the past provide any guidance, the upcoming principal energy source should already be around or will soon be emerging. Nobody has ever pulled energy technology out of a magic hat: in all Energy Eras of the past the successive critical energy technology had been deeply rooted in the previous Energy Era. The emergence of agriculture was based on the accumulation of gatherer-hunters' intimate knowledge of their natural environment, and farming was started slowly alongside traditional foraging. Coal-based steam technology emerged slowly in grain-based agrarian societies after coal had been adopted for metallurgical purposes (iron smelting) and as a fuel for selected industries as well as household use. (Besides, much water- and human/animal-powered machinery could be adopted to be driven by steam engines.) Oil-fueled internal combustion engines had their beginnings in the mid-Coal Age. Oil was first mined for use as a lamp fuel, was then fired in steam engines, and finally chosen as the ideal raw material to deliver liquid fuels for internal combustion engines.

Based on this analysis, there is one likely conclusion to be drawn: The next Energy Era will be the Nuclear Age. Think about it for a moment. Nuclear energy has emerged in the Oil Age and is fairly versatile, providing transport energy (currently principally for submarines), productive energy in the form of electricity, and enormous destructive energy for belligerent purposes: all indications that nuclear energy has the potential to truly define its own Energy Era. True, the thought of nuclear energy makes a lot of people uncomfortable. There is the danger of accidents or terrorist attacks with a high number of casualties spread over wide geographic areas and a long time-frame (cancer); there is the unresolved problem of long-term storage of radioactive waste from nuclear power plants; and there is the issue that weapons-grade nuclear material is proliferating together with nuclear power plants. However, people have always been uncomfortable with the emerging energy source of the upcoming Energy Era. Think about Coal Age Britain forcing self-propelled off-rail vehicles to be accompanied by a person with a red flag or lantern walking 60 yards ahead of the vehicle to warn horse riders and horse-drawn traffic. Besides, no one should expect the new energy source to be flawless: nuclear energy, if the Nuclear Age is truly to come, will be perfected in its own Energy Era, not in the previous one. And right now nuclear energy is on the rise again, as there is a certain consensus emerging that only nuclear energy will be able to bring anthropogenic climate change to a halt: it provides large amounts of concentrated (and non-intermittent) energy without releasing carbon dioxide into the atmosphere. The other nuclear wild card, of course, would be nuclear fusion. The technology is still too premature to predict if it is ever going to work safely and reliably, but if it does, it will undoubtedly define a new Energy Era: a nuclear fusion reactor would potentially yield very large amounts of energy from readily available materials.

Alternatively, the world might be heading into a Second Coal Age. Coal deposits are quite abundant and relatively widely spread over the continents; technology to capture carbon dioxide directly at the coal-fired power plants (to slow anthropogenic climate change) already exists (though this currently costs about a third of the energy gained); and coal liquefaction technology is in place to supply the transport sector with liquid fuels, though this is currently not economical. Besides, coal, like natural gas, can be easily used to produce hydrogen (H_2) or methanol (CH_3OH), which are both contestants to become energy carriers for transportation purposes in the future. For a while, it seemed as if the Oil Age would be succeeded by some sort of a Natural Gas Era, but gas deposits are nowhere nearly as abundant as coal deposits, and global natural gas production is expected to peak sometime between 2030 and 2040. (Otherwise, natural gas is actually a quite versatile fuel: It is used for heating, cooking, electricity generation, and vehicle propulsion, and it can be turned into liquid fuels.) In recent years, coal has been the fastest growing fossil fuel globally. In 2005 worldwide coal consumption grew by 5.7 percent, and in 2006 by 4.5 percent (compared to oil, 0.7 percent, and natural gas, 2.5 percent). Here it is worth noting that even an annual growth of just 2 percent would translate into a doubling of consumption within 35 years. So far, this rise in coal consumption has been driven by increasing demand for electricity. (China, where coal consumption grew by 8.7 percent in 2006, accounted for more than 70 percent of the global growth in coal consumption that year.) It will be a while before coal liquefaction becomes relevant. This technology will not only have to wait for even higher prices of conventional oil, it will then also have to compete against oil from quite abundant unconventional sources (heavy oil, tar sands, oil shales).

From the environmental standpoint the continuing dominance of fossil fuels is bad news, because all fossil fuels release climate-altering carbon dioxide gas when burned. Policy makers are therefore pushing for more environmentally friendly energy alternatives, such as water, wind, solar, and geothermal power. These energy sources will likely have a much larger share in a Second Coal Age or a Nuclear Age, and in the very long run they will necessarily become dominant, as they are the only truly renewable energy sources. However, renewables tend to deliver energy intermittently, and not in very concentrated form, which would be a problem for military strategists, for instance. Mature second generation bioethanol technology might be a wonderful solution for the transport sector, for instance, but in an ongoing war, nations would rather turn to coal liquefaction than to wait for the next harvest.

As long as oil remains the critical fuel for transportation and military might, the currently powerful nations will do all they can to secure access to the remaining oil deposits around the world. These nations have also long started to diversify their energy mix, with the oil crises of the 1970s setting into motion a distinct trend to substitute other fuels for oil in all applications where this is economically possible. At the beginning of the 21st century this diversification varied considerably between different countries, reflecting mainly their domestic resources and their attitude towards nuclear energy. (Leaving renewable energies other than hydropower aside, the U.S. energy mix consisted to 40 percent of oil, 25 percent coal, 25 percent natural gas, 8 percent nuclear, and 2 percent hydro. France, by comparison, used 37 percent oil, 37 percent nuclear, 14 percent gas, 7 percent hydro, and 4 percent coal. Britain used 38 percent gas, 34 percent oil, 18 percent coal, and 9 percent nuclear. Japan used 48 percent oil, 20 percent coal, 14 percent gas, 14 percent nuclear, and 4 percent hydro. China used 62 percent coal, 28 percent oil, 7 percent hydro, 3 percent gas.) However, oil remains the largest fuel overall. In terms of global consumption the distribution of the major commercial energy sources in 2006 was 36 percent oil, 28 percent coal, 24 percent gas, 6 percent hydro and 6 percent nuclear.

The International Energy Agency expects that oil will maintain its large share in the energy mix despite the enormous projected growth in global energy consumption, which is expected to rise by about a half within the 25-year period ending 2030. This would require that oil production until 2030 has to increase by about 40 percent (to reach 116 million barrels per day) over the 2006 level (of 84 million barrels per day). Given that oil production has peaked in many regions, this sounds quite optimistic, but it is safe to say that at least after 2030 the share of oil will decline substantially. The oil-importing nations will then have become very uncomfortable with the fact that more and more of the remaining oil reserves will be concentrated in the Middle East. Arguably, the main reason why politicians are currently promoting the development of alternative energy sources is not environmental consideration, but concern about energy security. Why are the United States and the European Union subsidizing bioethanol and biodiesel? Why are they setting ambitious goals in terms of future market shares of these fuels? Is it to militate against climate change or to decrease dependence on oil imports? One way or another, technological advances with respect to energy alternatives, combined with increasing oil prices, will likely decrease the share of oil in the energy mix before it comes to a true supply crisis. In short, the Oil Age will come to an end before we run entirely out of oil. Likely and hopefully, the transition will be smooth, with oil being slowly superseded in importance by another fuel, just as coal persisted when the Coal Age came to an end, and kept the largest share in the energy mix long after oil-fueled cars had taken the streets, and airplanes had taken the skies.

POLITICAL POWER IN THE NEW ENERGY ERA

This leaves us with the question of how global political, economic, and military power will be reshaped when the world is moving into the Nuclear Age or the Second Coal Age. Drawing again upon the lessons of the past, only those regions that have mastered the established energy technologies will manage a rapid transition into a new Energy Era. (The English, for instance, had the most efficient agricultural system in the world by the time they pioneered coal technology.) Currently this would mean the ability to drill for oil and to refine it, and to build internal combustion engines, automobiles and aircrafts. This is true for Europe, North America, and Japan,[184] but India and China have learned this, too, or are in the process of learning it. Brazil meets the criteria as well. (Brazilian aerospace conglomerate Embraer, the Empresa Brasileira de Aeronáutica S.A., ranks third only after Airbus and Boeing in terms of annual commercial aircraft production. At 192 million people as of 2008, Brazil is actually the fifth most populated country in the world.[185]) Another lesson from the past teaches that only nations rich in the energy resource that defines an Energy Era can become a superpower in that era. In this respect, a Second Coal Age would be bad news for the European Union. While the United States sits on 27 percent of the world's proven coal reserves, Russia on 17 percent, China on 13 percent, and India on 10 percent, the European Union (25 members) has no more than 4 percent. In terms of uranium for current nuclear technology, the European Union is likewise badly positioned. In fact, the EU does not have any uranium resources at all. The only prospect for this to change would be a membership of the Ukraine, but even the Ukrainian uranium resources are relatively meager, under 90,000 tons in terms of reasonably assured and inferred resources in 2005. (For comparison, world demand for uranium was around 68,000 tons in 2005, and may go up to 100,000 tons by 2030.) India (65,000 tons) and China (39,000 tons) do not have much uranium either, while Russia has around 172,000 tons in proved resources, and the United States as much as 340,000 tons (and its neighbor Canada another 440,000 tons). (Meanwhile, the largest resources are in Australia, 1.1 million tons, and Kazakhstan, 820,000 tons. South Africa, Namibia, and Brazil all have around 300,000 tons. Combined, Canada and Australia currently account for over half the world's primary uranium production.) On the other hand, the European Union has plenty of experience with nuclear energy. As of 2005, France, Britain, Germany, Sweden, Spain, and Belgium shared 125 operating nuclear power plants between them, while the United States had 104 (and Canada 18). This compares to 31 nuclear power plants operating in Russia, 15 in India, and 9 in China.[186]

In addition to energy technology, population size has always been a critical factor in the making of superpowers. For one thing, larger populations translate into larger internal markets, which allows for economic development without having to rely on intact political and trading relations with other regions. And then, larger populations have more brains and innovate more. This is especially true when larger shares of the physical work are being shifted towards inanimate energy (or towards slaves and beasts of burden during the earlier Energy Eras). Notably, larger economies are better positioned to funnel small, distributed surpluses into concentrated, capital-intensive projects that create knowledge and technology of all kinds. China and India still now have small per capita GDPs, but they have used substantial resources to establish space programs. It has been accepted for a while that capabilities to produce nuclear bombs, and missiles to deliver them, is conditional for any nation aspiring to become a superpower. Both China and India have achieved this, but future military strength will likely depend on space technology, which these two most populous nations in the world are also developing. In 2003 China has become the third nation, after the Soviet Union/Russia and the United States, to possess independent capability to put a person into space, and India is planning to achieve the same by 2014. India has been launching satellites since the 1970s, and has become a nation with independent satellite launch technology alongside Russia, the United States, the EU (whose European Space Agency will expand membership from the original western European members to the eastern European nations that joined the European Union), China, and Japan (where rockets free of U.S. technology have been used since 1994[187]). In 2007 India even demonstrated recovery technology, showing that it can safely make a space capsule reenter the atmosphere and bring it home[188].

Given such emerging competition, it is safe to say that U.S. power will decline in the future. This will happen even though the United States is well positioned in terms of energy resources that are likely to define the next Energy Era. (In case of coal, the term excellently positioned would actually be more appropriate.) It is simply not likely that the United States will ever again find itself in a position where it possesses and produces so much more of an Energy Era's defining fuel than any other nation, or any other but one, unless, perhaps, America develops its own functioning nuclear fusion reactor.

As things look now, global power will be more evenly distributed among North America, Europe, and Asia. If the United Nations is supposed to remain an institution that is to be taken seriously in terms of guiding the peaceful coexistence of the world's nations, it will have to be reformed to adequately reflect the current or emerging power set-up. Right now the Security Council, arguably the world's most powerful international institution, (as it investigates international disputes and decides whether or not a war is legal) reflects the power distribution at the end of World War II, with the United States, the Soviet Union, Britain, France, and China holding permanent seats with a veto right. It may be argued that such nations as Germany, Japan, or Italy (even though they currently do not possess nuclear weapons), may by this stage deserve a permanent seat just as much as Britain and France do, but this is a relatively minor issue. More importantly, India should soon be handed a permanent seat in the Security Council in order to reflect the realities of the early 21st century, and the seats of Britain and France should be merged into one EU seat. However, any restructuring of the United Nations Security Council will be difficult, as it would immediately start a discussion about fairness, justice, and righteousness. Much of the world would argue that the Council should not reflect military or economic power, but be restructured to evenly represent all continents or regions, with a permanent delegate each from South America and Africa, for instance.

NOTES

184. Nippon Oil Corporation is active internationally, and so is JAPEX, the Japan Petroleum Exploration Co., whose international focus is Canada, Indonesia, China, the Philippines, and Libya, for instance. JOGMEC, the Japan Oil, Gas and Metals National Corporation, was established in February 2004 with the mission to secure a stable supply of oil and natural gas (and nonferrous metals and minerals) for Japan. JAPEX, http://www.japex.co.jp/english/company/index.html. JOGMEC, http://www.jogmec.go.jp/english/index.html.

185. Only China, India, the United States, and Indonesia (238 million) have larger populations.

186. International Energy Agency, *World Energy Outlook 2006*, page 347, http://www.iea.org/textbase/nppdf/free/2006/weo2006.pdf.

187. "H-II Launch Vehicle," Launch Vehicles and Space Transportation Systems, Japan Aerospace Exploration Agency, http://www.jaxa.jp/projects/rockets/h2/index_e.html.

188. "Milestones," Indian Space Research Organization (ISRO), http://www.isro.org/mileston.htm.

CHAPTER 42

ENERGY FOR ALL

To be sure, the world in the early 21st century does not present itself as a democracy. The powerful nations, though many of them are themselves democracies, would never want the world as a whole to be democratic. Democracy means government by the people or rule of the majority. But the majority of people are now living in countries of low per capita income and meager political power. If any important decisions were ever to be decided on a one-vote-per-world-citizen or a one-vote-per-nation[189] basis, the global distribution of wealth and power would change rapidly. But de facto power today is not founded on large population size. It tends to result from high per capita energy consumption, which in itself may play a decisive role in triggering a fertility transition that eventually brings population growth to a halt. Such development has been witnessed in today's powerful nations a while ago, and wealth has in turn been distributed among a more moderate number of people. What is more, the shift towards the use of more (animal and) inanimate power in these societies released more people from physical work and allowed them to think and innovate more, and eventually to live longer and healthier lives. Today's lowest income nations, on the other hand, experienced a population explosion relatively recently, in the second half of the 20th century. A popular way to explain this rapid population growth in low-income nations is that industrialized nations "exported health, but forgot to export wealth."[190] This statement contains two claims. One, correctly, that introducing Western pharmaceuticals in developing countries has dramatically increased infant and childhood survival, which triggered fast population growth. The other, more disputed, that parents would have chosen to have fewer children before, or at least as a result of, experiencing better survival of their offspring, if only they had been wealthier. In reality, a delay in completing a full fertility transition may have many reasons, including such cultural issues as attitude towards birth control measures and regarding family (rather than the government) as responsible for retirement provisions. A direct influence towards smaller family size may be achieved by a larger share in inanimate energy consumption. In Western societies, urban parents eventually began viewing children as a cost factor, while the traditional large family size persisted for long in the less electrified and motorized countryside, where an additional child was viewed as an extra pair of hands to work in the near future.

Societies that consume large amounts of energy relative to their population size are generally wealthy. This correlation between energy consumption and standard of living can be illustrated by plotting energy consumption against per capita GDP, but the United Nations usually uses the "Human Development Index" (HDI), which is a better measure than per capita GDP to express living standards and human well-being, as it also takes into account such factors as life expectancy, mean years of schooling, and so on. Expressing all commercial and noncommercial energy consumption in terms of barrels of oil equivalent, it seems that a consumption of at least three quarters of a barrel of oil per person per year is required to provide all members of societies with at least the most basic needs. (Such consumption level was prevalent in parts of Western Europe before 1800, and in China before 1950.) Societies that experienced industrial advance and substantial improvements in quality of life (equivalent to a HDI of about 0.8.), consume 10 times this amount, about 7 barrels of oil equivalent per person a year. (This consumption level was reached in Western Europe and the United States around 1880, in Japan in the 1930s, and in China in the 1980s.) However, with additional energy consumption the achieved increase in human well-being, as measured by the HDI, becomes less pronounced: Most gains have already been achieved, and additional, relatively small, increases in living standards will cost relatively large additional energy inputs. Widespread affluence seems to require at least 15 barrels of oil equivalent per person per year, even if this energy is very efficiently used. (This consumption level was reached in France during the 1960s, and in Japan in the 1970s, for instance.)[191]

On global average, per capita energy consumption was nearly 13 barrels per year in 2004 (and expected to grow to nearly 14 barrels per year by 2015, a level at which it is projected to remain until 2030).[192] But this disguises large disparities. Per capita energy consumption averaged over developing countries was as low as 6 barrels per year in 2002, with the share of noncommercial energy being as large as a fifth. The latter is relevant, as the link between development and energy consumption is a lot stronger for commercial than noncommercial energy.[193] For comparison, U.S. commer-

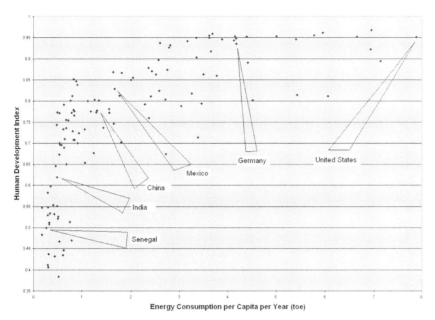

Correlation between Energy Consumption and Human Development Index
The amount of energy used is a key factor in human development. This has been observed in all countries. The link is especially strong during early stages of development. Once a certain level of human well-being has been achieved, further energy inputs do not seem to result in substantial further gains. (Based on data from International Energy Agency, "Key World Energy Statistics 2007," http://www.iea.org/textbase/nppdf/free/2007/key_stats_2007.pdf, and United Nations Development Programme, "Human Development Report 2007/2008," http://hdr.undp.org/en/media/hdr_20072008_en_complete.pdf.)

cial energy consumption was 59 barrels of oil equivalent per person in 2001 (based on the main energy sources oil, coal, natural gas, hydropower, wind power, and nuclear energy, but excluding the small contributions by commercial renewable energies other than hydro).[194] If all people in the world consumed energy as Americans do, the global energy consumption level would be some five times higher than it actually is. In terms of oil alone, if all countries would adopt U.S. per capita consumption habits, then the known oil reserves would last for no more than a few years.[195] It is thus obvious that, at least under the current energy regime, people in the industrialized world can only live the way they do because the majority of the world's population does not. If the Chinese or Indian population moved towards using as much energy per person as Americans and Europeans do, demand for (fossil) fuels would drive energy prices up to a level

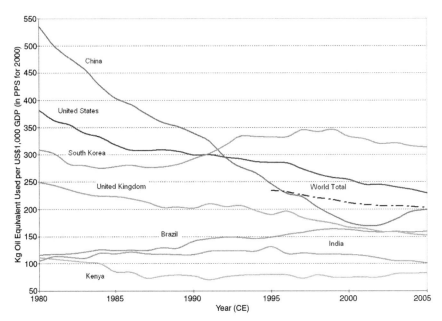

Energy Use per US$ 1,000 GDP (in PPP for 2000): Hope for Countries of Lower Energy Consumption? The creation of wealth has become less energy-intensive during the later Oil Age in Eurasia and North America, and thus on average for the world as a whole. Rather than by increased production efficiencies alone, this was to a good part achieved by a shift towards large contributions by the service sector to overall output. It is thus uncertain how much the trend towards lower energy intensity in mature Oil Age countries will help regions whose output is currently predominantly based on agricultural energy. Such areas will likely have to build an economy based on production before wealth creation can be based on services. (Based on data from Energy Information Administration, "World Energy Intensity—Total Primary Energy Consumption per Dollar of Gross Domestic Product Using Purchasing Power Parities, 1980–2005," http://www.eia.doe.gov/pub/international/iealf/tablee1p.xls.)

that would have a severe impact in the Western world. (This has already started, even though Chinese and Indian energy consumption is far below Western per capita consumption.) To be sure, compared to U.S. energy consumption patterns even energy use in the equivalently wealthy European Union appears meager: at "only" 29 barrels of oil equivalent per person per year Europeans (EU-15) consume half as much energy as Americans. A consumption level of around 30 barrels per person a year was actually quite typical for fully industrialized countries outside America at the beginning of the 21st century: It was around the same in Germany, France, Britain, Russia, Japan, South Korea, and Taiwan. Elsewhere people used a lot less energy. In South America, average commercial energy consumption was equivalent to 12 barrels of oil per person per year in Argentina, and 7 barrels per person per year in Brazil, the world's fifth most populous nation. The populous nations of developing Asia consumed even less commercial energy, 4.8 barrels of oil equivalent per person per year in China, 3.1 barrels per person per year in Indonesia (the world's fourth most populous nation), 2.2 barrels per person per year in India as well as Pakistan (the world's second and sixth most populous nations, respectively). In Bangladesh (the world's seventh most populous nation), people consumed even less, no more than 0.7 barrels of oil equivalent in commercial energy per year. In Africa as a whole, commercial energy consumption was equivalent to 2.5 barrels of oil per person per year, but relatively high consumption levels in such countries as Egypt (5.2 barrels per head per year) disguise the fact that per capita commercial energy consumption in large parts of sub-Saharan Africa is nearly negligible.

In terms of electrification, the picture looks similar. In countries such as Burkina Faso and Mozambique less than 7 percent of the population have access to electricity, and in sub-Saharan Africa as a whole the electrification rate is no more than 26 percent. For Westerners it might be hard to fully comprehend that in large parts of the world three-quarters of all people have to do without electric lights or other electric appliances, but in Ethiopia some 60 million people live without electricity, for instance, and in oil-exporting nation Nigeria over 70 million. The total number of Sub-Saharan Africans without electricity is around 550 million, roughly the same number as in (nuclear powers) India (487 million) and Pakistan (71 million) combined. The total figure is about 930 million for all of developing Asia, while globally some 1.6 billion people live without access to electricity.[196] This is more than a quarter of the world's total population. Affected are mainly people in the countryside: four out of five people without electricity live in rural areas. India's low electrification rate, at 56 percent, is actually in strong contrast with China's high electrification rate of 99 percent. Without China's rapid electrification programs, the global number of people without electricity would actually have increased steadily over the past three decades. Population growth tended to outpace electrification growth, and even in the decade between 1990 and 2000, the total number of people without electricity fell by no more than 500 million. This is problematic, as access to electricity is particularly crucial to human development. As with energy consumption in general, the correlation between per capita electricity consumption and Human Development Index ratings is strong and non-linear. At low levels of consumption, small increases in electricity consumption are associated with large improvements in human development. When a certain per capita electricity consumption level is reached, at about 5,000 kWh per year, HDI reaches a plateau. (To be sure, such averaged national data can be misleading, as a privileged urban minority may use a lot of electricity, while people in the countryside do not have access to electricity at all.)[197]

ENERGY AND THE UNITED NATIONS' MILLENNIUM DEVELOPMENT GOALS

Given the role that energy consumption plays in economic and societal development, it would be fair to assume that this issue is being prioritized in the context of helping developing countries. Nevertheless the Millennium Declaration, adopted in September 2000 by the member states of the United Nations in the largest gathering of world leaders in history, did not even mention energy.[198] Following consultations with the World Bank, the International Monetary Fund, the OECD, and the specialized agencies of the United Nations, the General Assembly recognized eight specific goals as part of the road map for implementing the declaration. These read as follows:

1. Eradicate extreme poverty and hunger.
2. Achieve universal primary education.
3. Promote gender equality and empower women.
4. Reduce child mortality.
5. Improve maternal health.
6. Combat HIV/AIDS, malaria, and other diseases.
7. Ensure environmental sustainability.
8. Develop a global partnership for development.

These 8 goals were defined in more detail through 18 targets to be achieved by 2015, but even these 18 targets did not mention energy at all.[199] Nevertheless, it should be obvious that increased energy consumption would help to meet nearly all the goals. Energy to pump water to the fields will help eradicate hunger (Goal 1), as drought is the main cause of food shortages in poor countries, and irrigation would boost crop yields by up to 400 percent. Energy for transport would make it easier for children to reach school (Goal 2). Energy for lighting generally improves quality of life, but specifically allows children to read and study for longer hours, thus improving school outcomes (Goal 2). Arguably, this is especially important in the tropics, where night falls early, around 6 p.m., throughout the year. Interestingly, for people now living in rural developing regions the first step into the Oil Age is still the use of kerosene, just as it was at the very beginning of the Oil Age in Western countries. Kerosene was initially used as lamp oil and is now generally the cheapest and most available fuel available for cooking, heating, and pumping water in regions that have not yet been electrified.[200] Given the experiences in terms of the empowerment and emancipation of Western women following household electrification with its domestic workload-reducing and time-saving effects, an increased electrification of developing countries would without doubt also help to achieve Goal 3. Similarly, energy for transport would make it easier to acquire fuel and water, a task usually performed by women and children.[201] (Empowered women would then be better able to make a choice on their family size, which would likely reduce fertility rates and slow population growth with an overall beneficial effect.) Energy to boil water that contains pathogens, a principal factor in child health and mortality, would help to achieve Goal 4, while substituting commercial fuels or electricity for traditional fuels would combat deadly lower respiratory infections,[202] which thrive on indoor air pollution that especially affects mothers and their children (Goals 4 and 5). Obviously, energy used for refrigeration would allow for storage of medicines and thus help battling all kinds of diseases (Goal 6), while commercial energy would bring to a halt environmental degeneration in form of deforestation (Goal 7).

Even though there is no specific Millennium Development Goal (MDG) on energy, the issue slowly received more attention within United Nations institutions.[203] The Economic and Social Council's Commission on Sustainable Development in its report on the ninth session (May 2000 and April 2001) stated under "Rural Energy" that in order "To implement the goal accepted by the international community to halve the proportion of people living on less than US$ 1 per day by 2015, access to affordable energy services is a prerequisite." It continues that "An effective strategy to address the energy needs of rural populations can be to promote the climbing of the energy ladder. This implies both improving ways of using biomass as well as moving from simple biomass fuels to the most convenient efficient form of energy appropriate to the task at hand, usually liquid or gaseous fuels for cooking and heating and electricity for most other uses."[204] At the United Nations Development Program's "Energy for Development Conference" in the Netherlands in December 2004 it was stressed that for the past few decades the number of people impoverished by a lack of modern energy services, roughly one-third of the world's population, has remained unchanged. The message, again, was: "No energy, no MDGs." Studies were presented that showed that access to modern energy services can have a decisive impact on the decline of poverty, whether through stimulating micro-enterprises, improving study conditions for children or reducing indoor air pollution from cooking. It was pointed out that nearly a third of the world's population depends on dung, firewood, and agricultural residues for cooking and heating, with women, particularly rural women and their children, bearing the brunt of this troubling reality, as they spend hours at a time gathering fuel wood, inefficiently processing food, and inhaling smoke from wood-fired cooking stoves. For these women and their families, dependence on traditional fuels and technologies barely allows fulfillment of the basic human needs of nutrition, warmth, and lighting, let alone the opportunity for more productive activities. It was concluded that meeting the MDGs will require the provision of modern energy services to over half a billion people by 2015, and that achieving all eight MDGs will require an annual average investment of $20 billion (roughly $20 per capita) in the development of energy infrastructure and delivery of energy services.[205]

In 2005, the Millennium Project, commissioned by the United Nations secretary-general to recommend a concrete action plan, presented its final report, "Investing in Development: A Practical Plan to Achieve the Millennium Development Goals."[206] This report states that "some areas important for development—and for achieving the Goals—are not included in the formal Goals framework. Energy services, sexual and reproductive health, and transport services are each vital to enabling and facilitating the achievement of the Goals." The report then continues that "improved energy services—including modern cooking fuels, access to electricity, and motive power—are necessary for meeting almost all the Goals." It further mentions that improved energy services can reduce the time and transport burden of women and young girls by reducing the need to collect biomass, for instance (thus facilitating school attendance), and that electricity is critical for providing basic social services, including health and education, and for powering machines that support income-generating opportunities, such as food processing, apparel production, and light manufacturing. In turn, the UN Millennium Project proposes that countries adopt the following specific targets for energy services to help achieve the Goals by 2015[207]:

- Reduce the number of people without effective access to modern cooking fuels by 50 percent and make improved cook-stoves widely available.
- Provide access to electricity for all schools, health facilities, and other key community facilities.
- Ensure access to motive power in each community.
- Provide access to electricity and modern energy services for all urban and periurban poor.

All these recommendations are obviously steps in the right direction, but even more important would be the definition of an overall minimum energy consumption goal. It has to be a principal priority to bring all the world's regions to a per capita commercial energy consumption level of at least 10 or so barrels of oil equivalent per year to make sure people everywhere will benefit from the same effects other regions experienced when stepping into the Coal Age and the Oil Age.

But this is not an easy task. First of all, population growth is continuing. According to current projections, the global population will reach 7.3 billion in 2015, 8.0 billion people by the year 2025, and 9.2 billion by 2050. This is substantially above the year 2007 level of 6.7 billion people (not to mention the 1950 figure of 2.5 billion people).[208] If the global energy consumption of 2005, around 84 billion barrels of oil equivalent,[209] was evenly distributed between all people on Earth, it would be sufficient to provide 7.3 billion people (as projected to live on the planet by 2015) with 7 barrels of oil equivalent per person per year, associated with a Human Development Index, HDI, of 0.8 (basic needs met). However, providing 7.3 billion people with 18 barrels of oil equivalent per person per year, which seems to be the minimum energy consumption requirement associated with a HDI of 0.9, would require 131 billion barrels of oil equivalent. Providing 9.2 billion people, as expected to live on the planet in 2050, with living standards reflected by a HDI of 0.8 would accordingly require 64 billion barrels of oil equivalent, while a HDI of 0.9 would require some 166 billion barrels of oil equivalent per year, twice the amount consumed globally in 2005, and all this under the unrealistic assumption that energy consumption would be equally distributed. If 9.2 billion people would consume energy at the current U.S. level of some 60 barrels of oil equivalent per person per year, the world would need 552 billion barrels of oil equivalent per year.

However, nearly all future population growth will occur in developing countries,[210] where, sadly, some two billion people currently do not have access to modern energy services. (Nearly 2.4 billion people use traditional biomass fuels for cooking and nearly 1.6 billion people do not have access to electricity.) Nobody expects people there to come anywhere close to Western living standards during the next half century, and even the very modest energy consumption and development goals may be difficult to achieve. The electrification of rural areas, in particular, is expensive and often technically difficult. It tends to fail to attract investment from the private sector and depends on a range of public and private partnerships.[211] World electricity demand increased by 30 percent within just 10 years between 1990 and 2000, but is expected to double again within less than 30 years. In terms of providing people with electricity it may be a welcome trend that global urbanization is continuing. In fact, the world reached a momentous milestone in 2008: for the first time in history, more than half of the human population, 3.3 billion people, have been living in urban areas. By 2030, this figure is expected to swell to almost 5 billion.[212] But even with so many living in urban areas, it would be unrealistic to hope that the number of people living without electricity could be halved by 2015 compared to the 2000 level. Much rather, the current number of some 1.6 billion people living without electricity will decrease by no more than 200 million by 2030, unless, of course, current policies are changed to put energy consumption into focus. In this context it is also important to realize that climate protection is less important than provision of cheap energy for the very poor, simply because the benefits will by far outweigh the negative impact of global warming expected in a relatively distant future.[213] People everywhere must be allowed to experience the beneficial technological and societal effects of increased energy consumption, with an acceleration of knowledge accumulation and innovation that will prepare them for future challenges.

NOTES

189. The closest the international community has come to this is the United Nations General Assembly, whose functions and powers are reduced to making recommendations. "Functions and Powers of the General Assembly," United Nations General Assembly, http://www.un.org/ga/about/background.shtml.

190. Joel E. Cohen, *How Many People Can the Earth Support?* (New York: W.W. Norton & Company, 1995). Even though the health situation is a lot better than it used to be, severe health problems remain. HIV/AIDS killed 1.6 million people in Africa in 2007, tuberculosis (TB) killed approximately 550,000, and at least 1 million people die from malaria each year. DATA, "HIV/AIDS, TB and Malaria," http://www.data.org/issues/aids.html.

191. Find a detailed discussion of this issue in Jose Goldemberg, "Energy and Human Well-Being," background paper to the Human Development Report 2001- Making New Technologies Work for Human Development, UNDP 2001, http://hdr.undp.org/en/reports/global/hdr2001/, hdr.undp.org/en/reports/global/hdr2001/papers/goldemberg-energy-1.pdf. The same concept is used by Vaclav Smil (*Essays in World History: Energy in World History*, 1994) and can generally be found in various publications by the International Energy Agency. See, for instance, *World Energy Outlook 2004*, which compares energy consumption to human development. *World Energy Outlook 2004*, Chapter 10: "Energy And Development," http://www.worldenergyoutlook.org/2004.asp, www.iea.org/textbase/nppdf/free/2004/weo2004.pdf. Conversion of tons of oil equivalent into barrels of oil: 1 metric ton of crude oil, based on worldwide average gravity, equals 7.33 barrels (BP Statistical Review on World Energy-June 2007). 1 toe/year=1.3kW. "BP Statistical Review of World Energy June 2007," http://www.bp.com/liveassets/bp_internet/globalbp/globalbp_uk_english/reports_and_publications/statistical_energy_review_2007/STAGING/local_assets/downloads/pdf/statistical_review_of_world_energy_full_report_2007.pdf.

192. International Energy Agency, *World Energy Outlook 2006*, http://www.iea.org/textbase/nppdf/free/2006/weo2006.pdf.

193. International Energy Agency, *World Energy Outlook 2004*, Chapter 10: "Energy and Development," http://www.worldenergyoutlook.org/2004.asp, www.iea.org/textbase/nppdf/free/2004/weo2004.pdf.

194. For the world as a whole, this figure was about 11 barrels of oil per person per year in 2001. (This is based on a worldwide commercial energy consumption equivalent to 66,885 million barrels of oil, and a population of about 6,157 million. All data in this paragraph based on the "BP Statistical Review of World Energy June 2002" (commercial energy consumption) and the "CIA—The World Factbook (population data). To be sure, by 2006 the figure for global energy consumption excluding renewables other than hydro had risen to 79.7 billion barrels of oil equivalent. "BP Statistical Review of World Energy June 2007," http://www.bp.com/liveassets/bp_internet/globalbp/globalbp_uk_english/reports_and_publications/statistical_energy_review_2007/STAGING/local_assets/downloads/pdf/statistical_review_of_world_energy_full_report_2007.pdf; BP's current Statistical Review of World Energy can be downloaded at www.bp.com/statisticalreview.

195. Based on the 2001 U.S. consumption level of 23.6 barrels of oil per person, the world population would have consumed 145.4 billion barrels. With known oil

reserves of 1,050 billion barrels in 2001, this would have lasted for just over seven years. Under the assumption of a total of 2,000 billion barrels of remaining oil resources (including 1,050 billion barrels in known reserves), oil would have lasted for less than 14 years at the U.S. consumption level.

196. All figures are for 2005. International Energy Agency, *World Energy Outlook 2006*, http://www.iea.org/textbase/nppdf/free/2006/weo2006.pdf.

197. International Energy Agency, *World Energy Outlook 2004*, Chapter 10: "Energy and Development," http://www.worldenergyoutlook.org/2004.asp, www.iea.org/textbase/nppdf/free/2004/weo2004.pdf.

198. General Assembly Resolution 55/2: United Nations Millennium Declaration (8 September 2000), http://www.un.org/millennium/declaration/ares552e.htm; "UN Millennium Development Goals," http://www.un.org/millenniumgoals/.

199. The framework of 8 goals and 18 targets was eventually complemented by 48 technical indicators to measure progress towards the Millennium Development Goals. Even within these 48 targets increased energy consumption as a means to achieve any of the goals was not mentioned. Under Goal 7 (Ensure Environmental Sustainability), Target 9 (Integrate the principles of sustainable development into country policies and programs and reverse the loss of environmental resources), Indicator 27 mentions "CO_2 emissions, total, per capita and per \$1 GDP (PPP)." This has also been expressed as "Energy use (kg oil equivalent) per \$1 GDP (PPP)," but here the idea is to use energy efficiently to create wealth in the context of sustainability. "Official list of MDG indicators," Millennium Development Goals Indicators—The official United Nations site for the MDG Indicators, United Nations Statistics Division-Development Indicators Unit, http://mdgs.un.org/unsd/mdg/Host.aspx?Content=Indicators/OfficialList.htm; "Energy use (Kg oil equivalent) per \$1,000 (PPP) GDP," Millennium Development Goals Indicators—The official United Nations site for the MDG Indicators, http://mdgs.un.org/unsd/mdg/SeriesDetail.aspx?srid=648&crid=; "Goals, targets and indicators," Millennium Project, http://www.unmillenniumproject.org/goals/gti.htm.

200. As incomes rise, households in developing countries typically switch to modern energy services for cooking, heating, lighting, and electric appliances. How quickly this occurs depends on the affordability of modern energy services, as well as their availability, and on cultural preferences. The process is in most cases a gradual one. People generally shift first from traditional fuels to intermediate modern fuels, such as coal and kerosene, and finally to advanced fuels, such as liquefied petroleum gas, natural gas, and electricity. International Energy Agency, *World Energy Outlook 2004*, Chapter 10: "Energy and Development," http://www.worldenergyoutlook.org/2004.asp, www.iea.org/textbase/nppdf/free/2004/weo2004.pdf.

201. Find more information on how increased energy consumption reduces the workload of women in developing countries in "Gender and Energy for Sustainable Development: A Toolkit and Resource Guide," United Nations Development Programme, 2004, http://www.undp.org/energy/gendereng.htm.

202. Lower respiratory infections kill more people in developing countries than HIV/AIDS, for instance. World Health Organization, "The top 10 causes of death," http://www.who.int/mediacentre/factsheets/fs310/en/.

203. Nowadays the World Bank is developing strategies to improve access to energy in developing countries, http://www.worldbank.org/energy.

204. "Commission on Sustainable Development-Report on the Ninth session 5 May 2000 and 16–27 April 2001," Economic and Social Council (United Nations Department of Economic and Social Affairs), E/2001/29, E/CN.17/2001/19, http://www.un.org/esa/sustdev/csd/ecn172001-19e.htm#Chapter%20IV. Commission on Sustainable Development (CSD), CSD-9, UN Department of Economic and Social Affairs-Division for Sustainable Development, http://www.un.org/esa/sustdev/csd/CSD9.htm.

205. Statement by Mr. Shoji Nishimoto, Assistant Administrator and Director, Bureau for Development Policy, United Nations Development Programme: Energy for Development Conference, Noordjwik-Netherlands, 12–14 December 2004, http://www.energyandenvironment.undp.org/undp/indexAction.cfm?module=Library&action=GetFile&DocumentAttachmentID=1620,http://www.energyandenvironment.undp.org/undp/index.cfm?module=Library&page=Document&DocumentID=5733.

206. MillenniumProject, Report to the United Nations Secretary-General, "Investing in Development: A Practical Plan to Achieve the Millennium Development Goals," 2005, http://www.unmillenniumproject.org/reports/fullreport.htm, http://www.unmillenniumproject.org/documents/MainReportComplete-lowres.pdf. "Energizing the Millennium Development Goals," Knowledge Expo: A UNDP co-hosted event, United Nations Development Programme, New York, May 2006, http://www.undp.org/energy/csd06-post_prog.htm.

207. See further recommendations in the following paper: Vijay Modi et al., "Energy Services for the Millennium Development Goals," Joint production of the Energy Sector Management Assistance Programme, United Nations Development Programme, UN Millennium Project, and the World Bank, 2005, http://www.unmillenniumproject.org/documents/MP_Energy_Low_Res.pdf.

208. "World Population Prospects: The 2006 Revision," United Nations, Department of Economic and Social Affairs-Population Division, http://www.un.org/esa/population/publications/wpp2006/wpp2006.htm; "Highlights," http://www.un.org/esa/population/publications/wpp2006/WPP2006_Highlights_rev.pdf; United Nations Population Fund, http://www.unfpa.org/index.htm.

209. The total global energy supply, a measure of commercial energy consumption including traditional biomass fuel (animal and plant waste, etc.) was 11,433.9 Mt of oil equivalent or ca. 83,8 billion barrels of oil equivalent in 2005. (Figure by IEA, 2007, quoted in: Human Development Report 2007/2008, United Nations Development Programme, 2007, http://hdr.undp.org/en/media/hdr_20072008_en_complete.pdf). It is a good enough assumption to use figures for commercial energy, as the share of noncommercial energy consumption tends to diminish with increasing development and energy consumption. However, a trend towards decentralized renewable energy consumption could change this reality: People may then consume increasingly more noncommercial renewable energy in form of solar energy or wind energy, for instance.

210. The United Nations differentiates between "More developed countries" and "Less developed countries." Less developed countries ("developing countries") include all countries in Africa, Asia (excluding Japan), and Latin America and the Caribbean, and the regions of Melanesia, Micronesia, and Polynesia. More developed countries include all countries in Europe, North America, Australia, New Zealand, and Japan.

211. In Chile, for instance, the government in 1994 used domestic resources and external aid to set up a special fund for rural electrification, competitively awarding subsidies to private electricity distribution companies that undertake projects. MillenniumProject, Report to the United Nations Secretary-General, "Investing in Development: A Practical Plan to Achieve the Millennium Development Goals," 2005, http://www.unmillenniumproject.org/reports/fullreport.htm, http://www.unmillenniumproject.org/documents/MainReportComplete-lowres.pdf.

212. "State of World Population 2007—Unleashing the Potential of Urban Growth," United Nations Population Fund, http://www.unfpa.org/swp/2007/english/introduction.html, http://www.unfpa.org/swp/2007/presskit/pdf/sowp2007_eng.pdf.

213. Obviously, inexpensive carbon-neutral energy would be the best solution. However, for the time being two traditional coal-fired power plants would provide more benefits than one super-modern, expensive coal-fired power plant with carbon capturing capabilities. Climate protection has to be in focus where it can be afforded, by those who have already entered the Oil Age.

BIBLIOGRAPHY TO PART V

A.S.L. & Associates database. http://www.asl-associates.com/sulfur1.htm.

ABC Television. "The Great Global Warming Swindle." http://www.abc.net.au/tv/swindle/.

American Coalition for Ethanol. http://www.ethanol.org/.

Ammann, Caspar M., Gerald A. Meehl, Warren M. Washington, and Charles S. Zender. "A Monthly and Latitudinally Varying Volcanic Forcing Dataset in Simulations of 20th Century Climate." *Geophysical Research Letters* 30, No. 12 (2003):1657. http://www.ncdc.noaa.gov/paleo/pubs/ammann2003/ammann2003.html.

Antelava, Natalia. "China's Increasing Hold over Kazakh Oil." *BBC News*, 20 August 2007. http://news.bbc.co.uk/2/hi/asia-pacific/6935292.stm.

Ballard Power Systems Inc. http://www.ballard.com/.

Barnola, J. M., D. Raynaud, C. Lorius, N. I. Barkov. "Historical CO_2 Record from the Vostok Ice Core." January 2003. Carbon Dioxide Information Analysis Center (CDIAC). http://cdiac.ornl.gov/ftp/trends/co2/vostok.icecore.co2.

Bartis, James T. "Policy Issues for Coal-to-Liquid Development." Testimony presented before the Senate Energy and Natural Resources Committee, May 2007. http://www.rand.org/pubs/testimonies/2007/RAND_CT281.pdf.

BASF Fuel Cells. http://www.basf.com/fuelcell.

Beckman, John E., and Terence J. Mahoney. "The Maunder Minimum and Climate Change: Have Historical Records Aided Current Research?" *Library and Information Services in Astronomy III, ASP Conference Series* 153 (1998). http://www.stsci.edu/stsci/meetings/lisa3/beckmanj.html.

Berg, C. "World Fuel Ethanol Analysis and Outlook." 2004. http://www.distill.com/World-Fuel-Ethanol-A&O-2004.html.

Berger, A., and M. F. Loutre. "Climate: An Exceptionally Long Interglacial Ahead?" *Science* 297 (5585) (2002):1287–1288. http://www.sciencemag.org/cgi/content/summary/297/5585/1287.

Berglen, Tore F., et al. "A Global Model of the Coupled Sulfur/Oxidant Chemistry in the Troposphere: The Sulfur Cycle." *Journal of Geophysical Research* 109, D19310, October 9, 2004. http://www.agu.org/pubs/crossref/2004/2003JD003948.shtml.

Black, Richard. "No Sun Link to Climate Change." *BBC News*, July 10, 2007. http://news.bbc.co.uk/2/hi/science/nature/6290228.stm.

Borenstein, Seth. "There Goes El Nino, Here Comes La Nina: Forecasters Expect Nino's 'Evil Twin Sister' To Bring More Hurricanes, More Drought." *The Associated Press*, February 28, 2007, Washington. http://www.cbsnews.com/stories/2007/02/28/tech/main2523483.shtml.

BP Statistical Review of World Energy. (Current edition.) http://www.bp.com/statisticalreview. (BP quotes OPEC (http://www.opec.org), the Oil & Gas Journal (http://www.ogj.com), and World Oil (http://www.worldoil.com/) for oil reserves data; the World Energy Council (Survey of Energy Resources, http://www.worldenergy.org) for coal reserves data; and the International Association for Natural Gas (http://www.cedigaz.org/) for natural gas reserves data.)

BP Statistical Review of World Energy, June 2002. http://www.bp.com/downloads/1087/statistical_review.pdf. (No longer active.)

BP Statistical Review of World Energy, June 2007. http://www.bp.com/liveassets/bp_internet/globalbp/globalbp_uk_english/reports_and_publications/statistical_energy_review_2007/STAGING/local_assets/downloads/pdf/statistical_review_of_world_energy_full_report_2007.pdf.

BP Statistical Review, June 2008. "Oil—Proved Reserves History." http://www.bp.com/liveassets/bp_internet/globalbp/globalbp_uk_english/reports_and_publications/statistical_energy_review_2008/STAGING/local_assets/downloads/spreadsheets/statistical_review_full_report_workbook_2008.xls.

BP. "Ethanol: The Global Ethanol Industry Continued to Grow Rapidly in 2006, with Output Increasing by 22%." http://www.bp.com/sectiongenericarticle.do?categoryId=9017929&contentId=7033479.

Budyko, M. I. *Climatic Changes*. Washington: American Geophysical Society, 1977.

Calik Enerji-Oil & Gas Projects. "Samsun—Ceyhan Crude Oil Pipeline Project." Natural Gas & Oil Transportation & Distribution. http://www.calikenerji.com.tr/oilgas.php?ID=5.

Carr, Mathew. "Global Climate in 2008 May Be Coolest Since 2000." *Bloomberg*, January 3, 2008. http://www.bloomberg.com/apps/news?pid=20601081&sid=aBBw.Bw5TOpg&refer=australia.

Carter, Bob. "There IS a Problem with Global Warming . . . It Stopped in 1998." *Telegraph*, April 9, 2006. http://www.telegraph.co.uk/opinion/main.jhtml?xml=/opinion/2006/04/09/do0907.xml&sSheet=/news/2006/04/09/ixworld.html.

Chandler, David L. "No Future for Fusion Power, Says Top Scientist." *New Scientist*, March 9, 2006. http://www.newscientist.com/channel/fundamentals/dn8827-no-future-for-fusion-power-says-top-scientist.html.

Channel 4. "The Great Global Warming Swindle: The Arguments." http://www.channel4.com/science/microsites/G/great_global_warming_swindle/arguments_2.html.

Channel 4. "Top Ten Eco Things You Can Do." September 2006. http://www.channel4.com/science/microsites/S/science/society/eco_topten.html.

Chicago Climate Exchange (CCX). "Overview." http://www.chicagoclimatex.com/content.jsf?id=821.

Chu, Steven, José Goldemberg, et al. "Lighting the Way: Toward a Sustainable Energy Future." InterAcademy Council, 2007. http://www.interacademycouncil.net/?id=12161.

Chylek, P., et al. "Greenland Warming of 1920–30 and 1995–2005." *Geophysical Research Letters* 33 (2006).

CIA—The World Factbook. https://www.cia.gov/library/publications/the-world-factbook/.

CIA-The World Factbook. "China." https://www.cia.gov/library/publications/the-world-factbook/geos/ch.html.

CIA-The World Factbook. "India." https://www.cia.gov/library/publications/the-world-factbook/geos/in.html.

CIA-The World Factbook. "Libya." https://www.cia.gov/library/publications/the-world-factbook/geos/ly.html.

CIA-The World Factbook. "Nigeria." https://www.cia.gov/library/publications/the-world-factbook/geos/ni.html.

CIA-The World Factbook. "Russia." https://www.cia.gov/library/publications/the-world-factbook/geos/rs.html.

CIA-The World Factbook. "Saudi Arabia." https://www.cia.gov/library/publications/the-world-factbook/geos/sa.html.

CIA-The World Factbook. "Venezuela." https://www.cia.gov/library/publications/the-world-factbook/geos/ve.html.

Climate Exchange Plc. "About CCX." http://climateexchangeplc.com/home/about-ccx.

Climate Prediction Center/Ncep, National Weather Service. "El Niño/Southern Oscillation (Enso) Diagnostic Discussion." National Oceanic and Atmospheric Administration, United States Department of Commerce, January 10, 2008. http://www.cpc.noaa.gov/products/analysis_monitoring/enso_advisory/ensodisc.html.

CNN. "Clinton Hails Global Warming Pact." December 11, 1997. http://www.cnn.com/ALLPOLITICS/1997/12/11/kyoto/.

Cook, Earl. *Man, Energy, Society.* San Francisco: W.H. Freeman, 1976.

Copenhagen Consensus 2004. http://www.copenhagenconsensus.com/Default.aspx?ID=158.

Craig, James R., David J. Vaughan, Brian J. Skinner. *Resources of the Earth: Origin, Use, and Environmental Impact.* Upper Saddle River: Prentice Hall, 1996.

Crutzen, P. J. "Albedo Enhancement by Stratospheric Sulfur Injections: A Contribution to Resolve a Policy Dilemma?" *Climatic Change* 77 (3–4) (2006): 211. http://www.springerlink.com/content/t1vn75m458373h63/.

Danish National Space Center, DTU. "Climate Change and Cosmic Rays." http://www.spacecenter.dk/research/sun-climate/other/global-warming.

DATA. "HIV/AIDS, TB and Malaria." http://www.data.org/issues/aids.html.

Deffeyes, K.N. *Hubbert's Peak: The Impending World Oil Shortage.* Princeton: Princeton University Press, 2001.

Deutch, John, et al. "The Future of Nuclear Energy." Massachusetts Institute of Technology, 2003. http://web.mit.edu/nuclearpower/.

DeYoung, Karen. "Falling on His Sword: Colin Powell's Most Significant Moment Turned Out To Be His Lowest." *Washington Post*, October 1, 2006, Page W12. http://www.washingtonpost.com/wp-dyn/content/article/2006/09/27/AR2006092700106.html.

Doran, P. T., et al. "Antarctic Climate Cooling and Terrestrial Ecosystem Response." *Nature* 415 (2002): 517–20.

Driessen, Paul. "Global Warming Insanity." *The Washington Times*, September 12, 2007. http://washingtontimes.com/article/20070912/COMMENTARY/109120009/1012.

Duffy, Gary. "Pressures Build on Amazon Jungle." *BBC News*, 14 January 2008. http://news.bbc.co.uk/2/hi/americas/7186776.stm.

Economist, The. "Can Coal Be Clean?" November 30, 2006. http://www.economist.com/printedition/displaystory.cfm?story_id=E1_RPTNPTN.

Economist, The. "China—The New Colonialists." March 13, 2008. http://www.economist.com/displaystory.cfm?story_id=10853534.

Economist, The. "China's Quest for Resources: A Ravenous Dragon." March 13, 2008. http://www.economist.com/displaystory.cfm?story_id=10795714.

Economist, The. "China's Quest For Resources: A Large Black Cloud." March 13, 2008. http://www.economist.com/specialreports/displaystory.cfm?story_id=10795813.

Economist, The. "Climatology: Grey-sky Thinking." July 5, 2007. http://www.economist.com/science/displaystory.cfm?story_id=9433721.

Economist, The. "Ethanol: Dirty as Well as Dear?" January 15, 2004. http://www.economist.com/displaystory.cfm?story_id=2352788.

Economist, The. "In Search of the Perfect Battery." March 6, 2008. http://www.economist.com/displaystory.cfm?story_id=10789409.

Economist, The. "Nuclear Fusion: Firing New Shots." April 19, 2007. http://www.economist.com/research/backgrounders/displaystory.cfm?story_id=9033026.

Economist, The. "The West and Russia: Lights, Camera and a Different Ending." February 14, 2008. http://www.economist.com/displayStory.cfm?story_id=10696195.

Economist Intelligence Unit, The. "China-Economic Data." *CountryData*, March 7, 2008. http://www.economist.com/countries/China/profile.cfm?folder=Profile%2DEconomic%20Data.

Economist Intelligence Unit, The. "China-Economic Structure." September 7, 2007. http://www.economist.com/countries/China/profile.cfm?folder=Profile%2DEconomic%20Structure.

EIC Online Newsletters. "Eni and Çalik Enerji—the Samsun-Ceyhan Oil Pipeline Project." June 2006. http://www.the-eic.com/News/Archive/2006/Jun/Article2232.htm.

Ellerman, A. Denny, and Barbara K. Buchner. "The European Union Emissions Trading Scheme: Origins, Allocation, and Early Results." *Rev Environmental Economics and Policy* 1(1) (2007): 66–87. http://reep.oxfordjournals.org/cgi/content/abstract/1/1/66, 2007.

Energy Information Administration. (The EIA is a statistical and analytical agency within the US Department of Energy.) http://www.eia.doe.gov.

Energy Information Administration. "Are We Running Out of Oil?" http://www.eia.doe.gov/oiaf/archive/ieo99/boxtext.html.

Energy Information Administration. "China-Oil." *Country Analysis Briefs*. http://www.eia.doe.gov/emeu/cabs/China/Oil.html.

Energy Information Administration. "Crude Oil Production and Crude Oil Well Productivity, 1954–2007." http://www.eia.doe.gov/emeu/aer/txt/ptb0502.html.

Energy Information Administration. Energy Topics from A to Z. http://www.eia.doe.gov/fueloverview.html.

Energy Information Administration. EIA Service Report: "Federal Energy Subsidies: Direct and Indirect Interventions in Energy Markets." DOE/EIA/SR/EMEU-92-02, November 20, 1992. Office of Energy Markets and End Use. http://tonto.eia.doe.gov/ftproot/service/emeu9202.pdf.

Energy Information Administration. *International Energy Outlook*. (Most recent edition.) Office of Integrated Analysis and Forecasting. http://www.eia.doe.gov/oiaf/ieo/.

Energy Information Administration. *International Energy Outlook 2007*, Office of Integrated Analysis and Forecasting, http://www.eia.doe.gov/oiaf/archive/ieo07/index.html.

Energy Information Administration. *International Energy Outlook 2007*. "Chapter 7: Energy-Related Carbon Dioxide Emissions." Office of Integrated Analysis and Forecasting, http://www.eia.doe.gov/oiaf/archive/ieo07/pdf/emissions.pdf.

Energy Information Administration. *International Energy Outlook*. (Archived editions.) Office of Integrated Analysis and Forecasting. http://www.eia.doe.gov/oiaf/ieo/ieoarchive.html.

Energy Information Administration. "Long Term World Oil Supply (A Resource Base/Production Path Analysis)," 2000. http://www.eia.doe.gov/pub/oil_gas/petroleum/presentations/2000/long_term_supply/sld001.htm.

Energy Information Administration. "Turkey." *Country Analysis Briefs*. http://www.eia.doe.gov/emeu/cabs/Turkey/Background.html.

Energy Information Administration. "World Energy Intensity—Total Primary Energy Consumption per Dollar of Gross Domestic Product Using Purchasing Power Parities, 1980–2005." http://www.eia.doe.gov/pub/international/iealf/tablee1p.xls

Energy Information Administration. "World Petroleum Consumption, 1960–2006." http://www.eia.doe.gov/emeu/aer/txt/ptb1110.html.

Energy Research Institute. (ERI is a national research organization conducting studies on China's energy issues.) http://www.eri.org.cn.

Energy and Resources Institute (TERI). (Formerly known as Tata Energy and Resources Institute, TERI is an India-based non-profit, scientific and policy research organization.) http://www.teriin.org.

Esper, J., E.R. Cook, and F.H. Schweingruber. "Low-Frequency Signals in Long Tree-Ring Chronologies for Reconstructing Past Temperature Variability." *Science* 295, 5563 (2002): 2250–2253. http://www.sciencemag.org/cgi/content/short/295/5563/2250. IGBP PAGES/World Data Center for Paleoclimatology Data Contribution Series #2003–036. NOAA/NGDC Paleoclimatology Program, Boulder, CO. ftp://ftp.ncdc.noaa.gov/pub/data/paleo/treering/reconstructions/n_hem_temp/esper2002_nhem_temp.txt.

Etheridge, D.M., L.P. Steele, R.J. Francey, and R.L. Langenfelds. "Historic CH4 Records from Antarctic and Greenland Ice Cores, Antarctic Firn Data, and Archived Air Samples from Cape Grim, Tasmania-Period of Record: 1008–1995." In *Trends: A Compendium of Data on Global Change*, CDIAC (Carbon Dioxide Information Analysis Center), Oak Ridge National Laboratory, U.S. Department of Energy, 2002. http://cdiac.ornl.gov/trends/atm_meth/lawdome_meth.html.

Etheridge, D.M., L.P. Steele, R.L. Langenfelds, R.J. Francey, J.-M. Barnola, V.I. Morgan. "Historical CO_2 Record Derived from a Spline Fit (20 year cutoff) of the Law Dome DE08 and DE08–2 Ice Cores." June 1998. Carbon Dioxide Information Analysis Center (CDIAC). http://cdiac.ornl.gov/ftp/trends/co2/lawdome.smoothed.yr20.

Etheridge, D.M., L.P. Steele, R.L. Langenfelds, R.J. Francey, J.-M. Barnola, V.I. Morgan. "Historical CO_2 Record Derived from a Spline Fit (75 year cutoff) of the Law Dome DSS, DE08, and DE08–2 Ice Cores." June 1998. Carbon Dioxide Information Analysis Center (CDIAC). http://cdiac.ornl.gov/ftp/trends/co2/lawdome.smoothed.yr75.

European Commission. "Questions and Answers on the Commission's Proposal to Revise the EU Emissions Trading System." *MEMO/08/35*, Brussels, January 23, 2008. http://europa.eu/rapid/pressReleasesAction.do?reference=MEMO/08/35&format=HTML&aged=0&language=EN&guiLanguage=en.

Fortson, Danny. "Oil Fields of Plenty?" *The Independent*, February 1, 2008. http://www.independent.co.uk/news/business/analysis-and-features/oil-fields-of-plenty-776776.html.

Friedli, H., et al. "Carbon-13/carbon-12 Ratios in Carbon Dioxide Extracted from Antarctic Ice." *Geophysical Research Letters* 11 (1984). http://chemport.cas.org/cgi-bin/sdcgi?APP=ftslink&action=reflink&origin=npg&version=1.0&coi=1:CAS:528:DyaL2MXivV2ltw%3D%3D&pissn=0028-0836&pyear=1985&md5=09da79795cc62c731c8e31d535ae29d8.

Friends of the Earth. "Analysis of the Bush Energy Plan." http://www.foe.org/camps/leg/bushwatch/energyplan.pdf.

Friis-Christensen, E., and K. Lassen. "Length of the Solar Cycle: An Indicator of Solar Activity Closely Associated with Climate." *Science* 254 (1991): 698–700.

Genencor—A Danisco Division. http://www.genencor.com/.

Gholz, Eugene, and Daryl G. Press. "Energy Alarmism: The Myths That Make Americans Worry about Oil." April 5, 2007. http://www.cato.org/pubs/pas/pa589.pdf.

Global Canopy Programme. http://www.globalcanopy.org/.

Global Volcanism Program, Department of Mineral Sciences. "Large Holocene Eruptions." National Museum of Natural History, Smithsonian Institution. http://www.volcano.si.edu/world/largeeruptions.cfm.

Goklany, Indur. "What to Do about Climate Change." *Policy Analysis* 609, February 5, (2008), Cato Institute. http://www.cato.org/pub_display.php?pub_id=9125.

Goldemberg, Jose. "Energy and Human Well-Being." Background paper to the Human Development Report 2001—Making New Technologies Work for Human Development, UNDP 2001. http://hdr.undp.org/en/reports/global/hdr2001/, hdr.undp.org/en/reports/global/hdr2001/papers/goldemberg-energy-1.pdf.

Greenpeace. "The Subsidy Scandal: The European Clash between Environmental Rhetoric and Public Spending." http://archive.greenpeace.org/~comms/97/climate/eusub.html.

Hansen, J. "Can We Defuse the Global Warming Time Bomb?" *NaturalScience* (2003). http://pubs.giss.nasa.gov/docs/2003/2003_Hansen.pdf.

Hansen, James E. "As Pure as Snow." *Science Briefs*, December 2003. http://www.giss.nasa.gov/research/briefs/hansen_10/.

Hansen, James et al. "Global Warming in the Twenty-First Century: An Alternative Scenario." *PNAS (Proceedings of the National Academy of Sciences of the United States of America)* 97, 18 (2000): 9875–9880. http://www.pnas.org/cgi/content/full/97/18/9875.

Harvard-Smithsonian Center for Astrophysics. "20th Century Climate Not So Hot." Press Release, March 31, 2003. http://cfa-www.harvard.edu/press/archive/pr0310.html.

Hebrew University of Jerusalem. "Karl Kordesch." Famous Scientists. Institute of Chemistry. http://chem.ch.huji.ac.il/history/kordesch.html.

Herrick, Catherine, and Bill Owens. "Rewriting The Science: Scientist Says Politicians Edit Global Warming Research." *60 Minutes*, CBS, March 19, 2006. http://www.cbsnews.com/stories/2006/03/17/60minutes/main1415985.shtml.

Herzog, Howard J., Ken Caldeira and Eric Adams. "Carbon Sequestration via Direct Injection." *The Carbon Capture and Sequestration Technologies Program at MIT.* http://sequestration.mit.edu/pdf/direct_injection.pdf.

Hirsch, Robert L., Roger Bezdek and Robert Wendling. "Peaking of World Oil Production: Impacts, Mitigation, & Risk Management." February 2005. http://www.netl.doe.gov/publications/others/pdf/Oil_Peaking_NETL.pdf.

Hoang, Gia Tuong, and Jean Jacquinot. "Controlled Fusion: The Next Step." *Physics World*, January 2004. http://physicsweb.org/article/world/17/1/6.

Holland, David M., Robert H. Thomas, Brad de Young, Mads H. Ribergaard, and Bjarne Lyberth. "Acceleration of Jakobshavn Isbræ Triggered by Warm Subsurface Ocean Waters." *Nature Geoscience* 1 (2008): 659–664. http://www.nature.com/ngeo/journal/v1/n10/full/ngeo316.html.

Holz, Robert Lee. "Tracking Carbon Dioxide." *The Wall Street Journal*, December 28–30, 2007, Science Journal.

Houghton, John. "The Great Global Warming Swindle. Programme directed by Martin Durkin, on Channel 4 on Thursday 8 March 2007—Critique." http://www.jri.org.uk/news/Critique_Channel4_Global_Warming_Swindle.pdf.

Houghton, John. *Global Warming: The Complete Briefing*. Cambridge: Cambridge University Press, 1997.

Howden, Daniel. "Deforestation: The hidden Cause of Global Warming." *The Independent*, May 14, 2007. http://www.independent.co.uk/environment/climate-change/deforestation-the-hidden-cause-of-global-warming-448734.html.

Hubbert, M. King. *Energy Resources: A Report to the Committee on Natural Resources*. Washington D.C.: National Academy of Sciences, 1962.

Hwang, Roland. "Money Down the Pipeline: The Hidden Subsidies to the Oil Industry." Union of Concerned Scientists. http://www.ucsusa.org/publication.cfm?publicationID=149.

Indian Space Research Organisation (ISRO). "Milestones." http://www.isro.org/mileston.htm.
International Association for Natural Gas Vehicles (IANGV). http://www.iangv.org/statistics.html.
International Energy Agency. http://iea.org/. (The Paris-based IEA is an autonomous body established in November 1974 within the framework of the Organisation for Economic Co-operation and Development (OECD) to implement an international energy programme.)
International Energy Agency. *Key World Energy Statistics*. http://www.iea.org/Textbase/publications/free_new_Desc.asp?PUBS_ID=1199.
International Energy Agency. "Key World Energy Statistics 2007." http://www.iea.org/textbase/nppdf/free/2007/key_stats_2007.pdf.
International Energy Agency. *World Energy Outlook*. (Current edition.) http://www.worldenergyoutlook.org/.
International Energy Agency. *World Energy Outlook 2004*. http://www.worldenergyoutlook.org/2004.asp. www.iea.org/textbase/nppdf/free/2004/weo2004.pdf.
International Energy Agency. *World Energy Outlook 2006*. http://www.iea.org/textbase/nppdf/free/2006/weo2006.pdf.
International Energy Agency. *World Energy Outlook 2007*. *http:*//www.iea.org/textbase/npsum/WEO2007SUM.pdf.
International Energy Agency. *World Energy Outlook 2008*. http://www.worldenergyoutlook.org/2008.asp.
International Energy Agency. *World Energy Outlook*. (Previous editions.) http://www.worldenergyoutlook.org/older.asp.
International Energy Agency. "World Energy Outlook 2007: Fact Sheet—Global Energy Demand." http://www.worldenergyoutlook.org/2007.asp. http://www.iea.org/Textbase/npsum/WEO2007SUM.pdf.
International Energy Agency. "World Energy Outlook 2007: Fact Sheet—Oil." http://www.iea.org//textbase/papers/2007/fs_oil.pdf.
International Thermonuclear Experimental Reactor (ITER). http://www.iter.org/.
Iogen. http://www.iogen.ca/.
Ivanhoe, L.F. "Get Ready For Another Oil Shock!" http://dieoff.org/page90.htm.
Japan Aerospace Exploration Agency. "H-II Launch Vehicle." Launch Vehicles and Space Transportation Systems,. http://www.jaxa.jp/projects/rockets/h2/index_e.html.
JAPEX, the Japan Petroleum Exploration Co. http://www.japex.co.jp/english/company/index.html.
JOGMEC, the Japan Oil, Gas and Metals National Corporation. http://www.jogmec.go.jp/english/index.html.
Jones, P. D., D. E. Parker, T. J. Osborn, K. R. Briffa. "Global Monthly and Annual Temperature Anomalies (degrees C), 1850–2007 (Relative to the 1961–1990 Mean)." September 2008. Carbon Dioxide Information Analysis Center (CDIAC). http://cdiac.ornl.gov/ftp/trends/temp/jonescru/global.dat.
Jørgensen, Torben Stockflet, and Aksel Walløe Hansen, "Comments on 'Variation of cosmic ray flux and global cloud coverage—a missing link in solar–climate relationships' by Henrik Svensmark and Eigil Friis-Christensen [Journal of Atmospheric

and Solar-Terrestrial Physics 59 (1997) 1225–1232]." *Journal of Atmospheric and Solar-Terrestrial Physics* 62, 1 (2000): 73–77. http://www.sciencedirect.com/science?_ob=ArticleURL&_udi=B6VHB-3YN9385-8&_user=10&_rdoc=1&_fmt=&_orig=search&_sort=d&view=c&_acct=C000050221&_version=1&_urlVersion=0&_userid=10&md5=be6aef97b4a3f3cc8e9bb63edc59d5e9.

Katzer, James, et al. "The Future of Coal." Massachusetts Institute of Technology, 2007. http://web.mit.edu/coal/.

Keeling, C. D., et al. "Atmospheric Carbon Dioxide Variations at Mauna Loa Observatory, Hawaii." *Tellus 28* (1976 and updates): 538–51.

Keeling, R. F., S. C. Piper, A. F. Bollenbacher, and S. J. Walker. "Atmospheric CO_2 Values (ppmv) Derived from In Situ Air Samples Collected at Mauna Loa, Hawaii, USA." May 2008. Carbon Dioxide Information Analysis Center (CDIAC). http://cdiac.ornl.gov/ftp/trends/co2/maunaloa.co2.

Kleveman, Lutz C. "Der Kampf ums Kaspische Öl." *Der Spiegel*, August 2002. http://www.spiegel.de/archiv/dossiers/0,1518,245770,00.html.

Koplow, Doug, and John Dernbach. "Federal Fossil Fuel Subsidies and Greenhouse Gas Emissions: A Case Study of Increasing Transparency for Fiscal Policy." *Annual Review of Energy and the Environment* 26 (2001): 361–389. http://arjournals.annualreviews.org/doi/full/10.1146/annurev.energy.26.1.361.

Koplow, Doug, and John Dernbach. *Fueling Global Warming—Federal Subsidies to Oil in the United States.* http://archive.greenpeace.org/~climate/oil/fdsub.html.

Lackner, Klaus. "Future of Coal Depends on CO_2 Disposal." *Statement at the U.S. Senate Energy and Natural Resources Committee Coal Conference*, March 10, 2005. http://energy.senate.gov/public_new/_files/Question2SelectedParticipantsSubmittals.doc.

Lambers, Hans, and Miquel Ribas-Carbo, eds. "Plant Respiration: From Cell to Ecosystem." In *Advances in Photosynthesis and Respiration Series* 18. Dordrecht: Springer, 2005.

Lassen, K., and E. Friis-Christensen. "Variability of the Solar Cycle Length during the Past Five Centuries and the Apparent Association with Terrestrial Climate." *Journal of Atmospheric and Terrestrial Physics* 57, 8 (1995): 835–845.

Lindzen, Richard S. "Global Warming: The Origin and Nature of the Alleged Scientific Consensus." *Regulation-The Cato Review of Business & Government* 15, No. 2 (1992). http://www.cato.org/pubs/regulation/regv15n2/reg15n2g.html.

Lindzen, Richard. "Climate of Fear: Global-warming Alarmists Intimidate Dissenting Scientists into Silence." The *Wall Street Journal*, April 12, 2006, The Opinionjournal Archives. http://www.opinionjournal.com/extra/?id=110008220.

Locke, John. *The Second Treatise of Civil Government.* (First published in 1690.) Sixth edition of 1764. http://ebooks.adelaide.edu.au/l/locke/john/l81s/.

Lockwood, Mike, and Claus Fröhlich. "Recent Oppositely Directed Trends in Solar Climate Forcings and the Global Mean Surface Air Temperature." *Proc. R. Soc. A* (2007). http://publishing.royalsociety.org/media/proceedings_a/rspa2007 1880.pdf.

Lomborg, Bjørn. "Stern Review: The Dodgy Numbers behind the Latest Warming Scare." *The Wall Street Journal*, November 2, 2006, The Opinionjournal Archives. http://www.opinionjournal.com/extra/?id=110009182.

Lomborg, Bjørn. *Verdens Sande Tilstand*. Viby: Centrum, 2002. English language edition: Bjørn Lomborg. *The Skeptical Environmentalist: Measuring the Real State of the World*. Cambridge: Cambridge University Press, 2001.

Lorius, C., et al. "A 150,000-year Climatic Record from Antarctic Ice." *Nature* 316 (1985): 591–596. http://www.nature.com/nature/journal/v316/n6029/abs/316591a0.html.

Lucas, D. D., and H. Akimoto. "Contributions of Anthropogenic and Natural Sources of Sulfur to SO_2, H_2SO_4(g) and Nanoparticle Formation." *Atmos. Chem. Phys. Discuss* 7 (2007): 7679–7721. www.atmos-chem-phys-discuss.net/7/7679/2007/. http://www.atmos-chem-phys-discuss.net/7/7679/2007/acpd-7-7679-2007-print.pdf.

Luscombe, Richard. "Blogger Gets Hot and Bothered over Nasa's Climate Data Error." *The Guardian*, August 16, 2007. http://www.guardian.co.uk/environment/2007/aug/16/1.

Marine Current Turbines Ltd. http://www.marineturbines.com/.

Maugeri, Leonardo. "Oil: Never Cry Wolf—Why the Petroleum Age Is Far from Over." *Science* 304 (2004): 1115.

Maugeri, Leonardo. "The Shell Game." *Newsweek*, February 16, 2004. http://www.msnbc.msn.com/id/4402707/site/newsweek/hl/da/.

McIntyre, Steve. "RMS and Sulphate Emissions." Climate Audit, May 6, 2007. http://www.climateaudit.org/?p=1536.

Meehl, Gerald A., et al. "Combinations of Natural and Anthropogenic Forcings in Twentieth-Century Climate." *Journal of Climate* 17 (2004) 3721–3727, American Meteorological Society. http://www.cgd.ucar.edu/ccr/publications/meehl_additivity.pdf.

Meehl, Gerald A., et al. "Solar and Greenhouse Gas Forcing and Climate Response in the Twentieth Century." *Journal of Climate* 16 (2003) 426. http://www.cgd.ucar.edu/ccr/publications/meehl_solar.pdf.

Michaels, Patrick J. "Carbon Copies." *American Spectator*, February 27, (2008). http://www.cato.org/pub_display.php?pub_id=9242.

Michaels, Patrick J. "Is the Sky Really Falling? A Review of Recent Global Warming Scare Stories." *Trade Policy Analysis* 576 (2006), Cato Institute. http://www.cato.org/pubs/pas/pa576.pdf.

Michaels, Patrick J. "More Ice Than Ever." *American Spectator*, February 5, 2008. http://www.cato.org/pub_display.php?pub_id=9136.

Millennium Project. "Goals, Targets and Indicators." http://www.unmillenniumproject.org/goals/gti.htm.

Millennium Project. Report to the United Nations Secretary-General. "Investing in Development: A Practical Plan to Achieve the Millennium Development Goals," 2005. http://www.unmillenniumproject.org/reports/fullreport.htm. http://www.unmillenniumproject.org/documents/MainReportComplete-lowres.pdf.

Milloy, Steven. "Al Gore's Inconvenient Electric Bill." *Fox News*, March 12, 2007. http://www.foxnews.com/story/0,2933,257958,00.html.

Milloy, Steven. "Must-See Global Warming TV." *Fox News*, March 19, 2007. http://www.foxnews.com/story/0,2933,258993,00.html.

MIT Fuel Cell Laboratory. "About Fuel Cells." Department of Mechanical Engineering, Massachusetts Institute of Technology. http://web.mit.edu/afs/athena.mit.edu/org/m/mecheng/fcp/about%20f%20cells.html.

Mitchell, Andrew W., Katherine Secoy and Niki Mardas. "Forests First in the Fight Against Climate Change-The Vivo Carbon Initiative." *Global Canopy Programme*, June (2007). http://www.globalcanopy.org/themedia/file/PDFs/Forests%20 First%20June%202007.pdf.

Modi, Vijay, Susan McDade, Dominique Lallement and Jamal Saghir. "Energy Services for the Millennium Development Goals." Joint production of the Energy Sector Management Assistance Programme, United Nations Development Programme, UN Millennium Project, and the World Bank, 2005. http://www.unmil lenniumproject.org/documents/MP_Energy_Low_Res.pdf.

Morgan, Michael D., Joseph M. Moran and James H. Wiersma. *Environmental Science: Managing Biological & Physical Resources*. Dubuque: Wm. C. Brown Publishers, 1993.

MSNBC News. "Storm over Pentagon Climate Scenario." February 26, 2004. http://msnbc.msn.com/id/4379905/.

Nakic'enovic', N., and R. Swart, ed. *Special Report on Emission Scenarios*. Cambridge: Cambridge University Press, 2000.

NASA Earth Observatory. "Milutin Milankovitch." http://earthobservatory.nasa.gov/Library/Giants/Milankovitch/.

NASA Goddard Institute for Space Studies. "GISS Surface Temperature Analysis-August 2007 Update and Effects." http://data.giss.nasa.gov/gistemp/updates/200708.html.

National Academy of Sciences-Board on Atmospheric Sciences and Climate (BASC). *Surface Temperature Reconstructions for the Last 2,000 Years*. Washington: The National Academies Press, 2006. http://books.nap.edu/openbook.php?record_id=11676&page=3.

Netherlands Environmental Assessment Agency. http://www.mnp.nl/en/index.html.

Netherlands Environmental Assessment Agency. "Emissions Data." http://www.mnp.nl/edgar/model/.

New Scientist. "Green Light for Nuclear Fusion Project." November 21, 2006. http://technology.newscientist.com/article/dn10633-green-light-for-nuclear-fusion-project.html.

Nobel Foundation. The Nobel Peace Prize 2007. http://nobelprize.org/nobel_prizes/peace/laureates/2007/index.html.

Nordhaus, William. "Count before You Leap: Economics of Climate Change." *The Economist*, July, 1990. http://www.uwmc.uwc.edu/geography/globcat/globwarm/nordhaus.htm.

Nordhaus, William. *Managing the Global Commons: The Economics of Climate Change*. Cambridge: MIT Press, 1994.

Novozymes. http://www.novozymes.com/.

OECD (Organization for Economic Cooperation and Development). http://www.oecd.org.

Ogino, Akifumi, Hideki Orito, Kazuhiro Shimada, and Hiroyuki Hirooka. "Evaluating Environmental Impacts of the Japanese Beef Cow-calf System by the Life Cycle Assessment Method." *Animal Science Journal* 78 (2007): 424. http://www.blackwell-synergy.com/doi/abs/10.1111/j.1740-0929.2007.00457.x.

Orn, G., U. Hansson and H. Rodhe, "Historical Worldwide Emissions of Anthropogenic Sulfur: 1860–1985." *Tech. Rep.* CM-91 (1996), Department of Meteorology/International Meteorological Institute, Stockholm University.

Østergaard, Nicolai. "Kinesiske kraftværkcr sodcr indlandsisen." *Ingeniøren*, September 19, 2008. http://ing.dk/artikel/91415?bund.

Parkins, William E. "Energy: Fusion Power: Will It Ever Come?" *Science* 311 (5766) (2006), 1380. http://www.sciencemag.org/cgi/content/summary/311/5766/1380.

Pearce, Fred. "Bert Bolin: Meteorologist and first chair of the IPCC who cajoled the world into action on climate change." *The Independent*, January 5, 2008. http://www.independent.co.uk/news/obituaries/bert-bolin-meteorologist-and-first-chair-of-the-ipcc-who-cajoled-the-world-into-action-on-climate-change-768355.html.

Penn State University. "Hydrogen-Future Fuel? An Infinite Charge: Fuel Cell Chemistry is a Simple Dance. The trick Is to Make It Cheaper." http://www.rps.psu.edu/hydrogen/infinite.html.

People for the Ethical Treatment of Animals (PETA). "Factsheets." http://peta.com/mc/facts.asp.

Petit, J.R., J. Jouzel, D. Raynaud, N.I. Barkov, J.M. Barnola, I. Basile, M. Bender, J. Chappellaz, J. Davis, G. Delaygue, M. Delmotte, V.M. Kotlyakov, M. Legrand, V. Lipenkov, C. Lorius, L. Pépin, C. Ritz, E. Saltzman, M. Stievenard. "Climate and Atmospheric History of the Past 420,000 years from the Vostok Ice Core, Antarctica." *Nature* 399 (1999): 429–436. IGBP PAGES/World Data Center for Paleoclimatology Data Contribution Series #2001-076. NOAA/NGDC Paleoclimatology Program, Boulder, CO. ftp://ftp.ncdc.noaa.gov/pub/data/paleo/icecore/antarctica/vostok/deutnat.txt.

Pinkerton, Frederick E., and Brian G. Wicke, "Bottling the Hydrogen Genie." *The Industrial Physicist*, American Institute of Physics, February/March 2004. http://www.aip.org/tip/INPHFA/vol-10/iss-1/p20.html.

Population Reference Bureau (PRB). DataFinder. (Database of demographic, health, economic, and environment variables for various world regions.) http://www.prb.org/Datafinder.aspx.

Princeton University. "ZEC-Zero Emission Coal Technology, CO_2 sequestration, chemistry of the environment." *CHM333 web page projects*, Fall 2002. http://www.princeton.edu/~chm333/2002/fall/co_two/minerals/zec.htm#_ftnref5.

Rasool, S. I., and S. H. Schneider, "Atmospheric Carbon Dioxide and Aerosols: Effects of Large Increases on Global Climate." *Science* 173 (3992) (1971):138–141. http://www.sciencemag.org/cgi/content/abstract/173/3992/138.

Raupach, Michael R., et al. "Global and Regional Drivers of Accelerating CO_2 Emissions." *PNAS* (*Proceedings of the National Academy of Sciences of the United States of America*) 104 (24) (2007): 10288–10293. http://www.pnas.org/cgi/reprint/0700609104v1.pdf. http://www.pnas.org/content/104/24/10288/suppl/DC1.

RealClimate-Climate science from climate scientists. "Tropical tropospheric trends." December 12, 2007. http://www.realclimate.org/index.php/archives/2007/12/tropical-troposphere-trends/.

Rosentrater, Kurt A. "Economics And Impacts Of Ethanol Manufacture, *BioCycle*, 47 (12) (2006): 44. http://www.jgpress.com/archives/_free/001205.html.

Schröder, K.-P., and Robert Connon Smith. "Distant Future of the Sun and Earth Revisited." *Monthly Notices of the Royal Astronomical Society* 386 (2008): 155–163. http://dx.doi.org/10.1111%2Fj.1365–2966.2008.13022.x.
Schwartz, Peter, and Doug Randall. "An Abrupt Climate Change Scenario and Its Implications for United States National Security—A Report Commissioned by the U.S. Defense Department." October 2003. http://www.ems.org/climate/pentagon_climatechange.pdf.
Serenergy. http://www.serenergy.dk/.
Smith, R. Jeffrey. "Army Files Cite Abuse of Afghans—Special Forces Unit Prompted Senior Officers' Complaints." Washington Post, February 18, 2005, Page A16. http://www.washingtonpost.com/wp-dyn/articles/A33178-2005Feb17.html.
Solanki, S.K., et al. "Unusual Activity of the Sun During Recent Decades Compared to the Previous 11,000 years." *Nature* 431 (2004): 1084–1087. http://www.ncdc.noaa.gov/paleo/pubs/solanki2004/solanki2004.html.
Specter, Michael. "Big Foot: In Measuring Carbon Emissions, It's Easy to Confuse Morality and Science." *The New Yorker*, February 25, 2008. http://www.newyorker.com/reporting/2008/02/25/080225fa_fact_specter.
SRI Consulting. "Carbon Capture From Natural Gas Power Generation." *Process Economics Program Report 180A*, December 2004. http://www.sriconsulting.com/PEP/Public/Reports/Phase_2004/RP180A/#xml=http://www.sriconsulting.com/cgi-bin/texis/webinator/search/pdfhi.txt?query=180A&pr=super&prox=page&rorder=500&rprox=500&rdfreq=500&rwfreq=500&rlead=500&sufs=0&order=r&id=47d2f15f7. http://www.sriconsulting.com/PEP/Private/Reports/Phase_2004/RP180A/RP180A.pdf.
Stanford University Global Climate and Energy Project (GCEP). "Exergy Flow Charts." http://gcep.stanford.edu/research/exergycharts.html.
Steinfeld, H., et al. "Livestock's Long Shadow: Environmental Issues and Options." *LEAD-Livestock Environment and Development Initiative*. Coordinated by the UN FAO's Animal Production and Health Division. Rome: FAO, 2006. http://www.virtualcentre.org/en/library/key_pub/longshad/A0701E00.htm. http://www.virtualcentre.org/en/library/key_pub/longshad/A0701E00.pdf.
Stern, D. I. "Global Sulfur Emissions from 1850 to 2000." *Chemosphere* 58 (2005): 163–175. http://www.ncbi.nlm.nih.gov/pubmed/15571748?ordinalpos=2&itool=EntrezSystem2.PEntrez.Pubmed.Pubmed_ResultsPanel.Pubmed_RVDocSum. http://ideas.repec.org/p/rpi/rpiwpe/0311.html.
Stern, N. *The Stern Review on the Economics of Climate Change*. Cambridge: Cambridge University Press, 2006. http://www.hm-treasury.gov.uk/stern_review_final_report.htm.
Stott, Peter A., Gareth S. Jones and John F. B. Mitchell. "Do Models Underestimate the Solar Contribution to Recent Climate Change?" *Journal of Climate* 16 (2003):4079–4093. http://climate.envsci.rutgers.edu/pdf/StottEtAl.pdf.
Study of Critical Environmental Problems. *Man's Impact on the Global Environment*. Cambridge, MA: MIT Press, 1970.
Study of Man's Impact on Climate. *Inadvertent Climate Modification*. Cambridge, MA: MIT Press, 1971.
Svensmark, Henrik, and Eigil Friis-Christensen. "Reply to Lockwood and Fröhlich—The Persistent Role of the Sun in Climate Forcing." *Scientific Report 3*, Danish

National Space Center, Danish Technical University, 2007. http://www.spacecenter.dk/publications/scientific-report-series/Scient_No._3.pdf.

Svensmark, Henrik, and Eigil Friis-Christensen. "Variation of Cosmic Ray Flux and Global Cloud Coverage—a Missing Link in Solar-Climate Relationships." *Journal of Atmospheric and Solar-Terrestrial Physics* 59, 11 (1997):1225–1232. http://www.sciencedirect.com/science?_ob=ArticleURL&_udi=B6VHB-3SW03B6-H&_user=10&_coverDate=07%2F31%2F1997&_rdoc=1&_fmt=summary&_orig=browse&_cdi=6062&_sort=d&_docanchor=&view=c&_ct=1&_acct=C000050221&_version=1&_urlVersion=0&_userid=10&md5=b2e3b6773a039b6b02396b1b02df3492.

Svensmark, Henrik. "A Brief Summary of Cosmoclimatology." *Astronomy & Geophysic* 48, 1 (2007), Danish National Space Center (DTU). http://www.spacecenter.dk/research/sun-climate/cosmoclimatology/a-brief-summary-on-cosmoclimatology.

Technisches Museum Wien. "Easy Rider." http://www.tmw.ac.at/default.asp?id=758&al=Deutsch.

Technisches Museum Wien. "Die Brennstoffzelle-Die 'kalte Verbrennung'." http://www.tmw.ac.at/default.asp?id=662&al=Deutsch.

Terranol. http://www.terranol.com/.

Tester, Jefferson, et al. "The Future of Geothermal Energy." Massachusetts Institute of Technology, 2006. http://geothermal.inel.gov/publications/future_of_geothermal_energy.pdf.

Ulrich, Lawrence. "Facing New Mileage Rules, Porsche Preps a Hybrid S.U.V." *New York Times*, January 6, 2008. http://www.nytimes.com/2008/01/06/automobiles/06HYBRID.html?_r=2&oref=slogin&ref=automobiles&pagewanted=print&oref=slogin.

Union of Concerned Scientists. "Biodiesel: FAQ." http://www.ucsusa.org/clean_vehicles/big_rig_cleanup/biodiesel.html.

United Nations Department of Economic and Social Affairs. (Economic and Social Council.) "Commission on Sustainable Development-Report on the Ninth session 5 May 2000 and 16–27 April 2001." E/2001/29, E/CN.17/2001/19. http://www.un.org/esa/sustdev/csd/ecn172001-19e.htm#Chapter%20IV. Commission on Sustainable Development (CSD), CSD-9, http://www.un.org/esa/sustdev/csd/CSD9.htm.

United Nations Department of Economic and Social Affairs. Population Division. Population Database. http://esa.un.org/unpp/.

United Nations Department of Economic and Social Affairs. Population Division. "World Population Prospects: The 2006 Revision." http://www.un.org/esa/population/publications/wpp2006/wpp2006.htm. "Highlights." http://www.un.org/esa/population/publications/wpp2006/WPP2006_Highlights_rev.pdf.

United Nations Development Programme. "Energizing the Millennium Development Goals." Knowledge Expo: A UNDP co-hosted event, New York, May 2006. http://www.undp.org/energy/csd06-post_prog.htm.

United Nations Development Programme. "Gender and Energy for Sustainable Development: A Toolkit and Resource Guide, 2004. http://www.undp.org/energy/gendereng.htm.

United Nations Development Programme. Human Development Report 2007/2008. http://hdr.undp.org/en/media/hdr_20072008_en_complete.pdf.

United Nations Development Programme. Statement by Mr. Shoji Nishimoto, Assistant Administrator and Director, Bureau for Development Policy. Energy for Development Conference, Noordjwik-Netherlands, 12–14 December 2004. http://www.energyandenvironment.undp.org/undp/indexAction.cfm?module=Library&action=GetFile&DocumentAttachmentID=1620. http://www.energyandenvironment.undp.org/undp/index.cfm?module=Library&page=Document&DocumentID=5733.

United Nations Environment Programme. "Alternative Fuels." http://www.unep.org/tnt-unep/toolkit/Actions/Tool14/index.html.

United Nations Environment Programme. *Global Environment Outlook, GEO Year Book 2003, International Environmental Agenda*, Box 4: Greening of the Biosphere. Nairobi: UNEP, 2003. http://new.unep.org/geo/yearbook/yb2003/box7a.htm.

United Nations Framework Convention on Climate Change. "Changes in GHG Emissions from 1990 to 2004 for Annex I Parties." http://unfccc.int/files/essential_background/background_publications_htmlpdf/application/pdf/ghg_table_06.pdf.

United Nations Framework Convention on Climate Change. "Kyoto Protocol to the United Nations Framework Convention on Climate Change." http://unfccc.int/kyoto_protocol/items/2830.php. http://unfccc.int/essential_background/kyoto_protocol/items/1678.php.

United Nations General Assembly. "Functions and Powers of the General Assembly." http://www.un.org/ga/about/background.shtml.

United Nations General Assembly. General Assembly Resolution 55/2: United Nations Millennium Declaration (8 September 2000). http://www.un.org/millennium/declaration/ares552e.htm.

United Nations Intergovernmental Panel on Climate Change. http://www.ipcc.ch/.

United Nations Intergovernmental Panel on Climate Change. P. Forster et al. "Changes in Atmospheric Constituents and in Radiative Forcing." In *Climate Change 2007: The Physical Science Basis. Contribution of Working Group I to the Fourth Assessment Report of the Intergovernmental Panel on Climate Change*, edited by S. Solomon et al. Cambridge/New York: Cambridge University Press, 2007. "Table 2.1. Present-day concentrations and RF for the measured LLGHGs," page 141. http://www.ipcc.ch/pdf/assessment-report/ar4/wg1/ar4-wg1-chapter2.pdf.

United Nations Intergovernmental Panel on Climate Change. "Climate Change 2007: The Physical Science Basis." Summary for Policymakers. http://www.ipcc.ch/SPM2feb07.pdf.

United Nations Intergovernmental Panel on Climate Change. "Equilibrium global mean temperature increase above pre-industrial (°C)." In *Contribution of Working Group III to the Fourth Assessment Report of the Intergovernmental Panel on Climate Change*, Summary for Policymakers, May 2007, page 17. http://www.ipcc.ch/pdf/assessment-report/ar4/wg3/ar4-wg3-spm.pdf.

United Nations Intergovernmental Panel on Climate Change. "Fourth Assessment Report: Climate Change 2007: Synthesis Report." Summary for Policymakers. http://www.ipcc.ch/pdf/assessment-report/ar4/syr/ar4_syr_spm.pdf.

United Nations Intergovernmental Panel on Climate Change. "IPCC Fourth Assessment Report." Working Group I Report "The Physical Science Basis." http://www.ipcc.ch/ipccreports/ar4-wg1.htm.

United Nations Intergovernmental Panel on Climate Change. "Fourth Assessment Report: Climate Change 2007: The Physical Science Basis, Summary for Policymakers." http://www.ipcc.ch/SPM2feb07.pdf.

United Nations Intergovernmental Panel on Climate Change. "IPCC Fourth Assessment Report, Second Draft chapter 2." February 28, 2006.

United Nations Intergovernmental Panel on Climate Change. Bert Metz et al., *IPCC Special Report on Carbon Dioxide Capture and Storage*, prepared by Working Group III of the Intergovernmental Panel on Climate Change. Cambridge: Cambridge University Press, 2005. http://www.ipcc.ch/pdf/special-reports/srccs/srccs_wholereport.pdf.

United Nations Intergovernmental Panel on Climate Change. S. Solomon et al. "Summary for Policymakers-Climate Change 2007: The Physical Science Basis," edited by H.L. Miller, Contribution of Working Group I to the Fourth Assessment Report of the Intergovernmental Panel on Climate Change, 2007. http://www.ipcc.ch/pdf/assessment-report/ar4/wg1/ar4-wg1-spm.pdf.

United Nations Intergovernmental Panel on Climate Change. "Trends in Aerosols." Climate Change 2001: Working Group I: The Scientific Basis. http://www.grida.no/climate/ipcc_tar/wg1/180.htm.

United Nations Organization. "UN Millennium Development Goals." http://www.un.org/millenniumgoals/.

United Nations Population Fund. http://www.unfpa.org/index.htm.

United Nations Population Fund. "State of World Population 2007—Unleashing the Potential of Urban Growth." http://www.unfpa.org/swp/2007/english/introduction.html. http://www.unfpa.org/swp/2007/presskit/pdf/sowp2007_eng.pdf.

United Nations Statistics Division—Development Indicators Unit. "Official list of MDG Indicators." Millennium Development Goals Indicators—The official United Nations site for the MDG Indicators. http://mdgs.un.org/unsd/mdg/Host.aspx?Content=Indicators/OfficialList.htm.

United Press International. "China Has Its Worst Spell of Acid Rain." September 22, 2006. http://www.upi.com/NewsTrack/Science/2006/09/22/china_has_its_worst_spell_of_acid_rain/7049/.

United States Department of Defense. ERDC-CERL Fuel Cell Project. "Fuel Cell Information Guide." http://www.dodfuelcell.com/fcdescriptions.html.

United States Department of Energy. "Energy Efficiency and Renewable Energy: Biodiesel Production." http://www.afdc.energy.gov/afdc/fuels/biodiesel_production.html.

United States Department of State. "U.S. Assistance to Kazakhstan—Fiscal Year 2003." http://www.state.gov/p/eur/rls/fs/29487.htm.

United States Environmental Protection Agency. "Sulfur Dioxide." http://www.epa.gov/air/airtrends/sulfur.html.

United States Geological Survey (USGS). Energy Resources Program. http://energy.usgs.gov/

United States Geological Survey (USGS). "Major Volcanic Eruptions Since 1900." Microsoft Encarta Online Encyclopedia. http://encarta.msn.com/media_701500557_761570122_-1_1/Major_Volcanic_Eruptions_Since_1900.html.

United States Geological Survey. "World Petroleum Assessment and Analysis." Washington, DC, various years.

United States Library of Congress. Sen. Joseph I. Lieberman et al. "America's Climate Security Act of 2007," 2191, 2007. http://thomas.loc.gov/cgi-bin/bdquery/z?d110:S.2191:.

United States Senate Energy and Natural Resources Committee. "Coal Conference 2005." http://energy.senate.gov/public_new/index.cfm?FuseAction=Conferences.Detail&Event_id=befd9d70-fcc9-4a81-8517-e3cd45c70ec5&Month=3&Year=2005. "Past Conferences." http://energy.senate.gov/public_new/index.cfm?FuseAction=Conferences.Home.

van der Veer, Jeroen. "Two Energy Futures-On Shell's New Energy Scenarios to 2050, Scramble and Blueprints." January 25, 2008, Speeches 2008. http://www.shell.com/home/content/media-en/news_and_library/speeches/2008/jvdv_two_energy_futures_25012008.html.

Vegetarian Society. "Information Sheets." http://www.vegsoc.org/info.

Vegetarian Society. "Why It's Green to Go Vegetarian." http://www.vegsoc.org/environment/why%20its%20green%20final%20small.pdf.

Veizer, Jan. "Celestial Climate Driver: A Perspective from Four Billion Years of the Carbon Cycle." *Geoscience Canada* 32, 1 (2005). http://www.gac.ca/publications/geoscience/GACV32No1Veizer.pdf.

Veizer, Jan, "Isotope Data," http://www.science.uottawa.ca/~veizer/isotope_data/.

Veizer, Jan, D. Ala, K. Azmy, P. Bruckschen, D. Buhl, F. Bruhn, G.A.F. Carden, A. Diener, S. Ebneth, Y. Godderis, T. Jasper, C. Korte, F. Pawellek, O. Podlaha, and H. Strauss. "87Sr/86Sr, $\delta 13C$ and $\delta 18O$ evolution of Phanerozoic seawater." *Chemical Geology* 161 (1999): 59–88.

Veizer, Jan, Y. Godderis, and L.M. Francois. "Evidence for Decoupling of Atmospheric CO_2 and Global Climate during the Phanerozoic Eon." *Nature* 408 (2000): 698–701.

Vidal, John, and David Adam. "China Overtakes US as World's Biggest CO_2 Emitter." *The Guardian*, June 19, 2007. http://www.guardian.co.uk/environment/2007/jun/19/china.usnews.

von Weizsäcker, Ernst Ulrich, Amory B. Lovins, and L. Hunter Lovins. *Faktor vier. Doppelter Wohlstand—halbierter Naturverbrauch.* München: Droemersche Verlagsanstalt Th. Knaur, 1995, 1996. English language edition: Ernst Ulrich von Weizsäcker, Amory B. Lovins, and L. Hunter Lovins. *Factor Four. Doubling Wealth—Halving Resource Use.* London: Earthscan, 1997.

Washington Post, The. "Tug of War Goes Beyond Chinese Oil Bid." July 10, 2005, Page F02. http://www.washingtonpost.com/wp-dyn/content/article/2005/07/09/AR2005070900109.html.

Wetlands International, "Peat CO_2." December 21, 2006. http://www.wetlands.org/publication.aspx?id=51a80e5f-4479-4200-9be0-66f1aa9f9ca9.

Wetlands International. "Shocking Climate Impact of Wetland Destruction in Indonesia." November 2006. http://www.wetlands.org/news.aspx?ID=2817de3d-7f6a-4eec-8fc4-7f9eb9d58828.

White House, The. "President Bush Discusses Global Climate Change." Press Release June 11, 2001. http://www.whitehouse.gov/news/releases/2001/06/20010611-2.html.

World Bank, The. "Energy." http://www.worldbank.org/energy.

World Energy Council. "Survey of Energy Resources." http://www.worldenergy.org/wec-geis/publications/reports/ser/overview.asp.

World Health Organization. "The Top 10 Causes of Death." Geneva: WHO, 2008. http://www.who.int/mediacentre/factsheets/fs310/en/.

World Health Organization. International Agency for Research on Cancer. "Tobacco Smoke and Involuntary Smoking: Summary of Data Reported and Evaluation, Tobacco smoking and tobacco smoke, Involuntary smoking." *IARC Monographs on the Evaluation of Carcinogenic Risks to Humans* 83. Geneva: WHO, 2002. http://monographs.iarc.fr/ENG/Monographs/vol83/volume83.pdf.

Wuppertal Institute for Climate, Environment and Energy. http://www.wupperinst.org/FactorFour/.

Zachos, James, et al. "Trends, Rhythms, and Aberrations in Global Climate 65 Ma to Present." *Science* 292 (5517) (2001): 686–693. http://www.sciencemag.org/cgi/content/abstract/292/5517/686. http://pangea.stanford.edu/research/Oceans/GES206/readings/Zachos2001.pdf. http://geosci.uchicago.edu/~archer/classes/GeoSci238/zachos.2001.since_65M.pdf.

Ziock, H.-J., and D. P. Harrison, "Zero Emission Coal Power, a New Concept." http://www.netl.doe.gov/publications/proceedings/01/carbon_seq/2b2.pdf.

EPILOGUE

People have been thinking about energy and its role in society for a long time. The first truly elaborate considerations were forwarded in the Coal Age, when it became clear that energy schemes rather than other technological advance set societies most markedly apart in terms of their progress. By the end of the Coal Age the topic of energy had attracted a host of all kinds of writers. Herbert George Wells, for instance, who is better known as H. G. Wells, is mostly noted for his science fiction novels such as *The Time Machine* (1895), *The Invisible Man* (1897), and *The War of the Worlds* (1898). However, in 1902 he published a non-fiction work titled *Anticipations of the Reaction of Mechanical and Scientific Progress upon Human Life and Thought*.[1] In this book Wells investigates the role of steam power and technological advance as a determinant of human progress. He concludes, for instance, that it was "not one cause, but a very complex and unprecedented series of causes, that set the steam locomotive going," and that "It was indirectly . . . that the introduction of coal became the decisive factor." He argues that the coal of England was actually more difficult to reach than elsewhere, which necessitated the introduction of steam power to pump water from mines. In contrast the Chinese, according to Wells, did not need to introduce steam power even though they had been using coal in iron smelting for a very long time: their coal was more easily accessible. On the importance of new energy technology, for transportation, for instance, Wells writes in *Anticipations*:

> No one who has studied the civil history of the nineteenth century will deny how far-reaching the consequences of changes in transit may be, and

no one who has studied the military performances of General Buller and General De Wet but will see that upon transport, upon locomotion, may also hang the most momentous issues of politics and war. The growth of our great cities, the rapid populating of America, the entry of China into the field of European politics are, for example, quite obviously and directly consequences of new methods of locomotion.

... The evolution of locomotion has a purely historical relation to the Western European peoples. It is no longer dependent upon them, or exclusively in their hands. The Malay nowadays sets out upon his pilgrimage to Mecca in an excursion steamship of iron, and the immemorial Hindoo goes a-shopping in a train, and in Japan and Australasia and America, there are plentiful hands and minds to take up the process now, even should the European let it fall.

The beginning of this twentieth century happens to coincide with a very interesting phase in that great development of means of land transit that has been the distinctive feature (speaking materially) of the nineteenth century. The nineteenth century, when it takes its place with the other centuries in the chronological charts of the future, will, if it needs a symbol, almost inevitably have as that symbol a steam engine running upon a railway. This period covers the first experiments, the first great developments, and the complete elaboration of that mode of transit, and the determination of nearly all the broad features of this century's history may be traced directly or indirectly to that process.

H. G. Wells soon returned to fiction-writing, but energy technology, of the productive as well as the destructive kind, and how it influences human development, remained a central theme of his work. In 1913 he wrote a novel *The World Set Free: A Story of Mankind*,[2] which describes atomic bombs, and how they will end the existence of society as it once was. In terms of nuclear physics, the book was based on the contemporary knowledge that radon releases huge amounts of energy, but does so very slowly, because its natural radioactive decay stretches over millennia. Thus, as Wells writes, "The problem ... was already being mooted by such scientific men as Ramsay, Rutherford, and Soddy, in the very beginning of the twentieth century, the problem of inducing radio-activity in the heavier elements and so tapping the internal energy of atoms." In terms of the plot as a whole, the novel provides an arena for Wells to reflect on humankind and energy technology, and to summarize their inter-woven development. He does so, brilliantly, by opening:

THE HISTORY OF MANKIND IS THE HISTORY OF THE ATTAINMENT OF EXTERNAL POWER. Man is the tool-using, fire-making animal. From the outset of his terrestrial career we find him supplementing the natural strength and bodily weapons of a beast by the heat of burning and the rough implement of stone. So he passed beyond the ape. From that he expands. Presently he added to himself the power of the horse and the ox, he borrowed the carrying strength of water and the driving force of the wind, he quickened his fire by blowing, and his simple tools, pointed first with copper and then with iron, increased and varied and became more elaborate and efficient. He sheltered his heat in houses and made his way easier by paths and roads. He complicated his social relationships and increased his efficiency by the division of labour. He began to store up knowledge. Contrivance followed contrivance, each making it possible for a man to do more. Always down the lengthening record, save for a set-back ever and again, he is doing more.... A quarter of a million years ago the utmost man was a savage, a being scarcely articulate, sheltering in holes in the rocks, armed with a rough-hewn flint or a fire-pointed stick, naked, living in small family groups, killed by some younger man so soon as his first virile activity declined. Over most of the great wildernesses of earth you would have sought him in vain; only in a few temperate and sub-tropical river valleys would you have found the squatting lairs of his little herds, a male, a few females, a child or so.

He knew no future then, no kind of life except the life he led. He fled the cave-bear over the rocks full of iron ore and the promise of sword and spear; he froze to death upon a ledge of coal; he drank water muddy with the clay that would one day make cups of porcelain; he chewed the ear of wild wheat he had plucked and gazed with a dim speculation in his eyes at the birds that soared beyond his reach. Or suddenly he became aware of the scent of another male and rose up roaring, his roars the formless precursors of moral admonitions. For he was a great individualist, that original, he suffered none other than himself.

So through the long generations, this heavy precursor, this ancestor of all of us, fought and bred and perished, changing almost imperceptibly.

Yet he changed. That keen chisel of necessity which sharpened the tiger's claw age by age and fined down the clumsy Orchippus to the swift grace of the horse, was at work upon him—is at work upon him still. The clumsier and more stupidly fierce among him were killed soonest and oftenest; the finer hand, the quicker eye, the bigger brain, the better balanced body prevailed; age by age, the implements were a little better made, the man a little more delicately adjusted to his possibilities. He became more social; his herd grew larger; no longer did each man kill or drive out his growing sons; a system of taboos made them tolerable to him, and they revered him alive and soon even after he was dead, and were his allies against the beasts and the rest of mankind. (But they were forbidden to touch the women of the tribe, they had to go out and capture women for themselves, and each son fled from his stepmother and hid from her lest the anger of the Old Man should be roused. All the world over, even to this day, these ancient inevitable taboos can be traced.) And now instead of caves came huts and hovels, and the fire was better tended and there were wrappings and garments; and so aided, the creature spread into colder climates, carrying food with him, storing food—until sometimes the neglected grass-seed sprouted again and gave a first hint of agriculture.

... For scores and hundreds of centuries, for myriads of generations that life of our fathers went on. From the beginning to the ripening of that phase of human life, from the first clumsy eolith of rudely chipped flint to the first implements of polished stone, was two or three thousand centuries, ten or fifteen thousand generations. So slowly, by human standards, did humanity gather itself together out of the dim intimations of the beast. And that first glimmering of speculation, that first story of achievement, that story-teller bright-eyed and flushed under his matted hair, gesticulating to his gaping, incredulous listener, gripping his wrist to keep him attentive, was the most marvellous beginning this world has ever seen. It doomed the mammoths, and it began the setting of that snare that shall catch the sun.

That dream was but a moment in a man's life, whose proper business it seemed was to get food and kill his fellows and beget after the manner of all that belongs to the fellowship of the beasts. About him, hidden from him by the thinnest of veils, were the untouched *sources of Power*, whose magnitude we scarcely do more than suspect even to-day, Power that could make his every conceivable dream come real. But the feet of the race were in the way of it, though he died blindly unknowing.

At last, in the generous levels of warm river valleys, where food is abundant and life very easy, the emerging human, overcoming his earlier jealousies, becoming, as necessity persecuted him less urgently, more social and tolerant and amenable, achieved a larger community. There began a division of labour, certain of the older men specialised in knowledge and direction, a strong man took the fatherly leadership in war, and priest and king began to develop their roles in the opening drama of man's history. The priest's solicitude was seed-time and harvest and fertility, and the king ruled peace and war. In a hundred river valleys about the warm, temperate zone of the earth there were already towns and temples, a score of thousand years ago. They flourished unrecorded, ignoring the past and unsuspicious of the future, for as yet writing had still to begin.

Very slowly did man increase his demand upon the illimitable wealth of Power that offered itself on every hand to him. He tamed certain animals, he developed his primordially haphazard agriculture into a ritual, he added first one metal to his resources and then another, until he had copper and tin and iron and lead and gold and silver to supplement his stone, he hewed and carved wood, made pottery, paddled down his river until he came to the sea, discovered the wheel and made the first roads. But his chief activity for a hundred centuries and more, was the subjugation of himself and others to larger and larger societies. The history of man is not simply the conquest of external power; it is first the conquest of those distrusts and fiercenesses, that self-concentration and intensity of animalism, that tie his hands from taking his inheritance. The ape in us still resents association. From the dawn of the age of polished stone to the achievement of the Peace of the World, man's dealings were chiefly with himself and his fellow man, trading, bargaining, law-making, propitiating, enslaving, conquering, exterminating, and every little increment in Power, he turned at once and always turns to the purposes of this confused elaborate struggle to socialise. To incorporate and comprehend his fellow men into a community of purpose became the last and greatest of his instincts. Already before the last polished phase of the stone age was over he had become a political animal. He made astonishingly far-reaching discoveries within himself, first of counting and then of writing and making records, and with that his town communities began to stretch out to dominion; in the valleys of the Nile, the Euphrates, and the great Chinese rivers, the first empires and the first written laws had their beginnings. Men specialised for fighting and rule as soldiers and knights. Later, as ships grew seaworthy, the Mediterranean which had been a barrier became a highway, and at last out of a tangle of pirate polities came the great struggle of Carthage and Rome. The history of Europe is the history of the victory and breaking up of the Roman Empire. Every ascendant monarch in Europe up to the last, aped Caesar and called himself Kaiser or Tsar or Imperator or Kasir-i-Hind. Measured by the duration of human life it is a vast space of time between that first dynasty in Egypt and the coming of the aeroplane, but by the scale that looks back to the makers of the eoliths, it is all of it a story of yesterday.

Now during this period of two hundred centuries or more, this period of the warring states, while men's minds were chiefly preoccupied by politics and mutual aggression, their progress in the acquirement of external Power was slow—rapid in comparison with the progress of the old stone age, but slow in comparison with this new age of systematic discovery in which we live. They did not very greatly alter the weapons and tactics of warfare, the methods of agriculture, seamanship, their knowledge of the habitable globe, or the devices and utensils of domestic life between the days of the early Egyptians and the days when Christopher Columbus was a child. Of course, there were inventions and changes, but there were also retrogressions; things were found out and then forgotten again; it was, on the whole, a progress, but it contained no steps; the peasant life was the same, there were already priests and lawyers and town craftsmen and territorial lords and rulers doctors, wise women, soldiers and sailors in Egypt and China and Assyria and south-eastern Europe at the beginning of that period, and they were doing much the same things and living much the same life as they were in Europe in A.D. 1500. The English excavators of the year A.D. 1900 could delve into the remains of Babylon and Egypt and disinter legal documents, domestic accounts, and family correspondence that they could read with the completest sympathy. There were great religious and moral changes throughout the period, empires and republics replaced one another, Italy tried a vast experiment in slavery, and indeed slavery was tried again and again and failed and failed and was still to be tested again and rejected again in the New World; Christianity and Mohammedanism swept away a thousand more specialised cults, but essentially these were progressive adaptations of mankind to material conditions that must have seemed fixed for ever. The idea of revolutionary changes in the material conditions of life would have been entirely strange to human thought through all that time.

. . . The latent energy of coal and the power of steam waited long on the verge of discovery, before they began to influence human lives. . . . The mining of coal for fuel, the smelting of iron upon a larger scale than men had ever done before, the steam pumping engine, the steam-engine and the steam-boat, followed one another in an order that had a kind of logical necessity. It is the most interesting and instructive chapter in the history of the human intelligence, the history of steam from its beginning as a fact in human consciousness to the perfection of the great turbine engines that preceded the utilisation of intra-molecular power. Nearly every human being must have seen steam, seen it incuriously for many thousands of years; the women in particular were always heating water, boiling it, seeing it boil away, seeing the lids of vessels dance with its fury; millions of people at different times must have watched steam pitching rocks out of volcanoes like cricket balls and blowing pumice into foam, and yet you may search the whole human record through, letters, books, inscriptions, pictures, for any glimmer of a realisation that here was force, here was strength to borrow and use. . . . Then suddenly man woke up to it, the railways spread like a network over the globe, the ever enlarging iron steamships began their staggering fight against wind and wave.

Steam was the first-comer in the new powers, it was the beginning of the Age of Energy that was to close the long history of the Warring States. But for a long time men did not realise the importance of this novelty. They would not recognise, they were not able to recognise that anything fundamental had happened to their immemorial necessities. They called the steam-engine the 'iron horse' and pretended that they had made the most partial of substitutions. Steam machinery and factory production were visibly revolutionising the conditions of industrial production, population was streaming steadily in from the country-side and concentrating in hitherto unthought-of masses about a few city centres, food was coming to them over enormous distances upon a scale that made the one sole precedent, the corn ships of imperial Rome, a petty incident; and a huge migration of peoples between Europe and Western Asia and America was in Progress, and—nobody seems to have realised that something new had come into human life, a strange swirl different altogether from any previous circling and mutation, a swirl like the swirl when at last the lock gates begin to open after a long phase of accumulating water and eddying inactivity. . . .

The sober Englishman at the close of the nineteenth century could sit at his breakfast-table, decide between tea from Ceylon or coffee from Brazil, devour an egg from France with some Danish ham, or eat a New Zealand chop, wind up his breakfast with a West Indian banana, glance at the latest telegrams from all the world, scrutinise the prices current of his geographically distributed investments in South Africa, Japan, and Egypt, and tell the two children he had begotten (in the place of his father's eight) that he thought the world changed very little. They must play cricket, keep their hair cut, go to the old school he had gone to, shirk the lessons he had shirked, learn a few scraps of Horace and Virgil and Homer for the confusion of cads, and all would be well with them. . . .

Electricity, though it was perhaps the earlier of the two to be studied, invaded the common life of men a few decades after the exploitation of steam. To electricity also, in spite of its provocative nearness all about him, mankind had been utterly blind for incalculable ages. Could anything be more emphatic than the appeal of electricity for attention? It thundered at man's ears, it signalled to him in blinding flashes, occasionally it killed him, and he could not see it as a thing that concerned him enough to merit study. It came into the house with the cat on any dry day and crackled insinuatingly whenever he stroked her fur. It rotted his metals when he put them together. . . . There is no single record that any one questioned why the cat's fur crackles or why hair is so unruly to brush on a frosty day, before the sixteenth century. For endless years man seems to have done his very successful best not to think about it at all; until this new spirit of the Seeker turned itself to these things.

. . . Then suddenly, in the half-century between 1880 and 1930, it ousted the steam-engine and took over traction, it ousted every other form of household heating, abolished distance with the perfected wireless telephone and the telephotograph. . . .

At the close of the nineteenth century, as a multitude of passages in the literature of that time witness, it was thought that the fact that man had at last had successful and profitable dealings with the steam that scalded him and the electricity that flashed and banged about the sky at him, was an amazing and perhaps a culminating exercise of his intelligence and his intellectual courage. . . . 'The great things are discovered,' wrote Gerald Brown in his summary of the nineteenth century. 'For us there remains little but the working out of detail.' . . . Yet now where there had been but a score or so of seekers, there were many thousands, and for one needle of speculation that had been probing the curtain of appearances in 1800, there were now hundreds. And already Chemistry, which had been content with her atoms and molecules for the better part of a century, was preparing herself for that vast next stride that was to revolutionise the whole life of man from top to bottom . . . 'Then that perpetual struggle for existence, that perpetual struggle to live on the bare surplus of Nature's energies will cease to be the lot of Man.'

Wells identifies and summarizes a great many aspects of human and societal development and energy use. He touches upon the fertility transition and the link between population size and idea-creation or the rate of innovation. Later on in the book, he also includes destructive energy in the overall energy command: "All through the nineteenth and twentieth centuries the amount of energy that men were able to command was continually increasing. Applied to warfare that meant that the power to inflict a blow, the power to destroy, was continually increasing." What is more, Wells identifies the problem of transit from one energy scheme to another. He describes that the introduction of atomic energy as a great new source of power also has adverse effects:

> Between these high lights accumulated disaster, social catastrophe. The coal mines were manifestly doomed to closure at no very distant date, the vast amount of capital invested in oil was becoming unsaleable, millions of coal miners, steel workers upon the old lines, vast swarms of unskilled or under-skilled labourers in innumerable occupations, were being flung out of employment by the superior efficiency of the new machinery, the rapid fall in the cost of transit was destroying high land values at every centre of population, the value of existing house property had become problematical.

On the other hand, and though he names domesticated animals a source of power, Wells does not seem to clearly identify agriculture as a source of energy. He actually warns about "the relapse of mankind to agricultural barbarism from which it had emerged so painfully." However, the book's most notable feature in terms of Energy Ages is that Wells sort of skips the Age of Oil by projecting a transition from the Coal Age straight into the Nuclear Age. But this is quite understandable, as Wells wrote this account in 1913, when coal energy was dominating, and much oil energy technology was still nascent.

More surprisingly, a great scholarly book by Fred Cottrell, *Energy and Society: The Relation between Energy, Social Change, and Economic Development*,[3] also puts the importance of coal well above the importance of oil, even though it was published 10 years after the end of World War II, in 1955. (This was incidentally the year before nuclear disaster had to be expected according to Wells's *The World Set Free*.) Cottrell points at the inability of oil to substitute for coal in iron smelting, and explains that "this, together with the relatively small deposits of oil in any one region as compared with coal, and low cost of transportation of oil," allows for the following conclusion: "petroleum is less likely to affect the distribution of population (that is, to cause the kind of relocation that resulted from the adoption of sail or coal) than the great surpluses it produces would at first seem to indicate." (With surpluses he means surplus energy.) Cottrell further explains:

> In the framework offered here energy sources have great significance. We hold that coal is the basic fuel for high-energy society. Any political system based on high-energy technology must, then, either be possessed of coal or operate in a world which guarantees continuous access to coal-bearing regions. Coal is the largest single source of energy being used in the world today; in 1950, for example, coal and lignite supplied 47.5 per cent of all the energy, exclusive of feed and food, which was consumed in the world.

To be sure, this statement ignores that in the United States the share of coal and oil in terms of energy consumption was even in 1950 (at 34 percent each), and that the trend observed during the previous three decades showed the decline of the importance of coal and the rise of the importance of oil in the United States as well as the rest of the world. Perhaps Cottrell viewed oil in terms of destructive more than constructive energy, as he was writing shortly after World War II, a conflict dominated by oil-fueled airplanes and tanks. Or he could not see past the decline of coal as he was an author born in the Coal Age (1903). But the general thoughts he developed about energy and its role in society are nonetheless excellent.

Defining plants and animals as "energy converters," Cottrell was well aware of the role of agriculture as both a source of energy and a driver of social change: "Food raising permits an increase in the number of persons who can be supported from a given piece of land and thus permits an increase in the surplus locally available." Such surplus energy can then "be either widely dispersed (to lead to a general increase in leisure, for instance) or concentrated, to be sacrificed to the gods, or destroyed at the death of a ruler, or expended in the military conquest of areas which themselves yield lower surpluses." Cottrell quotes Forde, who in his 1946 book *Habitat, Economy, and Society*[4] defines three categories of societies: food gatherers, cultivators, and pastoral nomads. According to Forde, "the range of economic and social variation among cultivators is greater than among food-gatherers," and Cottrell concludes that "it is generally true that as the energy available to man increases, the variety of his activities increases." In terms of these activities after the onset of agriculture, Cottrell cites Childe, who in his 1951 book *Social Evolution*[5] defines civilization as "the aggregation of large populations in cities; the differentiation within these of primary producers . . . full-time specialist artizans, merchants, officials, priests, and rulers; an effective concentration of economic and political power; the use of conventional symbols for recording and transmitting information (writing), and equally conventional standards of weights and of measures of time and space leading to some mathematical and calendrical science." Later on, the increased activities following a rise in energy available to people is exemplified by Cottrell through the discussion of the influence of sail ships (a manifestation of wind energy utilization) on the development of trade, and thus societies and the spread of (agricultural) technology. More generally, Cottrell stresses the importance of making choices, and disagrees with "those who believe in the inevitability of progress": "we take the position that the evidence can as well be used to deny as to support the inevitability of transition from low- to high-energy society." He also reflects on energy prices, wonderfully stating: "The great significance of a change in the cost of energy arises from the fact that energy is a part of the cost of achieving all values. It takes energy even to dream."

Cottrell includes various quantitative examples to explain the drivers behind energy choices. He demonstrates in terms of net energy returns why, for instance, it made sense to invest agricultural energy and manpower to mine "good bituminous coal" that "will produce about 3,500 Calories per pound:"

> A coal miner who consumes in his own body about 3,500 Calories a day will, if he mines 500 pounds of coal, produce coal with a heat value 500 times the heat value of the food which he consumed while mining it. At 20 per cent efficiency he expends about 1 horsepower-hour of mechanical energy to get the coal. Now, if the coal he mines is burned in a steam engine of even 1 per cent efficiency it will yield about 27 horsepower-hours of mechanical energy. The surplus of mechanical energy gained would thus be 26 horsepower-hours, or the equivalent of 26 man-days per man-day. A coal miner who consumed about 1/5 as much food as a horse, could thus deliver through the steam engine about 4 times the mechanical energy which the average horse in Watt's day was found to deliver. Little wonder that the iron horse began to replace his organic forebear!

To be sure, Cottrell's book also includes incorrect assumptions and conclusions, but these are largely founded in the circumstances and level of knowledge at the time when the book was written. In the early 1950s not all that much was known about the origins of agriculture, for instance. (Childe wrote in *Social Evolution* in 1951 that all civilizations derive from "the cultivation of the same cereals and the breeding of the same species of animals.") Perhaps more fundamentally, Cottrell wrote at a time when the achievements of the Green Revolution were not yet achieved or fully foreseeable. Already the foreword of *Energy and Society* explains that the introduction of more modern forms of agriculture (draft animals, tractors) forces people to move from the countryside to urban areas, but that "those left behind on farms tend to multiply to the point where they continue to require for themselves all the food the land will provide." This Malthusian[6] trap talk is then put into the context of food imports in western Europe, while later years showed that chemical inputs and modern farm machineries allowed for the rural population to thin out and still to produce milk lakes, butter mountains, and other agricultural oversupplies in the European Union.[7]

Twenty-one years later, in 1976, another energy classic was published:[8] *Man, Energy, Society*. In the foreword to this book, Earl Cook thanks Fred Cottrell, whom he never met, for the various ideas and concepts developed in *Energy and Society*, and M. King Hubbert, now of Hubbert Curve fame, for being a great teacher. Cook quotes M. King Hubbert's *Energy Resources: A Report to the Committee on Natural Resources* of 1962:[9]

> Since energy is an essential ingredient in all terrestrial activity, organic and inorganic, it follows that the history of the evolution of human culture

must also be a history of man's increasing ability to control and manipulate energy.

Naturally, Earl Cook then explains the methods of M. King Hubbert and others to estimate the time and volume of oil peak production, and Hubbert's success to predict the peak year for the U.S.-48 (1970). Cook shows that crude oil discovered in the United States between 1857 and 1934 amounted to 62.0 billion barrels, compared to 77.3 billion barrels between 1935 and 1974, and that the rate of discovery declined, with the only exception to this trend occurring in 1968, when the Prudhoe Bay oil field on Alaska's North Slope was discovered (and added to proved reserves in 1970).

And yet *Man, Energy, Society* is about much more than oil, as it analyzes the influence of animate and inanimate energy sources on human societies from their very beginnings. Cook investigates the work of Sahlins, for instance, who in *Stone Age Economies* (1972)[10] argued that gatherer-hunter societies should be called affluent, because they were satisfied with the products of a work-day of less than four hours and were not frustrated by unfulfilled wants. Moving through time towards ever more advanced energy regimes, Cook touches upon just about every aspect of energy-society interactions, sometimes including political consequences. He reflects on the issue that slavery declined in Europe during the Middle Ages, while inanimate energy technology (waterpower) advanced. He explores the question why Britain rather than China experienced an Industrial Revolution, offering a widely-accepted, but weak, explanation: "The British looked outward for material opportunity; the Chinese looked inward for the elements of a good life." He also observes that a coal-powered North defeated an agricultural South during the American Civil War; that Germany defeated France in 1870 on the basis of its coal technology; and that neither Britain nor the United States fought any high-energy society between 1815 and World War I.

Cook attempts to quantify the transition from one energy regime to the other. He offers a comparison of the energy mix in the United States in 1850 and 1950, for instance: work animals (mainly horses) accounted for 53 percent in 1850, but only one percent in 1950; human work for 13 percent in 1850, but also only one percent in 1950. The energy mix of 1850 further included 12 percent wind power, 9 percent waterpower, 6 percent fuel wood, and 6 percent coal, while coal accounted for 34 percent, oil for 34 percent, and gas for 21 percent by 1950. What is more, Cook reflects on the overall energy efficiency of societies. He maintains that the aggregate efficiency of the energy system of the United States was only about 36 percent by the time he was writing.

Man, Energy, Society also contains reflections on the deadly sides of energy. Cook discusses responsibility, or lack of it, in terms of administering destructive energy technology: "Sales to other countries of arms made in the United States grew from less than $1 billion in 1970 to $8.6 billion in 1974. One of the principal rationalizations was the need to improve the balance of trade, which was becoming increasingly adverse because of expensive oil imports. Such reasoning will lead to an abundance of lethal materiel in parts of the world where territorial conflict is at best merely dormant." Sadly, Cook here correctly foresaw such conflicts as the Iran-Iraq War (1980–1988), with at least one million casualties, but this statement also indicates that this extraordinary book, like any other book, was a product of its time. Cook clearly wrote under the impression of the first oil price shock. He had no way of knowing that oil in 1986, within ten years of the publication of *Man, Energy, Society* would become extraordinarily cheap to fuel an economic boom that lasted until the end of the 20th century. (The international crude oil price collapsed from about $22 to $6 a barrel in 1986). His Preface reads:

> It seems likely that the world will be plagued by energy problems for a long time to come. Such problems-especially if we count food as an energy source-are nothing new. However, there are some new and disturbing aspects to the contemporary energy scene. First, there is the dependence of a considerable part of the human population on a food subsidy provided by the fossil fuels, which are resources that can neither be renewed nor recycled. Second, there is a highly uneven geographical distribution of known fossil-energy resources, a distribution that is incongruent with the map of consumption and is becoming more so. Third, there is a momentum of population growth that threatens to overwhelm the productiveness of energy and food technology. And finally, an increasing rate of accumulation, in the environment, of waste material and unused heat derived directly and indirectly from the eager use of fossil and nuclear fuels.

Cook was writing before the global fertility transition had been achieved or at least before it was fully confirmed. In percentage terms, the peak of global population growth was reached in the years between 1965 and 1970 at 2.1 percent annually, while the peak in terms of the absolute number of people added per year (circa 87 million people) was not reached until the period 1985 to 1990. It was thus more of a concern in 1976 than it is today that global population growth would outstretch the achievements of the Green Revolution. And though Cook shows concern about environmental degradation due to energy use, his book was written at a time when anthropogenic climate change had not yet been fully accepted and defined as a problem. (Still in the early 1970s, many feared a climate cooling rather than warming due to fossil fuel burning. The concern was that an increase in global aerosol formation would be sufficient to reduce the surface temperature to trigger a new ice age.)

According to the insights contained in the works reviewed above, and even more so if we add other similar works,[11] and the additional knowledge that accumulated until the early 21st century, there should be little doubt that the history of the evolution of human culture must indeed also be a history of the increasing ability of people to control and manipulate energy, as M. King Hubbert put it in 1962. However, if we accept that energy technology did in fact impose a major influence on human societal development, and take into account that the progress towards more energy consumption was asymmetric in terms of time-line and geographical distribution, we must also conclude that energy technology had a major impact on political history. But how does the theory of energy technology as a main driver behind societal and political history compare to other theories of history?

Attempts to explain world history along its broadest patterns are rare. Why didn't West Africans sail up north to capture the people of Portugal for slave work? Why didn't the Aztecs conquer Spain? Why didn't Australian aborigines send convicts to England? And why didn't the Chinese colonize America or dump opium onto the British market? To answer such questions in one general theory of history seems to be impossible. The task is thus often dismissed, and those theories that prevail tend to be flawed one way or another. They typically apply to parts of the world only, rather than to all continents, or to certain centuries, rather than all of world history. And they often fail to integrate the story of the 20th century into the overall picture.

Theories that deal with the fates of peoples from different regions and continents target variations in intelligence, culture, environment, and resistance to disease. To claim that groups of people who are worse off or enjoy lower technological standards find themselves in that situation because they are not as bright as others is a simplistic and arrogant assertion that unfortunately had its adherents in many epochs. European Coal Age travelers, for instance, reported of African tribes that counted "one, two, many," which was widely interpreted as these foragers' inferior cognitive and lingual abilities rather than a sign that these groups apparently did not need a full counting system in their lifestyle. However, already back then voices were raised to proclaim that people are essentially the same all over the world in terms of their most fundamental characteristics. In *Descent of Man* (1871) Charles Darwin, for instance, insisted that all human populations exhibit the same "mental and moral faculties":[12]

> Although the existing races of man differ in many respects, as in color, hair, shape of skull, proportions of the body, etc., yet, if their whole structure be taken into consideration, they are found to resemble each other closely on a multitude of points. Many of these are so unimportant or of so singular a nature that it is extremely improbable that they should have been independently acquired by aboriginally distinct species or races. The same remark holds good with equal or greater force with respect to the numerous points of mental similarity between the most distinct races of man. The American aborigines, Negroes, and Europeans are as different from each other in mind as any three races that can be named; yet I was

constantly struck, while living with the Fuegians on board the "Beagle," with the many little traits of character showing how similar their minds were to ours; and so it was with a full-blooded negro with whom I happened once to be intimate.

Nevertheless, the view remained, sometimes more latent and sometimes more explicit, and sometimes in smaller, other times in larger circles, that intelligence is behind observed international and societal power patterns. In recent years such beliefs have been fueled by research that tied statistically significant variations in IQ (intelligence quotient) test results to ethnic background.[13] Hoping that the world had moved beyond the misfortunes of racism that persisted in the 20th century, the first question that comes to mind is what good could possibly come out of such research. This would then be connected to the question what causes these variations. It is widely agreed that the intelligence of individuals is the product of nature and nurture, that is, a combination of genetics and environment. Thus, if adverse environmental factors can be identified, it may be possible to eliminate them. But how, and how much, nature versus nurture contribute to the intelligence of individuals is not entirely clear. We know from the study of identical twins reared apart in different environments that the genetic component is there. In fact, such studies tend to conclude that the heritability of cognitive abilities is large.[14] Similarly, genome researchers reported that certain gene variations have been associated with the highly gifted.[15] On the other hand, it has been shown that environmental factors, from nutrition to all sorts of (other) socioeconomic realities, also influence intelligence as measured by IQ tests. It should thus be expected that people from different parts of the world, some overwhelmingly rural, others highly urbanized and schooled, would exhibit different patterns of intelligence, according to their environmental and societal surroundings and challenges.

In regions of similar standards of nutrition, health, schooling, and so on, variations *within* societies and ethnic groups are much larger than variations *between* them. It has in turn been claimed that even small differences in mean IQ between societies are critical, as they strongly influence the absolute number of very highly intelligent individuals available in a society. (Such individuals are rare according to the normal (bell-shaped) distribution of the overall societal IQ profile.) Highly intelligent individuals are essential for the pace of societal development, so the argument goes, because humans, like anthropoid primates in general, will copy the behavior of the innovators in the population as soon as a beneficial change is demonstrated to them. Opponents would answer, though, that the leaders of societies do not typically come from the pool of the extremely intelligent—they rather exhibit a complex mix of cognitive, communications, and other social skills, just as intelligence generally is but part of many qualities that contribute to life success.

Besides, if history should be explained according to IQ test results, we would have to expect that South Koreans, Taiwanese, or Japanese would rule the world: East Asians score slightly higher than Europeans, who in turn score somewhat higher than Americans on IQ tests.

If we accept that the environment influences people's intelligence, we would also have to conclude that people living in different environments will differ in terms of their intelligence. However, we have no way of knowing how intelligent people were in various regions at various points in time. Still now, intelligence levels (as measured by IQ tests) seem to change in societies that have long been highly schooled and genetically stable. Any given age group (such as, say, the 25-year-olds) scored a lot lower on the same IQ tests in the 1940s than in the 1990s in industrialized countries.[16] This effect is not well explained, but if something like it has been at work during other or all periods of the past, perhaps at times in concert with direct selective pressures, regional differences may well have developed. Mean intelligence levels may have started to diverge between societies migrating to different regions when some had to adapt to environments that allowed only the smartest to survive, or people turned out smarter one place than another because they had access to better nutrition. Alternatively, the life-style or social organization of societies may have differed in respect to exposing children to the kind of mental stimulation that leaves adults with more complex cognitive abilities.[17] But all this is one large pool of speculations that can hardly be verified and would contribute doubtfully to an explanation of world history.

Cultural explanations of world history are similarly speculative. Culture is even more difficult to define and measure than intelligence. If we return to Darwin's assertion of "mental and moral faculties" that are shared by all people on Earth, and thus must have evolved in "primeval times" within tribes ancestral to all of us, we hear why and how such faculties may have emerged (*Descent of Man*, 1871):[18]

> It must not be forgotten that although a high standard of morality gives but a slight or no advantage to each individual man and his children over other men of the same tribe, yet that an increase in the number of well-endowed men and an advancement in the standard of morality will certainly give an immense advantage to one tribe over another. A tribe including many members who, from possessing in a high degree the spirit of patriotism, fidelity, obedience, courage, and sympathy, were always ready to aid one another, and to sacrifice themselves for the common good, would be victorious over most other tribes; and this would be natural selection.

In reality we do not know how cultural features arose in societies, and we can only speculate how culture in turn may have influenced history. Christians, for instance, like to mention a positive attitude in the Judeo-Christian tradition towards work, and its sense of linear time: the latter is contrasted to Hindu or Buddhist cyclicality; the former to Islamic and other cultures' allegedly more pronounced association of work with low social status. And *within* Christian societies, the observation that Catholic southern Europe progressed more slowly than Protestant northern Europe after the Middle Ages has led Protestant writers to glorify a Protestant work ethic.

To be sure, the Judeo-Christian belief system, like that of the Greeks and Romans and all religions that emerged in agricultural societies employing slave work, traditionally put a low value on work and associated status with the kind and amount of work one had to do. (The Judeo-Christian religion actually viewed work as a curse—it was the punishment by God for Adam and Eve's original sin for which they were expelled from the Garden of Eden.) The attitude towards work somewhat improved in feudal, but slave-free, medieval Christian Europe, though work still held no intrinsic value when Protestantism emerged in the 16th century due to protests against Catholic Christian Church authorities. The split into Catholic and Protestant churches began with Martin Luther in Germany in 1517 and continued with Ulrich Zwingli in 1519 in Switzerland. Henry VIII renounced papal supremacy and proclaimed himself head of the Church of England in 1533 after the pope refused to recognize his divorce from Catherine of Aragon. John Calvin established Presbyterianism in Geneva (in the French-speaking part of Switzerland) in 1541. John Knox met Calvin in Geneva and in 1557 founded the Church of Scotland along Presbyterian lines.

Martin Luther put a positive value on every kind of work, teaching that God assigned each person to his own place in the social hierarchy, and that everybody, including the rich, should work diligently without attempting to change the profession to which one was born.[19] However, Luther disapproved of commerce as an occupation involving any real work. Calvin's teachings were similar, but he believed that profits were to be reinvested, and that it was appropriate to change the family trade or profession in search for the greatest profits. Calvinism also features a central theme of predestination, which at first glance would appear to promote laziness: only certain souls, the Elect, are chosen by God to inherit eternal life, while all others were damned, and there is nothing one could do to change her or his fate. However, Calvinists, although they had no way of knowing what their destiny was, supposedly devoted themselves to plain living and hard work in the hope that this would be signs for them to be among the Elect. This led to a new kind of work attitude, according to German Coal Age economic sociologist Max Weber, son of a Calvinist mother.[20] And because various Protestant sects, such as the English Puritans, the French Huguenots, and the Swiss and Dutch Reformed, subscribed to Calvinist theology, it is claimed that economic prosperity achieved in the regions where they settled (including the New World) were connected to their belief system and Protestant work ethic.

To be sure, all these speculations can be rejected. The German example shows that the country's Catholic south has always done better economically than many of Germany's more northern, Protestant regions. And Catholic Wallonia (in present-day Belgium) was the very heartland of coal-fueled industrialization in continental Europe. As far as England's Quakers, Baptists, and Presbyterians (Calvinists) are concerned, there were other reasons for their entrepreneurial spirit that seem just as plausible as the suggestion that they worked hard to show to be among the Elect: Following the restoration of monarchy and strict Anglican state religion in 1660, non-Anglican Protestants were excluded from holding civil and military offices and had to turn towards other occupations, such as banking and manufacturing, where they consequently prospered. (Similarly, Jews of medieval Europe became masters of trade after being excluded from crafts and production. They also excelled in banking and similar professions which were then often disdained by, or forbidden to, Christians. In India, many entrepreneurs have emerged from the Jain community because their strict vegetarianism and historical refusal to work with animal products excluded them from most conventional occupations, hence they turned to trading.[21]) In regards to the important role that various Protestant groups played in the buildup of New England and the United States of America, it is difficult to differentiate between religious-cultural motivations and the power of a hands-on second-chance immigrant mentality. In Britain, the idea that religious beliefs or doctrines helped the Industrial Revolution on its way is discouraged by the fact that many of the most important inventors and entrepreneurs of the time were entirely secular people. Some of them were active in the Lunar Society of Birmingham, which flourished between 1765 and 1809, and promoted the exchange of knowledge and technological know-how. It has also been pointed out that new Protestant belief systems may actually have been an adjustment to a new economic system (capitalism), rather than influencing its emergence. We may even speculate that new religious doctrines have been formulated, newly interpreted, or promoted, to serve the emerging factory system that needed diligent workers laboring under poor conditions for long hours. However, one way or another, it would be rather unlikely that the course of history would have been significantly different if Britain had stayed entirely Catholic.

There are also other, non-religious explanations to account for the emergence of a culture reflected in an entrepreneurial attitude or class that believed in progress and technology to facilitate the Industrial Revolution.[22] One rather strange theory maintains that the wealthy left more surviving offspring than the poor, which led to a downward social mobility that introduced, genetically or culturally, the urge to make and save money in the wider population.[23] Climatic conditions have also been claimed to have helped an allegedly especially progressive culture to emerge in Europe (or more generally in regions of temperate climate). Climate certainly has an impact on the physical work people can do. For physiological reasons, the same runner will perform differently in the thin air of the Andes than somewhere at sea level. Similarly, it has been argued that people will always be able to work harder in temperate climates than in hot tropical climates. However, such environmental differences have at least in part been countered by physical adaptations associated with human groups living in different environments: Wide Andean chests for increased oxygen intake; short, stocky build with a low surface-to-volume ratio to preserve body heat in Arctic regions; long, tender bodies or an increased number of sweat glands to disseminate heat in hot, arid regions.[24] But the impact of climate on culture has also been interpreted in different ways: The challenges of pronounced seasons at high latitudes allegedly shaped the attitude and inventiveness of people. (Think of saving and storing food, and producing warm clothing, for long, cold winters compared to finding mild, fertile conditions all year around.) Similarly, it is said that the extended indoor time during long winter nights provided ample time to think, plan, and innovate. Again, such claims are pure speculations. First of all, Europe imported most of its fundamentals, including farming technology, draft animals, metallurgy, writing systems, and religion, from warm southwestern Asian regions. In more recent centuries many innovations emerged in the relatively mild regions of coastal northwestern Europe, where the ocean (and more specifically the Gulf Stream) serves as a heat reservoir and buffer that obviates the large seasonal temperature variations typical for the continental climate of Europe's interior. You may also ask yourself what Scandinavians, who populate Europe's very north, with the longest winters and winter nights, have contributed to the world: But the relative lack of famous Nordic people should not be interpreted as some kind of genetic or environmental shortcoming. Even for highly developed economies it is difficult to pioneer much novel technology (or art) if the total population size is small. Norwegians, for instance, populate an area that is only slightly smaller than Germany, but they are only 4.6 million, compared to Germany's 82 million inhabitants (2008). The same line of argument would explain why Eurasia, more than other continents, would indeed have been expected to be a hotbed of invention and innovation: Eurasia is the planet's largest landmass and has throughout the Agricultural Age—and likely some time before—been home to a lot more people than other continents: more people, more brains, more innovation, and population centers close enough to one another to allow for ideas, inventions, domesticates, and more to spread from one end of the continent to the other.

Population size and density are also behind the germ theory of world history. Eurasia provided the most people as hosts for infectious diseases that thrived in settled societies. Centuries of devastating epidemics eventually left the surviving (and now rather rapidly growing) Eurasian population largely immune against diseases that were entirely unknown in the Americas and Australia. Hence it is argued that the rapid European conquest of these continents was mainly the effect of a sort of unconscious biological warfare. Africa, in contrast, evaded European domination for a long time. Africa was not isolated from Eurasia and many of its people had acquired immunity against the classic Old World diseases. It was actually more the other way around: tropical African diseases such as malaria and yellow fever seriously troubled Europeans attempting to enter the dark continent, as they called it.

The problem with the germ theory, though it sounds immediately plausible, is that it is based on very limited statistical data. We have no reliable information about Native American population size shortly before and after the arrival of the conquistadores. The Spanish loved to torture, kill, rape, or enslave Native Americans on the largest possible scale, which must have disrupted the social infrastructure and food supply of the primary survivors. So who is to say how much of the Native American death toll was due to Spanish cruelties and their aftermath, and how much to nasty Old World diseases? Besides, how much different would the course of history have been in the absence of smallpox, measles, and influenza? Very likely, Spanish iron, horses, and military skill would have been sufficient to annihilate the natives who were unfamiliar with hard metal and animal-back riding. In fact, the Andean Incas were allegedly a lot less susceptible to Old World diseases than the natives of the central Mexican valley. It did not help them much. They were defeated even faster than the Aztecs, and most of those who survived the initial bloodbath were in turn worked to death.

This is not to say that infectious diseases did not play a significant role in world history. They certainly did, sometimes more, and sometimes less, directly. The unprecedented wave of epidemics that *plagued* Eurasia in the early centuries C.E. contributed greatly to the downfall of the Roman Empire in the west and China's Han dynasty in the east. These epidemics are also associated with the rise of Christianity in the Mediterranean and of Buddhism in China. Both these religions found plenty of new followers looking for consolation when it became obvious that Chinese Confucianism and the established pantheon of Roman gods could do nothing to avert the disease death of millions. The spread of Islam was also facilitated by epidemics, which happened to weaken the Byzantine and Persian Empire in the 7th century. And the dramatic bubonic plague epidemic of the 14th century (which killed a third of all people living between India and Ireland) may have depopulated Europe just about enough to allow for the introduction of the kind of animal power that made the European agricultural system the most labor-productive in the world. In more recent times, the influenza epidemic of 1918 killed 400,000 civilian Germans: without it, the popular uprising against the government that led to Germany's World War I capitulation might never have taken place.[25]

No influenza epidemic, no peace Treaty of Versailles, no World War II. Who knows?

If we return to the size (and density) of Eurasia's population, which apparently contributed to both its inventiveness and its disease resistance, we need to ask why Eurasia became home to so many people in the first place. Foraging societies arriving from Africa may have multiplied according to the size and natural resources of Eurasia, but this would still leave the question open, why millennia-long foraging with small unconcentrated populations would give way to large, innovating agrarian societies, and why this happened in Eurasia a lot earlier than on other continents. Here the environmental explanation popularized by Jared Diamond's *Guns, Germs, and Steel* may be used: Eurasia was not just large, but also exceptionally well-endowed with species that lend themselves well to domestication. Besides, Eurasia has much territory in temperate climate zones, where agriculture is an entirely different trade than in tropical and subtropical regions, where the soil in itself does not tend to be as fertile, and the absence of a real winter season allows pests to thrive and proliferate uninterruptedly throughout the year. What is more, Eurasia shows a pronounced west-east orientation, especially when compared to the Americas, which allowed domesticates to spread rapidly far afield into climatically similar regions even before the Opening of the World.[26] But here we have already reached the limits of the explanatory power of this model of world history in terms of its time-frame. It explains the history of Eurasians in relation to people on other continents during the Agricultural Age, but it cannot shed light on the intra-Eurasian developments leading to, and following, the Opening of the World. Why did western Eurasians (Europeans) rather than eastern Eurasians (the Chinese) colonize the Americas (and other regions around the globe), and why was it *they* who experienced an Industrial Revolution that put them ahead even of the most productive eastern Eurasian agrarian societies?[27]

A BETTER FRAMEWORK OF HISTORY

And here we are back at the theory presented in *Sources of Power*. When world history is explained in terms of energy technology, the entire spectrum of human history (and prehistory), from the earliest foragers until the 21st century can be interpreted. The energy theory of history unfolds on different levels:

First, *Sources of Power* provides a framework to describe and categorize human history. Traditionally, historians divide human history into prehistory and history, which are then further subdivided. This system calls for reform. It contains various categories and subcategories that reflect outdated approaches to history, with some divisions being arbitrary and meaningless.

Let's start with "prehistory." This term, coined by French Coal Age scholars, literally means before history, even though it is defined as all of human history up until written records appeared (about 3000 B.C.E.). Prehistory is then subdivided into the Stone Age, the Bronze Age, and the Iron Age, which was the suggestion of Danish Coal Age archaeologist Christian Jürgensen Thomsen, who in the 1820s was confronted with the task of having to classify artifacts for a museum collection.[28] His approach may have been a fine idea at the time, and for his purpose, but is it still meaningful today? Scientists now have plenty of methods to investigate and categorize non-stone and non-metal (and other non-written) artifacts found at archaeological sites. Even more importantly, how does prehistory and the three-age system deal with agriculture? Scholars may not have been aware of it in the Coal Age, but the emergence of agriculture is now considered the most critical milestone in human history. Why would we hide it in the pre-literate history category, under Stone Age, combined with gatherer-hunters?[29] Based on the importance of energy technology in terms of human societal development, the categories "Foraging Age" and "Agricultural Age," as suggested in this book, are a lot more adequate to provide an overall framework that can in turn be subdivided to tune in on the more specific time-periods. It would also remove the severe problem that the three-age system does not meaningfully describe progression of societies outside (western) Eurasia. We should not, for instance, describe the Mesoamerican Aztecs and Maya, with their high population densities, irrigation systems, writing systems, calendar systems, complex religions, and so forth, as "Stone Age cultures" just because they did not know hard metals. They were, of course, agrarians.

When we proceed to what traditional historians call "history," we will find that it is, like prehistory, divided into three subcategories: Antiquity, the Middle Ages, and the Modern period.[30] Again, we will see that these categories are mainly based on events in western Eurasia. The start of ancient history (antiquity) is defined according to the appearance of writing systems, while its end, the beginning of the Middle Ages in Europe, is set to around 350 C.E., when the Roman Empire began falling apart.[31] It is less well defined when antiquity actually ended in other parts of the world. The term Middle Ages has been coined by European Renaissance scholars to describe pejoratively the period bridging classical Greek and Roman times, and their own era, which they perceived as culturally progressive. Later on, the border between the Middle Ages and the Modern Period has been set at circa 1450 C.E. (typically associated with the invention of printing in Germany, or alternatively with Columbus' first trip to America in 1492 CE.) However, the idea that "modernity" should have started in the 15th century can today only sound odd. (The meaning of modern refers to contemporary and will always reflect new standards.) As the Modern Period continues until this day, it comprises eras that are entirely different in terms of just about everything, from energy regimes and societal organization to population size and life expectancy. Again, it seems a lot more useful to apply the concept of Energy Eras also to these recent centuries: they simply provide a better framework to describe them.

That said, any attempt to categorize a continuous development into distinct phases will have its difficulties. Categorizations are introduced to handle complex or large matters, but such efforts can never be perfect. There is more than one way to set up an Energy Eras system of world history, and there are various challenges to be addressed. Some may argue, for instance, that the command of fire marked the start of a new Energy Era, while I decided that this event did not trigger social changes on a scale large enough to justify calling this the beginning of a new epoch. However, the command of fire would be a target point to subdivide the Foraging Age, especially when a wider consensus emerges in terms of when this actually happened, and what the consequences really were.[32] Similarly, the Agricultural Age calls for subdivisions beyond the Super-Agricultural Era. These may be based on any significant agricultural technology such as irrigation, hoe versus plow, new plow designs (such as the asymmetrical curved moldboard, which reached Europe from China in the 17th century), animal power in agriculture, harnessing technology (rigid horse collar), crop rotation systems, adoption of new domesticates (such as the rise of rye in more northern regions of medieval Europe); in short, anything that boosted agricultural output significantly enough to provide agrarian societies or regions with a competitive energy advantage.

In terms of the more recent Energy Eras, some may wish to bundle the Coal Age and Oil Age into a Fossil Fuel Age, as I have loosely done here and there in the book. However, this would disguise the transfer of global political and military power from coal-fueled Britain to oil-fueled America and Soviet Union. What is more, a Fossil Fuel Age may be said to have begun long before the Coal Age, when the Dutch started fueling their industries with peat, and the Netherlands emerged as the uncontested wealthiest region in the world. Then again, not everyone would agree that peat should be classified as a fossil fuel. Finland, for instance, with its widespread peat bogs, classifies peat as a slowly renewing biomass fuel. (This is in contrast to the general European Union view that categorizes peat as a fossil fuel.) More importantly, we need to ask how much of the Dutch wealth was actually due to peat usage. Did peat confer any military advantage? Did it gain global significance? We know that much income of the Dutch economy at the time was from trade, based on wind-powered shipping, and that thousands of windmills were operating in the Netherlands. Should we define a Wind Power Era? Or a WaterPower Era that pays tribute to England's early water-powered industrialization? The case against wind or waterpower to define a whole Energy Era is that they are not entirely universal, and rather intermittent, sources of power.³³ Both provided productive energy for stationary mechanical work, and wind power boosted waterborne mobility, but transport on land is supported by neither the one nor the other.³⁴ Consequently, the contribution of water or wind power to the total global energy mix has never been especially high.

Another problem with the Energy Era system is geographical. I defined the Energy Ages according to a global first-mover timescale. Accordingly, I ascertain that the Agricultural Age began when agriculture emerged in the Fertile Crescent, though most parts of the world remained foraging areas for millennia to come. Similarly, the Coal Age started when coal energy became relevant in Britain rather than the world as a whole. To be sure, there is also room for a difference in opinion when this actually happened. Some may claim the Coal Age started with Darby or Newcomen technology, many may want to choose the 1769 Watt patent of the external condenser, while others again may prefer the year 1776, when the first Boulton & Watt steam engine was completed, or when such engines spread into commerce during the following decade. I have, however, opted for circa 1800, the year the patent on the separated condenser finally expired, which allowed for the steam engine to quickly evolve into a more universal prime mover.

Arguably, it may be even more difficult to set a border between the Coal Age and the Oil Age. Here I tend to favor the time when the share of coal in the world's total energy mix began to decline in the second decade of the 20th century. However, it might also be argued that the production of the first proper gasoline-fueled cars or airplanes in the beginning of the 20th century mark the start of the Oil Age. Another option would be the year 1912, when the British Admiralty began to convert the Royal Navy from coal to oil, though the first of the new all-oil class of battleships was only completed in 1915. But anyone who has read the passage from Fred Cottrell's 1955 book cited above would probably hesitate putting the beginning of the Oil Age (much) before World War I. After all, oil did not pass coal globally before 1965 in terms of share in total energy consumption.

No matter which Energy Era is concerned, there have always been regions that achieved the transition into the new Energy Era later, or not at all. In case of such regions I have sometimes argued that "they had not yet entered the Coal Age," for instance, even though I described a time when the Coal Age had long begun according to the global (first mover) definition.³⁵ And when does a region truly enter a new Energy Era? Did the laying of railway tracks organized by the colonial overlords mean that India entered the Coal Age in the 19th century? Was the Oil Age entered by those African countries that use huge military budgets to operate oil-fueled imported fighter jets while their populations own hardly any cars per capita and a paved road network is practically non-existing outside urban areas? I would argue that a new Energy Era has only been entered when large parts of the population have also experienced the social transitions associated with the new Energy Era.

To be sure, it will not be easy to reform and replace the classic categories of history. Even though the advantages of the Energy Eras system seem obvious, and the listed problems can be, or have been, overcome, there will be opposition, as there are major vested interests at play. If you were running a well-established "Journal of World Prehistory," for instance, or a publication called "Post-Medieval Archaeology," you would probably not be amused about any attempts to change the classifications currently established in the study of history. Nevertheless, it needs to be done.

EXPLAINING HISTORY

At the next level, the energy theory of world history, as presented in this book, attempts to explain history. Energy technology, as an integral part of societal development, provided advantages that altered the competitive position of societies compared to one another. The role of agriculture as a technology to provide surplus food to trigger professional specialization (and a lot more) is well established (though agriculture is not always understood as an energy system). There was but one principal problem involved. As economies throughout the Agricultural Age were based on animate energy, their basic fuel was edible. Economic growth was thus connected to increased grain harvests that also fueled human population growth.[36] There is wide consensus that societies throughout the Agricultural Age were caught in the Malthusian trap as described by Thomas Malthus in 1798 in *An Essay on the Principle of Population*. Populations, so Malthus said, tend to grow without end, while food supply would eventually be limited by finite land. Put differently, any gains in agricultural productivity would be answered by accelerated population growth that leads people back to the edge of starvation. And indeed, the calorie intake of average people in England in 1800 was well below the levels enjoyed by average gatherer-hunters. Malthus thus named wars, diseases, and famines "positive checks" that increased mortality to adjust population numbers to fit the prevailing food supply.[37]

However, Malthus happened to write right at the transition into a new Energy Age. Historians and economists widely agree that the Industrial Revolution provided for an escape from the Malthusian trap, as it triggered productivity growth that outpaced population growth. This allowed average income to rise, which paved the road from ancient poverty to wide-spread affluence. Various explanations have been forwarded to explain why productivity would quite suddenly increase so much. They usually target economic or political transitions, and changes in institutions or people. However, the most obvious and plausible explanation is that the transformation of energy regimes was behind the productivity increase, causing a host of secondary effects. Some may also prefer to look at it this way: it became possible to escape from the Malthusian trap through a transition from edible to inanimate energy, first towards more waterpower, then, more critically, to coal energy. Growth of production output no longer had to be fueled by increases in agricultural output, which allowed for an accelerated rate of productivity growth without further population growth.[38] In turn living standards in different regions diverged radically. At the end of the Agricultural Age, in 1800, the wealth disparity between the richest and the poorest countries was no more than 4 to 1. Today, after some regions rushed, first, through the Coal Age, and then into the Oil Age, this disparity is more than 50 to 1 (or, depending on the way of measuring it, at least 20:1).

To be sure, the wider energy patterns of history are sometimes easier to identify, explain, and sometimes even to quantify, than the underlying energy realities. *Sources of Power* includes a concise political world history as an invitation to view world history through energy goggles, but not all energy aspects can be immediately identified as such or be verified or quantified in terms of their role and impact. There simply has not been that much focus on some of these issues, and more research is required. Is it true, for instance, that the introduction of a large share of animal power in European agriculture, which depends on surplus land, was the effect of large parts of the human population being wiped out by Black Death epidemics? Was the impact of the global agricultural revolution in terms of providing more nutritional energy a lot more pronounced because it helped agrarians to cope with climate change during the Little Ice Age? Had the use of coke in iron-smelting and the use of Newcomen steam engines in the mining districts spread enough by and during the Seven Years' War to give England an advantage in this conflict that set the stage for Britain to become a globally dominating power? Did the introduction of time-saving electrical household appliances really contribute much to the emancipation of women? Can it be coincidence that the dollar-based Bretton Woods exchange rate system collapsed in 1971, right after the U.S. oil production peak of 1970? Was the economic boom of the 1990s, which is widely associated with the rise of the Internet, not simply the result of low oil prices at that time?

To be sure, the theory of *Sources of Power* is not deterministic. Determinism as such is an outdated concept in many disciplines, including physics (since the advent of quantum mechanics), biology ("nature *and* nurture"), as well as history. The idea that all events or actions are inevitably predetermined ("fatalism") probably never had much appeal, accept for some religious groups. However, environmental determinism (be it climatic or geographical) is not especially popular either. It implies that people are guided strictly by stimulus-response, reacting in one particular way under given circumstances.[39] Historians nowadays, and rightly so, tend to reject this concept, and call any notion of determinism simplistic. However, to take a "decidedly non-deterministic"[40] view may be simplistic as well. For once, it may obstruct the view of causal connections. It also means to subscribe to a concept that is in itself simplistic. In complex situations, we cannot simply claim that a response will follow or will not follow a stimulus. There might be various responses and we have to view possible outcomes in terms of probabilities. Sometimes responses will depend on various stimuli. If all but one are in place, the occurrence of the last stimulus will indeed determine if the response will follow. Besides, certain aspects of determinism would probably appeal to all historians, as determinism maintains that the present would be different if the past were different. In philosophy, the meaning of determinism has changed. Only hard-liners would view determinism and free will as mutually exclusive. Compatibilists believe the two ideas can be coherently reconciled. We could thus argue that the theory of *Sources of Power* is not non-deterministic either. As I outlined in the introduction to this book, the standards of energy technology do indeed determine how much work can be done in a society, but it is up to people how they use their energy, how much of it, and what for. People would often want to maximize their energy consumption, or, especially if energy is not abundant, to improve their efficiencies to get the most use out of the available energy. Thus far, there have always been limits to the energy flows available to societies in useful, concentrated form. In the Agricultural Age these flows have been ultimately limited by solar output, and in turn by photosynthetic efficiency, for instance, or by the wind conditions in an area, or by the rivers flowing through a region. Environmental factors have thus indeed determined the limits of energy use and human activity. And though perspectives changed a lot in the Fossil Fuel Age, the limited energy content of mined fuels still imposed restrictions. If the realities of energy technology allowed it, we would, for instance, quite likely opt for some kind of flying vehicle for our daily individual mobility, rather than to drive around in cars.

Sometimes it can be an illustrative experiment of thought to take a theory to its extremes. This can typically be done at two ends. If we ask what would happen if humanity had no energy resources at all, the answer would be easy. We would not have to worry about anything any longer, as humankind would cease to exist. If, however, we ask what would happen if people had unlimited amounts of (environmentally friendly) energy in concentrated, useful form at their disposal, we may hope to discover new aspects of the theory of energy. With all energy restraints removed, what other factors would become most determining? Would people live happily side-by-side and enjoy the prosperity that comes with large energy supplies? Or would they start fighting, trying to regain historically lost territory, or gain control over the most habitable regions?

Answers to these questions would strongly depend on the precise scenario. If unlimited energy, useable for productive as well as destructive purposes, were to become available to some earlier than others, and the technology was difficult to copy or adopt by others, then the first mover may well attempt to dominate the rest of the world. We may, of course, hope that moral standards would prevent outright brutality. Historically, those in possession of more destructive energy have at best been as friendly as they could afford to be without compromising their standard of living. If the rest of the world would no longer in any way be a threat to the interests of the first mover, violence may certainly be avoided.

If we turn to peaceful scenarios, the large influx of energy command would likely further accelerate knowledge creation and technological advance. As this would include the health sciences, we might experience a kind of Malthusian backslide. Food would likely not be the problem in a world full of energy, but people would start to live even longer. The rise in record national life expectancy since 1840 was about three months per year on average. If this trend would continue only at this pace, the record would be 100 years in 2060. But obviously people restricted to one planet cannot do both, live forever *and* have children. Birth-rates have traditionally been relatively high in low-energy societies, and relatively low in high-energy societies. However, we have also observed that the very wealthy in high-energy societies tend to have more-than-average children. In the population context, even three surviving children per couple is a lot, because it is fifty percent above the replacement level of two children per two parents. As technology now allows it, we have in recent years observed a rising number of women giving birth in their 50s. If increasingly healthy, long-living women in an energy-rich world of affluence then decided to have more children, perhaps additional ones when they are over 60 years old, all areas of the world might eventually become so densely populated as to turn into unpleasant habitats. But then again, with unlimited amounts of energy, we would also be able to populate the oceans and put colonies into orbit or on other planets.

Returning to reality, insights gained from the energy theory should serve as a tool of analysis of historical and current events. Established explanations may be seen in a new light as soon as they are run through an analysis of energy restraints. What if, for instance, Europeans during the Coal Age had had weapons of mass destruction? They might well have cleared all natives out of the fertile areas of South and East Asia, just as they did in North America, where the task could easily be achieved with coal and superior agricultural energy. The current view is, however, that Europeans were merely attempting to dominate (and not to remove) the Asian population centers in order to sell their goods on these markets. This is an explanation based on what actually happened, not on what energy realities were dictating. In terms of the energy theory of history, the "market solution" may have simply been the second best choice, as the energy limitations of the time allowed Europeans to threaten and defeat Asians, but not to annihilate them (or to force eventual survivors to permanently live in reservations on marginal land).

RECONCILING OTHER THEORIES

To be sure, the Energy Theory of World History does not affront other theories of history. Quite on the contrary, it tends to explain them better and shows how they are embedded in a larger picture by integrating the different pieces of the puzzle into a grand view of world history. Those who emphasize the importance of agriculture, and the delays in the onset of farming on different continents, will see their concept in a wider framework of energy technologies. They will thus find their theory reconciled and merged with theories explaining the divergence between China and Europe through the Industrial Revolution. Those who like to focus on the role of climate may adopt the view that climatic conditions had the largest impact in terms of energy flows in societies (agriculture, waterpower, wind power), while those who keep insisting that intelligence and culture were involved may interpret these features in terms of responses to such flows.[41] Finally, those who stress the impact of diseases on world history will also find it useful to reanalyze their theory in light of energy realities. The 1918 influenza epidemic mentioned above may have killed more people than both World Wars combined, and yet it is often not even mentioned in traditional history books.[42] Similarly, the epidemics hassling Coal Age cities in the mid-19th century often get little attention. Why? Because the germ theory of history is mainly relevant to the Agricultural Age, when epidemics, while the human suffering was the same as later on, constituted the destruction of societies' main prime mover capacity. In his book edited by William H. McNeill and Ross E. Dunn, Vaclav Smil, for instance, makes the case against energy as the principal factor in history by pointing out that the fall of the Roman Empire—by far the most studied collapse in history—is explained by various different reasons such as "social dysfunction, internal conflict, invasions, epidemics, or climate change."[43] However, if we identify epidemics (and possibly climate change) as the factor that weakened the agrarian energy base, then social dysfunction, internal conflict, and inability to deal with invasions would certainly follow. In later Energy Eras epidemics would have never disrupted societies as much, because their economic foundation was increasingly based on prime movers indestructible by disease. Much work could still be done if only enough people survived to handle fossil-fueled apparatus, for instance, including the agricultural machinery used to produce food for the survivors. In principle, the argument also holds good for societies of the late Agricultural Age that shifted much work, especially in agriculture, to animals immune to epidemics affecting humans. One of the factors that J. R. McNeill names to explain human progress is "luck: in the eighteenth century a large part of the disease load that checked our numbers, and our productivity too, was lifted. Initially this had little to do with medicine or public health measures, but reflected a gradual adjustment between human hosts and some of our pathogens and parasites."[44] It is correct that the human population grew (about ten times) faster during the Super-Agricultural Era than before, and that this accelerated growth resulted mainly from a fall in death rates, and especially from a decrease in the frequency and severity of episodes of catastrophic mortality.[45] But even though the precise reasons for this trend are still debated, there was certainly more behind it than "luck," when it came to the acquiring of more immunity against disease. The view that medicine had little to do with the decreased mortality of the 18th century seems to be shared by many historians, who tend to point out that Edward Jenner published his experiments on vaccination by cowpox to protect against smallpox only in 1798. However, the technique of variolation had spread to Europe from China through the Ottoman Empire (and thus to all of Eurasia's principal population centers) already by the early 18th century, and was promoted in England as well as New England from the 1720s.[46] Meanwhile, the global agricultural revolution likely helped to improve the general health of people, as better nourished people are less susceptible to disease.[47] And then it may have helped the survival of people significantly in some regions in the Super-Agricultural Era that they had shifted quite a bit of work towards animal, water, and wind power. If we consider animal power in agriculture, even a relatively small such shift may have had a large impact to decrease the severity and impacts of epidemics. As more prime mover capacity survived to work the fields (assuming that horses and oxen were not affected by the epidemics), the extent to which famines wiped out a large number of primary survivors would have been significantly reduced. (The same argument holds for wind and waterpower to help primary survivors to mill grain, for instance. It also applies to other deadly secondary and tertiary effects that tended to follow epidemics.) Infectious diseases would thus have started to become less epidemic and more endemic, as envisioned by William H. McNeill.[48] And this energy aspect adds a facet to the views of Alfred W. Crosby, who considered the "population growth of the post-Columbian era . . . the most impressive single biological development of this millennium."[49]

EXPLAINING THE EMERGENCE OF ENERGY TECHNOLOGY

One question that remains is why energy technology emerged in the first place. Here, I would start by offering a general explanation. As I have pointed out in the early chapters of this book, humans, like all species, are subject to the principal energy imperative that they can only survive if they are able to extract more energy out of their environment than they expend in the process of doing so. The urge or instinct to be efficient, to maximize the return on any work effort, would thus be an intrinsic part of humans, evolutionarily installed, if you will. This attitude was likely a key factor in the emergence of agriculture, and may have later promoted efficiency improvements and the shift towards inanimate energy.

If we ask more specifically why agriculture and coal culture emerged,[50] why in particular places, and why at particular times, then the energy theory of history benefits from being somewhat of a synthesis of other theories: these issues have been studied, separately, for a long time by scholars of various fields. There are usually three categories of explanations. Either environmental conditions (let us include the societal and technological environment) are claimed to have been so rich or beneficial as to allow or prompt the next energy step to be taken; or environmental conditions were so poor as to push people towards taking the next step (necessity as the mother of invention); or environmental conditions were so beneficial in terms of the older, well-functioning energy regime that they prevented people from taking the step into a new era. All of these approaches can seem plausible, if skillfully put into the right context, and all of them can be discouraged, if tested under a different light. Applied to the emergence of agriculture, the different explanations may sound something like this, for instance: abundant large-seeded grasses tempted foragers to experiment actively with them and plant them; or a climatic downturn reduced edible plant growth in the wild, and perhaps led to the extinction of prey species, forcing foragers to become farmers; or a stable supply of uncultivated food species prevented spoiled foragers from investing into efforts that would have led to farming.[51] Applied to the emergence of coal culture, analogous reasoning is used: abundant and easily accessible coal invited agrarians to use it; or scarcity of fuel wood forced agrarians to use coal, and in turn the occurrence of coal in great depths forced them to invent coal-fired pumps to remove water from mines; or the established system of agriculture and production, perhaps with the availability of lots of virgin land as well as cheap labor, prevented efficient agrarians from being tempted to make investments into new energy regimes.

To be sure, it is sometimes difficult, if not impossible, within complex settings to identify which effects are merely coinciding and which are causally connected. It may also be difficult to differentiate between cause and effect. Increased energy flows may create a certain culture in a society, or a certain culture may seek to increase the harvested energy flows. The complex nature of things allows for the arguing in circles, and the chicken-and-egg problem permits everyone to pick whatever explanation she or he likes best. This has, for instance, become obvious in the discussions of why Europe rather than China experienced an indigenous Industrial Revolution.[52] This issue is often presented in an unfortunate way by asking, "Why did China fall behind?," which is certainly the wrong question to ask, at least, if it is not modified by "temporarily." We are just at one point in history, not at its end. You can ask why Europe had fallen behind China by the early Middle Ages just as much as you can ask why China had fallen behind the United States and western Europe by the Coal and Oil Age. In both cases one has been learning from the other's innovations to get ahead, and perhaps (or even likely?) the question will be reversed again by the end of the 21st century.

INTO THE FUTURE

It is by this stage unclear what the later decades of the 21st century will bring in terms of technology and social organization, but one thing is certain: the future, as the present and the past, will be affected by energy limitations. The trend towards more energy diversification will continue, as societies have learned the lessons from their energetic past. They will develop new energy regimes before the unsustainability of the old ones lead to collapse. A slow demise of the Mesopotamian kind will not repeat itself, as such threats as the finiteness of oil resources or global climate change are being taken seriously and are constantly reevaluated. Coal liquefaction technology is already in place, and nuclear fission energy holds much further potential, including the use of more abundant thorium instead of uranium as raw material. If a breakthrough in nuclear fusion technology can be achieved to allow its spread into commerce, the cards of the energy game will certainly be redistributed.

One of the consequences of the sort of semi-determinism inherent to the energy theory of history is that its predicting power diminishes whenever a grand new energy scheme enters the scene, because the overall patterns of energy command are redefined. While there were clear restrictions as to how far Napoleon could march his troops, and how long Hitler's Germany could keep fighting once it had failed to reach the oil resources of Azerbaijan, the course of history would have likely been different if Germany's, rather than America's international team of scientists had completed an atomic bomb during World War II. Similarly, the introduction of horses in warfare entirely changed the situation in the wider Fertile Crescent area. The good news in this respect is that energy technology tends to be in the making for a while before it becomes relevant. This should avoid sudden surprises. Besides, the H.G. Wells view, expressed in the foreword of the 1921 edition of his originally 1913 book *The World Set Free*, appears to be at least partly correct: "Certainly it seems now that nothing could have been more obvious to the people of the earlier twentieth century than the rapidity with which war was becoming impossible." Unfortunately, he was not entirely correct, but we can attest today that ever since the first atomic bomb was developed, nations in possession of such weapons have refrained from fighting one another openly. Instead, the Oil Age's nuclear powers have directed their might at those who possess no atomic bombs and happened to get in the way of their interests. This unleashed strong forces of nuclear proliferation, as just about every nation in the world would like to acquire nuclear weapons systems or will alternatively have to ally itself closely with at least one existing nuclear power. As their enormous destructive effect made nuclear weapons impractical to use in nearly all situations, various nations also started to downsize some of their arms systems (mini-nukes), and in the future the focus may change to more selective (bio)weapons that will destroy specific materials or people (by targeting genetic markers associated with particular ethnic groups). In a way, such weapons of knowledge imply that destructive energy will become less relevant in future years. However, delivery systems will still require quite a bit of energy, and so will the battlefield robots that might be the answer to ethnic bioweapons. And if nations start placing arms systems on satellites in the planet's orbit, the energy costs will be immense.

As pointed out earlier, we have to make sure there is plenty of constructive compared to destructive energy available at all times to maintain peace. Much constructive energy can still be gained or saved through efficiency improvements (and modern communications technology reduces the need for personal travel, for instance). Those who are uncomfortable with the enlarged prospects of (peaceful) nuclear energy in the future, I can understand: Nuclear fission technology does indeed involve the severe problems of atomic weapons proliferation and final waste storage, and it is a somewhat questionable technology in current form and under current circumstances. However, new energy regimes have always forced societies to change or improve their social organization (in this case in a global context to counter the spread of weapons-grade nuclear fuel), and technological difficulties have always been overcome as necessary (in this case to improve nuclear power plant decommissioning and waste storage). Besides, the enormous investments now made in renewable energy technology will enable it to take a good share of the burden off nuclear and coal energy. Exactly how people will meet the challenges of future years to produce and distribute productive energy for more people and regions to peacefully prosper in an ever more connected world, we cannot tell. The inability to predict the future and the inability of the energy theory to explain all aspects of world history are two sides of the same coin. Both history and future involve people, and people will make irregular choices. There will thus never be a single all-explanatory grand unification theory of history. But if there were, energy would undoubtedly be a main part of it.

NOTES

1. H. G. Wells, *Anticipations of the Reaction of Mechanical and Scientific Progress upon Human Life and Thought* (London: Chapman & Hall, 1902), http://www.gutenberg.org/files/19229/19229-h/19229-h.htm#FNanchor_3_3.

2. H. G. Wells, *The World Set Free: A Story of Mankind* (London: Macmillan & Co., 1914). Herbert George Wells, *The World Set Free*, The Project Gutenberg Ebook, produced by Charles Keller and David Widger, http://www.gutenberg.org/files/1059/1059-h/1059-h.htm.

3. Fred Cottrell, *Energy and Society: The Relation between Energy, Social Change, and Economic Development* (New York: McGraw-Hill Book Company, 1955).

4. C. Daryll Forde, *Habitat, Economy and Society: A Geographical Introduction to Ethnology* (New York: E.P. Dutton & Co., 1946).

5. V. Gordon Childe, *Social Evolution* (New York: Abelard Press, 1951).

6. An explanation of Thomas Malthus's theory follows further down in the text.

7. Besides, the foreword writer mentions "Christian fortitude" as one of the factors that "not only increased output per man-hour on the farms but also enabled factories and cities to develop with great rapidity."

8. Earl Cook, *Man, Energy, Society* (San Francisco: W.H. Freeman, 1976).

9. M. King Hubbert, *Energy Resources: A Report to the Committee on Natural Resources* (Washington D.C.: National Academy of Sciences, 1962).

10. Marshall Sahlins, *Stone Age Economies* (Chicago: Aldine-Atherton, 1972).

11. In recent years Vaclav Smil has emerged as the perhaps most renowned writer on broad energy issues. See, for instance, Vaclav Smil, *Energies: An Illustrated Guide to the Biosphere and Civilization* (Cambridge: MIT Press, 2000). One famous book published between those of Cottrell and Cook is Isaac Asimov's *Life and Energy* (Garden City, NY: Doubleday, 1962), which covers aspects of both animate and inanimate energies in human societies at different points in time.

12. Charles Darwin, *The Descent of Man, and Selection in Relation to Sex* (New York: Barnes & Noble Publishing, 2004), 152. Original edition: Charles Darwin, *The Descent of Man, and Selection in Relation to Sex* (London: John Murray, 1871), available online at Project Gutenberg, http://www.gutenberg.org/etext/2300.

13. Such variations have been known throughout much of the 20th century. The modern debate over racial differences was ignited by Arthur Jensen, a psychology professor at the University of California at Berkeley, when he attributed them to genetic factors in 1969. In 1994 the book *The Bell Curve* by American psychologist Richard Herrnstein and American social analyst Charles Murray renewed the debate. Critics have challenged the fundamentals of comparative global IQ studies, but apparently the "debate about racial and ethnic differences in IQ scores is not about if the differences exist but what causes them." Douglas K. Detterman, "Intelligence," Microsoft Encarta Online Encyclopedia, http://encarta.msn.com/encnet/refpages/RefArticle.aspx?refid=761570026.

14. R. Plomin et al., "Variability and Stability"; R. Plomin and T. S. Price, "The Relationship between Genetics and Intelligence," in N. Colangelo & G. A. Davis, *Handbook of Gifted Education* (Upper Saddle River: Pearson Education, Inc., 2003), 113–23.

Here is a word of warning in terms of applying the concept of heritability, which Arthur Jensen stated to be about 80 percent in terms of IQ scores, to intelligence studies at all: "Biological systems are complex, non-linear, and non-additive. Heritability estimates are attempts to impose a simplistic and reified dichotomy (nature/nurture) on non-dichotomous processes." Steven P. R. Rose, "Commentary: Heritability

Estimates—Long Past their Sell-by Date," *International Journal of Epidemiology* 35, 3 (2006): 525–27, http://ije.oxfordjournals.org/cgi/content/full/35/3/525.

15. D.M. Dick et al., "Association of CHRM2 with IQ: Converging evidence for a gene influencing intelligence," *Behav Genet.* 37(2) (2007): 265–72, http://www.ncbi.nlm.nih.gov/pubmed/17160701.

16. "When an IQ test is revised, it is restandardized with a new normative sample. The distribution of raw scores in the sample population determines the IQ that will be assigned to the raw scores of others who take the test. By analyzing the performance over the years of different normative samples on the same tests, researchers have concluded that performance on intelligence tests has risen significantly over time. This phenomenon, observed in industrialized countries around the world, is known as the Flynn effect, named after the researcher who discovered it, New Zealand philosopher James Flynn. Scores on some tests have increased dramatically. For example, scores on the Raven's Progressive Matrices, a widely used intelligence test, increased 15 points in 50 years when scored by the same norms. In other words, a representative sample of the population that took the test in 1992 scored an average of 15 points higher on the test than a representative sample that took the test in 1942." Douglas K. Detterman, "Intelligence," Microsoft Encarta Online Encyclopedia, http://encarta.msn.com/encyclopedia_761570026_3/Intelligence.html#p36.

17. One controversial concept of how lifestyle, language, and cognitive abilities are linked together has been revived by a recent study by Peter Gordon of Columbia University, who attempted to shed light on the connection between language and thinking by researching gatherer-hunters from the Brazilian Pirahã tribe, whose language only contains words for the numbers one and two as distinct from "many." The foragers were unable to reliably tell the difference between four objects placed in a row and five in the same configuration. (Obviously, there is no need for them to count to higher numbers in their daily life.) This result may support ideas of "linguistic determinism" in the sense that the language available to humans defines the complexity of their thoughts, which is a controversial issue. Gordon maintains that this is the case in terms of counting, while there are other things that we can think about that we cannot talk about. (This would come close to Michael Polanyi's ideas of tacit versus explicit knowing.) Celeste Biever, "Language May Shape Human Thought," *NewScientist.com*, August 19, 2004, http://www.newscientist.com/article.ns?id=dn6303.

18. Charles Darwin, *The Descent of Man*

19. Roger B. Hill, "History of Work Ethic: 4. Protestantism and the Protestant Ethic," Department of Workforce Education, Leadership & Social Foundations, The University of Georgia, http://www.coe.uga.edu/workethic/hpro.html; Roger B. Hill, "Historical Context of the Work Ethic, 1992–1996," http://www.coe.uga.edu/~rhill/workethic/hist.htm.

20. Max Weber, *Die protestantische Ethik und der "Geist" des Kapitalismus* (Weinheim: Beltz Athenäum, 2000). (Based on the original edition of 1904/05, with remarks on the changes made in the 1920 edition.)

21. Philip A. Wickham, *Strategic Entrepreneurship* (Harlow: Financial Times Prentice Hall, 2006), 167.

22. To be sure, entrepreneurship as such is considered a "pan-human" phenomenon. Philip A. Wickham, *Strategic Entrepreneurship*, (Harlow: Financial Times Prentice Hall, 2006), 170.
23. Gregory Clark, *A Farewell to Alms: A Brief Economic History of the World* (Princeton: Princeton University Press, 2007).
24. Luigi Luca Cavalli-Sforza, "Race," Microsoft Encarta Online Encyclopedia 2008, http://encarta.msn.com/encyclopedia_761576599_5/Race.html.
25. This is merely a speculation. The starved German population may have well revolted even in the absence of the influenza epidemic. We do not know how much exactly the epidemic contributed to the outbreak of the revolt.
26. To be sure, maize diffused over a long north-south stretch in the Americas, but this diffusion was slow compared to the spreading wheat in Eurasia.
27. In *Guns, Germs, and Steel*, Diamond attempts to find answers to this question, but they are weak. He mentions, for instance, that Europe's coastline is comparatively longer than China's, which may have ultimately been behind the more pronounced European efforts in overseas explorations.
28. Frenchman Nicholas Mahudel, in the Super-Agricultural Era, and others later on, had proposed similar systems before Thomsen.
29. It has been attempted to fix this by associating gatherer-hunters with the Stone Age subcategories "palaeolithic" and "mesolithic" (or "epipalaeolithic"), and people in transition to agriculture with subcategory "neolithic."
30. Sometimes the system is expanded to include five subcategories at this level by introducing "Cradle of civilization" before Antiquity, and splitting the Modern period into "Early Modern period" and "Modern period."
31. Often, the start of the Middle Ages is defined as the definite fall of the Western Roman Empire in the 5th century C.E., and sometimes a date as late as 600 C.E. is proposed.
32. Right now it is speculated that it was the command of fire that allowed people to migrate out of Africa, which would be a milestone indeed. A further fine tuning within the command of fire would be the ability to handle and conserve open fires, as opposed to the ability to actually start one.
33. Nevertheless, wind and waterpower would be well suited as subdivisions of the later Agricultural Age.
34. This might change, though. Electricity from hydroelectric stations and windmill parks may contribute substantially to the charging of electric vehicles sometime in the future.
35. To be sure, other systems are facing the same challenge. The Iron Age, for instance, describes different time periods in southwestern Asia and in Europe, and never took place in the Americas.
36. The direct connection to population growth was lacking when economic growth was based on animal, wind, or waterpower, or on efficiency gains.
37. Later on Malthus moderated his views by introducing "positive checks" that targeted birth rather than death rates, namely delayed marriage and fewer children.
38. Here it should be mentioned that this development during the transitional period was supported in the most unfortunate way by diseases that killed large shares of the population in the fast-growing unsanitary cities.

39. In his wonderfully politically correct book *Guns, Germs, and Steel*, Jared Diamond seems to attempt to demonstrate through ideas of environmental determinism that all people on Earth, regardless their ethnic background, are intrinsically equal in terms of their talents and abilities. Those who fell behind were in no way inferior to Europeans (or Eurasians), but happened to live in inferior environments. To be sure, equality among people cannot be verified by demonstrating differences in environments, even if we believe that environmental influence is important. So-called achievers would turn non-achievers if moved from favorable to unfavorable environments, but it cannot be claimed that non-achievers would turn achievers automatically when put into a more favorable environment, unless we assume there is really just one group of intrinsically equal people, which was the assumption we set out to test in the first place.

40. The expression "decidedly non-deterministic" comes from Vaclav Smil's own description of his book *Energy in World History* at "Complete Annotated Book List," http://home.cc.umanitoba.ca/~vsmil/complete_booklist.html. Smil equates the importance of energy in history with that of climate and disease. Vaclav Smil, *Essays in World History—Energy in World History* (Boulder, CO: Westview Press, 1994), 243.

41. The latter can think about it this way. If we take entirely equal people and put them in environments of different energies, then the group in a beneficial environment would not just progress faster, but the group may also change in its nature and culture in ways that may even give it an advantage when the original benefits of its environment no longer exist. This would then reconcile supporters of environmental determinism and believers that the answer to progress lies in people, culture, and institutions. To be sure, many would consider such view insulting, as it implies that the course of social evolution created better and worse cultures or people.

42. The death toll of the 1918 influenza epidemic has been repeatedly revised upward. The following paper suggests that it was of the order of 50 million, though it may have been 100 million. The higher figure would be even well above the total death toll of both World Wars, which was perhaps 75 million. (This figure is also uncertain.) N.P. Johnson and J. Mueller, "Updating the Accounts: Global mortality of the 1918–1920 "Spanish" influenza pandemic." *Bull Hist Med.* 76(1) (2002):105–15, http://www.ncbi.nlm.nih.gov/pubmed/11875246?dopt=Abstract.

43. Vaclav Smil, *Essays in World History*, 252.

44. John Robert McNeill, *Something New Under the Sun: An Environmental History of the Twentieth-century World* (New York: W. W. Norton & Company, 2001).

45. Joel E. Cohen, *How Many People Can the Earth Support?* (New York: W.W. Norton & Company, 1995), 42.

46. David A. Koplow, *Smallpox: The Fight to Eradicate a Global Scourge* (Berkeley: University of California Press, 2003), Chapter 1: "The Rise and Fall of Smallpox."

47. Some scholars have challenged the view that improved nutrition lowered death rates on the basis of data on decreasing body heights in Western Europe during the 18th century. Joel E. Cohen, *How Many People?*, 44.

48. William H. McNeill, *Populations and Politics since 1750*, (Charlottesville/London: University Press of Virginia, 1990), 4–5, cited in: Joel E. Cohen, *How Many People Can the Earth Support?*, 44.

49. Alfred W. Crosby, *The Columbian Exchange: Biological and Cultural Consequences of 1492* (Westport: Greenwood Press, 1972), 165.

50. The reasons behind the emergence of oil culture are less controversial. Coal Age engineers were well aware of the benefits of liquid fuels, and there were but few options around.

51. The latter explanation is used with respect to those resource-rich areas where sedentary foragers, such as the salmon-catching natives of the Pacific Northwest, never became true farmers.

52. Peter C. Perdue, "China in the Early Modern World: Shortcuts, Myths and Realities," *Education About Asia* 4:1 (1999): 21–26, http://web.mit.edu/21h.504/www/china_emod.htm, http://ocw.mit.edu/NR/rdonlyres/History/21H-504East-Asia-in-the-WorldSpring2003/70208DF1-1DAB-4B4A-97C8-6800A7D0EE81/0/china_emod.pdf.

BIBLIOGRAPHY

Asimov, Isaac. *Life and Energy.* Garden City, NY: Doubleday, 1962.

Biever, Celeste. "Language May Shape Human Thought." *NewScientist.com*, August 19, 2004. http://www.newscientist.com/article.ns?id=dn6303.

Cavalli-Sforza, Luigi Luca. "Race." Microsoft Encarta Online Encyclopedia. http://encarta.msn.com/encyclopedia_761576599_5/Race.html.

Childe, V. Gordon. *Social Evolution.* New York: Abelard Press, 1951.

Clark, Gregory. *A Farewell to Alms: A Brief Economic History of the World.* Princeton: Princeton University Press, 2007.

Cohen, Joel E. *How Many People Can the Earth Support?* New York: W.W. Norton & Company, 1995.

Cook, Earl. *Man, Energy, Society.* San Francisco: W.H. Freeman, 1976.

Cottrell, Fred. *Energy and Society: The Relation between Energy, Social Change, and Economic Development.* New York: McGraw-Hill Book Company, 1955.

Crosby, Alfred W. *The Columbian Exchange: Biological and Cultural Consequences of 1492.* Westport: Greenwood Press, 1972.

Darwin, Charles. *The Descent of Man, and Selection in Relation to Sex.* New York: Barnes & Noble Publishing, 2004. Original edition: Darwin, Charles *The Descent of Man, and Selection in Relation to Sex.* London: John Murray, 1871, available online at Project Gutenberg. http://www.gutenberg.org/etext/2300.

Detterman, Douglas K. "Intelligence." Microsoft Encarta Online Encyclopedia. http://encarta.msn.com/encnet/refpages/RefArticle.aspx?refid=761570026.

Diamond, Jared. *Guns, Germs, and Steel: The Fates of Human Societies.* New York: W.W. Norton & Company, 1997.

Dick, D.M., et al. "Association of CHRM2 with IQ: converging evidence for a gene influencing intelligence." *Behav Genet.* 37(2) (2007): 265–72. http://www.ncbi.nlm.nih.gov/pubmed/17160701.

Forde, C. Daryll. *Habitat, Economy and Society: A Geographical Introduction to Ethnology.* New York: E.P. Dutton & Co., Inc., 1946.

Herrnstein, Richard J., and Charles A. Murray. *The Bell Curve: Intelligence and Class Structure in American Life.* New York: Free Press, 1994.

Hill, Roger B. "Historical Context of the Work Ethic, 1992–1996." http://www.coe.uga.edu/~rhill/workethic/hist.htm.

Hill, Roger B. "History of Work Ethic: 4.Protestantism and the Protestant Ethic." Department of Workforce Education, Leadership & Social Foundations, The University of Georgia. http://www.coe.uga.edu/workethic/hpro.html.

Hubbert, M. King. *Energy Resources: A Report to the Committee on Natural Resources.* Washington D.C.: National Academy of Sciences, 1962.

Johnson, N.P., and J. Mueller, "Updating the Accounts: Global Mortality of the 1918–1920 "Spanish" Influenza Pandemic." *Bull Hist Med.* 76 (2002):105–15. http://www.ncbi.nlm.nih.gov/pubmed/11875246?dopt=Abstract.

Koplow, David A. *Smallpox: The Fight to Eradicate a Global Scourge.* Berkeley: University of California Press, 2003.

McNeill, John Robert. *Something New Under the Sun: An Environmental History of the Twentieth-century World.* New York: W. W. Norton & Company, 2001.

McNeill, William H. *Populations and Politics since 1750.* Charlottesville/London: University Press of Virginia, 1990.

Perdue, Peter C. "China in the Early Modern World: Shortcuts, Myths and Realities." *Education About Asia* 4:1 (1999): 21–26. http://web.mit.edu/21h.504/www/china_emod.htm. http://ocw.mit.edu/NR/rdonlyres/History/21H-504East-Asia-in-the-WorldSpring2003/70208DF1-1DAB-4B4A-97C8-6800A7D0EE81/0/china_emod.pdf.

Plomin, R., and T. S. Price. "The Relationship between Genetics and Intelligence." In N. Colangelo, and G. A. Davis. *Handbook of Gifted Education,* 113–123. Upper Saddle River: Pearson Education, Inc., 2003.

Plomin, R., N. L. Pedersen, P. Lichtenstein and G. E. McClearn. "Variability and Stability in Cognitive Abilities Are Largely Genetic Later in Life." *Behavior Genetics* 24 (3) (1994): 207. http://www.springerlink.com/content/t0844nw244473143/.

Rose, Steven P. R. "Commentary: Heritability Estimates—Long Past Their Sell-by Date," *International Journal of Epidemiology* 35, 3 (2006): 525–527. http://ije.oxfordjournals.org/cgi/content/full/35/3/525.

Sahlins, Marshall. *Stone Age Economies.* Chicago: Aldine-Atherton, 1972.

Smil, Vaclav. *Energies: An Illustrated Guide to the Biosphere and Civilization.* Cambridge: MIT Press, 2000.

Smil, Vaclav. *Essays in World History: Energy in World History.* Boulder, CO: Westview Press, 1994.

Weber, Max. *Die protestantische Ethik und der "Geist" des Kapitalismus.* Weinheim: Beltz Athenäum, 2000. (Based on the original edition of 1904/05, with remarks on the changes made in the 1920 edition.).

Wells, H. G. *Anticipations of the Reaction of Mechanical and Scientific Progress upon Human Life and Thought.* London: Chapman & Hall, 1902. http://www.gutenberg.org/files/19229/19229-h/19229-h.htm#FNanchor_3_3.

Wells, H. G. *The World Set Free: A Story of Mankind.* London: Macmillan & Co., 1914, Herbert George Wells, *The World Set Free,* The Project Gutenberg Ebook, produced by Charles Keller and David Widger. http://www.gutenberg.org/files/1059/1059-h/1059-h.htm.

Wickham, Philip A. *Strategic Entrepreneurship.* Harlow: Financial Times Prentice Hall, 2006.

Index

Abolition of slavery, 264–66
Accidents: nuclear, 472; road traffic, 417
Acidification, 716. *See also* Acid rain
Acid rain, 258
Acids, amino, 428–29, 467
Adaptations: of human body to different climates, 854
Ader, Clément, 390
Aerosols: role in climate change, 696–97
Afforestation, 716
Afghan Civil War, 620–21
Afghanistan, 341–42; British invasions, 341–42; civil war, 620–21; Soviet occupation, 617–18
Afghan Wars (British invasions), 341–42
Africa: colonization during Coal Age, 308–14; post–World War II, 525–26; post–World War II, U.S. and Soviet activities, 553–55
Agade. *See* Akkad (Agade)
Agricola, Georgius, 192. *See also* De Re Metallica (Agricola)
Agricultural Adjustment Act (AAA), 539
Agriculture, 17–21; British efficiency gains in Super-Agricultural Era, 163–64; emergence of, 23–35; energy inputs during Oil Age, 460–61; energy sink, 460–61; fertilization, 100–101; fruit trees, 48; influence on climate, 687; in the Oil Age, 451–63; original centers, 28–30; productivity gains during Coal Age, 284–85; spread of, 37–45; sustainability, 99–101; in the tropics, 29, 30, 43–44; work effort, 50–51
Aid, foreign, 486–87
AIDS, 495–96
Air: pollution, 418–19; 467, 692, 696; pollution, indoor, 419, 811; raids, 392–93, 435, 439, 500, 513, 601; travel, 393–94, 397
Airbus, 397, 411, 800
Airplanes: development in Word War I, 392–93; rocket-propelled, 440–41; with jet engines, 394–98; with piston engines, 388–94
Akkad (Agade), 106
Alaska: oil, 564; purchase, 321
Albedo, 692, 697
Alcohol. *See* Bioethanol
Alexander the Great, 102, 103
Algae, 719, 748
Algeria, 525
Al-Ghawar, 579
Alizarin, 274
Allende, Salvador, 550

Al-Qaeda, 623–25
Alternating current (AC), 221–23
Aluminum, 424; oxide: 427
Al Yamamah weapons deals, 611
Amazonia, 29
Amino acids, 428–29, 467
Ammonia, 436, 452, 469, 716; in refrigerators, 286. *See also* Haber-Bosch process
Anesthetic, 274
Anglo-Iranian Oil Company (AIOC), 592
Anglo-Persian Oil Company (APOC), 374–75, 571, 590–91. *See also* Anglo-Iranian Oil Company
Angola, 554
Animal: domestication, 24–27; domestication, suitability for, 66–69; feed requirements, 74–75, 285; genetically engineered, 457–58; husbandry, 19–20; husbandry, in the Oil Age, 455–58; power, 60–76; power, for field work, 69–71; power, for off-field work, 71–74
Antarctica: ice increasing, 707; ice melting, 688, 691, 706
Apollo (spacecraft), 444
Appalachian basin, 237
Aquaculture. *See* Fish farming
Arab expansion, 113–14
Arabian American Oil Company. *See* Aramco
Arab-Israeli War of 1967, 587
Arab-Israeli War of 1973, 588–90
Aramco (Arabian American Oil Company), 578
Archimedes' screw, 58, 214
Argentina: after World War II, 551
Arkwright, Richard, 159
Arrhenius, Svante August, 677
Asia: independence of nations during Oil Age, 516–25
Aspirin, 274
Atmosphere. *See* Climate change
Atomic bombs. *See* Nuclear bombs
Australia, 314–16
Austrian monarchy, 306–8; Galicia (oil), 363–64

Austronesian Expansion, 40–41
Automobile technology, spread of, 408–10
Aviation technology, spread of, 410–11
Azerbaijan, 505, 533; emergence of oil industry, 367–71
Aztecs, 130–31

Baby boom, 494
Babylon, 106
Bacteria, 49, 272–73, 416, 429, 448, 455
Baeyer, Adolf von, 274
Baku, 367–71, 505, 533; during World War II, 507, 508–10
Balfour declaration, 584–85
Balloon: manned flight, 386–88
Bantu Migrations, 39–40
Barley: resistance to salinity, 33 n.16
Barrel: volume definition in oil industry, 369
BASF (Badische Anilin und Soda Fabrik), 274, 436
Basic-oxygen process. *See* Linz-Donawitz process
Batteries, for electric vehicles, 756
Battle of Gallipoli, 373
BBC (British Broadcasting Corporation), 431
Beef: consumption, influence on climate change, 14, 715; production, energy requirements, 461; production, water requirements, 457
Beirut, U.S. embassy bombing, 612
Belge, Le, 241
Belgian Congo, 313–14; after World War II, 554–55. *See also* Congo, Democratic Republic of the
Belgium, 241
Bell, Alexander Graham, 269
Bell X-1, 441
Benz, Karl, 378–79
Berlin Conference (1884), 309
Bessemer, Henry, 195
Bessemer process, 195, 211
Best (company), 384
Bicycles, 267
Bin-Laden, Osama. *See* Osama bin Laden
Biobutanol, 751–52

Biodiesel, 752
Bioethanol, 748–751
Biofuels, liquid, 748–52
Biogas, 765
Biological weapons, 448–49
Biomass, 83–86; for electricity production, 765–66; modern, 724. *See also* Charcoal; Crops, genetically engineered; Wood
Bison, 30
Bitumen, 747
Black Death, 121
Blériot, Louis, 391
Blücher, 208–9
Boats. *See* Ships
Boeing, 394, 397, 800
Bohr, Nils, 438
Bombs, nuclear. *See* Nuclear bombs
Books, 122
Bosch, Carl. *See* Haber-Bosch process
Boulton, Matthew, 201–2, 206
Boulton & Watt, 201–4
Bow-and-Arrow (Bows), 87–88; composite, 88
Boxer Rebellion, 337–38
Braun, Wernher von, 442
Brazil, 275, 800; bioethanol production, 749; discovery, 135; post–World War II, 550
Breech-loading, 279
Bretton Woods system, 498, 515–16, 861
Bridges, 196–97
Bridgewater canal, 229
British Africa, 526
British Petroleum (BP), 374, 571, 592. *See also* Anglo-Persian Oil Company (APOC)
Buddhism, 139, 155
Buffon, Georges-Louis Leclerc de, 270
Bunsen, Robert Wilhelm, 217–18
Burma, 336; emergence of oil industry, 372
Burmah Oil Company, 372, 374
Bush, George H. W., 553, 601
Bush, George W., 102

Cable Tool Drilling. *See* Drilling
CAFE. *See* Corporate Average Fuel Economy
California: annexation by the United States, 320; gold rush of 1848, 323; oil production, 559–60; Russian expansion, 329–30
California Electric Light Company Inc., 220
Calvinism, 852
Cambodia, 545–46. *See also* Indochina
Camels, 64–65
Campodia, 336
Canada, 247
Canals, 74
Candles, 83
Canning, of food, 285–86
Cannons: naval, 91; siege, 90–91
Canola, 454
Cape of Good Hope, 134–35
Capitalism, 243
Car accidents, 417
Carbon-arc lamp, 220
Carbon dioxide: atmospheric, 677–79; atmospheric, future scenarios, 699–700; emissions permissions trading, 721–22; sequestration, 717–20. *See also* Climate change
Carbon dioxide in photosynthesis, 100, 693, 766
Carboniferous Crescent, 237
Carbon monoxide, 216, 469, 753. *See also* Coal gas; Synthesis gas
Carbon sinks, 692–94
Carnegie, Andrew, 247
Carnivores, 14, 19
Cartridges, metal, 279
Castro, Fidel, 549–50
Catapults, 89
Caterpillar Tractor Company, 384
Cattle: domestication, 20, 28; draft, 60, 70; Sahel, 30. *See also* Beef
Cells, fuel. *See* Fuel cells
Cells, photovoltaic. *See* Photovoltaic cells
Cellular telephones, 430
Cellulose 277–78, 426, 751
Celts, 119–20
Cement production, carbon dioxide release, 693
Central Intelligence Agency (CIA), 543

Central Treaty Organization (CENTO), 581, 595
Ceramics, advanced, 426–27
Cereal production, in the Oil Age, 451–52
Chanute, Octave, 390
Charcoal, 84–86; in gunpowder production, 89; in iron production, 192
Charlemagne, 102, 120
Chemical weapons, 447; use in Iran-Iraq War, 599–600
Chernobyl accident, 472
Childe, V. Gordon, 845
Chile: post–World War II, 550–51
Chile nitrates (saltpeter), 284, 436. *See also* Saltpeter
Chimpanzees, diet, 2
China: during the Oil Age, 517–18; early civilizations and dynasties, 140–41; end of dynastic rule, 338; European advance during Coal Age, 334–38; expansion into Tibet, 337; future power, 784–87; invasion by the Japanese, 512; naval explorations, 141–42; one-child policy, 494; Qing dynasty, 166–67
Chinese Exclusion Acts, 266, 539
Chlorofluorocarbons, 687
Christianity, 110–13, 155
Chuñu, 30
Churchill, Winston, 373, 591
Civetta, 215
Civil War (American), 245, 281
Climate: artificial cooling, 720–21; influence by clouds, 686, 689, 696–97; importance for progress, 853–54; modelling, 685–700; models, fitting of, 688–89; of the past, 679–85
Climate change, 677–725; consequences of, 689–92; costs of, 707–8; measures against, 713–25
Cloning, 457–58
Clouds, effect on climate, 686, 689, 696–97
CO_2. *See* Carbon dioxide
Coal: burning causing acid rain, 258; future consumption, 774–75; in iron production, 194–98; liquefaction, 752–53; mining, 193, 198; production, increase, 197–98; production, in Japan during Coal Age, 253; production, in various countries during Coal Age, 238; properties, 187–89; reserves, in various countries, 237–39, 800; use during Oil Age, 467–68
Coal Age: general education increase, 263; workers' rights, 263
Coal culture. *See* Coal technology
Coal gas, 216–18
Coal tar, 273–74
Coal technology: conditions during emergence, 228–32; emergence, 191–232; hardships during emergence, 257–58; spread, 235–56; spread to Japan, 250–53; spread to Russia, 248–50; transfer to colonies, 254–55
Cockerill, family, 241
Coke, 194–95, 216
Cold War, 531
Colombo. *See* Columbus, Christopher
Colonies during Coal Age, 301–45
Coltan (columbite-tantalite), 555
Columbus, Christopher, 128–29
Combine (harvester-thresher), 246
Commonwealth of Nations, 526
Communist Manifesto, 528
Compagnie Française des Pétroles (CFP), 573
Computers, 431–32
Condor, Operation, 552
Conference of Berlin (1884), 309
Confucionism, 155
Congo, Democratic Republic of the, 554–55. *See also* Belgian Congo
Congo, Republic of the, 553
Congress of Berlin (1878), 500
Congress of Vienna, 245
Contras, 551–52
Conventional oil, global reserves, 741–47
Cook, Earl, 846–48
Copernicus, Nicolaus, 155
Corn. *See* Maize
Coronary heart disease, 417
Corporate Average Fuel Economy (CAFE), 563, 701
Cort, Henry, 195
Cortés, Hernándo, 124, 131

Cotton, 159–63; genetically engineered, 454
Cotton gin, 160–61
Cottrell, Fred, 844–46
Crick, Francis, 428
Crop rotation, 101
Crops: genetically engineered, 453–55
Crossbows, 88
Crude oil. *See* Oil
Crusades, 114–15
Cuba, 326; post–World War II, 549–50
Cugnot, Nicolas-Joseph, 205–6
Cultural Revolution (China), 518
Culture: importance for progress, 851–54
Current. *See* Alternating current; Direct current

Da Gama, Vasco, 135
Daimler, Gottlieb, 378–79
Danzig (Gdansk), 507
Darby, Abraham, 194
D'Arcy, William Knox, 374
Darwin, Charles, 271, 849
Davy, Humphry, 220
DDT (insecticide), 453
Debt trap, 486
Deep offshore drilling, 746
Deforestation: role in climate change, 714–16
Deng Xiaoping, 518
De Re Metallica (Agricola), 192
Determinism, 861–62
Developing countries: future energy supply, 805–13
Diamond, Jared, 856
Diaz, Bartolomeu, 134
Diesel, Rudolf, 380–82
Diesel (fuel): from biomass, 752
Direct current (DC), 221–23
Direct-Methanol Fuel Cell (DMFC), 755
Dirty war (Argentina), 551
Diseases: infectious, 49; role in course of history, 854–56. *See also* Epidemics
DNA (deoxyribose nucleic acid), 428–29
Domestication of animals, 24–27
Donets basin, 249
Donkeys, 60–61
Dove. *See* Taube (airplane)
Drake, Edwin L., 366

Drilling: cable tool, 402–3 n.4; percussion (spring poles), 362, 398, 402–3 n.4; walking beam, 402–3 n.4
Dunlop, John B., 275
DuPont, 426
Dust Bowl, 684
Dutch: decline at the expense of England, 164; founding colony in Indonesia, 138; use of peat, 137–38
Dutch East Indies. *See* Indonesia
Dutch Golden Age, 137–39
Dyes: production during Coal Age, 273–74
Dynamite, 278
Dynamoelectric principle, 219

East India Company, 165
East Indies, Dutch. *See* Indonesia
Economic expansion: during Coal Age, 293–94; during Oil Age, 477–78
Edison, Thomas Alva, 220, 221, 222
Edo Bay, 250
Egypt: original civilization, 107
Ehrlich, Paul, 273
Einstein, Albert, 428
Eisenhower, Dwight, 591
Electric: motors, 221; train, 221; vehicles, 756
Electricity, 218–27; current production by fuel, 718; options for future generation, 761–70
Electrification in developing world, 807–10
Elephants, 65
El Niño, 688, 695
El Salvador, 551
Empires: to achieve Millennium Development Goals (MDG), 810–13; during Coal Age, 301–45; during Oil Age, 497–654; rise and fall, 97–103
Energy: for developing countries, 805–13; mix during Oil Age, 465–75; mix in different countries, 799; mix expected until 203, 773–77
Energy consumption: compared to economic output, 478, 809; importance for quality of life, 806; in the Oil Age, 413, 465–66; in the Oil Age, in different countries, 806–7

Energy density: ethanol compared to gasoline, 749; gas compared to gasoline, 377, 741; hydrogen ompared to gasoline, 755; methanol compared to ethanol, 755; oil products compared to coal, 372; peat compared to coal, 189; peat compared to wood, 137; wood compared to coal, 184, 189, 190 n.4
English activities in India: early on, 139–40; expansion, 164–65
Enlightenment, 155
Epidemics: in the Americas, 129, 134, 855; black death, 121; and Christianity, 110–11; general impact in Agricultural Age, 101; influenza (1918), 505; in the Roman empire, 110–11; smallpox, 271–72. *See also* Diseases, infectious
Ericsson, John, 214–15
Erie Canal, 242
Ethanol. *See* Bioethanol
Ethiopia, 573
Etrich, Igo, 391
Europe: early agriculture, 38–39; medieval technology, 121–23; national fragmentation, 123–24; rise of, 119–23
European agricultural expansion, 150–51
European Union: future power, 782–84
Explosives: cordite, 278; dynamite, 278; gun cotton, 277; gunpowder, 89–90; nitration, 277; nitroglycerin, 278; in the Oil Age, 435–436; Poudre B, 278; RDX (Research Department Explosive), 279; TNT (trinitrotoluene), 278

Fair Labor Standards Act, 415
Family size: in the Oil Age, 417
Faraday, Michael, 218
Farmer, Moses G., 220
Farming. *See* Agriculture
February Revolution (Russia, 1917), 527
Fertile Cresent, 28
Fertility transition: in Europe and Japan, 262; global, 493–95
Fertilizers: during the Agricultural Age, 100–101; during the Coal Age, 283–84; during the Oil Age, 452, 436
Feudalism, 120
Fire, Command of, 6

Firearms, hand-held, 91–95
First oil price shock, 588–90, 683, 848
Fish farming, 459–60
Fish production, 458
Fitch, John, 213
Flax. *See* Linen
Flexi Fuel Vehicles (FFV), 749
Flintlocks, 94–95
Flying shuttle, 158
Flynn effect, 851
Flywheels, 724–25
Food Canning, 285–86
Foragers, 1–9, 23, 25, 28, 47, 845; in Australia, 315; diet, 2–4; language, 849, 870 n.17; in New Zealand, 316–17; religion, 55; replacement by agrarians, 41–43; social organization, 5, 53; work hours, 1, 847
Ford, Gerald, 548
Ford, Henry, 409
Forde, C. Daryll, 845
Formosa. *See* Taiwan
Fourneyron, Benoit, 224
Four Tigers. *See* Tiger States
Francis, James B., 224
Franklin, Benjamin, 150
Franklin, Rosalind, 428
Franks, 102, 120
Franz Ferdinand (Archduke of Austria), 353–54, 499, 500
Frisch, Otto, 438
Fuel cells, 754–56
Fuel for transport: options for the future, 741–60
Fulton, Robert, 213
Future energy scenarios, 799

Galicia, 502; emergence of oil industry, 363–64
Galilei, Galileo, 156
Gallipoli, 503
Gas: lamps, 217–18; plants, 216–17; to power vehicles, 754; turbines for airplane propulsion, 394–98. *See also* Methane; Natural gas
Gas Light and Coke Company, 217
Gasoline, 377
Gas-to-liquids (GTL), 753–54
Gates, Bill, 420

Gatherer-hunters. *See* Foragers
Gathering. *See* Foragers
Gaza Strip, 587, 613
Gazprom, 536–37
Gdansk. *See* Danzig
General Electric Company, 222–23, 414
General Motors (GM), 563
Genetic engineering, 429; of crops, 453–55
Genghis Khan, 98–99
Geothermal energy, 724, 767–68
German Democratic Republic, 530
Ghawar Field, 578
Global Fertility Transition, 493–95
Global warming. *See* Climate change
Goats: domestication, 25–26, 28; herding, 44–45
Goodyear, Charles, 275
Gore, Al, 706
Gorillas, diet, 2
Gräf & Stift, 353–54
Gramme, Zénobe-Théophile, 219
Grand Coulee plant, 471
Gray, Elisha, 269
Greece: early civilizations, 107–8; post–World War II, 542
Greenhouse effect, 677
Greenhouse gas, 685–87; emissions permissions trading, 721–22. *See also* Radiative forcing
Greening, global, 690
Greenland, 679, 681
Green Revolution, 451–53, 492
Guano, 101
Guantánamo Bay, 627
Guatemala, 548
Guericke, Otto von, 199
Gulbenkian, Calouste 375, 376
Gulf Stream: role for climate change, 692
Gun cotton, 277
Gunpowder, 89–90
Gutenberg, 122

Haber, Fritz, 402, 436. *See also* Haber-Bosch process
Haber-Bosch process, 435–36, 460–61
Habsburgs, 124
Hadza, 1

Hahn, Otto, 437
Han China, 155
Hanford plant, 438
Hardships, when entering Oil Age, 418–20
Harnessing of horses. *See* Rigid horse collar
Harrison, John, 290
Heavy water reactors, 473
Heinkel (company), 394, 441
Henle, Jacob, 272
Henry's Law, 685
Hertz, Heinrich, 269
Herzl, Theodor, 584
Hewitt, Peter Cooper, 221
Hezbollah, 612
Highs, Thomas, 159
Hindenburg, 388
Hinduism, 139
Hiroshima, 438–39, 513
Hitler, Adolf, 507, 510
Hoe, 69
Hollerith, Herman, 431
Holocaust, 510–11
Holt (company), 383
Holy Roman emperor, 124
Hominids. *See* Humans
Homo erectus, 8, 43
Hong Kong, 521
Hooke, Robert, 199
Hornsby-Akroyd, 380–81, 383
Horses, 61–64; for field work, 70–71; for field work during Coal Age, 285; in medieval Europe, 121; population during Coal Age, 267; rigid collar 70–71
Hubbert, M. King, 743, 846–47
Human Development Index (HDI), 806
Humans: emergence and spread, 7–8
Humboldt, Alexander von, 270
Hunger in developing countries, 482–84
Huns, 102
Hunter-gatherers. *See* Foragers
Hunting: energy return, 2
Hussein, Saddam. *See* Saddam Hussein
Hutton, James, 270
Huygens, Christiaan, 199
Hydroelectricity. *See* Hydropower

Hydrogen: economy, 761; as energy carrier, 724; bomb, 439–40; for fuel cells, 754–55
Hydropower, 223–224, 470–71; future role, 762–64

Ibn Saud, 614
Ice: melting due to climate change, 691, 706, 707
Ice age cycles, 711–12
Ice cores, 678
Iceland, 679
Iceman. *See* Ötzi
Incandescent lamps, 220–21
Incas, 131–34
India: early civiliations, 139; English activities, early on, 139–40; English activities, increasing, 164–65; English advance during Coal Age, 332–34; future power, 787–88; Mughal (Mogul) empire, 341; post–World War II, 522–25
Indians, American. *See* Native Americans
Indochina, 336, 512; post–World War II, 545
Indonesia: emergence of oil industry, 371; independence, 517; invasion by Japanese, 507, 512; post–World War II, 547–48
Industrial Revolution: water-powered, 157–62
Infectious diseases. *See* Disease, infectious
Influenza epidemic (1918), 505
Intelligence: importance for progress, 850–51
Intercontinental ballistic missiles, 442
Intergovernmental Panel on Climate Change (IPCC), 682, 691
Internal combustion (IC) engines, 376–83
International Bank for Reconstruction and Development (IBRD), 516
International Criminal Court (ICC), 556
International Energy Outlook, 773
International Military Education and Training (IMET), 555–56
Internet, 433–34
Intifada, 613
Inuits, 2, 5

Iran, 590–94; revolution, 593–94; war against Iraq, 597–601. *See also* Oil; Persia
Iran-Contra Affair, 551–52, 598–99
Iranian revolution, 593–94
Iran-Iraq War, 597–601
Iraq: early history, 595; early oil prospecting, 375–76; invasion by United States (2003), 629–39; invasion by United States (2003), death toll, 638; invasion of Kuwait, 602–5; trade embargo, 606; war against Iran, 597–601
Iraq Petroleum Company, 573
Iron, wrought, 195
Iron production, 84–86, 191–98; in the Oil Age, 424
Irrigation, 99, 105
Islam, 113; split into Shiite (Shia) and Sunni branch, 114
Israel, 582–90; proclamation of state, 586
Israeli-Palestine conflict, 611–13, 640–41
Italy: role in World War I, 502

Japan: coal production, 253; during Coal Age, 338; early history, 143; invasion of China, 512; post–World War II, 518–20; rise of automobile industry, 519–20; shoguns, 250; spread of coalculture to, 250–53
Jenner, Edward, 272, 865
Jet planes. *See* Airplanes, with jet engines
Jews, 111, 582–84
Joint Combined Exercises and Training (JCET), 556
Judaism, 111
Junkers (company), 393, 441

Kai-shek, Chiang, 517
Kalahari desert, 1
Kaplan, Viktor, 224
Kazakhstan, 745, 786, 792
Kepler, Johannes, 155
Kerosene, 362–63
Khan, Abdul Qadeer, 524–25
Khoisan languages, 1
Khomeini, Ayatollah, 593–94
Kissinger, Henry, 546, 548, 551, 597
Knowledge Society, 415

Koch, Robert, 272–73
Königgrätz, Battle of, 279
Korea, 339
Korean War, 522, 544
Korolyov, Sergey, 444
Krupp, Alfred, 195, 246
Kurdistan, 572
Kurds, 596; insurrection during Iran-Iraq War, 600–601
Kuwait: early oil prospecting, 574–75; independence from Britain, 578; invasion by Iraq, 602–3; liberation after Iraqi invasion, 603–5
Kyoto Protocol, 702–4

Lamarck, Jeane Baptiste, 270
Lamp carbon-arc, 220
Lamp mercury-vapor, 221
Lamp neon, 221
lamps, gas, 217–18
Lamps, incandescent, 220–21
Landsteiner, Karl, 272
La Niña, 695
Laos, 336, 545–47. *See also* Indochina
Latin America: post–World War II U.S. imperialism, 548–53
Laval, Carl Gustaf P. de, 227
Law of the Minimum, 283
Lawrence, T. E. ("of Arabia"), 576
League of Nations, 514
Lebon, Phillipe, 216–17
Leibniz, Gottfried Wilhelm, 156
Lenin, Vladimir Ilyich, 527
Leninism, 528
Lenoir, Jean Joseph Etienne, 377
Leopold I (king of Belgium), 241
Leopold II (king of Belgium), 313
Libya, 525, 793
Liebig, Justus von, 283, 286
Life expectancy: increase in Britain between 1780 and 1913, 261; increase since 1840, 863; in the Oil Age, 415, 493
Life-style, in the Oil Age, 413–21
Light bulbs, 220–21
Lighting: kerosene, 362–63; whale oil, 362–63
Lilienthal, Otto, 389
Linde, Carl von, 435

Linen, 157–59
Linz-Donawitz process, 424
Lippisch, Alexander, 440
Liquefied natural gas (LNG), 469–70, 776
Little Ice Age, 147, 679, 681, 696
Liverpool & Manchester Railway 209–10
Livestock: role in climate change, 713–16. *See also* Animal, husbandry
Llama, 132
Locomotive Act of 1865, 211
Long Depression (1873 to 1896), 248
Longitude, determination of, 289–90
Los Alamos (New Mexico), 438
Louisiana purchase, 318
Lowell, Francis Cabot, 241
Lucas gusher, 558–59
Lukoil (LangepasUraiKogalymneft), 536, 636
Lumumba, Patrice, 554
Lusitania, 503
Luther, Martin, 852
Lyell, Charles, 271

Machine gun, 281
Magellan, Ferdinand, 135–36
Magnetism, 218–19
Maize: cultivation, 129; genetically engineered, 454
Malthus, Thomas, 860
Malthusian trap, 860–61, 863
Mamluks, 115
Mandela, Nelson, 526
Mao Zedong, 517–18
Marconi, Guglielmo, 269
Marcos, Ferdinand, 545
Marcus, Siegfried, 220, 378
Marshall Plan, 543
Marxism, 527–28
Matchlocks, 94
Maunder Minimum, 696
Maxim, Hiram Stevens, 281
Maxwell, James Clerk, 269
Mayas, 130
Meat consumption, 13; role in climate change, 713–16
Meat production in the Oil Age, 456
Medieval Warm Period, 147, 679, 681
Meiji Restoration, 251–52
Meitner, Lise, 437–38

Mendel, Gregor Johann, 428
Mesoamerican civilizations. *See* Aztecs; Incas; Mayas
Mesopotamia, 105
Messerschmitt, 396, 441
Metabolism, human, 57
Metallurgy, 84–86
Methane, 713–14; influence on climate, 686–87. *See also* Natural gas
Methanol: for fuel cells, 755–56
Meucci, Antonio, 269
Mexican War, 320–21
Mexico, oil industry, 566–67
Microprocessors, 432–33
Middle Ages, climate. *See* Medieval Warm Period
Middle East: during the Oil Age, 569–643, future power, 790–91
MIG (Mikoyan-Gurevich), 411
Migration of people, during Coal Age, 292–93
Millennium Development Goals (MDG), 810
Missiles: intercontinental ballistic, 442; modern Chinese, 518. *See also* Rockets
Mobile phones. *See* Cellular telephones
Mobutu, Joseph Désiré, 554–55
Mogul empire. *See* India, Mughal (Mogul) empire
Moluccas, 135–36
Mongols, 98–99
Monroe Doctrine (1823), 565
Montgolfier brothers, 386
Moon exploration: by Americans, 444–45; by Russians, 444
Morse, Samuel, 268
Mossadegh, Mohammed, 591
Motor vehicles, increase in numbers, 402
Motors, electric, 221
Mozambique, 554
MTBE (methyl tert-butyl ether), 750
Mughal empire. *See* India: Mughal (Mogul) empire
Mujahideen, support by the United States, 619
Murdock, William, 206, 217
Muscle Power, 57–65
Mussolini, Benito, 507
Myanmar. *See* Burma

Nagasaki, 438–39, 513
Napoleonic Wars, 229, 244–45, 306
NASA (National Aeronautics and Space Administration), 444–45, 704
National Defense Education Act (1958), 540
National Security Act (1947), 543
National Security Council (NSC), 543
Native Americans: eradication during Coal Age, 322–23
NATO (North Atlantic Treaty Organization), 530, 544
Natural gas: future consumption, 775–76; to make liquid fuel, 753–54; use during Oil Age, 468–70. *See also* Methane
Navigation at sea, 289–90
Nazi (Nationalsozialistische Deutsche Arbeiterpartei), 507
Neanderthals, 7, 43, 55
Neon lamp, 221
Net energy return, 846
Net primary production (NPP), 690
Newcomen, Thomas, 199–201
New Deal, 470, 539
New Economic Policy (NEP), 529
Newton, Isaac, 156
New Zealand, 316–18
Niagara Falls, power plant, 223
Nicaragua, 551
Nigeria, 526, 747, 792
Nightingale, Florence, 272
Nile, 107
Nine (September) 11, 2001, attacks, 616–17
Nitrate. *See* Saltpeter
Nitration (production of explosives), 277
Nitrogen, 436
Nitroglycerin, 278
Nitrous oxide (N_2O), 687
Nixon, Richard Milhous, 546
Nobel, Alfred, 278
Nobel, oil industry activities in Azerbaijan, 368–69
Nordhaus, William, 707–8
Noriega, Manuel, 552–53
Normandy, 510
North Atlantic Treaty Organization. *See* NATO
North Korea, 544; nuclear bomb, 524

Northwest European Coal Belt, 237
Northwest Passage, 691
Nuclear Age, 797–98
Nuclear bombs, 436–440
Nuclear energy, 471–73; for future electricity production, 768–70; for submarine propulsion, 471–72
Nuclear fusion, 769–70
Nuclear Nonproliferation Treaty of 1968, 446
Nuclear power for electricity production. *See* Nuclear energy
Nuclear powers, 446
Nuclear radiation tests on U.S. civilians, 446
Nuclear waste, final storage, 473

Oak Ridge National Laboratory, 438
Obesity, 417
Oceans, role in climate change, 687–88
October Revolution (Russia, 1917), 527
Ohain, Pabst von, 396
Oil: ancient uses, 361–62; drilling technology, 398–99, 746; future consumption, 776–77; future exporters, 788–93; heavy, 747; for lighting, 362–63; peak production, 743–45; production increase, 402; production in the United States, 557–65; properties 359–60; prospecting methods, 399–401; refining, 402; reserves, global (conventional), 741–47; reserves, in Persian Gulf nations, 569–70; shale, 748; for ship propulsion, 372–74, 376; transport of, 401; unconventional, 747–48
Oil and Gas Journal, 742
Oil Creek, 365–66
Oil culture. *See* Oil technology
Oil price: collapse in 1986, 532, 608–9; different types of oil, 641; OPEC target price band, 641; rise after 1999, 641–43; shock, first, 588–90, 683, 848; shock, second, 593–94, 705
Oil technology: emergence of, 361–405; spread of, 407–12
Okinawa, Battle of, 513
One-child policy, 494

OPEC. *See* Organisation of the Petroleum Exporting Countries
Open hearth process, 196, 216
Opening of the world, 127–44
Operation Condor, 552
Operation Paperclip, 397, 441, 443, 444, 541
Opium Wars, 335–36
Organisation of the Petroleum Exporting Countries (OPEC), 571, 580–81, 588, 641; reserves statements, 742
Organization: social, in agrarian societies, 53–55
Organization for Economic Cooperation and Development (OECD), 699
Ørsted, Hans Christian, 218
Osama bin Laden, 622–23
Oslo Accords, 613
Otto, Nikolaus, 377–78
Ottoman Empire, 115–16, 342–43, 375; Young Turks revolution, 500; collapse during World War I, 505; dissection after World War I, 571–73
Ötzi, 42–43
Overkill (of animals) theory, 67–68
Ozone hole, 687

Pacific, War in the, 511–13. *See also* World War II
Paddlewheels, 213
Pakistan, 522, 524–25
Palestine, 582, 584–90; conflict with Israel, 611–13, 640–41
Palestine Liberation Organization (PLO), 587, 612
Panama, post–World War II, 552–53
Panama Canal, 216, 291, 327–28
Pangaea, 237
Pan-Slavism, 500
Paperclip. *See* Operation Paperclip
Papin, Denis, 199
Parsons, Charles A., 227
Pastoralism, 44–45
Patents, role for progress, 231
Peak oil, 743–45
Pearl Harbor, 510, 513
Peat, use by the Dutch, 137–38
Pelton, Lester A., 224

PEM (proton exchange membrane) cells, 755
PEMEX (Petróleos Mexicanos), 567
Penicillin, 416
Pennsylvania oil, 364–67
Percussion drilling. *See* Drilling
Perkin, William, 274–75
Perry, Matthew (commodore), 250
Persia, 102, 341; early oil prospecting, 374–75; original empire, 106. *See also* Iran
Persian Gulf region, during the Oil Age, 569–643
Pesticides, 453
Peter I (the Great, of Russia), 248
Petroleum. *See* Oil
Pharmaceuticals, production during Coal Age, 274
Philippines, 327, post–World War II, 545
Phoenicians, 108
Photovoltaic cells, 724, 766–67
Pinatubo, Mount, 688
Pinochet, Augusto, 550–51
Pizarro, Francisco, 134
Plastics, 424–26
Ploesti (Ploiesti), 364, 502, 508
Plough. *See* Plow
Plow, 70
Plutonium, 438
Poland, 507
Polish Corridor, 507
Political power in the future, as derived from energy resources, 779–93
Pony Express, 64
Population: decline in rural areas in the Oil Age, 452; developments during Oil Age, 491–96; global status in later Oil Age, 495–96; of medieval Europe, 121; size, importance for progress, 854
Population growth: in Agricultural Age, 148–50; in Britain between 1750 and 1820, 230; in Britain between 1750 and 1901, 261; in Coal Age, 261–63
Porsche, Ferdinand, 391, 758
Portuguese naval expeditions, 127, 134–35
Potash (potassium oxide, K_2O), 101
Potatoes, 29–30, 132–33
Potatoes, Sweet, 316

Poudre B, 278
Poverty: in terms of energy consumption, 481–82; in terms of money, 482, 485
Powell, Colin, 546, 631
Power: human muscle, 58–60; animal, 60–76; hydro. *See* hydropower
Pressurized water reactor (PWR), 472
Printing press, steam-operated, 268
Project Sunshine, 446
Prospecting (oil), technology 399–401; international spread of, 407–8
Prussia, 307–8
Puerto Rico, 326
Puffing Billy, 208
Pulses, 48, 101
Putin, Vladimir, 537
Pyroscaphe, 213

Qatar, natural gas reserves, 775
Qing dynasty, 166–67
Quesnay, François, 295
Quinoa, 29, 132–33

Radiative forcing, 697–99, 713
Radioactivity, 436–37
Radio communication, 269
Railroad: early continental European lines, 242–43; early U.S. lines, 243. *See also* Train
Railway. *See* Railroad; Train
Rainhill Trials, 209–10
RDX (Research Department Explosive), 279
Reagan, Ronald, 551–52, 599
Red Flag Act. *See* Locomotive Act of 1865
Refining (oil), 402
Refrigeration, 286–87
Religion, 54–55; importance for progress, 851–53
Renaissance, 155
Renewable energy, 473–74, 722–25
Ressel, Josef, 214
Reza Shah Pahlavi, Mohammed, 591–93
Rifled barrels, 279
Rigid horse collar, 70–71
Rockefeller. *See* Standard Oil Company
Rockets, 442–45; Atlas, 444; Jupiter-C, 444; Redstone, 444; R-7 Semyorka,

443–44; Saturn, 444–45; V-1, 442; V-2, 442
Roe, George H., 220
Roman empire, 108–11
Romania: Ploesti (Ploiesti), 364, 502, 508; role in World War I, 502; role in World War II, 508
Roosevelt, Franklin, 539
Roosevelt, Theodore, 280
Roosevelt Corollary, 327
Rotary engines, 380
Rothschild, 583–84, 585; oil industry activities in Azerbaijan, 369
Royal Dutch Shell, 371; early activities in Iraq, 375, 376
Royal William (ship), 213
R-7 Semyorka, 443–44
Rubber production during Coal Age, 275
Ruhr valley, 246
Rumsfeld, Donald, 600, 637
Russia: coal and iron production in Coal Age, 249; expansion during Coal Age, 328–31; future power, 789–90; Peter, the Great, 248; revolution (1905), 527; serfdom, 248; spread of coal technology to, 248–50; surrender during World War I, 503; war against the Japanese, 339, 527; Western intervention in civil war, 528. *See also* Russian Federation; Soviet Union
Russian Federation, 532; natural gas reserves, 775; oil and gas production, 536–37. *See also* Russia; Soviet Union
Russo-Japanese War, 339, 527; effect on Azerbaijan, 533
Russo-Turkish Wars, 500
Rwanda, Tutsi extermination and role in war in Congo, 555

Saddam Hussein, 102, 595–97
Sailing, 80–81
Saint-Etienne, 242
salinization, 99
Saltpeter, 89; Chile, 284, 436
San, 1
Sandinistas, 551
San Juan Hill, Battle of, 280
Sarajevo, 353

Sargon, 106
Satellites, communication, 430
Saturn (rockets), 444–45
Saudi Arabia, 579–80; early history, 575–76; kings, 614; population and social development after World War II, 615; post-9/11, 627–29; relations to the West, 609–11; totalitarianism, 614–16
Savery, Thomas, 199
School of the Americas (SOA), 556
Schrödinger, Erwin, 428
Schumpeter, Joseph, 294
Sea level rise, 691
Second Coal Age, 798–99
Second oil price shock, 593–94, 705
Security Council, 514
Sediment Cores, 684
Sepoy Rebellion, 332–33
Serbia, 500–501
Seven Sisters, 578, 592
Seven Years' War, 167–69, 229, 244
Shale oil. *See* Oil, shale
Sharon, Ariel, 612
Shell. *See* Royal Dutch Shell
Ship propeller, 214–15
Ship screw, 214–15
Ships: diesel-powered, 381; steam-powered, 212–16; wind-powered, 80–81
Shoguns, 250
Siberian oil, 535
Siemens, Karl Wilhelm, 195, 216
Siemens, Werner, 219, 221
Siemens & Halske, 268
Siemens-Martin process, 196, 216
Silesia, 167, 168, 169, 506; Upper (coal fields), 237–38, 255 n.32
Singapore, 336, 521
Six-Day War. *See* Arab-Israeli War of 1967
Slater, Samuel, 241
Slavery, 127–28; abolition of, 264–66; in Caribbean sugar production, 152–54; in U.S. cotton production, 161–63; rights of slaves in Babylon, 54
Smallpox (variola), 49, 110, 111, 271–72; in the Americas, 129, 134, 855; in the Coal Age, 416; vaccination

(variolation), 271–72, 865; virus, 416–17
Smil, Vaclav, 864, 869 n.11
Smith Adam, 295–96
Smith, Francis, 214
Social disparities, in Oil Age, 419–20
Social organization. *See* Organization
Soil erosion, 100
Solar power, 724, 766–67; indirect electricity production, 767
Solar radiation, role in climate change, 696–96, 698
Somalia, 553
Somoza (García), Anastasio, 551
South Africa, 526
South Korea, 522, 544
Southern Cone, 151, 254
Soviet Union, 527–37; disintegration, 532; industrialization under Stalin, 529–30; occupation of Afghanistan, 617–18; oil production, 535–356; oil resources, 533–35; role in World War II, 530; space exploration, 531. *See also* Russia; Russian Federation
Space race, 531
Spears, 87
Speed: horse-drawn coaches, 212; steam trains, 210–11, 212
Spermaceti, 362–63
Spice Islands (the Moluccas), 135–36
Spice trade, 135–36
Spindletop (Texas), 558–59
Spinning, 157–59
Spinning jenny, 159
Springboard. *See* Drilling
Spring poles. *See* Drillling
Sputnik, 444; shock (1957), 540
SS *Great Britain*, 215
Stalin, Joseph V., 529
Stalingrad, Battle of, 508
Standard Oil Company, 560–62
Steam: coaches, 211; engines, 198–205; engines, high-pressure, 204; engines, in manufacturing, 204–5; ships, 212–16; shovel, 267; trains, 205–12; turbines, 225–28
Steel, 195–96
Stephenson, George, 208–9
Stephenson, Robert, 209–11

Stirrup, 64
Stockton & Darlington Railway, 209
Stone tools, 4
Storms, changing patterns due to climate change, 707, 708
Strassmann, Fritz, 437
Submarines, 381–82; nuclear-powered, 471–72
Suez Canal, 216, 291, 526
Suez Crises, 587
Sugar, 151–54; beet, 247
Suharto, Raden, 547–48
Sukarno, Achmed, 547
Sulfur, 90
Sun, 675; role in climate change, 695–96, 698
Super-Agricultural Era, 147–171
Swan, Joseph, 220
Sweet Potatoes, 316
Sykes-Picot Agreement, 571–73, 585
Synthesis gas, 753
Syphilis, 129; treatment, 274

Taiwan (Formosa), 517, 521–22, 544
Taliban, 621–22; being ousted, 625–26
Tank: development after World War I, 385–86; development during World War I, 384–85
Tankships, 401
Tar sands, 747–48
Tata, Jamshetji Nusserwanji, 333
Taube (airplane), 391
Technology advances: during Coal Age, 267–76; during Oil Age, 423–34
Telegraph, 268
Telephone, 268–69
Telescope, 156
Temperature, oceanic, record of the past, 679–85
Tennessee Valley Authority (TVA), 470–71, 539
Tesla, Nikola, 222–23, 269
Texas, oil production, 558–59
Thailand, 336
Thatcher, Margaret, 611, 612; role in climate change policy, 705
Theory of Relativity, 428
Three Gorges Dam, 471, 785
Three Mile Island accident, 472

Tibet, 337, 517
Tidal power, 723, 763–64
Tiger States, 480, 520–22
Titusville, Pennsylvania, 366
TNT (trinitrotoluene), 278
Tools: iron, 84; stone 4
Torrijos-Carter Canal Treaties (1977), 552
TOTAL, 573. *See also* Compagnie Française des Pétroles (CFP); TotalFinaElf
TotalFinaElf, 636
Town gas. *See* Coal gas
Tractor, 383–84; diesel-powered, 382; track-laying, 383
Trade: barriers, 480–81; increase during Coal Age, 289; increase during Oil Age, 478–81; theory, 295–96, 298 n.69
Train: diesel, 382; electric, 221; speed, 210–11, 212; steam, 205–12
Trans-Atlantic telephone cable, 429
Transcaucasian railway and pipeline, 369
Transgenic animals. *See* Animal, genetically engineered
Transgenic crops. *See* Crops
Transistors, 431
Transport fuel, options for the future, 741–60
Trans-Siberian Railroad, 249
Treaty of Versailles, 506–7
Trevithick, Richard, 206–8
Triple Alliance, 502
Triple Entente, 501
Truman Doctrine (1947), 543
Turbines: gas (airplane propulsion), 394–98; steam, 225–28; water, 224
Turbinia, 227–28
Turkey, 783–84
Turkish Petroleum Company, 375, 571, 573. *See also* Iraq Petroleum Company
Turks Ottoman, 115–16
Turks Seljuk, 114–15
Typewriters, 268

Ukraine, 249, 801
Unconventional oil, 747–48
Uniformitarianism, 270
Union of Soviet Socialist Republics (USSR). *See* Soviet Union

United Nations (UN), 514–15
United Nations Framework Convention on Climate Change (UNFCCC), 702
United States: activities in Latin America after World War II, 548–53; agriculture during Coal Age, 285; between the world wars, 539; Civil War, 245, 281; cooperation between government and oil companies, 562–63; during the Oil Age, 538–65; expansion during Coal Age, 318–24; future power, 779–82; Golden Era (1945 to 1970), 540–41; imperialism during Cold War, 541–56; oil imports, 564–65, 569; oil production decline, 563–64; oil production, 557–65; overseas imperialism during Coal Age, 324–28; overseas troops and military bases in late Oil Age, 556–57; post-1970 era, 541; post–World War II activities in Africa, 553–55; post–World War II activities in East and South Asia, 544–48; post–World War II global militarism, 555–57; rivalry with Britain for Middle East oil concessions, 576–77, role in World War I, 503; Saudi relations, 609–10; 2003 invasion of Iraq, 629–39
Unity of life, 429
Upper Silesia. *See* Silesia, Upper
Uranium, 437; deposits, 768–69; deposits, by country, 801
U.S. Geological Survey (USGS), 743
Usama bin Laden. *See* Osama bin Laden
USS *Cole*, 625
Utilitarian ethics, 700, 709

Vaccination, 416
Variola. *See* Smallpox
Variolation, 271–72, 865
Vegetarianism, 2, 13–14; role in climate change, 713–16, 751
Venera 1, 444
Venezuela, 793; oil industry, 567–69
Versailles, Treaty of, 506–7
Very First World War. *See* Seven Years' War
Vienna, Congress of, 245

Vietnam: French activities, 336; independence, 545–46. *See also* Indochina
Vietnam War, 545–47
Vikings, 123–24, 679, 681
Virus, 49, 416. *See also* Diseases, infectious
Volcanic activity: role in climate change, 697–98
Volga-Ural oil fields, 510, 535
Von Braun. *See* Braun, Wernher von
V-1 rocket, 442
Vostok, 680, 682–83, 685
V-2 rocket, 442

Wahhabism, 575
Walking beam. *See* Drilling
Wallace, Alfred Russel, 271
Wallonia, 93, 237, 241, 853
Walter, Hellmuth, 440
Wankel, Felix, 380
War in the Pacific (World War II), 511–13
War of 1812, 242, 319–20
War of the Pacific (1879 to 1884), 284
War on Terrorism, 626–27
Warsaw Pact, 531, 544
Water gas. *See* Coal gas
Water-logging, 99
Waterpower, 77–82; in textile production, 157–60. *See* Hydropower
Water turbines, 224
Waterwheels, 77–80; in medieval Europe, 122–23
Watson, Jim, 428
Watt, James, 201–4
Wave power, 764
Weapons: in the Agricultural Age, 87–95; in the Coal Age, 277–81; in the Oil Age, 435–50; nuclear, 436, 440
Weaving, 158–59
Weber, Max, 852
Weisskopf, Gustav Albin, 389–90
Weizmann, Chaim, 278, 585
Wells, Herbert George, 837–38, 844
Welsbach, Carl Auer von, 217–18, 220
Welsbach mantle, 218
West Bank, 587–88, 613

Western Union Company, 222
Westinghouse Electric Company, 222–23
Whale oil, 362–63
Wheat, resistance to salinity, 33 n.16
Wheelbarrow, 58
Wheel locks, 94
Wheels, spoked, 62
Whitehead. *See* Weisskopf, Gustav Albin
Whittle, Frank, 396
Wildcat well, 401
Wind energy, 77–82, 122, 473–74; future role, 764–65. *See also* Windmills
Windmills, 77–80; in medieval Europe, 122
Wind power. *See* Wind energy
Winsor, Frederic, 217
Witte, Sergey, 249–50
Woman suffrage, 264
Wood: energy density compared to coal, 184, 189; sustainable yield, 190 n.4
Wool, 157–59
Work hours: in foraging societies, 1, 847; in Oil Age societies, 415
World Bank, 516
World Energy Outlook, 773
The World Set Free: A Story of Mankind (Wells), 838
World Trade Center, bombing of 1993, 623. *See also* Nine (September) 11, 2001, attacks
World War I, 499–507; casualties, 505
World War II, 507–14; casualties, 435, 513–14
Wright, Wilbur and Orville, 390–91
Writing systems, of various civilizations 105–8
Wrought iron, 195

Xiaoping, Deng, 518

Young Turks, 500
Yukos, 536–37

Zaïre. *See* Belgian Congo; Congo, Democratic Republic of the
Zedong, Mao, 517–18
Zeppelin, 386–88
Zheng He, 142
Zionism, 584

ABOUT THE AUTHOR

Manfred Weissenbacher is an energy technology and strategy consultant working for governments and companies, and Associate Professor of Innovation Management at Copenhagen Business School's Department of Management, Politics and Philosophy. Previously with the Stanford Research Institute and the Danish Technological Institute, he developed energy storage systems at Battery Technologies, Inc. in Toronto, Canada, and at the Technical University of Graz, Austria, where he took his PhD. As a Fulbright scholar, he studied at the Monterey Institute of International Studies in California. His wide field of studies includes economics, history, political science, chemistry, environmental science, agricultural science, health science, and nonproliferation. He is the author of numerous chemistry journal papers, chemical economics market analyses, and the book *BSE and Creutzfeldt-Jakob Disease*, published in five languages. He invites his readers' comments at his Web site, www.manfredweissenbacher.com.